T0189450

Lecture Notes in Computer Science 14637

The series Lecture Notes in Computer Science (LNCS), including its subseries Lecture Notes in Artificial Intelligence (LNAI) and Lecture Notes in Bioinformatics (LNBI), has established itself as a medium for the publication of new developments in computer science and information technology research, teaching, and education.

LNCS enjoys close cooperation with the computer science R & D community, the series counts many renowned academics among its volume editors and paper authors, and collaborates with prestigious societies. Its mission is to serve this international community by providing an invaluable service, mainly focused on the publication of conference and workshop proceedings and postproceedings. LNCS commenced publication in 1973.

Xujin Chen · Bo Li
Editors

Theory and Applications of Models of Computation

18th Annual Conference, TAMC 2024
Hong Kong, China, May 13–15, 2024
Proceedings

 Springer

Editors
Xujin Chen (ID)
Chinese Academy of Sciences
Beijing, China

Bo Li
Hong Kong Polytechnic University
Hong Kong, China

ISSN 0302-9743   ISSN 1611-3349 (electronic)
Lecture Notes in Computer Science
ISBN 978-981-97-2339-3   ISBN 978-981-97-2340-9 (eBook)
https://doi.org/10.1007/978-981-97-2340-9

This Springer imprint is published by the registered company Springer Nature Singapore Pte Ltd.
The registered company address is: 152 Beach Road, #21-01/04 Gateway East, Singapore 189721, Singapore

Paper in this product is recyclable.

Preface

It is with great pleasure that we present the proceedings of the 2024 Annual Conference on Theory and Applications of Models of Computation (TAMC 2024), held in Hong Kong, China, from May 13–15, 2024. This volume contains all contributed papers presented at TAMC 2024. Over the last 18 years, the TAMC conference series has served as a valuable platform for researchers specializing in computational theory, information theory, and their applications. This year's conference centered around key themes including computability, complexity, algorithms, information theory, as well as their integration with machine learning theory and the foundations of artificial intelligence.

The Program Committee, composed of 28 active researchers from the field, reviewed 69 submissions and selected 30 papers for presentation at TAMC 2024. Every submission underwent a rigorous single-blind peer review process. Each paper had at least three reviews, with additional reviews solicited as needed.

We would like to express our heartfelt gratitude to everyone who dedicated time and effort to make TAMC 2024 a resounding success: the authors, whose innovative research enriches our collective knowledge, the Program Committee members, whose judgments are invaluable, and the reviewers, whose insights help to ensure the scientific quality. The authors of the papers and the Program Committee are from 23 countries and regions, reflecting the international status of the TAMC series of conferences.

Our gratitude is also directed to Springer for accepting TAMC 2024 for publication in this Lecture Notes in Computer Science (LNCS) proceedings volume. Our special appreciation goes to the general chairs Yixin Cao and Qing Li, and the local organization team for their outstanding dedication and exceptional work.

May 2024

Xujin Chen
Bo Li

Organization

Steering Committee

Manindra Agrawal   Indian Institute of Technology Kanpur, India
Jin-Yi Cai   University of Wisconsin, Madison, USA
John Hopcroft   Cornell University, USA
Angsheng Li   Beihang University, China
Zhiyong Liu   Institute of Computing Technology, Chinese Academy of Sciences, China

General Chairs

Yixin Cao   Hong Kong Polytechnic University, China
Qing Li   Hong Kong Polytechnic University, China

Program Committee Chairs

Xujin Chen   Chinese Academy of Sciences, China
Bo Li   Hong Kong Polytechnic University, China

Local Organizing Chairs

Ankang Sun   Hong Kong Polytechnic University, China
Mingqi Yuan   Hong Kong Polytechnic University, China
Yu Zhou   Hong Kong Polytechnic University, China

Program Committee

Hee-Kap Ahn   Pohang University of Science and Technology, South Korea
Hans L. Bodlaender   Utrecht University, The Netherlands
Anthony Bonato   Toronto Metropolitan University, Canada
Julian Bradfield   University of Edinburgh, UK
Xujin Chen   Chinese Academy of Sciences, China

Bhaskar DasGupta   University of Illinois, Chicago, USA
Qilong Feng   Central South University, China
Henning Fernau   Trier University, Germany
Ling Gai   University of Shanghai for Science and
    Technology, China
Nan Guan   City University of Hong Kong, China
Jesper Jansson   Kyoto University, Japan
Jan Kratochvil   Charles University, Czechia
Michael Lampis   LAMSADE, Université Paris Dauphine, France
Bo Li   Hong Kong Polytechnic University, China
Jian Li   Tsinghua University, China
Hsiang-Hsuan Liu   Utrecht University, The Netherlands
Klaus Meer   BTU Cottbus-Senftenberg, Germany
Mitsunori Ogihara   University of Miami, USA
Jose Rolim   University of Geneva, Switzerland
Shikha Singh   Williams College, USA
Warut Suksompong   National University of Singapore, Singapore
Xiaoming Sun   Chinese Academy of Sciences, China
Akira Suzuki   Tohoku University, Japan
Seeun William Umboh   University of Melbourne, Australia
Rohit Vaish   Indian Institute of Technology, Delhi, India
Chee Yap   New York University, USA
Yitong Yin   Nanjing University, China
Christos Zaroliagis   University of Patras, Greece

Additional Reviewers

Taehoon Ahn   Esther Galby
Hyung-Chan An   Jiarui Gan
Jiehua Chen   Loukas Georgiadis
Xin Chen   Tim A. Hartmann
Po-An Chen   Ararat Harutyunyan
Huairui Chu   Chris Heunen
Chaeyoon Chung   Po-Wei Huang
Jaehoon Chung   Lingxiao Huang
Argyrios Deligkas   Kai Jin
Frank Drewes   Mook Jung Kwon
Rüdiger Ehlers   Lawqueen Kanesh
Marek Elias   Byeonguk Kang
Taekang Eom   Geunho Kim
Rolf Fagerberg   Kei Kimura
Bernd Finkbeiner   Evangelos Kipouridis

Spyros Kontogiannis
Noleen Köhler
Alexander Lam
Jaegun Lee
Qian Li
Shi Li
Weian Li
Yaqiao Li
Wenjing Liu
Samuel McCauley
Nikolaos Melissinos
Anish Mukherjee
Torsten Mütze
Aidin Niaparast
Christos Nomikos
Panagiotis Patsilinakos
Matthias Pfretzschner
Giovanni Pighizzini
Aditya Potukuchi
Lev Reyzin
Burton Rosenberg
David Sankoff
Ramprasad Saptharishi
Xiaohan Shan

Florian Sikora
Ludwig Staiger
Frank Stephan
Ankang Sun
Takahiro Suzuki
Yuma Tamura
Wenjie Tang
Kostas Tsichlas
Manolis Vasilakis
Daniel Vaz
José Verschae
Chunyang Wang
Chenhao Wang
Adrian Wurm
Mingji Xia
Han Xiao
Qingjie Ye
Ismael González Yero
Sheung Man Yuen
Chihao Zhang
Ruilong Zhang
Zhijie Zhang
Chaodong Zheng

Contents

On Learning Families of Ideals in Lattices and Boolean Algebras

Nikolay Bazhenov[1,2]($\boxtimes$)(iD) and Manat Mustafa[3](iD)

[1] Sobolev Institute of Mathematics, 4 Acad. Koptyug Avenue,
Novosibirsk 630090, Russia
`nickbazh@yandex.ru`
[2] Kazakh-British Technical University, 59 Tole Bi Street, 050000 Almaty,
Kazakhstan
[3] Department of Mathematics, School of Sciences and Humanities,
Nazarbayev University, 53 Qabanbaybatyr Avenue, Astana 010000, Kazakhstan
`manat.mustafa@nu.edu.kz`

Abstract. The paper studies learnability from positive data for families of ideals in lattices and Boolean algebras. We established some connections between learnability and algebraic properties of the underlying structures. We prove that for a computable lattice L, the family of all its ideals is **BC**-learnable (i.e., learnable w.r.t. semantic convergence) if and only if each ideal of L is principal. In addition, **BC**-learnability for the family of all ideals is equivalent to **Ex**-learnability (learnability w.r.t. syntactic convergence).

In general, learnability depends on the choice of a computable isomorphic copy of L. Indeed, we show that for every infinite, computable atomic Boolean algebra B, there exist computable algebras A and C isomorphic to B such that the family of all computably enumerable ideals in A is **BC**-learnable, while the family of all computably enumerable ideals in C is not **BC**-learnable. Finally, we obtain a reverse-mathematical result about conservative **Ex**-learning for ideals.

Keywords: Algorithmic learning theory · Computable structure · Lattice · Boolean algebra · Linear order · Ideal · Inductive inference

1 Introduction

Algorithmic learning theory goes back to the works of Putnam [18] and Gold [8]. The theory investigates formal mathematical frameworks for inductive inference. As formulated in [22], inductive inference is the process of generating hypotheses for describing an unknown object from finitely many data points about the unknown object. The classical studies (up to the beginning of 2000s) mainly focused on learning for formal languages and for recursive functions. For a survey, the reader is referred to the monograph [14] and the papers [16, 22].

Let $\mathcal{S} = (A; f_1, f_2, \ldots, f_n)$ be an algebraic structure (for example, think about the abelian group of integers $(\mathbb{Z}; +)$ or the field of rationals $(\mathbb{Q}; +, \cdot)$). The

© The Author(s), under exclusive license to Springer Nature Singapore Pte Ltd. 2024
X. Chen and B. Li (Eds.): TAMC 2024, LNCS 14637, pp. 1–13, 2024.
https://doi.org/10.1007/978-981-97-2340-9_1

structure S is *computable* if A is a (Turing) computable set and the functions $f_1, f_2, \ldots, f_n$ are also computable.

Stephan and Ventsov [21] initiated systematic investigations of learnability for *classes of substructures* of a given computable structure S. Within their framework, the learner obtains only positive data (or text) about a substructure M of S, and the outputted sequence of hypotheses should converge to a program enumerating the substructure M to be learned. As usual, there are two main notions of convergence—semantical (**BC**) and syntactical (**Ex**). (The formal details are discussed in Sect. 2).

The paper [21] showed that learnability from text has strong connections to the algebraic properties of a given computable structure. For example, the class of all ideals of a computable ring R is **BC**-learnable from positive data if and only if the ring R is Noetherian (Theorem 3.1 in [21]). In addition, a computable ring R is Artinian if and only if the class of all its ideals is **Ex**-learnable from text with a constant bound on the number of mind changes (Theorem 4.2 in [21]).

While the work [21] was more focused on the learnability of the full class of substructures for a structure S, Harizanov and Stephan [12] restricted their attention only to computably enumerable (c.e.) substructures of S—they studied learnability for the class of all c.e. subspaces of a computable vector space. Following this approach, the paper [7] considered learnability for some families of closed sets in a c.e. matroid. We also refer to [5,6] for related results.

The current paper continues investigations of learnability from positive data for classes of substructures [7,12,21]. Here we focus on lattices $(L; \vee, \wedge)$ and Boolean algebras $(B; \vee, \wedge, \overline{(\cdot)}, 0, 1)$. We study learnability for families of ideals in these structures.

The paper is arranged as follows. Section 2 gives the necessary preliminaries. Recall that a lattice $\mathcal{L} = (L; \vee, \wedge)$ is computable if its domain L is a computable subset of $\mathbb{N}$, and the operations $\vee$ and $\wedge$ are computable functions. Theorem 2 (in Sect. 3) establishes the following. For a computable lattice $\mathcal{L}$, the family of all its ideals $AId(\mathcal{L})$ is **BC**-learnable if and only if each ideal of $\mathcal{L}$ is principal. In addition, for $AId(\mathcal{L})$, the notions of **BC**-learnability and **Ex**-learnability coincide. A similar result is obtained for the family $CEId(\mathcal{L})$ containing all c.e. ideals of $\mathcal{L}$. Subsection 3.1 gives some applications of Theorem 2.

Let B be a computable, infinite Boolean algebra. In Sect. 4, we show that if B is atomic, then B always has two computable copies A and C isomorphic to B such that $CEId(A)$ is **BC**-learnable, but $CEId(C)$ is not **BC**-learnable (Theorem 3 and Proposition 4). In addition, if B is not atomic, then $CEId(B)$ is not **BC**-learnable. In Sect. 5, we follow the approach of [13] and give a reverse-mathematical result about conservative **Ex**-learnability of ideals (Theorem 4).

2   Preliminaries

As usual, $(\varphi_e)_{e \in \mathbb{N}}$ denotes the standard list of all unary partial computable functions, and $W_e = \mathrm{dom}(\varphi_e)$, for $e \in \mathbb{N}$. For the background on computability theory, we refer to [17].

Preliminaries on computable structure theory can be found in [2]. As usual in this theory, if $\mathcal{S}$ is an infinite computable structure, then, without loss of generality, one may assume that the domain of $\mathcal{S}$ is equal to $\mathbb{N}$. If $\mathcal{A}$ is an arbitrary structure and $\mathcal{C}$ is a computable structure isomorphic to $\mathcal{A}$, then one says that $\mathcal{C}$ is a *computable isomorphic copy* of $\mathcal{A}$.

For a finite string σ, by $|\sigma|$ we denote the length of σ. By Λ we denote the empty string. By $\leq_{\mathbb{N}}$ we denote the standard ordering of natural numbers.

2.1 Algorithmic Learning

Our learning framework mainly follows the conventions of [12, 21] which, in turn, follow the Gold-style learning for families of c.e. sets [8, 14]. Note that in this paper, we consider only learnability from *positive data* (or text). A *text* for a non-empty set $V \subseteq \mathbb{N}$ is an arbitrary function $t \colon \mathbb{N} \to \mathbb{N}$ such that range$(t) = V$. (Here we will work only with non-empty c.e. subsets of $\mathbb{N}$, so we do not need to use a special pause symbol '#' in texts t).

Given a finite sequence $a_0, a_1, \ldots, a_n$ of natural numbers, a learner M (algorithmically) outputs a conjecture $e_n \in \mathbb{N}$. More formally, a *learner* M is a computable function acting from $\mathbb{N}^{<\mathbb{N}}$, the set of all finite sequences of natural numbers, into $\mathbb{N}$. Without loss of generality, we may always assume that $M(\Lambda) = 0$.

If t is a text and $n \in \mathbb{N}$, then by $t \upharpoonright n$ we denote the finite string $t(0)t(1)\ldots t(n-1)$ from $\mathbb{N}^{<\mathbb{N}}$. If $\sigma \in \mathbb{N}^{<\mathbb{N}}$, then by $\sigma \subset t$ we denote the fact that σ is an initial segment of t (that is, $\sigma = t \upharpoonright |\sigma|$).

We use the following two main notions of learnability:

(i) A learner M **BC**-*learns* a c.e. set V if for every text t of V, there exists $n_0 \in \mathbb{N}$ such that
$$(\forall n \geq n_0)(W_{M(t \upharpoonright n)} = V).$$

(ii) A learner M **Ex**-*learns* a c.e. V if for every text t of V, there exists $n_0 \in \mathbb{N}$ such that
$$(\forall n \geq n_0)(M(t \upharpoonright n) = M(t \upharpoonright n_0) \text{ and } W_{M(t \upharpoonright n_0)} = V).$$

Here **BC** stands for 'behaviorally correct', and **Ex** stands for 'explanatory'. Informally speaking, for **Ex**-learnability, the sequence of hypotheses must converge syntactically, while **BC**-learnability allows semantical convergence.

A family of sets $\mathcal{K}$ is **BC**-*learnable* (from text) if there exists a learner M that **BC**-learns each set from $\mathcal{K}$. The notion of an **Ex**-learnable family is defined in a similar way. It is clear that **Ex**-learnability always implies **BC**-learnability.

We also consider the natural constraint of *conservativeness* [1]. A learner M *conservatively* **Ex**-*learns* a c.e. set V if M **Ex**-learns V and, in addition, for every text t of V and all $n \in \mathbb{N}$, $k \geq 1$, we have:
$$M(t \upharpoonright (n+k)) \neq M(t \upharpoonright n) \;\Rightarrow\; \text{range}(t \upharpoonright (n+k)) \nsubseteq W_{M(t \upharpoonright n)}.$$

Intuitively, a conservative learner does not make unjustified changes: a conjecture changes only if the 'old' $W_{M(t\restriction n)}$ becomes unsatisfactory.

One of the key ingredients of our proofs is the notion of a *locking sequence*. Let M be a learner, and let $V \subseteq \mathbb{N}$. A string $\sigma \in \mathbb{N}^{<\mathbb{N}}$ is called a *locking sequence for M on V* if range(σ) $\subseteq V$ and the following holds:

$$(\forall \tau \in \mathbb{N}^{<\mathbb{N}})[\text{range}(\tau) \subseteq V \;\to\; (M(\sigma\tau) = M(\sigma) \;\&\; W_{M(\sigma)} = V)].$$

A string σ is a **BC**-*locking sequence for M on V* if range(σ) $\subseteq V$ and

$$(\forall \tau \in \mathbb{N}^{<\mathbb{N}})[\text{range}(\tau) \subseteq V \;\to\; W_{M(\sigma\tau)} = V]. \tag{1}$$

Theorem 1 ([3]). *If a learner M **Ex**-learns a set V, then there exists a locking sequence for M on V. A similar fact is true for **BC**-learning and **BC**-locking sequences.*

2.2   Lattices and Boolean Algebras

We treat lattices as structures in the signature $\{\vee, \wedge\}$. For a lattice $\mathcal{L} = (L; \vee, \wedge)$, its induced order $\leq_{\mathcal{L}}$ is defined as follows: $x \leq_{\mathcal{L}} y$ if and only if $x \vee y = y$. If the subscript is clear from the context, then we write $\leq$ in place of $\leq_{\mathcal{L}}$. Notice the following: if the operation $\vee$ is computable, then the order relation $\leq$ is also computable.

A non-empty set $I \subseteq L$ is an *ideal* of the lattice $\mathcal{L}$ if I satisfies the following:

– if $a, b \in I$, then $a \vee b \in I$;
– if $a \leq b$ and $b \in I$, then $a \in I$.

For a non-empty set $A \subseteq L$, by id(A) we denote the *ideal of $\mathcal{L}$ generated by the set A* (i.e., the least ideal containing A). For an element $a \in \mathcal{L}$, we will write id(a) in place of id($\{a\}$).

An ideal I is *principal* if $I = \text{id}(a)$ for some element $a \in \mathcal{L}$. The following fact is well-known:

Lemma 1 (see, e.g., Lemma 5 in [10]). *Let A be a non-empty subset of L, and let I be an ideal of $\mathcal{L}$. Then $I = \text{id}(A)$ if and only if*

$$I = \{x : x \leq (a_0 \vee a_1 \vee \cdots \vee a_n)\} \text{ for some } n \in \mathbb{N} \text{ and } a_i \in A\}. \tag{2}$$

Consequently, finitely generated ideals of $\mathcal{L}$ are precisely the principal ideals. In addition, for an element $a \in \mathcal{L}$, we have id(a) = $\{x : x \leq a\}$.

Definition 1. *For a computable lattice $\mathcal{L}$, we use the following notations:*

– *$AId(\mathcal{L})$ is the family of all ideals of $\mathcal{L}$;*
– *$CEId(\mathcal{L})$ is the family of all c.e. ideals of $\mathcal{L}$.*

Notice that $CEId(\mathcal{L})$ contains all principal ideals of $\mathcal{L}$. We refer to [10] for further background on lattice theory.

Boolean algebras are viewed as structures in the signature $\{\vee, \wedge, \overline{(\cdot)}, 0, 1\}$. Let $\mathcal{B} = (B; \vee, \wedge, \overline{(\cdot)}, 0, 1)$ be a Boolean algebra. Recall that the reduct $(B; \vee, \wedge)$ is a distributive lattice.

An element $a \in \mathcal{B}$ is an *atom* if a is a minimal non-zero element in $\mathcal{B}$, i.e., $0 <_{\mathcal{B}} a$ and there is no b with $0 <_{\mathcal{B}} b <_{\mathcal{B}} a$. A Boolean algebra $\mathcal{B}$ is *atomic* if for every element $c \in \mathcal{B}$ such that $c \neq 0$, there exists an atom a satisfying $a \leq_{\mathcal{B}} c$.

An element b is *atomless* if there are no atoms a such that $a \leq_{\mathcal{B}} b$. It is clear that an algebra $\mathcal{B}$ is atomic if and only if it does not contain non-zero atomless elements.

If $\mathcal{M} = (M; \leq)$ is a linear order with a least element, then by $\mathrm{Intalg}(\mathcal{M})$ we denote the *interval Boolean algebra* associated with $\mathcal{M}$. Roughly speaking, this is the subalgebra $\mathcal{C}$ of the algebra of all subsets of M such that $\mathcal{C}$ is generated by the set of half-open intervals of $\mathcal{M}$. See Example 1.11 in [15] for the formal details.

The *Fréchet ideal* $Fr(\mathcal{B})$ is the lattice-ideal of $\mathcal{B}$ generated by the set of all atoms of $\mathcal{B}$. We refer to [9,15] for further preliminaries on Boolean algebras.

3   Learning Ideals of a Lattice

Here we obtain a general criterion for learnability of the family $AId(\mathcal{L})$ containing all ideals of a lattice $\mathcal{L}$.

Theorem 2. *Let $\mathcal{L}$ be a computable lattice. Then the following conditions are equivalent:*

(a) the family of all ideals $AId(\mathcal{L})$ is **Ex***-learnable;*
(b) the family $AId(\mathcal{L})$ is **BC***-learnable;*
(c) every ideal of $\mathcal{L}$ is principal.

Proof. The direction (a)⇒(b) is trivial.

(b)⇒(c). Suppose that the family $AId(\mathcal{L})$ is **BC**-learnable by a learner M. Towards a contradiction, assume that $\mathcal{L}$ has a non-principal ideal I.

By Theorem 1, one can choose a non-empty **BC**-locking sequence σ for the learner M on I. Consider the finite set $F = \mathrm{range}(\sigma)$, and define a finitely generated ideal $J = \mathrm{id}(F)$. By Lemma 1, J is a principal ideal. It is clear that $J \subset I$. Consider an arbitrary text t for the ideal J such that $\sigma \subset t$. Applying the definition of a **BC**-locking sequence for M on I (see Eq. (1)), we deduce that for every $n \geq |\sigma|$, we have $W_{M(t \restriction n)} = I$. Since $\mathrm{range}(t) = J$ and M **BC**-learns I, we obtain that $I = J$, which contradicts the non-principality of I. Therefore, every ideal of $\mathcal{L}$ must be principal.

(c)⇒(a). Suppose that every ideal of $\mathcal{L}$ is principal. Without loss of generality, we may assume that the lattice $\mathcal{L}$ is infinite and $\mathrm{dom}(\mathcal{L}) = \mathbb{N}$. (If $\mathcal{L}$ is finite, then the family $AId(\mathcal{L})$ is finite and, hence, it is trivially **Ex**-learnable).

We choose a computable function $\psi \colon \mathbb{N}^{<\mathbb{N}} \to \mathbb{N}$ such that for any non-empty $\sigma \in \mathbb{N}^{<\mathbb{N}}$, we have $W_{\psi(\sigma)} = \mathrm{id}(\mathrm{range}(\sigma))$. Such a function ψ exists: indeed, given a string σ with $\mathrm{range}(\sigma) = \{a_0 <_{\mathbb{N}} a_1 <_{\mathbb{N}} \cdots <_{\mathbb{N}} a_m\}$, we collect into $W_{\psi(\sigma)}$ precisely those $x \in \mathbb{N}$ such that $x \leq_{\mathcal{L}} (a_0 \vee a_1 \vee \cdots \vee a_m)$, see Eq. (2).

We construct a learner M. For a given $\sigma = a_0 a_1 \ldots a_n \in \mathbb{N}^{n+1}$, we define

$$M(\sigma) = \begin{cases} \psi(\sigma), & \text{if } n = 0 \text{ or } a_n \not\leq_{\mathcal{L}} (a_0 \vee a_1 \vee \cdots \vee a_{n-1}), \\ \psi(a_0 a_1 \ldots a_{n-1}), & \text{if } n \geq 1 \text{ and } a_n \leq_{\mathcal{L}} (a_0 \vee a_1 \vee \cdots \vee a_{n-1}). \end{cases}$$

Observe that the function M is computable, since the relation $\leq_{\mathcal{L}}$ is computable.

Let I be an arbitrary ideal of $\mathcal{L}$. Since I is principal, there exists b such that $I = \mathrm{id}(b)$. Let t be an arbitrary text for the ideal I. We find the least index $n_0 \in \mathbb{N}$ such that $(t(0) \vee t(1) \vee \cdots \vee t(n_0)) = b$. Then for any $n > n_0$, we have $t(n) \leq_{\mathcal{L}} b = (t(0) \vee t(1) \vee \cdots \vee t(n-1))$. Hence, $M(t \restriction n) = M(t \restriction n_0)$, and the set $W_{M(t \restriction n_0)} = W_{\psi(t \restriction n_0)}$ is equal to $\mathrm{id}(b) = I$. We conclude that all ideals I are **Ex**-learnable by the learner M.   □

We note that the **Ex**-learner M constructed above is conservative.

Recall that every principal ideal of a computable lattice is c.e. If we use this fact, then essentially the same proof as above gives us the following:

Corollary 1. *For a computable lattice $\mathcal{L}$, the following conditions are equivalent:*

*(a) the family of all c.e. ideals $CEId(\mathcal{L})$ is **Ex**-learnable;*
*(b) the family $CEId(\mathcal{L})$ is **BC**-learnable;*
(c) every c.e. ideal of $\mathcal{L}$ is principal.

3.1   Applications and Examples

Here we apply the characterizations of learnability obtained in Theorem 2 and Corollary 1 to some familiar computable lattices. Note that these characterizations allow us to state our results only for the case of **BC**-learnability.

Firstly, recall that any linear order $\mathcal{M} = (M; \leq)$ can be viewed as a distributive lattice with respect to the following operations: $x \vee y = \max(x, y)$ and $x \wedge y = \min(x, y)$. In addition, lattice-ideals of $\mathcal{M}$ are precisely the *down-sets* (i.e., non-empty sets D satisfying the following: if $a \leq b$ and $b \in D$, then $a \in D$).

As usual, by $\mathcal{M}^*$ we denote the *reversal* of the order $\mathcal{M}$—that is, $x \leq_{\mathcal{M}^*} y$ if and only if $y \leq_{\mathcal{M}} x$. By ω we denote the (order-type of the) standard order of natural numbers.

Proposition 1. *Let $\mathcal{M}$ be a computable linear order.*

*(i) The family $AId(\mathcal{M})$ of all down-sets of $\mathcal{M}$ is **BC**-learnable if and only if $\mathcal{M}$ is isomorphic to the reversal α^* of some ordinal α.*
*(ii) If $\mathcal{M}$ is isomorphic to an infinite ordinal, then the family $CEId(\mathcal{M})$ of all c.e. down-sets of $\mathcal{M}$ is not **BC**-learnable.*

*(iii) There exist two computable linear orders $\mathcal{A}$ and $\mathcal{B}$ such that both $\mathcal{A}$ and $\mathcal{B}$ are isomorphic to $\omega + \omega^*$, the family $CEId(\mathcal{A})$ is not **BC**-learnable, and the family $CEId(\mathcal{B})$ is **BC**-learnable.*

Remark 1. Notice that item (iii) of Proposition 1 shows that in general, learnability of the family of all c.e. ideals of a lattice $\mathcal{L}$ depends on a particular choice of a computable isomorphic copy $\mathcal{A} \cong \mathcal{L}$.

Proof (of Proposition 1). **(i)** Firstly, assume that $\mathcal{M}$ is isomorphic to α^*, where α is an ordinal. Then every non-empty down-set D in $\mathcal{M}$ has a greatest element $b \in D$. Thus, $D = \mathrm{id}(b)$ inside $\mathcal{M}$, and by Theorem 2, the family $AId(\mathcal{M})$ is **BC**-learnable.

Secondly, assume that $\mathcal{M}$ is not isomorphic to the reversal of an ordinal. Then there exists a non-empty set S inside $\mathcal{M}$ with the following property: if $x \in S$, then there exists $y \in S$ such that $x <_\mathcal{M} y$. Consider the down-set $\mathrm{id}(S)$ and observe that $\mathrm{id}(S)$ cannot be a principal ideal. Thus, the family $AId(\mathcal{M})$ is not **BC**-learnable.

(ii) If the order $\mathcal{M}$ is isomorphic to ω, then the whole domain of $\mathcal{M}$ forms a computable non-principal ideal. Otherwise, there is a (unique) element a such that the down-set $\mathrm{id}(a) = \{x : x \leq_\mathcal{M} a\}$ forms an order isomorphic to $\omega + 1$. Then the down-set $\{x : x <_\mathcal{M} a\}$ is a computable non-principal ideal, since its order-type is precisely ω. We conclude that by Corollary 1, the family $CEId(\mathcal{M})$ is not **BC**-learnable.

(iii) We choose the order $\mathcal{A}$ as a 'standard' computable isomorphic copy of $\omega + \omega^*$: say, define $2k <_\mathcal{A} 2\ell + 1$ for all $k, \ell \in \mathbb{N}$, and

$$0 <_\mathcal{A} 2 <_\mathcal{A} 4 <_\mathcal{A} \cdots <_\mathcal{A} 2n <_\mathcal{A} \cdots ,$$
$$1 >_\mathcal{A} 3 >_\mathcal{A} 5 >_\mathcal{A} \cdots >_\mathcal{A} 2n + 1 >_\mathcal{A} \cdots .$$

Then the set of all even numbers is a computable down-set inside $\mathcal{A}$, and this set is not a principal ideal. By Corollary 1, the family $CEId(\mathcal{A})$ is not **BC**-learnable.

Harizanov [11, Proposition 3.1] proved that for a given Δ_2^0 Turing degree **d**, there exists a computable linear order $\mathcal{B}$ isomorphic to $\omega + \omega^*$ such that the set of all elements from the "ω-part" of $\mathcal{B}$ has degree **d**. We denote this set by $\Omega(\mathcal{B})$.

To finish the proof, it is sufficient to observe two things. Firstly, the set $\Omega(\mathcal{B})$ is the only non-principal ideal of $\mathcal{B}$. Secondly, if one chooses a degree **d** which does not contain c.e. sets (e.g., take a properly d.c.e. degree that exists by the classical result of Cooper [4]), then the set $\Omega(\mathcal{B})$ cannot be c.e. By Corollary 1, we deduce that in this case, the family $CEId(\mathcal{B})$ is **BC**-learnable. $\square$

Now we give another simple application of Theorem 2. Let $\mathcal{L}$ be a lattice. An element $c \in \mathcal{L}$ is *compact* if c satisfies the following: for every subset S of $\mathrm{dom}(\mathcal{L})$, if S has a supremum $\sup S$ and $c \leq_\mathcal{L} \sup S$, then $c \leq_\mathcal{L} \sup F$ for some finite subset $F \subseteq S$.

Proposition 2. *If a computable lattice $\mathcal{L}$ contains a non-compact element, then the family of all ideals $AId(\mathcal{L})$ is not **BC**-learnable.*

Proof. Let c be a non-compact element of $\mathcal{L}$. Choose an infinite set S with the following properties: S has a supremum $\sup S$, $c \leq \sup S$, and for every finite subset $F \subset S$, we have $c \not\leq \sup F$. We show that the ideal $\mathrm{id}(S)$ is not principal.

Towards a contradiction, assume that $\mathrm{id}(S) = \mathrm{id}(e)$ for some element e. Then we have $c \leq \sup S \leq e$. On the other hand, since $e \in \mathrm{id}(S)$, Lemma 1 implies that $e \leq (a_0 \vee a_1 \vee \cdots \vee a_n)$ for some elements $a_i \in S$. Thus, $c \leq (a_0 \vee a_1 \vee \cdots \vee a_n)$, which contradicts the choice of c and S. We deduce that the ideal $\mathrm{id}(S)$ is not principal, and hence, by Theorem 2, the family $AId(\mathcal{L})$ is not **BC**-learnable. $\square$

Notice that a computable lattice $\mathcal{L}$ can have all its elements compact, but still the family $AId(\mathcal{L})$ could be not **BC**-learnable. Indeed, by Proposition 1, the ordinal ω is an example of such $\mathcal{L}$.

In conclusion of this section, we give another example of an infinite lattice $\mathcal{L}$ with learnable $AId(\mathcal{L})$.

Example 1. Consider the following computable lattice $\mathcal{M}$. The lattice has a least element $\perp$ and a greatest element $\top$. All the other elements a_i, $i \in \mathbb{N}$, are pairwise incomparable, and they satisfy $\perp <_{\mathcal{M}} a_i <_{\mathcal{M}} \top$. We observe that the lattice $\mathcal{M}$ is modular and non-distributive, and every ideal of $\mathcal{M}$ is principal. Hence, the family $AId(\mathcal{M})$ is **BC**-learnable.

4   Learning Ideals of a Boolean Algebra

Suppose that a lattice $\mathcal{L}$ is given. Recall that we have noticed that in general, learnability properties of ideals depend on the choice of a computable isomorphic copy of the lattice $\mathcal{L}$ (see Remark 1). Motivated by this phenomenon, this section considers the following question:

Problem 1. Let $\mathcal{B}$ be a Boolean algebra. How does the choice of a computable structure $\mathcal{A}$ isomorphic to $\mathcal{B}$ affect the learnability of the family $CEId(\mathcal{A})$ of all c.e. ideals of $\mathcal{A}$?

Firstly, we establish the following result.

Theorem 3. *For a computable Boolean algebra $\mathcal{B}$, the following conditions are equivalent:*

(a) *There exists a computable structure $\mathcal{A} \cong \mathcal{B}$ such that the family $CEId(\mathcal{A})$ is **BC**-learnable.*
(b) *The algebra $\mathcal{B}$ is atomic.*

Proof. **(b)$\Rightarrow$(a).** This is a direct consequence of Corollary 1 and the following unpublished result of Pavel Alaev:

Proposition 3 (Alaev, private communication). *If $\mathcal{B}$ is a computable, atomic Boolean algebra, then there exists a computable $\mathcal{A}$ isomorphic to $\mathcal{B}$ such that every c.e. ideal of $\mathcal{A}$ is principal.*

(a)⇒(b). Suppose that a Boolean algebra $\mathcal{B}$ is not atomic, and let $\mathcal{C}$ be an arbitrary computable isomorphic copy of $\mathcal{B}$. By Corollary 1, it is sufficient to show that $\mathcal{C}$ has a c.e. non-principal ideal I.

Since $\mathcal{C}$ is not atomic, one can choose a non-zero atomless element c in $\mathcal{C}$. We construct a computable sequence $(d_i)_{i \in \mathbb{N}}$ as follows. Put $d_0 := c$.

Suppose that d_i is already defined, and we have $0 <_\mathcal{C} d_i <_\mathcal{C} d_{i-1}$ (if $i \geq 1$). Since the element c is atomless and $d_i \leq_\mathcal{C} c$, there exists an element e such that $0 <_\mathcal{C} e <_\mathcal{C} d_i$. We find the $\leq_\mathbb{N}$-least such element e and declare $d_{i+1} := e$.

We define an ideal $I := \mathrm{id}(\{d_i \wedge \overline{d_{i+1}} : i \in \mathbb{N}\})$. Since the sequence $(d_i)_{i \in \mathbb{N}}$ is computable, Eq. (2) implies that the ideal I is c.e. Towards a contradiction, assume that $I = \mathrm{id}(e)$ for some element e. Then by Eq. (2), for some $N \in \mathbb{N}$, we have

$$e \leq_\mathcal{C} \left(\bigvee_{i \leq N} d_i \wedge \overline{d_{i+1}} \right) = d_0 \wedge \overline{d_{N+1}}.$$

On the other hand, $d_{N+1} \leq_\mathcal{C} e \leq_\mathcal{C} d_0 \wedge \overline{d_{N+1}}$—and this is impossible, since $d_{N+1} \wedge (d_0 \wedge \overline{d_{N+1}}) = 0 \neq d_{N+1}$. We conclude that I is a c.e. non-principal ideal of $\mathcal{C}$, and hence, the family $CEId(\mathcal{C})$ is not **BC**-learnable.   □

Theorem 3 shows that every computable atomic algebra $\mathcal{B}$ has a 'nice' (from the learnability point of view) isomorphic copy. On the other hand, the next result shows that one could also construct a 'bad' isomorphic copy of $\mathcal{B}$.

Proposition 4. *Let $\mathcal{B}$ be a computable, atomic Boolean algebra such that $\mathcal{B}$ is infinite. Then there exists a computable structure $\mathcal{C} \cong \mathcal{B}$ such that the family $CEId(\mathcal{C})$ is not **BC**-learnable.*

Proof (Sketch). Notice that the Fréchet ideal $Fr(\mathcal{B})$ is non-principal. By η we denote the (order-type of the) standard order of rationals. It is known (see, e.g., Proposition 1.5 in [19]) that the algebra $\mathcal{B}$ satisfies exactly one of the following two conditions:

1. $\mathcal{B} \cong \mathrm{Intalg}(1 + \omega \cdot \eta)$;
2. $\mathcal{B} \cong \mathcal{D} \times \mathrm{Intalg}(\omega)$ for some computable Boolean algebra $\mathcal{D}$.

For these two cases, standard constructions of (computable) Boolean algebra theory allow us to obtain the following facts.

- There exists a computable isomorphic copy $\mathcal{C}$ of $\mathrm{Intalg}(1 + \omega \cdot \eta)$ such that its Fréchet ideal $Fr(\mathcal{C})$ is computable.
- There exist a computable isomorphic copy $\mathcal{C}$ of $\mathcal{D} \times \mathrm{Intalg}(\omega)$ and an element $e \in \mathcal{C}$ such that the ideal $I = \{x : x \in Fr(\mathcal{C})$ and $x <_\mathcal{C} e\}$ is computable and non-principal.

By Corollary 1, the family $CEId(\mathcal{C})$ is not **BC**-learnable.   □

5   Conservative Learning, and Reverse Mathematics

Hölzl, Jain, and Stephan [13] initiated systematic investigations of inductive inference within the framework of reverse mathematics. Inspired by their results, here we give one application of reverse mathematics to our setting of learnability for classes of ideals.

In reverse mathematics, one works within subsystems of second-order arithmetic Z_2. One can verify that, given two theorems, either one implies the other (provided some tools offered by the base theory), or the two are incomparable. Some common subsystems of Z_2 are obtained by limiting the comprehension and induction axioms to specific classes of formulae. Here we use the following two subsystems: RCA_0 is the weak base theory with comprehension for Δ_1^0-definable sets and with induction limited to Σ_1^0-formulae; ACA_0 extends RCA_0 allowing for definitions of sets by arithmetical comprehension. We refer to the monograph [20] for the background on reverse mathematics.

A model $(M, \mathcal{S}) \models RCA_0$ consists of two parts: M is its first-order part, and $\mathcal{S}$ is a class of subsets of M. Note that $\mathcal{S}$ is not necessarily equal to the set of all subsets of M. Typically, one uses $\mathbb{N}$ to denote M—so, in what follows, a reader should remember that $\mathbb{N}$ contains (*not necessarily standard*) natural numbers belonging to the model.

Note that RCA_0 allows standard computability-theoretic encodings, so one could view a finite tuple $(a_1, \ldots, a_n)$ of natural numbers as its code $\langle a_1, \ldots, a_n \rangle \in \mathbb{N}$. As in [13], a sequence $(A_i)_{i \in \mathbb{N}}$ is a *uniformly represented family of sets* if the set $\{\langle i, x \rangle : x \in A_i\}$ belongs to $\mathcal{S}$.

For reverse-mathematical considerations, the paper [13] introduced several formal versions of the famous Angluin's tell-tale criterion [1]. One of them is the following:

Definition 2 (see Definition 5 in [13]). *A uniformly represented family of sets $(A_i)_{i \in \mathbb{N}}$ satisfies the tell-tale criterion in general if for each $i \in \mathbb{N}$ there is $b_i \in \mathbb{N}$ such that*
$$\neg \exists j (A_i \cap \{0, 1, \ldots, b_i\} \subseteq A_j \subset A_i).$$

Corollary 11 of [13] proves that over RCA_0, the system ACA_0 is equivalent to the following statement: every uniformly represented family satisfying the tell-tale criterion in general is conservatively **Ex**-learnable.

In what follows, we establish a version of this result for the case of learnability for classes of ideals. Firstly, we discuss our conventions (within RCA_0). The omitted details could be recovered from [13].

We will work with the Boolean algebra $\mathcal{B}_{\mathbb{N}}$ containing all finite and all cofinite subsets of $\mathbb{N}$. The atoms of $\mathcal{B}_{\mathbb{N}}$ are precisely the one-element sets $\{i\}$, for $i \in \mathbb{N}$. If an ideal I is a subset of the Fréchet ideal $Fr(\mathcal{B}_{\mathbb{N}})$, then it is clear that

$$I = \mathrm{id}(\{\{j\} : \{j\} \in I\}). \tag{3}$$

By $SFr(\mathcal{B}_{\mathbb{N}})$ we denote the set of all ideals I such that $I \subseteq Fr(\mathcal{B}_{\mathbb{N}})$ and I is not the least ideal of $\mathcal{B}_{\mathbb{N}}$.

For convenience, we assume that a text for an ideal $I \in SFr(\mathcal{B}_\mathbb{N})$ is a function $t \colon \mathbb{N} \to \mathbb{N}$ such that $t \in \mathcal{S}$ and range$(t) = \{j : \{j\} \in I\}$. (Here we identify a function t with its graph, and hence we may say that t belongs to $\mathcal{S}$, i.e., to the second-order part of a model.) In other words, instead of enumerating the ideal I 'directly', a text enumerates (indices of) all the atoms belonging to I. By Eq. (3), this notion of a text is well-defined. So, in our representations, we will identify the ideal I itself with the set range(t) for a text t for I.

A learner is a function $M \colon \mathbb{N}^{<\mathbb{N}} \to \mathbb{N}$ such that $M \in \mathcal{S}$. If one fixes a uniformly represented family of sets $(A_i)_{i \in \mathbb{N}}$, then the outputted conjecture i corresponds to A_i.

Theorem 4. *Over* RCA$_0$, *ACA$_0$ is equivalent to the following statement: every uniformly represented family of ideals from $SFr(\mathcal{B}_\mathbb{N})$ satisfying the tell-tale criterion in general is conservatively* **Ex**-*learnable.*

Proof. ($\Rightarrow$). Follows directly from Corollary 11 of [13].

($\Leftarrow$). Let $g \colon \mathbb{N} \to \mathbb{N}$ be an injective function belonging to $\mathcal{S}$. By Lemma III.1.3 in [20], to deduce ACA$_0$, it is sufficient to prove that the set range(g) also belongs to $\mathcal{S}$.

We define a uniformly represented family of ideals $(I_{\langle i,m \rangle})_{i,m \in \mathbb{N}}$ as follows:

- $I_{\langle i,0 \rangle} := \mathrm{id}(\{\{0\}, \{\langle i+1, k \rangle\} : k \in \mathbb{N}\});$
- $I_{\langle i,m+1 \rangle} := \mathrm{id}(\{\{0\}, \{\langle i+1, k \rangle\} : k \leq m\}),$ if $g(m) = i;$
- $I_{\langle i,m+1 \rangle} := \mathrm{id}(\{\{0\}\}),$ if $g(m) \neq i.$

Observe that the sequence $(I_{\langle i,m \rangle})_{i,m \in \mathbb{N}}$ satisfies the tell-tale criterion in general.

Let M be a conservative **Ex**-learner for the family $(I_{\langle i,m \rangle})_{i,m \in \omega}$. For $i \in \mathbb{N}$, consider a text t_i such that $t_i(0) = 0$ and $t_i(k+1) = \langle i+1, k \rangle$. Since t_i is a text for $I_{\langle i,0 \rangle}$ (and M **Ex**-learns $I_{\langle i,0 \rangle}$), there exists an index l such that $M(t_i \upharpoonright l) = \langle i,0 \rangle$. By applying minimization (Theorem II.3.5 in [20]), we get a function $h(i) = \mu l[M(t_i \upharpoonright l) = \langle i,0 \rangle]$, and this h belongs to $\mathcal{S}$.

We prove that $i \in \mathrm{range}(g)$ if and only if $(\exists m \leq h(i))(g(m) = i)$. Indeed, towards a contradiction, assume that $g(m) = i$ for some $m > h(i)$ (recall that g is injective). Then the ideal $J := I_{\langle i,m+1 \rangle}$ has a text t' such that $t' \supset t_i \upharpoonright h(i)$. Since $J \subset I_{\langle i,0 \rangle}$ and the learner M is conservative, we have $M(t' \upharpoonright l) = \langle i,0 \rangle$ for all $l \geq h(i)$. Thus, M does not learn the ideal J, which gives a contradiction.

Notice that the set range(g) is Σ_1^0-definable. On the other hand, $i \notin \mathrm{range}(g)$ if and only if $(\forall m \leq h(i))(g(m) \neq i)$; this implies Π_1^0-definability of range(g). By applying Δ_1^0-comprehension, we obtain that range(g) belongs to $\mathcal{S}$. $\square$

Acknowledgements. The work was supported by Nazarbayev University Faculty Development Competitive Research Grants 201223FD8823. The research of Bazhenov is funded by the Science Committee of the Ministry of Science and Higher Education of the Republic of Kazakhstan (Grant No. AP19576325). The work of Bazhenov was also carried out within the framework of the state contract of the Sobolev Institute of Mathematics (project no. FWNF-2022-0011).

References

1. Angluin, D.: Inductive inference of formal languages from positive data. Inf. Control **45**(2), 117–135 (1980). https://doi.org/10.1016/S0019-9958(80)90285-5
2. Ash, C.J., Knight, J.: Computable Structures and the Hyperarithmetical Hierarchy. Studies in Logic and the Foundations of Mathematics, vol. 144. Elsevier Science B.V., Amsterdam (2000)
3. Blum, L., Blum, M.: Toward a mathematical theory of inductive inference. Inf. Control **28**(2), 125–155 (1975). https://doi.org/10.1016/S0019-9958(75)90261-2
4. Cooper, S.B.: Degrees of unsolvability. Ph.D. thesis, Leicester University (1971)
5. de Brecht, M., Kobayashi, M., Tokunaga, H., Yamamoto, A.: Inferability of closed set systems from positive data. In: Washio, T., Satoh, K., Takeda, H., Inokuchi, A. (eds.) JSAI 2006. LNCS (LNAI), vol. 4384, pp. 265–275. Springer, Heidelberg (2007). https://doi.org/10.1007/978-3-540-69902-6_23
6. de Brecht, M., Yamamoto, A.: Topological properties of concept spaces (full version). Inf. Comput. **208**(4), 327–340 (2010). https://doi.org/10.1016/J.IC.2009.08.001
7. Gao, Z., Stephan, F., Wu, G., Yamamoto, A.: Learning families of closed sets in matroids. In: Dinneen, M.J., Khoussainov, B., Nies, A. (eds.) WTCS 2012. LNCS, vol. 7160, pp. 120–139. Springer, Heidelberg (2012). https://doi.org/10.1007/978-3-642-27654-5_10
8. Gold, E.M.: Language identification in the limit. Inf. Control **10**(5), 447–474 (1967). https://doi.org/10.1016/S0019-9958(67)91165-5
9. Goncharov, S.S.: Countable Boolean Algebras and Decidability. Consultants Bureau, New York (1997)
10. Grätzer, G.: Lattice Theory: Foundation. Birkhäuser, Basel (2011)
11. Harizanov, V.S.: Turing degrees of certain isomorphic images of computable relations. Ann. Pure Appl. Logic **93**(1–3), 103–113 (1998). https://doi.org/10.1016/S0168-0072(97)00056-0
12. Harizanov, V.S., Stephan, F.: On the learnability of vector spaces. J. Comput. Syst. Sci. **73**(1), 109–122 (2007). https://doi.org/10.1016/j.jcss.2006.09.001
13. Hölzl, R., Jain, S., Stephan, F.: Inductive inference and reverse mathematics. Ann. Pure Appl. Logic **167**(12), 1242–1266 (2016). https://doi.org/10.1016/j.apal.2016.06.002
14. Jain, S., Osherson, D., Royer, J.S., Sharma, A.: Systems that Learn. MIT Press, Cambridge (1999)
15. Koppelberg, S.: Handbook of Boolean Algebras, vol. 1. North-Holland, Amsterdam (1989)
16. Lange, S., Zeugmann, T., Zilles, S.: Learning indexed families of recursive languages from positive data: a survey. Theor. Comput. Sci. **397**(1–3), 194–232 (2008). https://doi.org/10.1016/j.tcs.2008.02.030
17. Odifreddi, P.: Classical Recursion Theory. Studies in Logic and the Foundations of Mathematics, vol. 125. Elsevier Science B.V., Amsterdam (1992)
18. Putnam, H.: Trial and error predicates and the solution to a problem of Mostowski. J. Symb. Log. **30**(1), 49–57 (1965). https://doi.org/10.2307/2270581
19. Remmel, J.B.: Recursive Boolean algebras. In: Monk, J.D., Bonnet, R. (eds.) Handbook of Boolean Algebras, vol. 3, pp. 1097–1165. North-Holland, Amsterdam (1989)

20. Simpson, S.G.: Subsystems of Second Order Arithmetic, 2nd edn. Cambridge University Press, Cambridge (2009)
21. Stephan, F., Ventsov, Y.: Learning algebraic structures from text. Theor. Comput. Sci. **268**(2), 221–273 (2001). https://doi.org/10.1016/S0304-3975(00)00272-3
22. Zeugmann, T., Zilles, S.: Learning recursive functions: a survey. Theor. Comput. Sci. **397**(1–3), 4–56 (2008). https://doi.org/10.1016/j.tcs.2008.02.021

Unambiguous and Co-nondeterministic Computations of Finite Automata and Pushdown Automata Families and the Effects of Multiple Counters

Tomoyuki Yamakami[(✉)]

Faculty of Engineering, University of Fukui, 3-9-1 Bunkyo, Fukui 910-8507, Japan
TomoyukiYamakami@gmail.com

Abstract. Nonuniform families of polynomial-size finite automata and pushdown automata respectively have strong connections to nonuniform-NL and nonuniform-LOGCFL. We examine the behaviors of unambiguous and co-nondeterministic computations produced by such families of automata operating multiple counters. As its consequences, we obtain various collapses of the complexity classes of families of promise problems solvable by finite and pushdown automata families when all valid instances are limited to either polynomially long strings or unary strings. A key technical ingredient of our proofs is an inductive counting of reachable vertices of each computation graph of finite and pushdown automata that operate multiple counters simultaneously.

1 Background and Challenging Questions

1.1 Two Important Open Questions in Nonuniform Polynomial State Complexity Theory

Nondeterministic computation has played an important role in computational complexity theory as well as automata theory. Associated with such computation, there are two central and crucial questions to resolve. (i) Can any co-nondeterministic computation be simulated on an appropriate nondeterministic machine? (ii) Can any nondeterministic machine be made unambiguous? In the polynomial-time setting, these questions correspond to the famous NP =?co-NP and NP =?UP questions. In this work, we attempt to resolve these questions in the setting of nonuniform polynomial state complexity classes.

We quickly review the origin and the latest progress of the study of nonuniform state complexity classes. Apart from a standard uniform model of finite automata, Berman and Lingas [1] and Sakoda and Sipser [13] considered, as a "collective" model of computations, nonuniform families of two-way finite automata indexed by natural numbers and they studied the computational power of these families of finite automata having *polynomial size* (i.e., having polynomially many inner states). Unlike Boolean circuit families, these automata are

X. Chen and B. Li (Eds.): TAMC 2024, LNCS 14637, pp. 14–25, 2024.
https://doi.org/10.1007/978-981-97-2340-9_2

allowed to take "arbitrarily" long inputs. Of those families of polynomial-size finite automata, Sakoda and Sipser focused on the models of two-way deterministic finite automata (or 2dfa's, for short) and two-way nondeterministic finite automata (or 2nfa's). They introduced the complexity classes, dubbed as 2D and 2N, which consist of all families of "promise" problems[1] solvable respectively by nonuniform families of polynomial-size 2dfa's and 2nfa's. As their natural extensions, nonuniform families of two-way deterministic pushdown automata (or 2dpda's) and two-way nondeterministic pushdown automata (or 2npda's) running in polynomial time have also been studied lately [16,19]. Similarly to 2D and 2N, these pushdown automata models induce two corresponding complexity classes of promise problem families, denoted respectively 2DPD and 2NPD. Since an introduction of nonuniform polynomial-size finite automata families, various machine types (such as deterministic, nondeterministic, alternating, probabilistic, and quantum) have been studied in depth [3,7–10,15–19].

An importance of polynomial-size finite automata families comes from their close connection to decision problems in the nonuniform variants of the log-space complexity classes L and NL, when all promised (or valid) instances given to underlying finite automata are limited to polynomially long strings (where this condition is referred to as a *polynomial ceiling*) [9], or when all valid instances are limited to unary strings [10]. In a similar fashion, 2DPD and 2NPD are closely related to the nonuniform versions of LOGDCFL and LOGCFL [16] when all valid instances are only polynomially long, where LOGCFL (resp., LOGDCFL) is the collection of all languages log-space many-one reducible to context-free (resp., deterministic context-free) languages. These strong correspondences to standard computational complexity theory provide one of the good reasons to investigate the fundamental properties of various types of nonuniform automata families in hopes of achieving a better understanding of parallel complexity classes, such as L, NL, LOGDCFL, and LOGCFL, in the nonuniform setting.

For nonuniform finite and pushdown automata families, nevertheless, the aforementioned two central open questions (i)–(ii) correspond to the 2N =?2U, 2N =?co-2N, 2NPD =?2UPD, and 2NPD =?co-2NPD questions. Unfortunately, these four equalities are not known to hold at this moment. It is therefore desirable to continue the study on the behaviors of finite and pushdown automata families in order to deepen our understandings of these machine families and to eventually resolve those four central questions.

1.2   New Challenges and Main Contributions

The primary purpose of this work is to present "partial" solutions to the four important questions raised in Sect. 1.1 associated with 2N and 2NPD by studying the computational power of nonuniform families of polynomial-size finite and pushdown automata in depth.

---

[1] A *promise problem* over alphabet Σ is a pair (A, R) satisfying that $A, R \subseteq \Sigma^*$ and $A \cap R = \varnothing$. A *language* L over Σ can be identified with a unique promise problem having the form $(L, \Sigma^* - L)$.

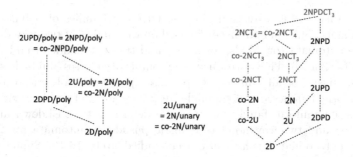

Fig. 1. Containments among nonuniform polynomial state complexity classes shown in this work. Remark that the collapse 2U/poly = 2N/poly comes from [18] and 2U/unary = 2N/unary = co-2N/unary are drawn from [4,5] in Sect. 4.

In the course of our study, we further look into the key role of "counters", each of which is essentially a stack manipulating only a single symbol, say, "1" except for the bottom marker $\bot$. Since the total number of 1s in a counter can be viewed as a natural number, the counter is able to "count", as its name suggests. For the first time, we supplement multiple counters to the existing models of nonuniform finite and pushdown automata families. We remark that, for short runtime computation, counters are significantly weaker[2] in functionality than full-scale stacks. Even though, the proper use of counters can help us not only trace the tape head location but also count the number of steps.

By appending multiple counters to finite and pushdown automata, we obtain the machine models of *counter automata* and *counter pushdown automata*. Two additional complexity classes, $2NCT_k$ and $2NPDCT_k$ are naturally obtained by taking families of polynomial-size nondeterministic counter automata and counter pushdown automata operating k counters.

The use of multiple counters makes it possible to show in Sect. 3 that 2N and co-2N coincide and that 4 counters are enough (namely, $2NCT_4 = \text{co-}2NCT_4$). This result further leads to the equivalence between co-2N and 2N in Sect. 4.3 when all promise problem families are restricted to having polynomial ceilings. Under the same restriction, we will show that co-2NPD and 2NPD also coincide. Our results are briefly summarized in Fig. 1, in which the suffix "/poly" refers to the polynomial ceiling restriction and the suffix "/unary" refers to the restriction to unary input strings. To obtain some of the equalities in the figure, we will exploit a close connection between *parameterized decision problems* and families of promise problems, which was first observed in [17] and then fully developed in [16,18,19].

---

[2] With exponential overhead, 2-counter automata can simulate a Turing machine [11].

2 Preliminaries: Notions and Notation

2.1 Numbers, Languages, and Pushdown Automata

The set of all *natural numbers* (including 0) is denoted $\mathbb{N}$ and the positive-integer set $\mathbb{N} - \{0\}$ is expressed as $\mathbb{N}^+$. Given two integers m and n with $m \leq n$, the notation $[m, n]_{\mathbb{Z}}$ denotes the *integer set* $\{m, m + 1, m + 2, \ldots, n\}$. As a special case, we write $[n]$ for $[1, n]_{\mathbb{Z}}$ when $n \in \mathbb{N}^+$. In this work, all *polynomials* must have nonnegative coefficients and all *logarithms* are taken to the base 2 with the notation $\log 0$ being treated as 0. The *power set* of a set S is denoted $\mathcal{P}(S)$.

Given an alphabet Σ and a number $n \in \mathbb{N}$, the notation Σ^n (resp., $\Sigma^{\leq n}$) denotes the set of all strings over Σ of length exactly n (resp., at most n). The *empty string* is always denoted ε.

As a basic machine model, we use *two-way nondeterministic finite automata* (or 2nfa's, for short) that make neither ε-moves[3] nor stationary moves. This means that input tape heads of 2nfa's always move to adjacent tape cells without stopping at any tape cells. Given a finite automaton M, the *state complexity* $sc(M)$ of M is the total number of inner states used for M.

Another important machine model is *two-way nondeterministic pushdown automata* (or 2npda's) N over alphabet Σ with a stack alphabet Γ including the bottom marker $\perp$. Notice that N is allowed to make ε-moves whereas 2nfa's make no ε-moves. The *stack-state complexity* $ssc(N)$ of N denotes the product $|Q| \cdot |\Gamma^{\leq e}|$, which turns out to be a useful complexity measure [16,19], where Q is a set of inner states and e is the *push size* (i.e., the maximum length of pushed strings into the stack at any single push operation).

A *counter* is a special kind of (pushdown) stack whose alphabet consists only of a single symbol, say, "1" except for $\perp$. In this work, we freely equip multiple counters to finite automata and pushdown automata. These machines are respectively called *counter automata* and *counter pushdown automata*. We conveniently abbreviate a *two-way nondeterministic counter automaton* as a 2ncta and a *two-way nondeterministic counter pushdown automaton* as a 2npdcta.

2.2 Promise Problems and Nonuniform Families

Unlike the notion of languages, *promise problems* over alphabet Σ are formally of the form (A, R) satisfying that $A, R \subseteq \Sigma^*$ and $A \cap R = \varnothing$. We say that a 1nfa (1ncta, 1npda, or 1npdcta) M *solves* (A, R) if (i) for any $x \in A$, M accepts x (i.e., there exists an accepting computation path of M on x) and (ii) for any $x \in R$, M rejects x (i.e., all computation paths of M on x are rejecting). Any string in $A \cup R$ is said to be *promised* or *valid*. Since we do not impose any further condition on all strings outside of $A \cup R$, it suffices to focus only on the promised strings in our later discussion.

For any nondeterministic machine models discussed in Sect. 2.1, a machine is said to be *unambiguous* if it has at most one accepting computation path on each

---

[3] An ε-move of an automaton refers to the case where the automaton makes a transition without reading any input symbol.

promised instance. For other instances, there is no restriction on the number of accepting/rejecting computation paths.

Throughout this work, we consider any "family" $\mathcal{L}$ of promise problems $(L_n^{(+)}, L_n^{(-)})$ over a common fixed alphabet Σ indexed by natural numbers $n \in \mathbb{N}$. Such a family $\mathcal{L}$ is said to have a *polynomial ceiling* if there exists a polynomial p such that $L_n^{(+)} \cup L_n^{(-)} \subseteq \Sigma^{\leq p(n)}$ holds for all indices $n \in \mathbb{N}$. Given a complexity class $\mathcal{C}$ of families of promise problems, if we restrict our attention to only promise problem families in $\mathcal{C}$ having a polynomial ceiling, then we obtain the subclass of $\mathcal{C}$, expressed as $\mathcal{C}/\mathrm{poly}$. Moreover, when all promise problems are restricted to the ones over unary alphabets, we obtain the subclass $\mathcal{C}/\mathrm{unary}$. Those exotic notations come from [7,8] and are adopted in [10,15–19].

A family $\mathcal{M} = \{M_n\}_{n \in \mathbb{N}}$ of 2nfa's (resp., 2npda's) over alphabet Σ is of *polynomial size* if there exists a polynomial p satisfying $sc(M_n) \leq p(n)$ (resp., $ssc(M_n) \leq p(n)$) for all $n \in \mathbb{N}$. Similar notions are definable for other machine models, such as 2dfa's, 2ncta's, 2npda's, 2npdcta's, and 2dpdcta's.

Given a family $\mathcal{L} = \{(L_n^{(+)}, L_n^{(-)})\}_{n \in \mathbb{N}}$ of promise problems, a family $\mathcal{M} = \{M_n\}_{n \in \mathbb{N}}$ of nondeterministic machines, and a polynomial p, we say that M_n *solves* $(L_n^{(+)}, L_n^{(-)})$ *within time* $p(n, |x|)$ if (1) for any $x \in L_n^{(+)}$, there exists an accepting computation path of M_n on x having length at most $p(n, |x|)$ and (2) for any $x \in L_n^{(-)}$, there is no accepting computation path of M_n on x but there is a rejecting computation path of having length at most $p(n, |x|)$. Moreover, $\mathcal{M}$ is said to *solve* $\mathcal{L}$ *in polynomial time* if there is a polynomial p such that, for all indices $n \in \mathbb{N}$, M_n solves $(L_n^{(+)}, L_n^{(-)})$ within time $p(n, |x|)$. We define 2N as the collection of all families of promise problems solvable by nonuniform families of polynomial-size 2nfa's in polynomial time.[4] By replacing these 2nfa's with 2dfa's, 2dpda's, and 2npda's, we respectively obtain 2D, 2DPD, and 2NPD. For any $k \in \mathbb{N}^+$, we further define $2\mathrm{NCT}_k$ and $2\mathrm{NPDCT}_k$ using k-counter 2ncta's and k-counter 2npdcta's,[5] respectively. When $k = 1$, we tend to drop the subscript "k" and write, e.g., 2NCT instead of $2\mathrm{NCT}_1$. The notation co-$\mathcal{L}$ denotes $\{(L_n^{(-)}, L_n^{(+)})\}_{n \in \mathbb{N}}$. Given a complexity class $\mathcal{C}$ of promise problem families, such as 2N and 2NPD, co-$\mathcal{C}$ expresses the class $\{\mathrm{co}\text{-}\mathcal{L} \mid \mathcal{L} \in \mathcal{C}\}$. It follows that $2\mathrm{D} = \mathrm{co}\text{-}2\mathrm{D}$ and $2\mathrm{DPD} = \mathrm{co}\text{-}2\mathrm{DPD}$.

In addition, the use of unambiguous 1nfa's and unambiguous 2npda's introduces the complexity classes 2U and 2UPD, respectively, and their multi-counter variants $2\mathrm{UCT}_k$ and $2\mathrm{UPDCT}_k$. It then follows that $2\mathrm{D} \subseteq 2\mathrm{DPD} \subseteq 2\mathrm{UPD} \subseteq 2\mathrm{UPDCT}$ and $2\mathrm{U} \subseteq 2\mathrm{N} \subseteq 2\mathrm{NCT} \subseteq 2\mathrm{NPD} \subseteq 2\mathrm{NPDCT}$.

2.3 Reducing the Number of Counters in Use

It is possible to reduce the number of counters in use on multi-counter automata and multi-counter pushdown automata. Minsky [11] earlier demonstrated how to

---

[4] As shown by Geffert et al. [4], in the case of 2dfa's and 2nfa's, removing the "polynomial time" requirement does not change the definitions of 2D and 2N.

[5] We remark that, with the use of multiple counters, it is possible to force 2ncta's and 2npdcta's to halt within polynomial time on *all computation paths*.

simulate a Turing machine on a 2-counter automaton with exponential overhead. Since we cannot use the same simulation technique due to its large overhead, we need to take another, more direct approach toward $2\mathrm{NCT}_k$ and $2\mathrm{NPDCT}_k$. Even without the requirement of polynomial ceiling, it is possible in general to reduce the number of counters in use down to "4" for 2ncta's and "3" for 2npdcta's as shown below.

Proposition 1. *For any constants* $k, k' \in \mathbb{N}$ *with* $k \geq 4$ *and* $k' \geq 3$, $2\mathrm{NCT}_k = 2\mathrm{NCT}_4$ *and* $2\mathrm{NPDCT}_{k'} = 2\mathrm{NPDCT}_3$. *The same holds for the deterministic case.*

A core of the proof of Proposition 1 is the following lemma on the simulation of every pair of counters by a single counter with the heavy use of an appropriately defined "pairing" function. Let $\mathcal{M} = \{M_n\}_{n \in \mathbb{N}}$ denote any polynomial-size family of 2ncta's or of 2npdcta's running in time, in particular, $(n|x|)^t$ for a fixed constant $t \in \mathbb{N}^+$. This $\mathcal{M}$ satisfies the following lemma.

Lemma 2. *There exists a fixed deterministic procedure by which any single move of push/pop operations of two counters of M_n can be simulated by a series of operations with one counter with the help of 3 extra counters. These extra 3 counters are emptied after each simulation and thus they are reusable for any other purposes. If we freely use a stack during this simulation procedure, then we need only two extra counters instead of three. The state complexity of the procedure is $n^{O(1)}$.*

Proof Sketch. For each index $j \in [2]$, the notation i_j denotes the current content (viewed as a natural number) of the jth counter of M_n on input x. Let us consider the pair (i_1, i_2) and encode it into a number defined as $i_1 \cdot p + i_2$, which is succinctly expressed as $\langle i_1, i_2 \rangle_p$, where $p = (n|x|)^t$. We wish to simulate the behavior of the two counters of the 2npda M_n using four counters CT1, CT2, CT3, and CT4. (a) We produce the number p in CT2 using CT1, CT4, and a tape head. (b) We then check, in the current value $\langle i_1, i_2 \rangle_p$ stored in CT3, whether $i_j = 0$ or not for each $j \in [2]$. (c) We next simulate a single push/pop operation of the two counters.

In the case where M_n is a multi-counter 2npda's, we can use its stack as a reusable counter during the processes (a)–(c) by first writing a separator, say, # viewed as a new bottom marker.   □

It is not clear that "4" and "3" are the smallest numbers supporting Proposition 1 for 2ncta's and 2npdcta's, respectively. We may conjecture that $2\mathrm{NCT}_i \neq 2\mathrm{NCT}_{i+1}$ and $2\mathrm{NPDCT}_j \neq 2\mathrm{NPDCT}_{j+1}$ for all $i \in [3]$ and $j \in [2]$.

3   Reachability with No Polynomial Ceiling Bounds

Let us consider the question raised in Sect. 1.1 on the closure property under complementation, namely, the 2N =?co-2N question. Unfortunately, we do not know its answer. With the presence of "counters", however, it is possible to provide a complete solution to this question.

Theorem 3. *For any constant $k \geq 4$, co-$2\mathrm{NCT}_k \subseteq 2\mathrm{NCT}_4$.*

Corollary 4. $2\mathrm{NCT}_4 = $ co-$2\mathrm{NCT}_4$.

A key to the proof of Theorem 3 is the following lemma.

Lemma 5. *For any constant $k \in \mathbb{N}^+$, co-$2\mathrm{NCT}_k \subseteq 2\mathrm{NCT}_{5k+12}$.*

The proof of Theorem 3 is described as follows. By Proposition 1, it suffices to consider the case of $k = 4$. We then obtain co-$2\mathrm{NCT}_4 \subseteq 2\mathrm{NCT}_{32}$ by Lemma 5. Proposition 1 again leads to $2\mathrm{NCT}_{32} \subseteq 2\mathrm{NCT}_4$. Therefore, co-$2\mathrm{NCT}_4 \subseteq 2\mathrm{NCT}_4$ follows. By taking the "complementation" of the both sides of this inclusion, we also obtain $2\mathrm{NCT}_4 \subseteq$ co-$2\mathrm{NCT}_4$. Corollary 4 is thus obtained.

To prove Lemma 5, we wish to use an algorithmic technique known as *inductive counting*. This intriguing technique was discovered independently by Immerman [6] and Szelepcsényi [14] in order to prove that $\mathrm{NL} = $ co-NL.

For the description of the proof of Lemma 5, we introduce the following special notion of (internal) configurations for a family $\{M_n\}_{n \in \mathbb{N}}$ of the nth k-counter 2ncta M_n running in time polynomial, say, $r(n, |x|)$. A *configuration* of M_n on input x is of the form $(q, l, \vec{m})$ with $q \in Q$, $l \in [0, |x| + 1]_{\mathbb{Z}}$, and $\vec{m} = (m_1, m_2, \ldots, m_k) \in \mathbb{N}^k$. This form indicates that M_n is in inner state q, scanning the lth tape cell, and M_n's ith counter holds a number m_i for each index $i \in [k]$. We abbreviate as $CONF_{n,|x|}$ the configuration space $Q_n \times [0, |x| + 1]_{\mathbb{Z}} \times [0, r(n, |x|)]_{\mathbb{Z}}^k$. Given two configurations c_1 and c_2 of M_n on x, the notation $c_1 \vdash_x c_2$ means that c_2 is "reachable" from c_1 by making a single move of M_n on x. We abbreviate as $c_1 \vdash_x^{t-1} c_t$ a chain of transitions $c_1 \vdash_x c_2 \vdash_x \cdots \vdash_x c_t$. Moreover, $\vdash_x^*$ denotes the transitive closure of $\vdash_x$.

Proof Sketch of Lemma 5. Given a family $\mathcal{L} = \{(L_n^{(+)}, L_n^{(-)})\}_{n \in \mathbb{N}}$ in co-$2\mathrm{NCT}_k$, we consider a nonuniform family $\{M_n\}_{n \in \mathbb{N}}$ of polynomial-size k-counter 2ncta's that solves co-$\mathcal{L}$ within $r(n, |x|)$ time for a polynomial r.

Our goal is to build a $(5k + 12)$-counter 2ncta P_n, which solves the promise problem $(L_n^{(+)}, L_n^{(-)})$ for each index $n \in \mathbb{N}$ in polynomial time. A basic idea of constructing P_n is to provide a procedure of deciding nondeterministically whether M_n rejects input x; in other words, whether all computation paths of M_n on x reach non-accepting inner states. For this purpose, we need to "count" the number of non-accepting computation paths of M_n on x. If this number matches the total number of computation paths, then we know that M_n rejects x. From this follows $\mathcal{L} \in 2\mathrm{NCT}_{5k+12}$.

We arbitrarily fix a number n and a valid input x. In what follows, we deal with configurations of the form $(q, l, \vec{m})$ in $CONF_{n,|x|}$. It is important to note that, with the use of additional $k + 1$ counters, we can "enumerate" all elements in $CONF_{n,|x|}$, ensuring a linear order on $CONF_{n,|x|}$. This fact makes it possible for us to select the elements of $CONF_{n,|x|}$ sequentially one by one in the construction of P_n. For each number $i \in [0, r(n, |x|)]_{\mathbb{Z}}$, we define $V_i = \{(q, l, \vec{m}) \in CONF_{n,|x|} \mid (q_0, 0, \vec{0}) \vdash_x^i (q, l, \vec{m})\}$ and set $N_i = |V_i|$. Clearly, N_0 equals 1.

We calculate the value $N_{r(n,|x|)}$ by inductively calculating N_i for each $i \in [0, r(n,|x|)]_{\mathbb{Z}}$ using extra reusable counters. We use a counter to remember the value i in $[0, r(n,|x|)]_{\mathbb{Z}}$. Since $0 \leq N_i \leq |CONF_{n,|x|}| = |Q_n|(r(n,|x|)+1)^k(|x|+2)$, we also need to remember the value N_i using another counter. Furthermore, we can remember the current location of M_n's tape head using an additional counter. To hold $\vec{m}$, we also need extra k counters.

After calculating $N_{r(n,|x|)}$, we sequentially pick all elements $(z, t, \vec{e})$ from $(Q_n - Q_{acc,n}) \times [0, |x|+1]_{\mathbb{Z}} \times [0, r(n,|x|)]_{\mathbb{Z}}^k$ one by one and check whether $N_{r(n,|x|)}$ matches the number of non-accepting computation paths. If so, then we accept x; otherwise, we reject x.

It is possible to prove that, in a certain computation path of P_n, the value N_i is correctly calculable for each index $i \in [0, r(n,|x|)]_{\mathbb{Z}}$. Therefore, P_n correctly solves $(L_n^{(+)}, L_n^{(-)})$. The above procedure can be implemented on an appropriate 2ncta with the total of $5(k+1) + 7$ counters. Thus, $\mathcal{L}$ belongs to $2\mathrm{NCT}_{5k+12}$. □

4   Effects of Polynomial Ceiling Bounds

4.1   Elimination of Counters

In Sect. 2.3, we have discussed how to reduce the number of counters down to three or four. By the presence of polynomial ceilings, it is further possible to eliminate all counters from counter automata and counter pushdown automata.

Theorem 6. *Let k be an arbitrary constant in $\mathbb{N}^+$. (1) $2\mathrm{NCT}_k/\mathrm{poly} = 2\mathrm{N}/\mathrm{poly}$. (2) $2\mathrm{NPDCT}_k/\mathrm{poly} = 2\mathrm{NPD}/\mathrm{poly}$. The same statement holds even if underlying nondeterministic machines are changed to deterministic ones.*

Proof Sketch. Here, we prove only (2) since the proof of (1) is in essence similar. Since $2\mathrm{NPD} \subseteq 2\mathrm{NPDCT}_k$ for all $k \geq 1$, we obtain $2\mathrm{NPD}/\mathrm{poly} \subseteq 2\mathrm{NPDCT}_k/\mathrm{poly}$. We next show that $2\mathrm{NPDCT}_k/\mathrm{poly} \subseteq 2\mathrm{NPD}/\mathrm{poly}$. Consider any family $\mathcal{L} = \{(L_n^{(+)}, L_n^{(-)})\}_{n \in \mathbb{N}}$ of promise problems in $2\mathrm{NPDCT}_k/\mathrm{poly}$ over alphabet Σ. There exists a polynomial p satisfying $L_n^{(+)} \cup L_n^{(-)} \subseteq \Sigma^{\leq p(n)}$ for all $n \in \mathbb{N}$. Take any family $\{M_n\}_{n \in \mathbb{N}}$ of polynomial-size 2npdcta's that solves $\mathcal{L}$ with k counters in time polynomial in $(n, |x|)$. Let q denote a polynomial such that, for any $n \in \mathbb{N}$ and any promised input $x \in \Sigma^*$, $q(n, |x|)$ upper-bounds the runtime of M_n on x. Since p is a ceiling, all the k counters of M_n hold the number at most $q(n, p(n))$ on all "valid" instances. Define $r(n) = q(n, p(n))$. Since r is a polynomial, it is possible to express the contents of the k counters in the form of inner states. Therefore, without using any counter, we can simulate M_n on the input x by running an appropriate 2npda whose stack-state complexity is polynomially bounded. This implies that $\mathcal{L} \in 2\mathrm{NPD}/\mathrm{poly}$.

The last part of the theorem follows in a similar way as described above. □

Corollary 7. *(1) If $2\mathrm{DCT}_4 = 2\mathrm{NCT}_4$, then $\mathrm{L}/\mathrm{poly} = \mathrm{NL}/\mathrm{poly}$. (2) If $2\mathrm{DPDCT}_3 = 2\mathrm{NPDCT}_3$, then $\mathrm{LOGDCFL}/\mathrm{poly} = \mathrm{LOGCFL}/\mathrm{poly}$.*

The proof of this corollary is obtainable from Theorem 6 as well as the following two results. (i) $2\mathrm{D}/\mathrm{poly} = 2\mathrm{N}/\mathrm{poly}$ implies $\mathrm{L}/\mathrm{poly} = \mathrm{NL}/\mathrm{poly}$ [9]. (ii) $2\mathrm{DPD}/\mathrm{poly} = 2\mathrm{NPD}/\mathrm{poly}$ implies $\mathrm{LOGDCFL}/\mathrm{poly} = \mathrm{LOGCFL}/\mathrm{poly}$ [16].

4.2 Unambiguity of 2N and 2NPD

We turn our attention to the question of whether we can make 2nfa's and 2npda's unambiguous. In the simple case of unary inputs, let us recall the result of Geffert and Pighizzini [5], who showed that any 2nfa can be simulated by an appropriate 2ufa with a polynomial increase of the state complexity of the 2nfa. From this follows the collapse of 2N/unary to 2U/unary as stated in Fig. 1.

Next, we look into a relationship between 2NPD and 2UPD when all promise problems are limited to having polynomial ceilings.

Theorem 8. *(1)* 2N/poly = 2U/poly. *(2)* 2NPD/poly = 2UPD/poly.

The statement (1) of Theorem 8 was already proven in [18]. However, we do not know whether 2N = 2U or $2\text{NCT}_k = 2\text{UCT}_k$ even though (1) holds.

Hereafter, we focus on proving the statement (2) of Theorem 8. For this purpose, we use the nonuniform (i.e., the Karp-Lipton style polynomial-size advice-enhanced extension) computational model[6] of *two-way auxiliary unambiguous pushdown automata* (or aux-2upda's), which are 2upda equipped with auxiliary work tapes, and the complexity class $\text{UAuxPDA,TISP}(n^{O(1)}, \log n)$/poly induced by aux-2upda's that run in time $n^{O(1)}$ using work space $O(\log n)$ together with polynomially-bounded advice functions. Reinhardt and Allender [12] demonstrated that LOGCFL/poly coincides with $\text{UAuxPDA,TISP}(n^{O(1)}, \log n)$/poly.

Proof Sketch of Theorem 8(2). By definition, 2UPD is contained in 2NPD. We then intend to show that 2NPD/poly $\subseteq$ 2UPD (because this statement is equivalent to 2NPD/poly = 2UPD/poly).

Discovered in [15] was a close connection between parameterized decision problems and families of promise problems solvable by certain finite automata families. We quickly review necessary terminology, introduced in [20] and fully developed in [15–19]. A *parameterized decision problem* over alphabet Σ is of the form (L, m), where $L \subseteq \Sigma^*$ and $m(\cdot)$ is a size parameter (i.e., a mapping of Σ^* to $\mathbb{N}$). Any size parameter computable by an appropriate log-space deterministic Turing machine (DTM) is called a *logspace size parameter*. A typical example is $m_{\parallel}$ defined as $m_{\parallel}(x) = |x|$ for all $x \in \Sigma^*$. All parameterized decision problems whose size parameters m satisfy $|x| \leq q(m(x))$ for all $x \in \Sigma^*$ form the complexity class PHSP, where q is an appropriate polynomial depending only on m. See [17] for more information. The notation para-LOGCFL/poly denotes the collection of all parameterized decision problems (L, m) with logspace size parameters m solvable by *two-way auxiliary nondeterministic pushdown automata* (or aux-2npda's) running in time $m(x)^{O(1)}$ and space $O(\log m(x))$ with the use of advice strings of length polynomial in $m(x)$, where x represents an "arbitrary" input. As a special case, if m is fixed to $m_{\parallel}$, we obtain LOGCFL/poly.

Let $\mathcal{L} = \{(L_n^{(+)}, L_n^{(-)})\}_{n \in \mathbb{N}}$ and $\mathcal{K} = \{(\mathcal{K}_n^{(+)}, \mathcal{K}_n^{(-)})\}_{n \in \mathbb{N}}$ be any two families of promise problems over a common alphabet Σ. Given a parameterized decision

---

[6] This means aux-2upda's running in $n^{O(1)}$ time using $O(\log n)$ work space on any promised instance of the form $(x, h(|x|))$, where h is a polynomially-bounded advice function.

problem (L, m) over Σ, we say that $\mathcal{L}$ is *induced from* (L, m) if, for any index $n \in \mathbb{N}$, $L_n^{(+)} = L \cap \Sigma_{(n)}$ and $L_n^{(-)} = \overline{L} \cap \Sigma_{(n)}$, where $\Sigma_{(n)} = \{x \in \Sigma^* \mid m(x) = n\}$. The family $\mathcal{K}$ is said to be an *extension* of $\mathcal{L}$ if $L_n^{(+)} \subseteq K_n^{(+)}$ and $L_n^{(-)} \subseteq K_n^{(-)}$ hold for any $n \in \mathbb{N}$. Moreover, $\mathcal{L}$ is *L-good* if the set $\{1^n \# x \mid n \in \mathbb{N}, x \in L_n^{(+)} \cup L_n^{(-)}\}$ belongs to L. A collection $\mathcal{F}$ of promise problem families is *L-good* if all elements of $\mathcal{F}$ has an L-good extension in $\mathcal{F}$. From $\mathcal{L}$, we define $K_n^{(+)} = \{1^n \# x \mid x \in L_n^{(+)}\}$ and $K_n^{(-)} = \{1^n \# x \mid x \notin L_n^{(+)}\} \cup S_n$ for any $n \in \mathbb{N}$, where $\Sigma_\# = \Sigma \cup \{\#\}$ and $S_n = \{z \# x \mid z \in \Sigma^n - \{1^n\}, x \in (\Sigma_\#)^*\} \cup \Sigma^n$. Finally, we say that (K, m) is *induced from* $\mathcal{L}$ if $K = \bigcup_{n \in \mathbb{N}} K_n^{(+)}$ and $m : (\Sigma_\#)^* \to \mathbb{N}$ satisfies that (i) $m(w) = n$ holds for any string w of the form $1^n \# x$ with $x \in L_n^{(+)} \cup L_n^{(-)}$ and (ii) $m(w) = |w|$ holds for all other strings w. Note that $\overline{K}$ equals $\bigcup_{n \in \mathbb{N}} K_n^{(-)}$. See [17] for more information. The L-goodness can be proven for 2NPD/poly, 2UPD/poly, and co-2NPD/poly in a way similar to [16].

In the rest of this proof, we succinctly write LOGUCFL/poly for UAuxPDA,TISP$(n^{O(1)}, \log n)$/poly. Let us assert the following claim concerning LOGUCFL/poly and 2UPD. A similar claim was proven first in [17] and later proven in [18] for UL/poly and in [16] for LOGCFL/poly.

Claim 9. *(i)* LOGCFL $\subseteq$ LOGUCFL/poly *implies* para-LOGCFL/poly $\cap$ PHSP $\subseteq$ para-LOGUCFL/poly. *(ii)* para-LOGCFL/poly $\cap$ PHSP $\subseteq$ para-LOGUCFL/poly *implies* 2NPD/poly $\subseteq$ 2UPD.

Since it is known in [12] that LOGCFL/poly $\subseteq$ LOGUCFL/poly, Claim 9 then concludes that 2NPD/poly $\subseteq$ 2UPD, as requested.

A key to the proof of Claim 9(ii) is the following two statements (a)–(b), which are analogous to [17, Proposition 5.1]. Let $\mathcal{L}$ denote any family $\{(L_n^{(+)}, L_n^{(-)})\}_{n \in \mathbb{N}}$ of promise problems, let L and K be any decision problems over the common alphabet Σ, and let m denote any logspace size parameter over Σ. We write $(\mathcal{C}, \mathcal{D})$ for any element of the set $\{(\text{LOGCFL}, 2\text{NPD}), (\text{LOGUCFL}, 2\text{UPD})\}$.

(a) If $\mathcal{L}$ is induces from (L, m), then $(L, m) \in$ para-$\mathcal{C}$/poly $\cap$ PHSP iff $\mathcal{L} \in \mathcal{D}$/poly.

(b) If $\mathcal{L}$ is L-good and (K, m) is induced from $\mathcal{L}$, then $(K, m) \in$ para-$\mathcal{C}$/poly $\cap$ PHSP iff $\mathcal{L} \in \mathcal{D}$/poly.

Note that the validity of the above statements (a)–(b) for the first case of $\mathcal{C} = \text{LOGCFL}$ and $\mathcal{D} = 2\text{NPD}$ comes from [16, Proposition 4.3]. The second case of $\mathcal{C} = \text{LOGUCFL}$ and $\mathcal{D} = 2\text{UPD}$ can be handled by a similar argument.

Let us assume that para-LOGCFL/poly $\cap$ PHSP $\subseteq$ para-LOGUCFL/poly. Our goal is to verify that 2NPD/poly $\subseteq$ 2UPD. For any family $\mathcal{L}$ of promise problems in 2NPD/poly over alphabet Σ, we take the parameterized decision problem (K, m) induced from $\mathcal{L}$. By (b), it follows that $(K, m) \in$ para-LOGCFL/poly $\cap$ PHSP iff $\mathcal{L} \in 2\text{NPD}$/poly. Thus, we obtain $(K, m) \in$ para-LOGCFL/poly. Our assumption implies that $(K, m) \in$ para-LOGUCFL/poly. By (b) follows the membership of $\mathcal{L}$ to 2UPD/poly. Therefore, Claim 9(ii) holds. □

4.3   Complementation of 2N and 2NPD

Let us look into the question raised in Sect. 1.1 concerning the closure property under complementation. Geffert, Mereghetti, and Pighizzini [4] earlier demonstrated a simulation of a "complementary" 2nfa (i.e., a two-way finite automaton making co-nondeterministic moves) on unary inputs by another 2nfa with a polynomial increase of the state complexity. From this fact, we instantly conclude that co-2N/unary coincides with 2N/unary as stated in Fig. 1.

Hereafter, we discuss the complementation closures of 2N and 2NPD when all valid instances are limited to polynomially long strings.

Theorem 10. *(1)* 2N/poly = co-2N/poly. *(2)* 2NPD/poly = co-2NPD/poly

It is important to note that the statement (1) does not require the use of parameterized version of NL. This is rather a direct consequence of Theorem 3 (and Corollary 4) together with Theorem 6(1). This fact exemplifies the strength of the additional use of counters provided to underlying finite automata.

Proof Sketch of Theorem 10. (1) By Corollary 2, $2NCT_4 = co\text{-}2NCT_4$ follows. We thus conclude that $2NCT_4$/poly = co-$2NCT_4$/poly. Theorem 6(1) then implies that 2N/poly = co-2N/poly.

(2) The proof for LOGCFL = co-LOGCFL in [2] can be carried over to the advised setting, and thus we immediately obtain LOGCFL/poly = co-LOGCFL/poly. Similarly to Claim 9, we assert the following claim. Here, we obtain para-co-LOGCFL/poly from the definition of para-LOGCFL/poly by simply replacing (L, m) with $(\overline{L}, m)$, where $\overline{L}$ is the *complement* of L.

Claim 11. *(i)* co-LOGCFL $\subseteq$ LOGCFL/poly *implies* para-co-LOGCFL/poly$\cap$ PHSP $\subseteq$ para-LOGCFL/poly. *(ii)* para-co-LOGCFL/poly $\cap$ PHSP $\subseteq$ para-LOGCFL/poly *implies* co-2NPD/poly $\subseteq$ 2NPD.

Claim 11 leads to the desired equality 2NPD/poly = co-2NPD/poly from LOGCFL/poly = co-LOGCFL/poly, which is a direct consequence of LOGCFL = co-LOGCFL [2].

Different from the proof of Claim 9, the proof of Claim 11 exploits the fact that 2NPD and co-2NPD (as well as para-LOGCFL and para-co-LOGCFL) are "complementary" in nature. The proof of Claim 11(ii), in particular, requires the statements similar to the two statements (a)–(b) given in the proof of Claim 9(ii) with the choice of $\mathcal{C} = $ co-LOGCFL and $\mathcal{D} = $ co-2NPD. These statements can be verified by exploring the "complementary" property of para-co-LOGCFL. $\square$

References

1. Berman, P., Lingas, A.: On complexity of regular languages in terms of finite automata. Report 304. Institute of Computer Science. Polish Academy of Science, Warsaw (1977)

2. Borodin, A., Cook, S.A., Dymond, P.W., Ruzzo, W.L., Tompa, M.: Two applications of inductive counting for complementation problems. SIAM J. Comput. **18**, 559–578 (1989)
3. Geffert, V.: An alternating hierarchy for finite automata. Theor. Comput. Sci. **445**, 1–24 (2012)
4. Geffert, V., Mereghetti, C., Pighizzini, G.: Complementing two-way finite automata. Inf. Comput. **205**, 1173–1187 (2007)
5. Geffert, V., Pighizzini, G.: Two-way unary automata versus logarithmic space. Inf. Comput. **209**, 1016–1025 (2011)
6. Immerman, N.: Nondeterministic space is closed under complement. SIAM J. Comput. **17**, 935–938 (1988)
7. Kapoutsis, C.A.: Size complexity of two-way finite automata. In: Diekert, V., Nowotka, D. (eds.) DLT 2009. LNCS, vol. 5583, pp. 47–66. Springer, Heidelberg (2009). https://doi.org/10.1007/978-3-642-02737-6_4
8. Kapoutsis, C.A.: Minicomplexity. J. Automat. Lang. Combin. **17**, 205–224 (2012)
9. Kapoutsis, C.A.: Two-way automata versus logarithmic space. Theory Comput. Syst. **55**, 421–447 (2014)
10. Kapoutsis, C.A., Pighizzini, G.: Two-way automata characterizations of L/poly versus NL. Theory Comput. Syst. **56**, 662–685 (2015)
11. Minsky, M.L.: Computation: Finite and Infinite Machines. Prentice-Hall, Englewood Cliff (1967)
12. Reinhardt, K., Allender, E.: Making nondeterminism unambiguous. SIAM J. Comput. **29**, 1118–1181 (2000)
13. Sakoda, W.J., Sipser, M.: Nondeterminism and the size of two-way finite automata. In: The Proceedings of STOC 1978, pp. 275–286 (1978)
14. Szelepcsényi, R.: The method of forced enumeration for nondeterministic automata. Acta Inform. **26**, 279–284 (1988)
15. Yamakami, T.: State complexity characterizations of parameterized degree-bounded graph connectivity, sub-linear space computation, and the linear space hypothesis. Theor. Comput. Sci. **798**, 2–22 (2019)
16. Yamakami, T.: Parameterizations of logarithmic-space reductions, stack-state complexity of nonuniform families of pushdown automata, and a road to the LOGCFL⊆LOGDCFL/poly question. arXiv:2108.12779 (2021)
17. Yamakami, T.: Nonuniform families of polynomial-size quantum finite automata and quantum logarithmic-space computation with polynomial-size advice. Inf. Comput. **286**, Article no. 104783 (2022)
18. Yamakami, T.: Unambiguity and fewness for nonuniform families of polynomial-size nondeterministic finite automata. In: Lin, A.W., Zetzsche, G., Potapov, I. (eds.) RP 2022. LNCS, vol. 13608, pp. 77–92. Springer, Cham (2022). https://doi.org/10.1007/978-3-031-19135-0_6 A corrected version is at arXiv:2311.09979
19. Yamakami, T.: Power of counting by nonuniform families of polynomial-size finite automata. In: Fernau, H., Jansen, K. (eds.) FCT 2023. LNCS, vol. 14292, pp. 421–435. Springer, Cham (2023). https://doi.org/10.1007/978-3-031-43587-4_30 A corrected and expanded version is available at arXiv:2310.18965
20. Yamakami, T.: The 2CNF Boolean formula satsifiability problem and the linear space hypothesis. J. Comput. Syst. Sci. **136**, 88–112 (2023)

A Gray Code of Ordered Trees

Shin-ichi Nakano[✉] [ID]

Gunma University, Kiryu 376-8515, Japan
nakano@gunma-u.ac.jp

Abstract. A combinatorial Gray code for a set of combinatorial objects is a sequence of all combinatorial objects in the set so that each object is derived from the preceding object by changing a small part.

In this paper we design a Gray code for ordered trees with n vertices such that each ordered tree is derived from the preceding ordered tree by removing a leaf then appending a leaf elsewhere. Thus, the change is just remove-and-append a leaf, which is of minimum size.

1 Introduction

A classical Gray code for n-bit binary numbers is a sequence of all n-bit binary numbers so that each number is derived from the preceding number by changing exactly one bit. A combinatorial Gray code for a set of combinatorial objects is a sequence of all combinatorial objects in the set so that each object is derived from the preceding object by changing a small (constant) part.

When we generate all combinatorial objects and the number of such objects is huge if we can compute them as a combinatorial Gray code then we can output (or store) each object as the difference from the preceding object and we may compute each object in a constant time. Also, when we repeatedly solve some problem for a class of objects, a solution for an object may help to compute a solution for a similar successive object. See surveys for combinatorial Gray codes [5,8].

For binary trees with n vertices one can generate all binary trees so that each binary tree is derived from the preceding binary tree by a rotation operation at a vertex [3,4]. The number of change of edges in a rotation operation is three [1, p9]. Also one can generate all binary trees with n vertices so that each tree is derived from the preceding tree by removing a subtree and place it elsewhere [1, Exercise 25]. However the levels of many vertices may be changed, where the level of a vertex is the number of vertices on the path from the vertex to the root.

For ordered tree with n vertices, by slightly modifying the algorithm in [6], one can generate all ordered trees with n vertices so that each ordered tree is derived from the preceding ordered tree by either (1) removing at most two leaves then appending one leaf, or (2) removing one leaf then appending at most two leaves [7]. The number of change of edges is at most three. Also another Gray code using pointer-based representations is known [2].

X. Chen and B. Li (Eds.): TAMC 2024, LNCS 14637, pp. 26–37, 2024.
https://doi.org/10.1007/978-981-97-2340-9_3

In this paper we design a Gray code for ordered trees with n vertices such that each ordered tree is derived from the preceding ordered tree by removing a leaf then appending a leaf elsewhere. Thus the change is just remove-and-append a leaf, which is minimum, and other vertices remain as they were, including their levels. Our Gray code is based on a tree structure among the ordered trees.

The remainder of this paper is organized as follows. Section 2 gives some definitions and basic lemmas. In Sect. 3 we design our algorithm to construct a Gray code for the ordered trees with n vertices. Finally Sect. 4 is a conclusion.

2   Preliminaries

A *tree* is a connected graph with no cycle. A *rooted tree* is a tree with a designated vertex as *the root*. *The level of a vertex* v in a rooted tree is the number of vertices on the path from v to the root. The level of the root is 1. For each vertex v except the root if the neighbor vertex of v on the path from v to the root is p then p is *the parent* of v and v is *a child* of p. The root has no parent. In this paper we always draw each child vertex below its parent. A vertex with no child is called *a leaf*. An *ordered tree* is a rooted tree in which the left-to-right order of the child vertices of each vertex is defined. The number of ordered trees with exactly $n+1$ vertices is known as the n-th Catalan number $_{2n}C_n/(n+1)$ [1, p12], where $_{2n}C_n$ is the number of subsets of n elements of a set of $2n$ elements.

Given an ordered tree T, let $P_r(T) = (v_0, v_1, \cdots, v_k)$ be the path from the root v_0 to a leaf v_k such that, for each $i = 1, 2, \cdots, k$, v_i is the rightmost child of v_{i-1}. $P_r(T)$ is called *the rightmost path* of T and v_k is called *the rightmost leaf* of T. The number of edges in $P_r(T)$ is denoted by $rpl(T)$.

For an ordered tree T if the rightmost child of the root has exactly one child as a leaf then we say T has *the pony-tail*.

For two distinct ordered trees T and T', if T' is derived from T by appending a new leaf as the rightmost leaf then removing any other leaf, then we say T *is copying* T'. When T is copying T' if the parent of the rightmost leaf of T' has two or more child vertices then $rpl(T) \geq rpl(T')$ holds, otherwise, the parent of the rightmost leaf of T' has exactly one child vertex, which is the rightmost leaf, and $rpl(T) = rpl(T')-1$ holds. So if T is copying T', $rpl(T) = 1$ and $rpl(T') > 1$ then T' has the pony-tail and $rpl(T') = 2$.

Let S_k be the set of the ordered trees with exactly k vertices. In this paper we design, for each $k = 1, 2, \cdots, n$, a combinatorial Gray code for S_k, that is a sequence of all ordered trees in S_k such that each ordered tree is derived from the preceding ordered tree by removing a leaf then appending a leaf elsewhere. We call the change *delete-and-append a leaf*.

For an ordered tree T with $n \geq 2$ vertices let $p(T)$ be the ordered tree derived from T by removing the rightmost leaf. We say $p(T)$ is *the parent* of T, and T is a child of $p(T)$. For any ordered tree T in S_n if we repeatedly compute the parent of the derived ordered tree we obtain the sequence $T, p(T), p(p(T)), \cdots$ of ordered trees, which ends with the trivial ordered tree consisting of exactly one vertex. We call the sequence *the removing sequence of* T [6].

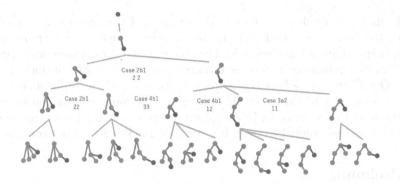

Fig. 1. The family tree F_n of S_n.

By merging the removing sequences of the ordered trees in S_n one can obtain an (unordered) tree F_n of ordered trees [6] (See an example for $n = 5$ in Fig. 1) in which the root corresponds to the trivial ordered tree with exactly one vertex, each vertex at level k corresponds to some ordered tree in S_k, and each edge corresponds to some ordered tree and its parent. We call the tree *the family tree*. Note that we have not determined yet the left-to-right order of the child ordered trees of each order tree in F_n. We have the following three lemmas.

Lemma 1. *There is a bijection between the ordered trees in S_k and the vertices at level k in F_n.*

Proof. Given an ordered tree T with exactly k vertices, by repeatedly appending a new leaf as the rightmost child of the root, one can obtain a descendant tree $T' \in S_n$ in F_n. Thus every ordered tree in S_k appears in the removing sequence of some tree in S_n and so corresponds to a vertex at level k in F_n.

Clearly every vertex at level k in F_n corresponds to an ordered tree with exactly k vertices.   □

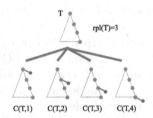

Fig. 2. An illustration for Lemma 2.

Lemma 2. *Let T be an ordered tree in S_k with $k < n$. T has $rpl(T) + 1$ child ordered trees in F_n.*

Proof. For each $i = 1, 2, \cdots, rpl(T)+1$, by appending a new leaf as the rightmost child leaf of the vertex on $P_r(T)$ at level i, one can obtain a distinct child ordered tree. See Fig. 2.   □

We denote by $C(T, i)$ the child ordered tree of T derived from T by appending a new leaf as the rightmost child leaf of the vertex on $P_r(T)$ at level i. Thus $rpl(C(T, i)) = i$.

By Lemma 2, every ordered tree T in S_k with $k < n$ except the ordered tree with exactly one vertex has two or more child ordered trees in F_n since $rpl(T) \geq 1$. Clearly the ordered tree with exactly one vertex has exactly one child ordered tree in F_n.

Lemma 3. *Any ordered tree is derived from its sibling ordered tree in F_n by delete-and-append a leaf.*

Proof. Any ordered tree is derived from its sibling ordered tree by deleting the rightmost leaf then appending a leaf as the rightmost leaf at the suitable level.   □

In this paper we show that, by suitably defining the left-to-right order of the child ordered trees of each ordered tree in F_n, we can define an ordered tree F_n^O such that, for each k, a Gray code for S_k is appeared as the left-to-right sequence of the ordered trees corresponding to the vertices at level k of F_n^O. Thus a Gray code for S_n is appeared as the left-to-right sequence of the ordered trees corresponding to the leaves of F_n^O. See an example for $n = 5$ in Fig. 1.

3   Algorithm

In this section we design a Gray code for S_k for each $k = 1, 2, \cdots, n$, where S_k is the set of the ordered trees with exactly k vertices.

Induction on Levels
We proceed by induction on levels. Let F_k be the subtree of F_n induced by $S_1 \cup S_2 \cup \cdots \cup S_k$. The Gray code for S_1 is trivial and unique since $|S_1| = 1$. Similar for S_2 since $|S_2| = 1$. Assume that, for an integer $k < n$, we have defined a left-to-right order of child ordered trees of each ordered tree in $S_1 \cup S_2 \cup \cdots \cup S_{k-1}$, we have obtained an ordered tree F_k^O corresponding to F_k, and we have constructed a Gray code for S_k as the left-to-right sequence of the ordered trees corresponding to the leaves of F_k^O. Then we are going to define a left-to-right order of the child ordered trees of each ordered tree in S_k so that it extends F_k^O to an ordered tree F_{k+1}^O and a Gray code for S_{k+1} is appeared as the left-to-right sequence of the ordered trees corresponding to the leaves of F_{k+1}^O.

Basic Strategy of Algorithm
Let $(T_1, T_2, \cdots)$ be our Gray code for S_k. We are going to define a left-to-right order of the child ordered trees of each T_i in S_k, then we obtain a sequence of ordered trees in S_{k+1}, which is a Gray code for S_{k+1}, say $(T_1', T_2', \cdots)$.

If two consecutive ordered trees T'_j and T'_{j+1} in the sequence are siblings in F^O_{k+1}, then one can be derived from the other by delete-and-append a leaf by Lemma 3. However if two consecutive ordered trees T'_j and T'_{j+1} are not siblings in F^O_{k+1}, that is, T'_j is the rightmost child ordered tree of T_i and T'_{j+1} is the leftmost child ordered tree of T_{i+1} for some i, then we have ten cases to consider. We have the following lemma for T_i and T_{i+1}.

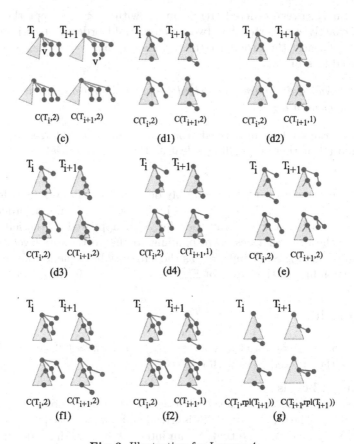

Fig. 3. Illustration for Lemma 4.

Lemma 4. *Assume that T_i can be derived from T_{i+1} by delete-and-append a leaf. Then the following hold. (Note that (c)-(f) are cases in which exactly one of T_i and T_{i+1} has rpl 1 and (g)-(g'') are cases in which both T_i and T_{i+1} have rpl more than 1.)*

(a) $C(T_i, 1)$ *can be derived from* $C(T_{i+1}, 1)$ *by delete-and-append a leaf.*

(b) *If* $rpl(T_i) = rpl(T_{i+1}) = 1$, *then* $C(T_i, 2)$ *can be derived from* $C(T_{i+1}, 2)$ *by delete-and-append a leaf.*

(c) *If* T_i *has the pony-tail,* $rpl(T_{i+1}) = 1$, T_i *is copying* T_{i+1} *and* T_{i+1} *is copying* T_i, *then* $C(T_i, 2)$ *can be derived from* $C(T_{i+1}, 2)$ *by delete-and-append a leaf.*

(d) If $rpl(T_i) = 1$, $rpl(T_{i+1}) > 1$, and T_{i+1} has no pony-tail (so T_{i+1} is copying T_i), then $C(T_i, 2)$ cannot be derived from $C(T_{i+1}, 2)$ by delete-and-append a leaf (See Fig. 3 (d1) and (d3)), however $C(T_i, 2)$ can be derived from $C(T_{i+1}, 1)$ by delete-and-append a leaf. (See Fig. 3 (d2) and (d4).)

(e) If $rpl(T_i) = 1$, $rpl(T_{i+1}) > 1$, T_{i+1} has the pony-tail, and T_i is copying T_{i+1}, then $C(T_i, 2)$ can be derived from $C(T_{i+1}, 2)$ by delete-and-append a leaf. (See Fig. 3 (e).)

(e') If $rpl(T_i) > 1$, $rpl(T_{i+1}) = 1$, T_i has the pony-tail, and T_{i+1} is copying T_i, then $C(T_{i+1}, 2)$ can be derived from $C(T_i, 2)$ by delete-and-append a leaf.

(f) If $rpl(T_i) = 1$, $rpl(T_{i+1}) > 1$, T_{i+1} has the pony-tail, T_{i+1} is copying T_i, then $C(T_i, 2)$ cannot be derived from $C(T_{i+1}, 2)$ by delete-and-append a leaf (See Fig. 3 (f1)), however $C(T_i, 2)$ can be derived from $C(T_{i+1}, 1)$ by delete-and-append a leaf. (See Fig. 3 (f2).)

(g) If $rpl(T_i) > rpl(T_{i+1}) \geq 2$, then $C(T_i, rpl(T_{i+1}))$ can be derived from $C(T_{i+1}, rpl(T_{i+1}))$ by delete-and-append a leaf. (See Fig. 3 (g).)

(g') If $rpl(T_i) = rpl(T_{i+1}) \geq 2$, then $C(T_i, rpl(T_i))$ can be derived from $C(T_{i+1}, rpl(T_i))$ by delete-and-append a leaf, and $C(T_i, rpl(T_i) + 1)$ can be derived from $C(T_{i+1}, rpl(T_i) + 1)$ by delete-and-append a leaf.

(g'') If $rpl(T_{i+1}) > rpl(T_i) \geq 2$, then $C(T_{i+1}, rpl(T_i))$ can be derived from $C(T_i, rpl(T_i))$. by delete-and-append a leaf. Also if $rpl(T_{i+1}) > rpl(T_i) \geq 2$, then $C(T_i, 1)$ can be derived from $C(T_{i+1}, rpl(T_i))$ by delete-and-append a leaf.

Proof. (a) (b) We have the following two cases. Case 1: T_i is derived from T_{i+1} by removing the rightmost leaf then appending a new leaf elsewhere. Case 2: T_i is derived from T_{i+1} by removing a leaf which is not the rightmost leaf then appending a new leaf elsewhere. For both cases the claim holds.

(c) Choose v and v' so that T_{i+1} is derived from T_i by appending the rightmost leaf at level 1 then deleting a leaf v (since T_i is copying T_{i+1}), and T_i is derived from T_{i+1} by appending the rightmost leaf at level 2 then deleting a leaf v' (since T_{i+1} is copying T_i).

We can show that exactly one of v or v' is a child (leaf) of the root, as follows. If v is a child of the root of T_i and v' is a child of the root of T_{i+1} then, since T_i is copying T_{i+1}, the degree of the root of T_i is equal to the degree of the root of T_{i+1}, and, since T_{i+1} is copying T_i, the degree of the root of T_{i+1} minus 1 is equal to the degree of the root of T_i, a contradiction. Also if v is not a child of the root of T_i and v' is not a child of the root of T_{i+1} then, since T_i is copying T_{i+1}, the degree of the root of T_i plus 1 is equal to the degree of the root of T_{i+1}, and, since T_{i+1} is copying T_i, the degree of the root of T_{i+1} is equal to the degree of the root of T_i, a contradiction. Thus exactly one of v or v' is a child (leaf) of the root.

Assume first that v is a child (leaf) of the root of T_i. Let $x_1, x_2, \cdots, x_d$ be the child vertices of the root in T_i except v in right-to-left order, and $y_1, y_2, \cdots, y_{d+1}$ the child vertices of the root in T_{i+1} in right-to-left order and v' belong to a subtree rooted at y_j. Since T_i is copying T_{i+1}, after removing v from T_i, the subtrees rooted at $x_1, x_2, \cdots, x_d$ are identical to the subtrees rooted at $y_2, y_3, \cdots, y_{d+1}$, respectively. Also since T_{i+1} is copying T_i, after removing v'

from T_{i+1}, the subtrees rooted at $y_2, y_3, \cdots, y_{d+1}$ are identical to the subtrees rooted at $x_2, x_3, \cdots, x_d$, including the trivial subtree rooted at v, respectively. Then the subtree rooted at y_j has only one child v', and the subtrees rooted at $y_2, y_3, \cdots, y_j$ are identical (to the pony tail). See Fig. 3 (c). Now $C(T_i, 2)$ is derived from $C(T_{i+1}, 2)$ by delete-and-append a leaf.

Similar for the case where v' is a child (leaf) of the root of T_{i+1}.

(d) Since T_{i+1} has no pony-tail, either (Case 1) the rightmost child vertex of the root of T_{i+1} has two or more child vertices (See Fig. 3 (d1)), or (Case 2) $rpl(T_{i+1}) > 2$. (See Fig. 3 (d3)). Since $rpl(T_i) = 1$ the rightmost child vertex of the root of T_i has no child vertex. For Case 1, the rightmost child vertex of the root of $C(T_{i+1}, 2)$ has three or more child vertices, while the rightmost child vertex of the root of $C(T_i, 2)$ has exactly one child vertex. Thus $C(T_i, 2)$ cannot be derived from $C(T_{i+1}, 2)$ by delete-and-append a leaf. See Fig. 3 (d1). For Case 2 we need to remove at least two vertices and append at least two vertices to obtain $C(T_i, 2)$ from $C(T_{i+1}, 2)$. Thus $C(T_i, 2)$ cannot be derived from $C(T_{i+1}, 2)$ by delete-and-append a leaf. See Fig. 3 (d3). However $C(T_i, 2)$ can be derived from $C(T_{i+1}, 1)$ by delete-and-append a leaf. See Fig. 3 (d2) and (d4).

(e) Omitted. See Fig. 3 (e). (e′) Similar to (e). (f) Omitted. See Fig. 3 (f1) and (f2). (g) Omitted. See Fig. 3 (g). (g′)(g″) Similar to (g).   □

Step of Algorithm

Let $(T_1, T_2, \cdots)$ be a Gray code for S_k corresponding to the leaves of F_k^O and we are going to define a left-to-right order of the child ordered trees of each ordered tree in S_k and construct a Gray code $(T_1', T_2', \cdots)$ for S_{k+1} corresponding to the leaves of F_{k+1}^O. When we start step i assume that we have already defined the left-to-right order of the child ordered trees of $T_1, T_2, \cdots, T_{i-1}$ and the leftmost child ordered tree of T_i, and in step i we are going to define the left-to-right order of the child ordered trees of T_i except the leftmost one, and the leftmost child ordered trees of T_{i+1}. See Fig. 4. The part we are going to define in the current step i is depicted as a grey rectangle. We proceed with several cases based on $rpl(T_i), rpl(T_{i+1})$ and the leftmost child of T_i, as explained later.

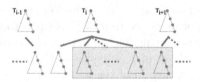

Fig. 4. An illustration for step i of the algorithm.

Loop Invariant

Our algorithm satisfies the following condition

(co1) For consecutive three ordered trees T_{u-1}, T_u, T_{u+1} at level k, if $rpl(T_{u-1}) = rpl(T_{u+1}) = 1$ and $rpl(T_u) > 1$ then T_u has the pony-tail and T_{u+1} is copying T_u. Also if $rpl(T_{u-1}) = rpl(T_{u+1}) \geq 2$ then $rpl(T_{u-1}) > rpl(T_u)$.

Our algorithm also satisfies the following condition at each step i.

(co2) Let $T'_{i'}$ be the leftmost child ordered tree of T_i. For consecutive three ordered trees $T'_{u'-1}, T'_{u'}, T'_{u'+1}$ at level $k+1$ with $u'+1 \le i'$, if $rpl(T'_{u'-1}) = rpl(T'_{u'+1}) = 1$ and $rpl(T'_{u'}) > 1$ then $T'_{u'}$ has the pony-tail and $T'_{u'+1}$ is copying $T'_{u'}$. Also if $rpl(T'_{u'-1}) = rpl(T'_{u'+1}) \ge 2$ then $rpl(T'_{u'-1}) > rpl(T'_{u'})$.

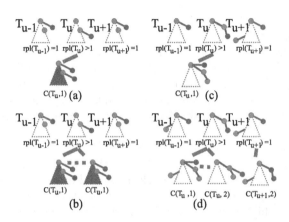

Fig. 5. Illustrations for the loop invariant.

The intuitive reason why we need those condition is as follows.
Assume that there are T_{u-1}, T_u, T_{u+1} with $rpl(T_{u-1}) = rpl(T_{u+1}) = 1, rpl(T_u) > 1$, T_u has no pony-tail, and $C(T_u, 1)$ is the leftmost child of T_u (see Fig. 5(a)). If we try to set $C(T_u, 1)$ at the rightmost child of T_u, then we fail to construct a Gray code for S_{k+1} since the same tree appear twice. (See Fig. 5(b).) So our algorithm tries to exclude any occurrence of such consecutive three ordered trees. Note that even when $rpl(T_{u-1}) = rpl(T_{u+1}) = 1$, $rpl(T_u) > 1$ and $C(T_u, 1)$ is the leftmost child of T_u, if T_u has the pony-tail and T_{u+1} is copying T_u (see Fig. 5(c)), then we can set $C(T_u, 2)$ at the rightmost child of T_u and $C(T_{u+1}, 2)$ at the leftmost child of T_{u+1} (by Lemma 4(e')) and we can proceed successfully. (See an example in Fig. 5(d).)

Assume that there are T_{u-1}, T_u, T_{u+1} with $rpl(T_{u-1}) = rpl(T_{u+1}) \ge 2$, $rpl(T_u) > rpl(T_{u-1})$, and $C(T_u, rpl(T_{u-1}))$ is the leftmost child of T_u. If we try to set $C(T_u, rpl(T_{u-1}))$ at the rightmost child of T_u, then we fail. So our algorithm tries to exclude any occurrence of such consecutive three ordered trees.

Algorithm
First we set $C(T_1, 1)$ as the leftmost child of T_1.

Assume that we have done step $1, 2, \cdots, i-1$. Now we execute the next step i of our algorithm if T_{i+1} exists. (If T_i is the last ordered tree in the Gray code of S_k then we order the remaining child of T_i with decreasing order of rpl from left to right.) (Also note that, in the following cases, if $rpl(T_i) \geq 3$ then $C(T_i, 1)$ never appear at the second leftmost child of T_i.)

We have the following four cases for step i.

Case 1: $rpl(T_i) = 1$ and $rpl(T_{i+1}) = 1$.

Case 1a: If $C(T_i, 1)$ is the leftmost child of T_i then we set $C(T_i, 2)$ as the rightmost child of T_i and $C(T_{i+1}, 2)$ as the leftmost child of T_{i+1} (by Lemma 4(b)).

Case 1b: Otherwise, $C(T_i, 1)$ is not the leftmost child of T_i then we set $C(T_i, 1)$ as the rightmost child of T_i and $C(T_{i+1}, 1)$ as the leftmost child of T_{i+1} (by Lemma 4(a)).

Case 2: $rpl(T_i) = 1$ and $rpl(T_{i+1}) > 1$.

We have two subcases.

Case 2a: T_{i+1} has no pony-tail. (So T_{i+1} is copying T_i.)

Case 2a1: If $C(T_i, 1)$ is the leftmost child of T_i then we set $C(T_i, 2)$ as the rightmost child of T_i and $C(T_{i+1}, 1)$ as the leftmost child of T_{i+1} (by Lemma 4(d)).

Case 2a2: If $C(T_i, 1)$ is not the leftmost child of T_i then we set $C(T_i, 1)$ as the rightmost child of T_i and $C(T_{i+1}, 1)$ as the leftmost child of T_{i+1} (by Lemma 4(a)).

Case 2b: T_{i+1} has the pony-tail and T_i is copying T_{i+1}.

Case 2b1: If $C(T_i, 1)$ is the leftmost child of T_i then we set $C(T_i, 2)$ as the rightmost child of T_i and $C(T_{i+1}, 2)$ as the leftmost child of T_{i+1} (by Lemma 4(e)).

Case 2b2: If $C(T_i, 1)$ is not the leftmost child of T_i then we set $C(T_i, 1)$ as the rightmost child of T_i and $C(T_{i+1}, 1)$ as the leftmost child of T_{i+1} (by Lemma 4(a)).

Case 2c: T_{i+1} has the pony-tail and T_{i+1} is copying T_i.

Case 2c1: If $C(T_i, 1)$ is the leftmost child of T_i then we set $C(T_i, 2)$ as the rightmost child of T_i and $C(T_{i+1}, 1)$ as the leftmost child of T_{i+1} (by Lemma 4(f)).

Case 2c2: If $C(T_i, 1)$ is not the leftmost child of T_i then we set $C(T_i, 1)$ as the rightmost child of T_i and $C(T_{i+1}, 1)$ as the leftmost child of T_{i+1} (by Lemma 4(a)).

Case 3: $rpl(T_i) > 1$ and $rpl(T_{i+1}) = 1$.

We have two subcases.

Case 3a: T_i has no pony-tail. (So T_i is copying T_{i+1}.)

Case 3a1: If $C(T_i, 1)$ is the leftmost child of T_i then we can prove that this case never occurs, as follows.

We have set $C(T_i, 1)$ as the leftmost child of T_i with $rpl(T_i) > 1$ in the preceding step of either Case 2a1, 2a2, 2b2, 2c1 or 2c2. (In Case 4 we never set $C(T_i, 1)$ as the leftmost child of T_i.) In those cases $rpl(T_{i-1}) = 1$ holds, and in Case 3a1 $rpl(T_i) > 1$ and $rpl(T_{i+1}) = 1$ hold and T_i has no pony-tail. This contradicts (co1).

Case 3a2: If $C(T_i, 1)$ is not the leftmost child of T_i then we set $C(T_i, 1)$ as the rightmost child of T_i and $C(T_{i+1}, 1)$ as the leftmost child of T_{i+1} (by Lemma 4(a)). Set other child ordered trees of T_i between the leftmost child and the rightmost child with decreasing order of rpl from left to right.

Case 3b: T_i has the pony-tail and T_{i+1} is copying T_i.

Case 3b1: If $C(T_i, 1)$ is the leftmost child of T_i then we set $C(T_i, 2)$ as the rightmost child of T_i and $C(T_{i+1}, 2)$ as the leftmost child of T_{i+1} (by Lemma 4(e$'$)). Set the remaining child $C(T_i, 3)$ of T_i as the middle child of T_i.

Case 3b2: If $C(T_i, 1)$ is not the leftmost child of T_i then we set $C(T_i, 1)$ as the rightmost child of T_i and $C(T_{i+1}, 1)$ as the leftmost child of T_{i+1} (by Lemma 4(a)). Set the remaining child as the middle child of T_i.

Case 3c: T_i has the pony-tail and T_i is copying T_{i+1}.

Case 3c1: $C(T_i, 1)$ is the leftmost child of T_i. If T_{i+1} is also copying T_i then we set $C(T_i, 2)$ as the rightmost child of T_i and $C(T_{i+1}, 2)$ as the leftmost child of T_{i+1} (by Lemma 4(c)) and set the remaining child as the middle child of T_i. Otherwise one can prove that this case never occurs. Similar to Case 3a1.

Case 3c2: If $C(T_i, 1)$ is not the leftmost child of T_i then we set $C(T_i, 1)$ as the rightmost child of T_i and $C(T_{i+1}, 1)$ as the leftmost child of T_{i+1} (by Lemma 4(a)). Set the remaining child as the middle child of T_i

Case 4: $rpl(T_i) > 1$ and $rpl(T_{i+1}) > 1$.

Case 4a: $C(T_i, 1)$ is the leftmost child of T_i.

Case 4a1: $rpl(T_i) \leq rpl(T_{i+1})$.

We set $C(T_i, rpl(T_i))$ as the rightmost child of T_i and $C(T_{i+1}, rpl(T_i))$ as the leftmost child of T_{i+1} (by Lemma 4(g$'$)(g$''$)).

Set other child ordered trees of T_i between the leftmost child $C(T_i, 1)$ and the rightmost child $C(T_i, rpl(T_i))$ with increasing order of rpl from left to right.

Case 4a2: $rpl(T_i) > rpl(T_{i+1})$.

We set $C(T_i, rpl(T_{i+1}))$ as the rightmost child of T_i and $C(T_{i+1}, rpl(T_{i+1}))$ as the leftmost child of T_{i+1} (by Lemma 4(g)).

Set other child ordered trees of T_i between the leftmost child $C(T_i, 1)$ and the rightmost child $C(T_i, rpl(T_{i+1}))$ with increasing order of rpl from left to right.

Case 4b: $C(T_i, 1)$ is not the leftmost child of T_i.

Let T be the leftmost child of T_i.

Case 4b1: $rpl(T_i) \leq rpl(T_{i+1})$.

If $rpl(T_i) < rpl(T_{i+1})$ then we set $C(T_i, 1)$ as the rightmost child of T_i and $C(T_{i+1}, rpl(T_i))$ as the leftmost child of T_{i+1} (by Lemma 4(g$''$)).

Otherwise $rpl(T_i) = rpl(T_{i+1})$ holds. If $rpl(T) = rpl(T_i)$ then we set $C(T_i, rpl(T_i) + 1)$ as the rightmost child of T_i and $C(T_{i+1}, rpl(T_i) + 1)$ as the leftmost child of T_{i+1} (by Lemma 4(g$'$)), and if $rpl(T) \neq rpl(T_i)$ then we set $C(T_i, rpl(T_i))$ as the rightmost child of T_i and $C(T_{i+1}, rpl(T_i))$ as the leftmost child of T_{i+1} (by Lemma 4(g$'$)).

Set other child ordered trees of T_i between the leftmost child and the rightmost child with decreasing order of rpl from left to right. (Note that if $rpl(T_i) \geq 3$ then $C(T_i, 1)$ never appears at the second leftmost child of T_i.)

Case 4b2: $rpl(T_i) > rpl(T_{i+1})$ and $rpl(T) \neq rpl(T_{i+1})$.

We set $C(T_i, rpl(T_{i+1}))$ as the rightmost child of T_i and $C(T_{i+1}, rpl(T_{i+1}))$ as the leftmost child of T_{i+1} (by Lemma 4(g)). Set other child ordered trees of T_i between the leftmost child and the rightmost child with decreasing order of rpl from left to right. (Note that $C(T_i, 1)$ never appears at the second leftmost child of T_i since $rpl(T_i) \geq 3$ holds.)

Case 4b3: $rpl(T_i) > rpl(T_{i+1})$ and $rpl(T) = rpl(T_{i+1})$.

We show this case never occurs in Lemma 5 below.

The description of the four cases for step i is completed. We have the following three lemmas.

Lemma 5. Case 4b3 *never occurs*.

Proof. Assume for a contradiction that the case occurs. In Case 4b we have defined T as the leftmost child of T_i.

If $rpl(T) > 2$, then we have set T in Case 4 of the preceding step $i - 1$. If $rpl(T_{i-1}) < rpl(T_i)$ then we set $C(T_i, rpl(T_{i-1}))$ as T in either Case 4a1 or Case 4b1, then $rpl(T_{i-1}) = rpl(T) = rpl(T_{i+1}) < rpl(T_i)$ holds, which contradicts (co1). Otherwise if $rpl(T_{i-1}) = rpl(T_i)$ then we set $C(T_i, rpl(T_i))$ or $C(T_i, rpl(T_i) + 1)$ as T in Case 4b1, then it contradict the condition $rpl(T_i) > rpl(T_{i+1})$ and $rpl(T) = rpl(T_{i+1})$ of Case 4b3. Otherwise, $rpl(T_{i-1}) > rpl(T_i)$ holds, then we set $C(T_i, rpl(T_i))$ as T in either Case 4a2 or Case 4b2, so $rpl(T) = rpl(T_i)$ holds, which contradicts Case 4b3.

If $rpl(T) = 2$, then we set T in either Case 2b1, 4a1, 4a2, 4b1 or 4b2 of the preceding step $i - 1$. If we set T in Case 2b1 then T_i has the pony-tail and $rpl(T_i) = 2$, which contradicts $rpl(T_i) > rpl(T_{i+1}) > 1$. If we set T in Case 4a1 or Case 4b1 then $rpl(T_{i-1}) = rpl(T) = rpl(T_{i+1}) < rpl(T_i)$, which contradicts (co1). If we set T in either Case 4a2 or Case 4b2 then $rpl(T_{i-1}) > rpl(T_i) = rpl(T)$ which contradicts Case 4b3. □

Lemma 6. *Assume that T_s is derived from T_{s-1} by delete-and-append a leaf. If $rpl(T_{s-1}) = 1$ (so T_s is copying T_{s-1}) then $C(T_s, 1)$ is copying $C(T_{s-1}, 2)$.*

Proof. We have two cases. Either T_s has the pony tail or does not. For each case the claim holds. See Fig. 6. □

We need the above lemma in the proof of the next lemma.

Lemma 7. *Assume that (co1) is satisfied. If (co2) is satisfied for $i = 1, 2, \cdots, s$ then, after executing step $i = s$, (co2) is satisfied for $i = s + 1$.*

Proof. Omitted. See full version.

Theorem 1. *There is a Gray code for ordered trees with n vertices such that each ordered tree is derived from the preceding ordered tree by removing a leaf then appending a leaf.*

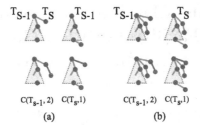

Fig. 6. Illustrations for Lemma 6.

By constructing the necessary part of F_n^O on the fly one can generate each ordered tree in a Gray code for S_n in $O(n^2)$ time for each ordered tree.

4 Conclusion

In this paper we have designed a Gray code for ordered trees with n vertices such that each ordered tree is derived from the preceding ordered tree by removing a leaf then appending a leaf.

Can we design a faster algorithm to compute the Gray code?

Is it possible to simplify the construction of the Gray code, possibly based on other choice of a tree structure of the ordered trees with n vertices?

Can we design a Gray code for other combinatorial objects based on a tree structure of the objects?

References

1. Knuth, D.E.: The Art of Computer Programming, vol. 4. Generating All Trees, History of Combinatorial Generation. Addison-Wesley (2006)
2. Lapey, P., Williams, A.: Pop & push: ordered tree iteration in O(1)-time. In: 33rd International Symposium on Algorithms and Computation, ISAAC 2022, Seoul, Korea, 19–21 December 2022. LIPIcs, vol. 248, pp. 53:1–53:16 (2022)
3. Lucas, J.M.: The rotation graph of binary trees is Hamiltonian. J. Algorithms **8**(4), 503–535 (1987)
4. Lucas, J.M., van Baronaigien, D.R., Ruskey, F.: On rotations and the generation of binary trees. J. Algorithms **15**(3), 343–366 (1993)
5. Mütze, T.: Combinatorial Gray codes - an updated survey. CoRR, abs/2202.01280 (2022)
6. Nakano, S.-I.: Efficient generation of plane trees. Inf. Process. Lett. **84**(3), 167–172 (2002)
7. Nakano, S.-I.: Family trees for enumeration. Int. J. Found. Comput. Sci. **34**, 715–736 (2023)
8. Savage, C.D.: A survey of combinatorial gray codes. SIAM Rev. **39**(4), 605–629 (1997)

Mechanism Design with Predictions for Facility Location Games with Candidate Locations

Jiazhu Fang⬤, Qizhi Fang, Wenjing Liu(✉)⬤, and Qingqin Nong⬤

Ocean University of China, Qingdao 266100, China
`fjz@stu.ouc.edu.cn`, {`qfang,liuwj,qqnong`}`@ouc.edu.cn`

Abstract. We study mechanism design with predictions in the single (obnoxious) facility location games with candidate locations on the real line, which complements the existing literature on mechanism design with predictions. We first consider the single facility location games with candidate locations, where each agent prefers the facility (e.g., a school) to be located as close to her as possible. We study two social objectives: minimizing the maximum cost and the social cost, and provide deterministic, anonymous, and group strategy-proof mechanisms with predictions that achieve good trade-offs between consistency and robustness, respectively. Furthermore, we represent the approximation ratio as a function of the prediction error, indicating that mechanisms can achieve better performance even when predictions are not fully accurate. We also consider the single obnoxious facility location games with candidate locations, where each agent prefers the facility (e.g., a garbage transfer station) to be located as far away from her as possible. For the objective of maximizing the minimum utility, we show that any strategy-proof mechanism with predictions is unbounded robust. For the objective of maximizing the social utility, we provide a deterministic, anonymous, and group strategy-proof mechanism with prediction that achieves a good trade-off between consistency and robustness.

Keywords: Mechanism design with predictions · Facility location · Candidate locations · Consistency · Robustness

1 Introduction

Designing *algorithms with imperfect predictions* is a recently popular research direction, providing a basis for finding methods that go beyond *worst-case analysis*, which is a traditional method for algorithm analysis in computer science. While worst-case analysis can ensure the robustness of the algorithm to some extent, the worst-case scenario often does not occur, which hinders the design

Supplementary Information The online version contains supplementary material available at https://doi.org/10.1007/978-981-97-2340-9_4.

of algorithms with better performance. The lower bounds of the worst-case are often determined by uncertain information, such as unknown future information in online problems [5] and private information of agents in mechanism design problems [23]. Introducing prediction provides a possibility to break through the lower bounds.

In the framework of algorithmic design with imperfect predictions, it is usually assumed that a machine learning method can provide predictions about unknown information, and the algorithm designer can use the imperfect predictions to design the algorithm. The goal of such algorithms is to achieve good performance that surpasses the worst-case bound, when the predictions are accurate, and to never perform much worse than the optimal algorithm in the worst-case, even when the predictions are arbitrarily inaccurate. Lykouris and Vassilvitskii [20] formally introduced *consistency* and *robustness* to represent the competitive ratios of an online algorithm with respect to accurate and arbitrarily inaccurate predictions respectively, which have now become standard terminologies for analyzing the performance of algorithms with imperfect predictions. Much of the related work has focused on online algorithm design, and many classic online problems have been considered, such as ski rental [15,24], caching [25], and scheduling [16,18].

Mechanism design and online algorithm design are similar in that they both involve unknown information. Due to the private information held by strategic agents, the strategy-proof (SP, no agent can benefit from misreporting) mechanism usually produces only approximately optimal solutions. Introducing predictions may allow us to obtain an optimal SP mechanism. Therefore, we hope to obtain an ideal mechanism through predictions, one that can elegantly transform from optimal performance when predictions are accurate to the best-known worst-case performance when predictions are arbitrarily inaccurate, thus achieving the best of both worlds. However, in many cases, mechanisms may rely on predicted values to achieve the optimal performance, and when predictions are inaccurate, unbounded robustness may arise. Alternatively, we hope that mechanisms can approach the above-mentioned ideal performance as closely as possible, that is, achieve the best possible trade-off between consistency and robustness. The application of predictions to mechanism design was first introduced by Agrawal et al. [1], who considered the classical mechanism design problem without money – *facility location*, and subsequently, Xu and Lu [27] studied other mechanism design problems under predictions, both showing that mechanisms with predictions can achieve a good (optimal) trade-off between robustness and consistency.

In the framework of mechanism design with predictions, we study the facility location games with candidate locations [13,26], which complements the results of mechanism design with predictions. In the classic facility location problem, facilities can be placed anywhere in the metric space. However, in reality, most facilities can only be built in specific locations due to restrictions on land use and area. For example, schools can only be built in certain spacious areas, while waste transfer stations can only be built in specific locations due to wind direction and surrounding residential areas. Note that the facility location problem with candidate locations is a more general formulation than its counterpart without candidate locations, as the latter corresponds to the special case where all points in the metric space are considered as candidate locations.

1.1 Related Work

Algorithm (Mechanism) with Predictions. Early relevant work on algorithm design with predictions can be found in this survey [21]. Lykouris and Vassilvitskii [20] formally introduced consistency and robustness as two main indicators for measuring the performance of algorithms with predictions. Subsequently, a series of classic problems have been studied, such as secretary problems [8], online graph problems [2], maximum flow [22], shortest path [10] as well as k-means clustering [9] and Nash social welfare maximization [4]. Here, we mainly introduce the literature on mechanism design with predictions. Agrawal et al. [1] first introduced this framework into the field of mechanism design and studied facility location problems in two-dimensional Euclidean space. For the maximum cost objective, they provided a 1-consistent and $(1 + \sqrt{2})$-robust SP mechanism, and for the social cost objective, they introduced a confidence value parameter $c \in (0, 1)$ and provided a $\frac{\sqrt{2c^2+2}}{1+c}$-consistent and $\frac{\sqrt{2c^2+2}}{1-c}$-robust SP mechanism, achieving an optimal trade-off between consistency and robustness. They also provided the optimal 1-consistent and 2-robust SP mechanism for the one-dimensional case under the maximum cost objective. Xu and Lu [27] studied four classic mechanism design problems, and improved the known approximation ratio using predictions for the two facility location game on the real line. Istrate and Bonchis [17] considered the obnoxious facility location problem and achieved trade-offs between robustness and consistency on line segments, squares, circles, and trees. Subsequently, Gkatzelis et al. [14] extended the prediction framework to decentralized mechanism design in strategic settings. Balkanski et al. [3] studied classic strategic scheduling problems, employing predictions to provide a 6-consistent and $2n$-robust SP mechanism. For more related literature, please refer to the webpage of algorithms with predictions [19].

Strategic Facility Location. Due to the extensive literature on facility location problems, we only provide a brief review of the most relevant literature. Procaccia and Tennenholtz [23] proposed approximate mechanism design without money for facility location games, which uses the approximation ratio to study SP mechanisms for the optimization problem, thereby initiating a research trend in strategic facility location. Tang et al. [26] considered the single and two facility location games with candidate locations under social cost and maximum cost objective. For the single facility problem with maximum cost objective, they presented a deterministic 3-approximate group strategy-proof (GSP, no coalition of agents can benefit from misreporting) mechanism and proved that there does not exist any deterministic (or randomized) SP mechanism with approximation ratio better than 3 (or 2). Feldman et al. [11] studied the relationship among three types of candidate selection mechanisms: voting, ranking and location mechanisms. They proved that the median mechanism, which places the facility at the candidate location closest to the median agent, is SP and 3-approximate under social cost, and no deterministic SP mechanism can do better. Gai et al. [13] studied the single obnoxious and two facility location games with social utility objective on the real line. For the single facility problem, they provided a deterministic 3-approximate GSP mechanism which is also the best possible SP mechanism. For further literature on facility location, please refer to [7].

1.2 Our Contribution

Facility Location Game with Candidate Locations. In Sect. 3, we study the single facility location game with candidate locations on the real line under two social objectives: minimizing the maximum cost and minimizing the social cost. For the maximum cost objective, we present a deterministic, anonymous and GSP mechanism with predictions, which is 1-consistent and 3-robust, achieving the best trade-off between consistency and robustness. More generally, we represent the approximation ratio of the mechanism as a function of the prediction error $\delta \geq 0$, obtaining that it is $\min\{1 + \delta, 3\}$-approximate, which corresponds to the optimal mechanism when $\delta = 0$ and gradually increases linearly to 3-approximate. We also analyze its performance under the social cost objective, which is 1-consistent and $(2n - 1)$-robust. By reducing the reliance on predictions, we present a deterministic, anonymous and GSP mechanism which is $\frac{3-\gamma}{1+\gamma}$-consistent and $\frac{3+\gamma}{1-\gamma}$-robust under social cost, where $\gamma \in [0, 1)$ can be adjusted based on the degree of trust in the prediction accuracy. If the designer is not confident in the prediction, setting $\gamma = 0$ will yield a 3-consistent and 3-robust mechanism that matches the best performance guarantee without predictions [11]. As γ gradually increases from 0 to 1, the consistency decreases from 3 to 1, achieving the optimal performance, but at the cost of increasing the robustness. We also represent the approximation ratio of the mechanism as a function of the prediction error δ, obtaining that it is $\min\{\frac{3-\gamma}{1+\gamma} + \delta, \frac{3+\gamma}{1-\gamma}\}$-approximate, which starts at $\frac{3-\gamma}{1+\gamma}$-approximate when $\delta = 0$, corresponding to the consistency guarantee, and gradually increases linearly to $\frac{3+\gamma}{1-\gamma}$-approximate, corresponding to the robustness guarantee.

Obnoxious Facility Location Game with Candidate Locations. In Sect. 4, we study the single obnoxious facility location game with candidate locations on the real line under two social objectives: maximizing the minimum utility and maximizing the social utility. For the minimum utility objective, we show that any SP mechanism with predictions is unbounded robust. We mainly focus on the social utility objective. By regarding the prediction as the location chosen by γn virtual voters and adopting a voting mechanism for the leftmost and rightmost candidate points, we obtain a deterministic, anonymous and GSP mechanism which is $\frac{3-\gamma}{1+\gamma}$-consistent and $\frac{3+\gamma}{1-\gamma}$-robust.

2   Preliminaries

In the single (obnoxious) facility location game with candidate locations, let $N = \{1, \ldots, n\}$ be the set of agents located on the real line, and each agent $i \in N$ has a private location $x_i \in \mathbb{R}$. We use $\mathbf{x} = (x_1, \ldots, x_n) \in \mathbb{R}^n$ to denote the *location profile* of the n agents. Let $M \subseteq \mathbb{R}$ be a compact set representing the set of candidate locations for the facility. Let $a, b \in M$ be the leftmost candidate point and the rightmost candidate point, respectively. The distance between any two points $x, y \in \mathbb{R}$ is $d(x, y) = |x - y|$. Denote an instance by $I(\mathbf{x}, M)$ or simply by I. A *(deterministic) mechanism* is a function $f : \mathbb{R}^n \to M$ which maps a given location profile $\mathbf{x}$ to a facility location. A mechanism f is *anonymous*, if its outcome does not depend on identities of the agents, i.e., for every location profile $\mathbf{x}$ and every permutation of agents $\pi : N \to N$, $f(x_1, \ldots, x_n) = f(x_{\pi(1)}, \ldots, x_{\pi(n)})$. Here, we focus on anonymous mechanisms and assume w.l.o.g. that $x_1 \le \ldots \le x_n$ for every location profile $\mathbf{x}$.

In the single facility location setting, each agent prefers the facility to be located as close to her as possible. Given an instance I and a facility location $y \in M$, the distance from the facility to agent $i \in N$ is considered as her cost, denoted by $\text{cost}(x_i, y) = d(x_i, y)$. We study two standard social objectives: minimizing the maximum cost and minimizing the social cost. The *maximum cost* of a facility location $y \in M$ with respect to the profile $\mathbf{x} \in \mathbb{R}^n$ is the maximum distance to all agents, denoted by

$$MC(\mathbf{x}, y) = \max_{i \in N} \text{cost}(x_i, y) = \max_{i \in N} d(x_i, y),$$

and the *social cost* of y with respect to $\mathbf{x}$ is the total distance to all agents, denoted by

$$SC(\mathbf{x}, y) = \sum_{i \in N} \text{cost}(x_i, y) = \sum_{i \in N} d(x_i, y).$$

In the single obnoxious facility location setting, each agent prefers the facility to be located as far away from her as possible, and the distance from the facility to agent $i \in N$ is interpreted as her utility, denoted by $\text{utility}(x_i, y) = d(x_i, y)$. We also consider two social objectives: maximizing the minimum utility and maximizing the social utility. The *minimum utility* of a facility location $y \in M$ with respect to the profile $\mathbf{x} \in \mathbb{R}^n$ is the minimum distance to all agents, denoted by

$$MU(\mathbf{x}, y) = \min_{i \in N} \text{utility}(x_i, y) = \min_{i \in N} d(x_i, y),$$

and the *social utility* of y with respect to $\mathbf{x}$ is the total distance to all agents, denoted by

$$SU(\mathbf{x}, y) = \sum_{i \in N} \text{utility}(x_i, y) = \sum_{i \in N} d(x_i, y).$$

The following definitions are only explicitly given in the single facility location setting and they are analogous in the obnoxious setting.

In the strategic setting, each agent may be incentivized to misreport to reduce her cost. A mechanism f is *strategy-proof* (SP) if no agent can benefit from misreporting, regardless of the reports of the others, that is, for every $i \in N$,

$$\text{cost}\left(x_i, f(x_i, \mathbf{x}_{-i})\right) \leq \text{cost}\left(x_i, f(x'_i, \mathbf{x}_{-i})\right),$$

for every $x_i, x'_i \in \mathbb{R}$ and $\mathbf{x}_{-i} = (x_1, \ldots, x_{i-1}, x_{i+1}, \ldots, x_n) \in \mathbb{R}^{n-1}$.

A mechanism f is *group strategy-proof* (GSP), if no coalition of agents can decrease each members cost from misreporting, regardless of the reports of the other agents, that is, for every $G \subseteq N$, every $\mathbf{x}_G = (x_i)_{i \in G}, \mathbf{x}'_G = (x'_i)_{i \in G} \in \mathbb{R}^{|G|}$ and $\mathbf{x}_{-G} = (x_i)_{i \notin G} \in \mathbb{R}^{|N \setminus G|}$, there exists $i \in N$ such that

$$\text{cost}\left(x_i, f(\mathbf{x}_G, \mathbf{x}_{-G})\right) \leq \text{cost}\left(x_i, f(\mathbf{x}'_G, \mathbf{x}_{-G})\right).$$

Given a social objective function C to be minimized and an instance $I(\mathbf{x}, M)$, denote by $o(I)$ and $OPT(I)$ the optimal facility location and the optimum value respectively, or simply by o and OPT when I is clear from the context. A mechanism f achieves an *approximation ratio* of $\alpha \geq 1$ if for every instance $I(\mathbf{x}, M)$,

$$C(\mathbf{x}, f(\mathbf{x})) \leq \alpha \cdot C(\mathbf{x}, o) = \alpha \cdot OPT(I).$$

In the problem of mechanism design with predictions, the private location for each agent is predicted. When the predicted location profile $\hat{\mathbf{x}} = (\hat{x}_1, \ldots, \hat{x}_n)$ is obtained, it is easy to compute the optimal location $\hat{o}$ for $\hat{\mathbf{x}}$. Therefore, we directly design mechanisms using the predicted value $\hat{o}$, corresponding to the optimal location o for $\mathbf{x}$. A (deterministic) *mechanism with predictions* is a function $f : R^n \times M \to M$ which maps a given location profile $\mathbf{x}$ and a predicted optimal location $\hat{o}$ to a facility location. Similarly, we stipulate that the mechanism with predictions, denoted as $f(\mathbf{x}, \hat{o})$, needs to be (G)SP, and use consistency and robustness to measure its performance. Given a social objective function C to be minimized, a mechanism f is α-consistent if it achieves an approximation ratio of α when the prediction is accurate ($\hat{o} = o$), i.e., for every instance $I(\mathbf{x}, M)$,

$$C(\mathbf{x}, f(\mathbf{x}, o)) \leq \alpha \cdot C(\mathbf{x}, o) = \alpha \cdot OPT(I).$$

A mechanism f is β-*robust* if it achieves an approximation ratio of β even when the prediction is arbitrarily inaccurate, i.e., for every instance $I(\mathbf{x}, M)$,

$$\max_{\hat{o}} C(\mathbf{x}, f(\mathbf{x}, \hat{o})) \leq \beta \cdot C(\mathbf{x}, o) = \beta \cdot OPT(I).$$

To achieve a smooth trade-off between consistency and robustness, we represent the mechanism's approximation ratio as a function of the *prediction error* to observe the performance of mechanisms when the prediction is not fully accurate. Given an instance $I(\mathbf{x}, M)$, we define the *prediction error* as $\delta(\hat{o}, I) = \frac{d(\hat{o}, o)}{OPT(I)}$ when the social objective function is minimizing the maximum cost, representing the distance between the predicted optimal value $\hat{o}$ and the true optimal value o, normalized by the optimal value $OPT(I)$. For the social cost objective, define the *prediction error* as $\delta(\hat{o}, I) = \frac{d(\hat{o}, o)}{OPT(I)/n}$, representing the distance between the predicted optimal value $\hat{o}$ and the true optimal value o, normalized by $OPT(I)/n$.

A mechanism f is called $\alpha(\delta)$-*approximate*, if given an upper bound δ on the prediction error, it achieves an approximation ratio of $\alpha(\delta)$ when the prediction error of every instance does not exceed δ, i.e., for every instance $I(\mathbf{x}, M)$,

$$\max_{I, \hat{o}:\ \delta(\hat{o}, I) \leq \delta} C(\mathbf{x}, f(\mathbf{x}, \hat{o})) \leq \alpha(\delta) \cdot C(\mathbf{x}, o) = \alpha(\delta) \cdot OPT(I).$$

It can be observed that when $\delta = 0$, the mechanism's approximation ratio corresponds to the consistency guarantee, and when $\delta \to \infty$, it corresponds to the robustness guarantee.

The goal is to design (G)SP mechanisms that can capture an optimal solution when the prediction is accurate and matches the best-known results without predictions when the prediction is arbitrarily inaccurate. Furthermore, we also consider the mechanism's approximation guarantees when the prediction is not fully accurate.

3   Single Facility Location Game

In this section, we study the single facility location game with candidate locations on the real line, where each agent wants the facility (e.g., a supermarket) to be built as close to her as possible, and the distance from the facility to her is considered as her cost. We focus on two social objectives: minimizing the maximum cost and minimizing the social cost.

3.1   Maximum Cost

For the objective of minimizing the maximum cost, we first present a deterministic, anonymous and GSP mechanism with predictions that is 1-consistent, capturing an optimal solution when the prediction is accurate, and 3-robust, matching the best approximation guarantee without predictions, thereby achieving the best of both worlds.

Mechanism 1

Input: A reported location profile $\mathbf{x} = (x_1, \ldots, x_n) \in \mathbb{R}^n$; Candidate locations $M \subseteq \mathbb{R}$; A prediction $\hat{o} \in M$.
Output: A facility location $y \in M$.
if $\hat{o} \in [x_1, x_n]$ then
return $y = \hat{o}$.
else if $\hat{o} < x_1$ then
return $y \in \arg\min_{g \in M} |x_1 - g|$, breaking ties in any deterministic way.
else
return $y \in \arg\min_{g \in M} |x_n - g|$, breaking ties in any deterministic way.
end if

Given the predicted optimal location $\hat{o}$ and the reported location profile $\mathbf{x} = (x_1, \ldots, x_n)$, we determine the facility location of by comparing the location relationship between $\hat{o}$ and $\mathbf{x}$, as illustrated by Mechanism 1. Specifically, when $\hat{o}$ falls within the range $[x_1, x_n]$, indicating that the prediction is not significantly inaccurate, and we directly output location $\hat{o}$. When $\hat{o}$ falls outside the range $[x_1, x_n]$, signifying a highly inaccurate prediction, we output the candidate location closest to either x_1 or x_n based on the location of $\hat{o}$, breaking ties in any deterministic way. In the following, we analyze the performance of the Mechanism 1.

Lemma 1. *Mechanism 1 is anonymous and GSP[1].*

Theorem 1. *For the single facility location game on the real line, Mechanism 1 is 1-consistent and 3-robust under the maximum cost objective.*

In accordance with the lower bound of any deterministic SP mechanism provided by Tang et al. [26], it is evident that the trade-off between consistency and robustness in our results is optimal. We now consider the performance of Mechanism 1 when the prediction is not fully accurate via the prediction error.

Theorem 2. *For the single facility location game on the real line, Mechanism 1 achieves a $\min\{1+\delta, 3\}$-approximation under the maximum cost objective, where δ is an upper bound on the prediction error.*

3.2   Social Cost

In this subsection, we focus on the objective of minimizing the social cost. For the single facility location game with candidate locations on the real line, Feldman et al. [11] proved that no deterministic SP mechanism without predictions can achieve an approximation ratio better than 3. Can we break through this lower bound by introducing predictions, as under the maximum cost objective?

We first analyze the performance of Mechanism 1 under social cost and show that it is 1-consistent and $(2n-1)$-robust which is tight (see *Supplementary Material*). Note that Mechanism 1 is equivalent to select the candidate location which is closest to the median of $\mathbf{x}$ and $(n-1)$ virtual points $\hat{o}$. It provides a poor robustness guarantee under social cost, probably because it relies heavily on the prediction to achieve 1-consistency. Motivated by this, we introduce a parameter $\gamma \in [0, 1)$ to regulate the reliance on the prediction and obtain the following Mechanism 2[2].

---

[1] Due to length limitation, all the proofs are included in *Supplementary Material*.

[2] Due to length limitation, both the equivalent statement of Mechanism 1 and the high-level idea of Mechanism 2 are given in *Supplementary Material*.

Mechanism 2

Input: A reported location profile $\mathbf{x} = (x_1, \ldots, x_n) \in \mathbb{R}^n$; Candidate locations $M \subseteq \mathbb{R}$; A prediction $\hat{o} \in M$; $\gamma \in [0, 1)$ such that γn is an integer.

Output: A facility location $y \in M$.

Create a γn-dimensional virtual location profile $\mathbf{x}^v = (x_1^v, \ldots, x_{\gamma n}^v)$, where $x_i^v = \hat{o}$ for any $i = 1, \ldots, \gamma n$.

Let x_m be the median of $(\mathbf{x}, \mathbf{x}^v)$, breaking ties in any deterministic way.

return $y \in \arg\min_{g \in M} |x_m - g|$, breaking ties in any deterministic way.

Theorem 3. *For the single facility location game on the real line, Mechanism 2 with some constant $\gamma \in [0, 1)$ is deterministic, anonymous and GSP, and it is $\frac{3-\gamma}{1+\gamma}$-consistent and $\frac{3+\gamma}{1-\gamma}$-robust under the social cost objective*[3].

We now analyze the performance of Mechanism 2 when the prediction is not fully correct via the prediction error.

Theorem 4. *For the single facility location game on the real line, Mechanism 2 with some constant $\gamma \in [0, 1)$ achieves a $\min\{\frac{3-\gamma}{1+\gamma} + \delta, \frac{3+\gamma}{1-\gamma}\}$-approximation under the social cost objective, where δ is an upper bound on the prediction error.*

4   Single Obnoxious Facility Location Game

In this section, we study the single obnoxious facility location game with candidate locations on the real line, where each agent wants the facility (e.g., a garbage dump) to be built as far away from her as possible, and the distance from her location to the facility can be interpreted as her utility. For the objective of maximizing the minimum utility, Tang et al. [26] showed that all SP mechanisms without predictions have unbounded approximation ratios, which implies that any SP mechanism with predictions is unbounded robust[4]. We now focus on the objective of maximizing the social utility.

Obviously, the optimal solution to the problem is either the leftmost candidate location a or the rightmost one b. Therefore, the predicted optimal solution $\hat{o}$ is also in $\{a, b\}$. We introduce the reliance parameter $\gamma \in [0, 1)$ just like in Sect. 3.2 and adopt the voting idea to determine the facility location. Specifically, there are $(1 + \gamma)n$ voters voting for two candidate locations $\{a, b\}$, where each agent $i \in N$ supports the candidate location farther from her, and the other γn virtual voters support $\hat{o}$. Then the facility location is determined by the majority rule. For this mechanism (denoted as Mechanism 3), we have the following conclusion.

Theorem 5. *For the single obnoxious facility location game on the real line, Mechanism 3 is deterministic, anonymous, GSP, and it is $\frac{3-\gamma}{1+\gamma}$-consistent and $\frac{3+\gamma}{1-\gamma}$-robust under the social utility objective*[5].

---

[3] Mechanism 2's consistency and robustness will change with the reliance parameter γ and there is a tradeoff between them; see Fig. 1 in *Supplementary Material*.

[4] For completeness, we also give a specific description; see *Supplementary Material*.

[5] See Fig. 1 in *Supplementary Material* for an intuitive interpretation.

While implementing Mechanism 3, the mechanism designer can choose the value of γ based on her confidence in the predicted value $\hat{o}$. If the designer is not confident in the prediction $\hat{o}$, setting $\gamma = 0$ will yield a 3-consistent and 3-robust mechanism that matches the best performance guarantee without predictions [13].

Here, we need not to analyze the performance of Mechanism 3 when predictions are not fully accurate via the prediction error, since the predicted value $\hat{o}$ is either o or $\{a, b\}\backslash o$.

Mechanism 3

Input: A reported location profile $\mathbf{x} = (x_1, \ldots, x_n) \in \mathbb{R}^n$; Candidate locations $M \subseteq \mathbb{R}$; A prediction $\hat{o} \in \{a, b\}$; $\gamma \in [0, 1)$.
Output: A facility location $y \in \{a, b\}$.
Let n_1 be the number of agents with $x_i \leq \frac{a+b}{2}$, and n_2 be the number of agents with $x_i > \frac{a+b}{2}$.
if $(\hat{o} = b \,\&\&\, n_1 + \gamma n > n_2) \| (\hat{o} = a \,\&\&\, n_1 > n_2 + \gamma n)$ **then**
return $y = b$.
else
return $y = a$.
end if

---

5 Conclusions and Open Problems

In this paper, we studied the single (obnoxious) facility location game with candidate locations on the real line under the framework of mechanism design with predictions. We first considered the single facility location game. For the objective of minimizing the maximum cost, we gave a deterministic, anonymous, and GSP mechanism, and proved that it is 1-consistent and 3-robust, achieving the best of both worlds. We also analyzed its performance under the social cost objective, but it provides poor robustness guarantee. By reducing its reliance on the prediction, we presented a deterministic, anonymous, and GSP mechanism, and proved that it is $\frac{3-\gamma}{1+\gamma}$-consistent and $\frac{3+\gamma}{1-\gamma}$-robust. We respectively represented the approximation ratios of the two mechanisms as functions of the prediction error, demonstrating that the mechanisms can elegantly transition from consistency guarantees to robustness guarantees. We also considered the single obnoxious facility location game under the objective of maximizing the social utility, and provided a deterministic, anonymous, and GSP mechanism, which is $\frac{3-\gamma}{1+\gamma}$-consistent and $\frac{3+\gamma}{1-\gamma}$-robust, achieving a good trade-off between consistency and robustness. However, there are still many open questions in this direction:

1. For the case of locating two facilities, can we provide a GSP mechanism that achieves a good trade-off between consistency and robustness? We have attempted to do so, but the results are unsatisfactory.

2. Can we extend the line to other metric spaces such as trees or general graphs and use predictions to achieve good results? What about facility location games in other settings, such as under the double-peaked preference [12] or under other social objectives (e.g., the minimax envy [6])?
3. Randomized mechanisms can usually achieve better approximation ratios than the deterministic. Can we design randomized mechanisms with predictions to achieve a better performance?

Acknowledgements. This research was supported in part by the National Natural Science Foundation of China (12201590, 12171444, 11971447).

References

1. Agrawal, P., Balkanski, E., Gkatzelis, V., Ou, T., Tan, X.: Learning-augmented mechanism design: leveraging predictions for facility location. In: Proceedings of the 23rd ACM Conference on Economics and Computation, pp. 497–528 (2022). https://doi.org/10.1145/3490486.3538306
2. Azar, Y., Panigrahi, D., Touitou, N.: Online graph algorithms with predictions. In: Proceedings of the 2022 Annual ACM-SIAM Symposium on Discrete Algorithms, pp. 35–66 (2022). https://doi.org/10.1137/1.9781611977073.3
3. Balkanski, E., Gkatzelis, V., Tan, X.: Strategyproof scheduling with predictions. arXiv preprint arXiv:2209.04058 (2022). https://doi.org/10.48550/arXiv.2209.04058
4. Banerjee, S., Gkatzelis, V., Gorokh, A., Jin, B.: Online nash social welfare maximization with predictions. In: Proceedings of the 2022 Annual ACM-SIAM Symposium on Discrete Algorithms, pp. 1–19 (2022). https://doi.org/10.1137/1.9781611977073.1
5. Borodin, A., El-Yaniv, R.: Online Computation and Competitive Analysis. Cambridge University Press, Cambridge (2005)
6. Cai, Q., Filos-Ratsikas, A., Tang, P.: Facility location with minimax envy. In: Proceedings of the 25th International Joint Conference on Artificial Intelligence, pp. 137—143 (2016). https://doi.org/10.5555/3060621.3060641
7. Chan, H., Filos-Ratsikas, A., Li, B., Li, M., Wang, C.: Mechanism design for facility location problems: a survey. In: Proceedings of the 30th International Joint Conference on Artificial Intelligence, pp. 4356–4365 (2021). https://doi.org/10.24963/ijcai.2021/596
8. Dütting, P., Lattanzi, S., Paes Leme, R., Vassilvitskii, S.: Secretaries with advice. In: Proceedings of the 22nd ACM Conference on Economics and Computation, pp. 409–429 (2021). https://doi.org/10.1145/3465456.3467623
9. Ergun, J.C., Feng, Z., Silwal, S., Woodruff, D.P., Zhou, S.: Learning-augmented k-means clustering. arXiv preprint arXiv:2110.14094 (2021). https://doi.org/10.48550/arXiv.2210.17028
10. Feijen, W., Schäfer, G.: Dijkstras algorithm with predictions to solve the single-source many-targets shortest-path problem. arXiv preprint arXiv:2112.11927 (2021). https://doi.org/10.48550/arXiv.2112.11927
11. Feldman, M., Fiat, A., Golomb, I.: On voting and facility location. In: Proceedings of the 2016 ACM Conference on Economics and Computation, pp. 269–286 (2016). https://doi.org/10.1145/2940716.2940725

12. Filos-Ratsikas, A., Li, M., Zhang, J., Zhang, Q.: Facility location with double-peaked preferences. Auton. Agents Mulit-Agent Syst. **31**, 1209–1235 (2017). https://doi.org/10.1007/s10458-017-9361-0

13. Gai, L., Liang, M., Wang, C.: Obnoxious facility location games with candidate locations. In: Ni, Q., Wu, W. (eds.) AAIM 2022. LNCS, vol. 13513, pp. 96–105. Springer, Cham (2022). https://doi.org/10.1007/978-3-031-16081-3_9

14. Gkatzelis, V., Kollias, K., Sgouritsa, A., Tan, X.: Improved price of anarchy via predictions. In: Proceedings of the 23rd ACM Conference on Economics and Computation, pp. 529–557 (2022). https://doi.org/10.1145/3490486.3538296

15. Gollapudi, S., Panigrahi, D.: Online algorithms for rent-or-buy with expert advice. In: Chaudhuri, K., Salakhutdinov, R. (eds.) ICML 2019, Long Beach, California, USA, vol. 97, pp. 2319–2327 (2019)

16. Im, S., Kumar, R., Montazer Qaem, M., Purohit, M.: Non-clairvoyant scheduling with predictions. In: Proceedings of the 33rd ACM Symposium on Parallelism in Algorithms and Architectures, pp. 285–294 (2021). https://doi.org/10.1145/3409964.3461790

17. Istrate, G., Bonchis, C.: Mechanism design with predictions for obnoxious facility location. arXiv preprint arXiv:2212.09521 (2022). https://doi.org/10.48550/arXiv.2212.09521

18. Lattanzi, S., Lavastida, T., Moseley, B., Vassilvitskii, S.: Online scheduling via learned weights. In: Proceedings of the 2020 ACM-SIAM Symposium on Discrete Algorithms, pp. 1859–1877 (2020). https://doi.org/10.1137/1.9781611975994.11

19. Lindermayr, A., Megow, N.: Algorithms with predictions. https://algorithms-with-predictions.github.io. Accessed 17 Apr 2023

20. Lykouris, T., Vassilvitskii, S.: Competitive caching with machine learned advice. J. ACM **68**(4), 1–25 (2021). https://doi.org/10.1145/3447579

21. Mitzenmacher, M., Vassilvitskii, S.: Algorithms with predictions. Commum. ACM **65**(7), 33–35 (2022). https://doi.org/10.1145/3528087

22. Polak, A., Zub, M.: Learning-augmented maximum flow. arXiv preprint arXiv:2207.12911 (2022). https://doi.org/10.48550/arXiv.2207.12911

23. Procaccia, A.D., Tennenholtz, M.: Approximate mechanism design without money. ACM Trans. Econ. Comput. **1**(4), 1–26 (2013). https://doi.org/10.1145/2542174.2542175

24. Purohit, M., Svitkina, Z., Kumar, R.: Improving online algorithms via ml predictions. In: Bengio, S., Wallach, H., Larochelle, H., Grauman, K., Cesa-Bianchi, N., Garnett, R. (eds.) NeurIPS 2018, Montréal, Canada, vol. 31, pp. 9661–9670 (2018)

25. Rohatgi, D.: Near-optimal bounds for online caching with machine learned advice. In: Proceedings of the 14th Annual ACM-SIAM Symposium on Discrete Algorithms, pp. 1834–1845 (2020). https://doi.org/10.1137/1.9781611975994.112

26. Tang, Z., Wang, C., Zhang, M., Zhao, Y.: Strategyproof facility location with limited locations. J. Oper. Res. Soc. China **11**, 553–567 (2023). https://doi.org/10.1007/s40305-021-00378-1

27. Xu, C., Lu, P.: Mechanism design with predictions. In: Proceedings of the 31st International Joint Conference on Artificial Intelligence, pp. 571–577 (2022). https://doi.org/10.24963/ijcai.2022/81

An Improved Approximation Algorithm for Metric Triangle Packing

Jingyang Zhao and Mingyu Xiao[✉]

University of Electronic Science and Technology of China, Chengdu, China
myxiao@gmail.com

Abstract. Given an edge-weighted metric complete graph with n vertices, the maximum weight metric triangle packing problem is to find a set of $n/3$ vertex-disjoint triangles with the total weight of all triangles in the packing maximized. Several simple methods can lead to a 2/3-approximation ratio. However, this barrier is not easy to break. Chen et al. proposed a randomized approximation algorithm with an expected ratio of $(0.66768-\varepsilon)$ for any constant $\varepsilon > 0$. In this paper, we improve the approximation ratio to $(0.66835 - \varepsilon)$. Furthermore, we can derandomize our algorithm.

1 Introduction

In a graph with n vertices, a *triangle packing* is a set of vertex-disjoint triangles (i.e., a simple cycle on three different vertices). The triangle packing is called *perfect* if its size is $n/3$ (i.e., it can cover all vertices). Given an unweighted graph, the Maximum Triangle Packing (MTP) problem is to find a triangle packing of maximum cardinality. In an edge-weighted graph, every edge has a non-negative weight. There are two natural variants. If the graph is an edge-weighted complete graph, a perfect triangle packing will exist. The Maximum Weight Triangle Packing (MWTP) problem is to find a perfect triangle packing such that the total weight of all triangles is maximized. Furthermore, if the graph is an edge-weighted metric complete graph (i.e., the weight of edges satisfies the symmetric and triangle inequality properties), the problem is called the Maximum Weight Metric Triangle Packing (MWMTP) problem.

In this paper, we mainly study approximation algorithms of MWMTP.

1.1 Related Work

It is known [20] that even deciding whether an unweighted graph contains a perfect triangle packing is NP-hard. Hence, MTP, MWTP and MWMTP are all NP-hard. MTP also includes the well-known 3-dimensional matching problem as a special case [11]. Guruswam et al. [12] showed that MTP remains NP-hard on chordal, planar, line, and total graphs. Moreover, MTP has been proved to be APX-hard even on graphs with maximum degree 4 [19,25]. Chlebík and Chlebíková [7] showed that MTP is NP-hard to approximate better than 0.9929.

X. Chen and B. Li (Eds.): TAMC 2024, LNCS 14637, pp. 50–62, 2024.
https://doi.org/10.1007/978-981-97-2340-9_5

MTP is a special case of the unweighted 3-Set Packing problem, which admits an approximation ratio of $(2/3 - \varepsilon)$ [13,18] and $(3/4 - \varepsilon)$ [8,9]. For MTP on graphs with maximum degree 4, Manic and Wakabayashi [23] proposed a 0.8333-approximation algorithm.

Similarly, MWTP can be seen as a special case of the weighted 3-Set Packing problem, which admits an approximation ratio of $(1/2 - \varepsilon)$ [1,3], $1/(2 - 1/63700992 + \varepsilon)$ [24], and $(1/1.786)$ [26]. For MWTP, there are some independent approximation algorithms. Hassin and Rubinstein [15,16] proposed a randomized $(0.518 - \varepsilon)$-approximation algorithm. Chen et al. [5,6] proposed an improved randomized $(0.523 - \varepsilon)$-approximation algorithm. Using the method of pessimistic estimator, van Zuylen [29] proposed a deterministic algorithm with the same approximation ratio. The current best ratio is due to the $1/1.786$-approximation algorithm for the weighted 3-Set Packing problem [26]. For MWTP on $\{0,1\}$-weighted graphs (i.e., a complete graph with edge weights 0 and 1), Bar-Noy et al. [2] proposed a 3/5-approximation algorithm.

For MWMTP, Hassin et al. [17] gave the first deterministic 2/3-approximation algorithm. Note that one can see that it is easy to design a 2/3-approximation algorithm. Chen et al. [4] proposed a nontrivial randomized approximation algorithm with an expected ratio of $(0.66768 - \varepsilon)$.

1.2 Our Results

In this paper, we propose a deterministic $(0.66835 - \varepsilon)$-approximation algorithm, which improves the deterministic 2/3-approximation algorithm [17] and the randomized $(0.66768 - \varepsilon)$-approximation algorithm [4]. Our algorithm is based on the randomized algorithm in [4], but the framework of our analysis is completely different. The main differences are shown as follows.

Firstly, the previous algorithm considers so-called balanced/unbalanced triangles in an optimal solution of MWMTP, while in our algorithm we use novel definitions to make tighter analysis and do not need to separate them. Secondly, we also consider orientations of cycles in a cycle packing, which can simplify the structure significantly and enable us to design better algorithms. For example, the previous decomposition algorithm could only lead to a probability of 1/27 (for one specific event) while our new decomposition algorithm enables us to obtain a probability of at least 97/1215 (see details in Sect. 4.3). Lastly, our algorithm is deterministic, which is obtained by derandomizing our randomized algorithm: we first propose a new randomized algorithm such that if one specific matching is given our algorithm is deterministic, and then we further derandomize our algorithm by finding a desirable deterministic matching by the method of conditional exceptions (see details in Sect. 5).

Due to limited space, the proofs of lemmas and theorems marked with "*" were omitted and they can be found in the full version of this paper [28].

2   Preliminaries

We use $G = (V, E)$ to denote a complete graph with n vertices, where $n \bmod 3 = 0$. There is a non-negative weight function $w : E \to \mathbb{R}_{\geq 0}$ on the edges in E. The weight function w is a semi-metric function, i.e., it satisfies the symmetric and triangle inequality properties. For any weight function $w : X \to \mathbb{R}_{\geq 0}$, we extend it to subsets of X, i.e., we define $w(Y) = \sum_{x \in Y} w(x)$ for $Y \subseteq X$.

A triangle $t = xyz$ is a simple cycle on three different vertices $\{x, y, z\}$. It contains exactly three edges $\{xy, xz, yz\}$. We may also use $\{a_t, b_t, c_t\}$ to denote them such that $w(a_t) \leq w(b_t) \leq w(c_t)$.

Two subgraphs or sets of edges are *vertex-disjoint* if they do not share a common vertex. As mentioned, a *perfect triangle packing* in graph G is a set of vertex-disjoint $n/3$ triangles, and all vertices are covered. It can be seen as the edges of $n/3$ vertex-disjoint triangles. In the following, we will always consider a triangle packing as a perfect triangle packing. We will use $\mathcal{B}^*$ to denote the maximum weight triangle packing.

A *matching* is a set of vertex-disjoint edges. In this paper, we often consider a matching of size $n/3$ and use $\mathcal{M}^*$ to denote the maximum weight matching, which can be found in $O(n^3)$ time [10,21]. A *cycle packing* is a set of vertex-disjoint simple cycles, the length of each cycle is at least 3, and all vertices of the graph are covered. We use $\mathcal{C}^*$ to denote the maximum weight cycle packing, which can be found in $O(n^3)$ time [14]. We can get $w(\mathcal{C}^*) \geq w(\mathcal{B}^*)$ since a triangle packing is also a cycle packing. Given a set of edges $\mathcal{X}$ and a vertex x, we may simply use $x \in \mathcal{X}$ to denote that there is an edge in $\mathcal{X}$ that contains the vertex x.

Let ε be a constant such that $0 < \varepsilon \leq 2/5$. A cycle C is *short* if $|C| \leq \frac{2}{\varepsilon}$; otherwise, the cycle is *long*. For each long cycle $C \in \mathcal{C}^*$, it is easy to see that we can delete at most $\lceil \frac{\varepsilon}{2}|C| \rceil < \frac{\varepsilon}{2}|C| + 1 < \varepsilon|C|$ edges to get a set of paths such that the length of each path, $|P|$, satisfies that $3 \leq |P| \leq \frac{2}{\varepsilon}$. Moreover, the total weight of these paths is at least $(1 - \varepsilon)w(C)$. By connecting the endpoints of each path, we can get a set of short cycles. Hence, we can get a short cycle packing (i.e., a cycle packing containing only short cycles) in polynomial time, denoted by $\mathcal{C}$, such that $w(\mathcal{C}) \geq (1 - \varepsilon)w(\mathcal{C}^*)$. Note that the constant ε needs to be at most 2/5. Otherwise, even the cycle of length 5 is a long cycle, and then the short cycle packing may contain a cycle of length less than 3.

We first define several kinds of triangles in $\mathcal{B}^*$. Fix the short cycle packing $\mathcal{C}$. An edge is an *internal edge* if both of its endpoints fall on the same cycle in $\mathcal{C}$; otherwise, it is an *external edge*. Consider a triangle t in $\mathcal{B}^*$. There are three cases: 1) t is an *internal* triangle if it contains three internal edges; 2) t is a *partial-external* triangle if it contains one internal edge and two external edges; 3) t is an *external* triangle if it contains three external edges. See Fig. 1 for an illustration.

Given the short cycle packing $\mathcal{C}$, for the sake of analysis, we orient each cycle with an arbitrary direction (note that the cycles in [4] are not oriented). Then, $\mathcal{C}$ becomes a set of directed cycles. For a vertex $x \in t \in \mathcal{B}^*$, it is an *external vertex* if it is incident to two external edges of t. Hence, a partial-

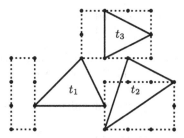

Fig. 1. An illustration of the three kinds of triangles, where there are three cycles (the dotted edges) in $\mathcal{C}$ and three triangles (the solid edges) in $\mathcal{B}^*$: t_1 is an external triangle, t_2 is a partial-external triangle, and t_3 is an internal triangle

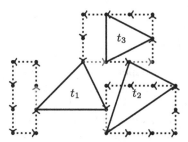

Fig. 2. An illustration of the oriented cycles (the dotted directed edges) and out-edges (the red dotted directed edges), where there are three out-edges from t_1 (an external triangle), one out-edge from t_2 (a partial-external triangle), and no out-edges from t_3 (an internal triangle) (Color figure online)

external triangle contains only one external vertex, and an external triangle contains three external vertices. For an external vertex $x \in t \in \mathcal{B}^*$, assume $x \in C \in \mathcal{C}$, the neighbor of x on the directed cycle C is denoted by x', and then the directed edge xx' is called an *out-edge* of triangle t. The set of out-edges of triangle t is denoted by E_t. Analogously, for each partial-external triangle $|E_t| = 1$, for each external triangle $|E_t| = 3$, and for each internal triangle $|E_t| = 0$. See Fig. 2 for an illustration.

Next, we further define type-1 and type-2 triangles. Fix a constant τ with $0 \leq \tau \leq 1/3$. For each triangle $t \in \mathcal{B}^*$, it is a *type-1* triangle if it holds that $w(e) > \frac{1}{2}(1-\tau)w(t)$ for every edge $e \in E_t$; otherwise, it is a *type-2* triangle (note that our definition is different from the definition in [4]). For this definition, we only consider partial-external triangles and external triangles since internal triangles have no out-edges. Intuitively, the out-edges of type-1 triangles are all heavy edges.

We split $\mathcal{B}^*$ into five disjoint sets $\mathcal{B}_1^*$, $\mathcal{B}_2^*$, $\mathcal{B}_3^*$, $\mathcal{B}_4^*$, and $\mathcal{B}_5^*$ such that

$\mathcal{B}_1^*$: the set of all internal triangles in $\mathcal{B}^*$;
$\mathcal{B}_2^*$: the set of all type-1 partial-external triangles in $\mathcal{B}^*$;
$\mathcal{B}_3^*$: the set of all type-2 partial-external triangles in $\mathcal{B}^*$;

$\mathcal{B}_4^*$: the set of all type-1 external triangles in $\mathcal{B}^*$;
$\mathcal{B}_5^*$: the set of all type-2 external triangles in $\mathcal{B}^*$.

Note that $\mathcal{B}_1^*$ is the set of all internal triangles in $\mathcal{B}^*$, $\mathcal{B}_2^* \cup \mathcal{B}_3^*$ is the set of all partial-external triangles in $\mathcal{B}^*$, and $\mathcal{B}_4^* \cup \mathcal{B}_5^*$ is the set of all external triangles in $\mathcal{B}^*$. Hence, we have $\mathcal{B}^* = \mathcal{B}_1^* \cup \mathcal{B}_2^* \cup \mathcal{B}_3^* \cup \mathcal{B}_4^* \cup \mathcal{B}_5^*$.

We also propose two new important definitions, which will be used frequently:

$$u_i = \sum_{t \in \mathcal{B}_i^*} w(a_t) / \sum_{t \in \mathcal{B}_i^*} w(c_t) \quad \text{and} \quad v_i = \sum_{t \in \mathcal{B}_i^*} w(b_t) / \sum_{t \in \mathcal{B}_i^*} w(c_t).$$

So, we can get $\sum_{t \in \mathcal{B}_i^*} w(a_t) = \frac{u_i}{u_i + v_i + 1} w(\mathcal{B}_i^*)$, $\sum_{t \in \mathcal{B}_i^*} w(b_t) = \frac{v_i}{u_i + v_i + 1} w(\mathcal{B}_i^*)$, and $\sum_{t \in \mathcal{B}_i^*} w(c_t) = \frac{1}{u_i + v_i + 1} w(\mathcal{B}_i^*)$.

2.1   Paper Organization

In the remaining parts of the paper, we will describe the approximation algorithm. It will compute three triangle packings: $\mathcal{T}_1$, $\mathcal{T}_2$, and $\mathcal{T}_3$. The computation of $\mathcal{T}_1$, which can be seen in Sect. 3, is based on the short cycle packing $\mathcal{C}$ via a dynamic method. In Sect. 4, we will use the maximum weight matching $\mathcal{M}^*$ to compute $\mathcal{T}_2$. In Sect. 5, we will first compute a randomized matching on the short cycle packing $\mathcal{C}$, and then use it to compute $\mathcal{T}_3$. The algorithm of $\mathcal{T}_3$ is randomized but we show that it can be derandomized efficiently by the method of conditional expectations. The approximation algorithm will return the best one. Hence, the approximation ratio is $\frac{\max\{w(\mathcal{T}_1),\ w(\mathcal{T}_2),\ w(\mathcal{T}_3)\}}{w(\mathcal{B}^*)}$. The trade-off among these three triangle packings will be shown in Sect. 6.

3   The First Triangle Packing

3.1   The Algorithm

A *partial triangle packing*, denoted by $\mathcal{P}$, is a set of triangle-components and edge-components such that the total number of components is at most $n/3$ and the edges of each component are all internal edges. The augmented weight of $\mathcal{P}$ is defined as $\widetilde{w}(\mathcal{P}) = \sum_{t \in \mathcal{P}} w(t) + \sum_{e \in \mathcal{P}} 2w(e)$. Suppose there are p_1 triangle-components and p_2 edge-components in $\mathcal{P}$, we have $n - 3p_1 - 2p_2 \geq p_2$ unused vertices (i.e., vertices not in $\mathcal{P}$). Hence there are at least as many unused vertices as edge-components. Given a partial triangle packing $\mathcal{P}$, one can construct a triangle packing $\mathcal{T}$ as follows:

Step 1. Pick all triangle-components into the packing;
Step 2. Arbitrarily assign an unused vertex for each edge-component and complete them into a triangle;
Step 3. Arbitrarily construct a set of triangles if there are still unused vertices.

The weight of triangles in Step 1 is $\sum_{t \in \mathcal{P}} w(t)$ and the weight of triangles in Step 2 is at least $\sum_{e \in \mathcal{P}} 2w(e)$ by the triangle inequality. So, $w(\mathcal{T}) \geq \tilde{w}(\mathcal{P})$.

Since the length of each cycle in $\mathcal{C}$ is bounded by a constant, the maximum augmented weight partial triangle packing, denoted by $\mathcal{P}^*$, can be found in polynomial time via a dynamic programming method [4].

Lemma 1 ([4]). *There is a polynomial-time algorithm that can compute a triangle packing $\mathcal{T}_1$ such that $w(\mathcal{T}_1) \geq \tilde{w}(\mathcal{P}^*)$.*

3.2 The Analysis

Let p_1 and p_2 be the number of internal triangles and partial-external triangles, respectively. After deleting all external edges of all triangles in $\mathcal{B}^*$, we can get a set of components, denoted by $\mathcal{P}$, where there are exactly p_1 triangle-components and p_2 edge-components. We can get $p_1 + p_2 \leq n/3$ since a triangle-component corresponds to an internal triangle and an edge-component corresponds to a partial-external triangle. Hence, $\mathcal{P}$ is a partial triangle packing. We can get that $w(\mathcal{T}_1) \geq \tilde{w}(\mathcal{P}^*) \geq \tilde{w}(\mathcal{P}) = \sum_{t \in \mathcal{P}} w(t) + \sum_{e \in \mathcal{P}} 2w(e)$. It is easy to see that $\sum_{t \in \mathcal{P}} w(t) = w(\mathcal{B}_1^*)$ since the triangle-components in $\mathcal{P}$ contains all internal triangles and $\mathcal{B}_1$ is the set of all internal triangles. However, for edge-components, it contains only one (internal) edge of each partial-external triangle. In the worst case, the contained edge in each partial-external triangle t is the least weighted edge a_t. So, we have $\sum_{e \in \mathcal{P}} 2w(e) \geq \sum_{t \in \mathcal{B}_2^* \cup \mathcal{B}_3^*} 2w(a_t)$. Then, we can get that $w(\mathcal{T}_1) \geq w(\mathcal{B}_1^*) + \sum_{t \in \mathcal{B}_2^* \cup \mathcal{B}_3^*} 2w(a_t)$. Recall that $\sum_{t \in \mathcal{B}_i^*} w(a_t) = \frac{u_i}{u_i + v_i + 1} w(\mathcal{B}_i^*)$. By Lemma 1, we can get

Lemma 2. *There is a polynomial-time algorithm that can compute a triangle packing $\mathcal{T}_1$ such that $w(\mathcal{T}_1) \geq w(\mathcal{B}_1^*) + \sum_{i=2}^{3} \frac{2u_i}{u_i + v_i + 1} w(\mathcal{B}_i^*)$.*

4 The Second Triangle Packing

4.1 The Algorithm

The second triangle packing $\mathcal{T}_2$ is generated using the maximum weight matching $\mathcal{M}^*$. The algorithm is simple and contains two following steps.

Step 1. Find the maximum weight matching $\mathcal{M}^*$ (of size $n/3$) in $O(n^3)$ time [10, 21];

Step 2. Arbitrarily assign an unused vertex for each edge of $\mathcal{M}^*$ and complete them into a triangle.

The running time is $O(n^3)$, dominated by computing $\mathcal{M}^*$.

4.2 The Analysis

Consider a triangle $t = xyz \in \mathcal{T}_2$, and assume w.l.o.g. that $xy \in \mathcal{M}^*$. By the triangle inequality, we can get that $w(t) = w(xy) + w(xz) + w(yz) \geq 2w(xy)$. Hence, it is easy to get $w(\mathcal{T}_2) \geq 2w(\mathcal{M}^*)$. Take the most weighted edge c_t for each triangle $t \in \mathcal{B}^*$. Then, the set of edges forms a matching of size $n/3$. Since $\mathcal{M}^*$ is the maximum weight matching of size $n/3$, we have $w(\mathcal{M}^*) \geq \sum_{t \in \mathcal{B}^*} w(c_t)$. Recall that $\sum_{t \in \mathcal{B}_i^*} w(c_t) = \frac{1}{u_i + v_i + 1} w(\mathcal{B}_i^*)$. We can get the following lemma.

Lemma 3. *There is a polynomial-time algorithm that can compute a triangle packing $\mathcal{T}_2$ such that $w(\mathcal{T}_2) \geq \sum_{t \in \mathcal{B}^*} 2w(c_t) = \sum_{i=1}^{5} \frac{2}{u_i + v_i + 1} w(\mathcal{B}_i^*)$.*

It is worth noting that that the algorithm of $\mathcal{T}_2$ is a simple 2/3-approximation algorithm. Since $u_i \leq v_i \leq 1$, we can get $w(\mathcal{T}_2) \geq \frac{2}{3} w(\mathcal{B}^*)$ by Lemma 3. Next, we will construct a randomized matching $\mathcal{Z}$ of size $n/3$, which can be used to prove that $w(\mathcal{T}_2) \geq 2w(\mathcal{M}^*) \geq 2 \cdot \mathbb{E}[w(\mathcal{Z})]$. A good advantage of $\mathcal{Z}$ is that if type-1 triangles in $\mathcal{B}^*$ has a nonzero weight, the matching will have a strictly larger weight than $\frac{1}{3} w(\mathcal{B}^*)$. In this case, $\mathcal{T}_2$ will have a strictly larger weight than $\frac{2}{3} w(\mathcal{B}^*)$ since $\mathcal{M}^*$ is the maximum weight matching.

4.3 The Randomized Matching Algorithm

Our randomized matching algorithm is a refined version of the algorithm presented in [4], which mainly contains four following steps.

Step 1. For each triangle $t \in \mathcal{B}^*$, select an edge e_t uniformly at random. The set of selected edges is denoted by $\mathcal{X}$.

Step 2. For each type-1 triangle $t \in \mathcal{B}^*$, if there is an out-edge that does not share a common vertex with any edge of $\mathcal{X}$, select it. The set of selected edges is denoted by $\mathcal{Y}$.

Step 3. Initialize $\mathcal{Z} = \emptyset$. Consider the out-edges in $\mathcal{Y}$, which contains a set of edge-components, path-components, and cycle-components. We consider three following cases.

- For each edge-component, select the edge into $\mathcal{Z}$.
- For each path-component or even cycle-component (i.e., the number of vertices in it is even), partition it into two matchings, select one matching uniformly at random, and then select the edges in the chosen matching into $\mathcal{Z}$.
- For each odd cycle-component (i.e., the number of vertices in it is odd), select one edge uniformly at random, delete it, partition it into two matchings, select one matching uniformly at random, and then select the edges in the chosen matching into $\mathcal{Z}$.

Note that $\mathcal{Z}$ is a matching containing only out-edges. Moreover, the out-edges are vertex-disjoint with the edges in $\mathcal{X}$.

Step 4. Consider each triangle $t \in \mathcal{B}^*$. If $\mathcal{Z}$ contains no out-edges of t, select the edge e_t in $\mathcal{X}$ into $\mathcal{Z}$.

Roughly speaking, the main idea of the randomized matching algorithm in [4] is to decompose the out-edges in $\mathcal{Y}$ into three matchings and put one into $\mathcal{Z}$, while we can put more out-edges into $\mathcal{Z}$ in Step 3 in our algorithm. Their decomposition is simple because in their algorithm the out-edges may form a multi-graph where parallel edges exist. As a result, the structure is very complicated, and it is not even obvious to get a better decomposition. Although our algorithm is also complicated, we can quickly analyze the expected weight of the randomized matching $\mathcal{Z}$ because the cycles in $\mathcal{C}$ are all oriented.

Lemma 4 (*). *The edge set $\mathcal{Z}$ is a matching of size $n/3$.*

Next, we analyze the expected weight of the randomized matching $\mathcal{Z}$.

Lemma 5 (*). *It holds that $\mathbb{E}[w(\mathcal{Z})] > \frac{1}{3}w(\mathcal{B}^*) + \sum_{t \in \mathcal{B}_2^*, e \in E_t} (\frac{1-3\tau}{6}w(t) \cdot Pr[e \in \mathcal{Z}]) + \sum_{t \in \mathcal{B}_4^*} (\frac{1-3\tau}{6}w(t) \cdot \sum_{e \in E_t} Pr[e \in \mathcal{Z}]).$*

We give a lower bound of $Pr[e \in \mathcal{Z}]$ for each type-1 triangle t with $e \in E_t$.

Lemma 6 (*). *We have $Pr[e \in \mathcal{Z}] \geq \frac{97}{1215}$ for each type-1 triangle t with any $e \in E_t$.*

Lemma 7. *The triangle packing $\mathcal{T}_2$ satisfies that $w(\mathcal{T}_2) \geq \frac{2}{3}w(\mathcal{B}^*) + \frac{97(1-3\tau)}{3645}w(\mathcal{B}_2^*) + \frac{97(1-3\tau)}{1215}w(\mathcal{B}_4^*).$*

5 The Third Triangle Packing

Recall that we split triangles into type-1 and type-2 triangles. Especially, the randomized matching in Sect. 4 is based on type-1 triangles. However, there may not exist type-1 triangles in the worst case. Then, it comes to the third triangle packing $\mathcal{T}_3$, which uses the property of type-2 triangles. The ideas of $\mathcal{T}_3$ are mainly from the randomized algorithm in [4]. We will first propose a new randomized algorithm and then derandomize it.

5.1 The Randomized Algorithm

Our new randomized algorithm of $\mathcal{T}_3$ is based on two edge-disjoint matchings $\mathcal{X}$ and $\mathcal{Y}$, where $\mathcal{X}$ is a randomized matching of size $n/3$ on the short cycle packing and $\mathcal{Y}$ is a matching of size at most $n/3$ determined by $\mathcal{X}$. The algorithm of $\mathcal{X}$ in our algorithm is obtained directly from [4] but the algorithm of $\mathcal{Y}$ is different: the triangle packing algorithm in [4] is randomized even if $\mathcal{X}$ is determined, while our algorithm of $\mathcal{T}_3$ would be deterministic. So, our new randomized algorithm can be derandomized simply by finding a desirable deterministic matching $\mathcal{X}$.

The algorithm of $\mathcal{X}$ in [4] mainly contains four steps.

Step 1. Initialize $\mathcal{L} = \mathcal{R} = \emptyset$.
Step 2. For each even cycle $C \in \mathcal{C}$, partition it into two matchings, select one matching uniformly at random, and select the edges in the chosen matching into $\mathcal{L}$.

Step 3. For each odd cycle $C \in \mathcal{C}$, select one edge uniformly at random, delete it, partition the rest edges into two matchings, select one matching uniformly at random, and select the edges in the chosen matching into $\mathcal{L}$ and $\mathcal{R}$ in the following way: select one edge uniformly at random, put the edge into $\mathcal{R}$, and put the rest edges into $\mathcal{L}$.

Step 4. Select $\frac{2}{3}|\mathcal{L}|$ edges of $\mathcal{L}$ uniformly at random, and put them into $\mathcal{X}$, and put all edges of $\mathcal{R}$ into $\mathcal{X}$.

The running time is dominated by computing the short cycle packing $\mathcal{C}$, which is polynomial. The randomized matching $\mathcal{X}$ has some good properties.

Lemma 8 ([4]). *The size of $\mathcal{X}$ is $n/3$. For every edge $e \in C \in \mathcal{C}$ and every vertex $v \notin C$, we have $Pr[e \in \mathcal{X}] = \frac{1}{3}$ and $Pr[e \in \mathcal{X} \wedge v \notin \mathcal{X}] \geq \frac{1}{9}$.*

Next, we introduce the algorithm of the third triangle packing T_3, where we will first compute the matching $\mathcal{Y}$ and then obtain T_3 using $\mathcal{X}$ and $\mathcal{Y}$. We first give some definitions.

For some edge $e = xy \in C \in \mathcal{C}$ and vertex $z \notin C$, we call $(x, y; z)$ a *triplet*. Recall that τ is a fixed constant defined before. A triplet $(x, y; z)$ is *good* if it satisfies that $w(xy) \leq (1 - \tau)(w(xz) + w(yz))$.

We construct a multi-graph H. Initially, graph H contains n vertices only and no edges. For each good triplet $(x, y; z)$, we add two edges xz and yz with an augmented weight $\widetilde{w}(xz) = \widetilde{w}(yz) = w(xz) + w(yz)$, and each of them has a label corresponding to the good triplet $(x, y; z)$. Hence, graph H contains only external edges. Two edges of H are called *conflicting* if their corresponding triplets share a common vertex.

The algorithm of T_3 is shown as follows.

Step 1. Find a maximum augmented weight matching $\mathcal{Y}^*$ in graph H in $O(n^3)$ time [10,21]. Note that there is no constraint on its size, and hence we have $0 \leq |\mathcal{Y}^*| \leq n/2$.

Step 2. Let $\mathcal{Y}_{\mathcal{X}}^*$ denote the set of edges $\{zx \in \mathcal{Y}^* \mid xy \in \mathcal{X}, z \notin \mathcal{X}\}$. We claim that for each edge in $\mathcal{Y}_{\mathcal{X}}^*$ there is at most one different edge in $\mathcal{Y}_{\mathcal{X}}^*$ that are conflicting with it (see Lemma 9).

Step 3. For each pair of conflicting edges of $\mathcal{Y}_{\mathcal{X}}^*$, delete the edge with the less augmented weight (if their augmented weights are the same we may simply delete one of them arbitrarily). The remained matching is denoted by $\mathcal{Y}$.

Step 4. Note that $\mathcal{X} \cup \mathcal{Y}$ is a set of components of size $n/3$, which contains $|\mathcal{Y}|$ path-components (each contains 3 vertices) and $|\mathcal{X}| - |\mathcal{Y}|$ edge-components. For each path-component xyz, complete it into a triangle. For each edge-component, arbitrarily assign an unused vertex, and complete it into a triangle.

We need to prove the claim in Step 2.

Lemma 9 (*). *For any edge $zx \in \mathcal{Y}_{\mathcal{X}}^*$, there is at most one different edge in $\mathcal{Y}_{\mathcal{X}}^*$ that are conflicting with it.*

5.2 The Analysis

Lemma 10 (*). $\mathbb{E}[w(\mathcal{T}_3)] \geq \frac{\tau}{18}\widetilde{w}(\mathcal{Y}^*) + \frac{2}{3}w(\mathcal{C})$.

Next, we give a lower bound of $\widetilde{w}(\mathcal{Y}^*)$.

Lemma 11 (*). *It holds that* $\widetilde{w}(\mathcal{Y}^*) \geq \sum_{i\in\{3,5\}} \frac{1}{2}w(\mathcal{B}_i^*)$.

Recall that $w(\mathcal{C}) \geq (1-\varepsilon)w(\mathcal{C}^*) \geq (1-\varepsilon)w(\mathcal{B}^*)$ and $\mathbb{E}[w(\mathcal{T}_3)] \geq \frac{\tau}{18}\widetilde{w}(\mathcal{Y}^*) + \frac{2}{3}w(\mathcal{C})$. Then, we have the following lemma.

Lemma 12. *There is a polynomial-time randomized algorithm that can compute a triangle packing* $\mathcal{T}_3$ *such that* $\mathbb{E}[w(\mathcal{T}_3)] \geq \frac{2}{3}(1-\varepsilon)w(\mathcal{B}^*) + \sum_{i\in\{3,5\}} \frac{\tau}{36}w(\mathcal{B}_i^*)$.

Next, we show that the algorithm can be derandomized efficiently by the method of conditional expectations [27].

5.3 The Derandomization

The third triangle packing $\mathcal{T}_3$ is based on the randomized matching $\mathcal{X}$ on $\mathcal{C}$. Let $f(\mathcal{X}) = \frac{\tau}{2}\widetilde{w}(\mathcal{Y}_{\mathcal{X}}^*) + 2w(\mathcal{X})$. By the proof of Lemma 10, we have $w(\mathcal{T}_3) \geq f(\mathcal{X})$ and $\mathbb{E}[f(\mathcal{X})] \geq \frac{\tau}{18}\widetilde{w}(\mathcal{Y}^*) + \frac{2}{3}w(\mathcal{C})$. If we derandomize the matching $\mathcal{X}$ such that $f(\mathcal{X}) \geq \frac{\tau}{18}\widetilde{w}(\mathcal{Y}^*) + \frac{2}{3}w(\mathcal{C})$, then we will obtain a deterministic algorithm of $\mathcal{T}_3$.

Recall that the randomized matching algorithm mainly contains two phases. In the first phase, the algorithm obtains two edge sets $\mathcal{L}$ and $\mathcal{R}$ by making random decisions on each cycle in $\mathcal{C}$. After that, the algorithm obtains the matching $\mathcal{X}$ by choosing $\frac{2}{3}|\mathcal{L}|$ edges in $\mathcal{L}$ uniformly at random and all edges in $\mathcal{R}$, which can be seen as making random decisions on each edge in $\mathcal{L}$. Using the method of conditional expectations, these two phases can be derandomized efficiently. The main idea is to consider random decisions in the algorithm sequentially. For each random decision, we can compute the expected weight of $f(\mathcal{X})$ conditioned on each possible outcome of the random decision. Since at least one of these outcomes has an expected weight of at least $\mathbb{E}[f(\mathcal{X})]$, we fix this outcome and continue to the next random decision. By repeating this procedure, we can get a deterministic solution with a weight of at least $\mathbb{E}[f(\mathcal{X})]$.

In our derandomization (see the full version), we consider the two phrases, respectively. And, we show that there is a polynomial number of outcomes for each random decision in the algorithm, and moreover, the conditional expected weight of $f(\mathcal{X})$ can be computed in polynomial time. So, the derandomization can be done in polynomial time.

Lemma 13 (*). *There is a polynomial-time algorithm that can compute a triangle packing* $\mathcal{T}_3$ *such that* $w(\mathcal{T}_3) \geq \frac{2}{3}(1-\varepsilon)w(\mathcal{B}^*) + \sum_{i\in\{3,5\}} \frac{\tau}{36}w(\mathcal{B}_i^*)$.

6   The Trade-Off

We first introduce some new parameters. We define $\alpha_i = w(\mathcal{B}_i^*)/w(\mathcal{B}^*)$, which measures the weight proportion of the triangles in $\mathcal{B}_i^*$ compared to the triangles in $\mathcal{B}^*$. Note that

$$\alpha_1 + \alpha_2 + \alpha_3 + \alpha_4 + \alpha_5 = 1. \tag{1}$$

Then, we define $\rho_i = \sum_{t\in\mathcal{B}_i^*} w(a_t)/w(\mathcal{B}^*)$, $\sigma_i = \sum_{t\in\mathcal{B}_i^*} w(b_t)/w(\mathcal{B}^*)$, and $\theta_i = \sum_{t\in\mathcal{B}_i^*} w(c_t)/w(\mathcal{B}^*)$. Recall that $\sum_{t\in\mathcal{B}_i^*} w(a_t) = \frac{u_i}{u_i+v_i+1}w(\mathcal{B}_i^*) = \frac{u_i}{u_i+v_i+1}\alpha_i w(\mathcal{B}^*)$. So, $\rho_i = \frac{u_i}{u_i+v_i+1}\alpha_i$. Analogously, we get $\sigma_i = \frac{v_i}{u_i+v_i+1}\alpha_i$ and $\theta_i = \frac{1}{u_i+v_i+1}\alpha_i$. So,

$$\rho_i + \sigma_i + \theta_i = \alpha_i, \quad i \in \{1,2,3,4,5\}. \tag{2}$$

By Lemmas 2, 3, 7, and 13, we can get that

$$w(\mathcal{T}_1)/w(\mathcal{B}^*) \geq \alpha_1 + 2\rho_2 + 2\rho_3, \tag{3}$$

$$w(\mathcal{T}_2)/w(\mathcal{B}^*) \geq 2\theta_1 + 2\theta_2 + 2\theta_3 + 2\theta_4 + 2\theta_5, \tag{4}$$

$$w(\mathcal{T}_2)/w(\mathcal{B}^*) \geq \frac{2}{3} + \frac{97(1-3\tau)}{3645}\alpha_2 + \frac{97(1-3\tau)}{1215}\alpha_4, \tag{5}$$

$$w(\mathcal{T}_3)/w(\mathcal{B}^*) \geq \frac{2}{3}(1-\varepsilon) + \frac{\tau}{36}\alpha_3 + \frac{\tau}{36}\alpha_5. \tag{6}$$

For each triangle t, we have $w(c_t) \geq w(b_t) \geq w(a_t)$ by the definition and $w(a_t) + w(b_t) \geq w(c_t)$ by the triangle inequality. Hence, we also have

$$\rho_i + \sigma_i \geq \theta_i \geq \sigma_i \geq \rho_i \geq 0, \quad i \in \{1,2,3,4,5\}. \tag{7}$$

If τ is fixed, using (1)–(7) the ratio $\frac{\max\{w(\mathcal{T}_1),\, w(\mathcal{T}_2),\, w(\mathcal{T}_3)\}}{w(\mathcal{B}^*)}$ can be obtained via solving a linear program (see the full version). Setting $\tau = 0.25$, we can get

Theorem 1. *For MWMTP with any constant $\varepsilon > 0$, there is a polynomial-time $(0.66835 - \varepsilon)$-approximation algorithm.*

7   Conclusion

In this paper, we consider approximation algorithms for the maximum weight metric triangle packing problem. This problem admits an almost-trivial 2/3-approximation algorithm [17]. The first nontrivial result, given by Chen et al. [4], is a randomized $(0.66768 - \varepsilon)$-approximation algorithm. Based on novel modifications, deep analysis, and conditional expectations, we propose a deterministic $(0.66835 - \varepsilon)$-approximation algorithm. Whether it admits a simple algorithm with a better-than-2/3-approximation ratio is still unknown.

In the future, it would be interesting to study the well-related maximum weight metric 3-path packing problem, where we need to find a set of $n/3$ vertex-disjoint 3-paths with the total weight maximized. This problem admits a similar almost-trivial 3/4-approximation algorithm (see [22]). However, it is still unknown to obtain a nontrivial approximation algorithm for this problem.

Acknowledgments. The work is supported by the National Natural Science Foundation of China, under grant 62372095.

References

1. Arkin, E.M., Hassin, R.: On local search for weighted k-set packing. Math. Oper. Res. **23**(3), 640–648 (1998)
2. Bar-Noy, A., Peleg, D., Rabanca, G., Vigan, I.: Improved approximation algorithms for weighted 2-path partitions. Discrete Appl. Math. **239**, 15–37 (2018)
3. Berman, P.: A $d/2$ approximation for maximum weight independent set in d-claw free graphs. Nord. J. Comput. **7**(3), 178–184 (2000)
4. Chen, Y., Chen, Z., Lin, G., Wang, L., Zhang, A.: A randomized approximation algorithm for metric triangle packing. J. Comb. Optim. **41**(1), 12–27 (2021)
5. Chen, Z., Tanahashi, R., Wang, L.: An improved randomized approximation algorithm for maximum triangle packing. Discrete Appl. Math. **157**(7), 1640–1646 (2009)
6. Chen, Z., Tanahashi, R., Wang, L.: Erratum to "an improved randomized approximation algorithm for maximum triangle packing" [Discrete Appl. Math. 157 (2009) 1640–1646]. Discrete Appl. Math. **158**(9), 1045–1047 (2010)
7. Chlebík, M., Chlebíková, J.: Approximation hardness for small occurrence instances of NP-hard problems. In: CIAC 2003, vol. 2653, pp. 152–164 (2003)
8. Cygan, M.: Improved approximation for 3-dimensional matching via bounded pathwidth local search. In: FOCS 2013, pp. 509–518. IEEE Computer Society (2013)
9. Fürer, M., Yu, H.: Approximating the k-set packing problem by local improvements. In: Fouilhoux, P., Gouveia, L., Mahjoub, A., Paschos, V. (eds.) ISCO 2014. LNCS, vol. 8596, pp. 408–420. Springer, Cham (2014). https://doi.org/10.1007/978-3-319-09174-7_35
10. Gabow, H.N.: Implementation of algorithms for maximum matching on nonbipartite graphs. Ph.D. thesis, Stanford University (1974)
11. Garey, M.R., Johnson, D.S.: Computers and Intractability: A Guide to the Theory of NP-Completeness. W. H. Freeman (1979)
12. Guruswami, V., Rangan, C.P., Chang, M., Chang, G.J., Wong, C.K.: The vertex-disjoint triangles problem. In: Hromkovič, J., Sýkora, O. (eds.) WG 1998. LNCS, vol. 1517, pp. 26–37. Springer, Heidelberg (1998). https://doi.org/10.1007/10692760_3
13. Halldórsson, M.M.: Approximating discrete collections via local improvements. In: SODA 1995, pp. 160–169. ACM/SIAM (1995)
14. Hartvigsen, D.: Extensions of matching theory. Ph.D. thesis, Carnegie-Mellon University (1984)
15. Hassin, R., Rubinstein, S.: An approximation algorithm for maximum triangle packing. Discrete Appl. Math. **154**(6), 971–979 (2006)
16. Hassin, R., Rubinstein, S.: Erratum to "an approximation algorithm for maximum triangle packing": [Discrete Applied Mathematics 154 (2006) 971–979]. Discrete Appl. Math. **154**(18), 2620 (2006)
17. Hassin, R., Rubinstein, S., Tamir, A.: Approximation algorithms for maximum dispersion. Oper. Res. Lett. **21**(3), 133–137 (1997)
18. Hurkens, C.A.J., Schrijver, A.: On the size of systems of sets every t of which have an SDR, with an application to the worst-case ratio of heuristics for packing problems. SIAM J. Discrete Math. **2**(1), 68–72 (1989)

19. Kann, V.: Maximum bounded 3-dimensional matching is MAX SNP-complete. Inf. Process. Lett. **37**(1), 27–35 (1991)
20. Kirkpatrick, D.G., Hell, P.: On the completeness of a generalized matching problem. In: STOC 1978, pp. 240–245. ACM (1978)
21. Lawler, E.: Combinatorial Optimization: Networks and Matroids. Holt, Rinehart and Winston (1976)
22. Li, S., Yu, W.: Approximation algorithms for the maximum-weight cycle/path packing problems. Asia Pac. J. Oper. Res. **40**(4), 2340003:1–2340003:16 (2023)
23. Manic, G., Wakabayashi, Y.: Packing triangles in low degree graphs and indifference graphs. Discrete Math. **308**(8), 1455–1471 (2008)
24. Neuwohner, M.: An improved approximation algorithm for the maximum weight independent set problem in d-claw free graphs. In: STACS 2021. LIPIcs, vol. 187, pp. 53:1–53:20. Schloss Dagstuhl - Leibniz-Zentrum für Informatik (2021)
25. van Rooij, J.M.M., van Kooten Niekerk, M.E., Bodlaender, H.L.: Partition into triangles on bounded degree graphs. Theory Comput. Syst. **52**(4), 687–718 (2013)
26. Thiery, T., Ward, J.: An improved approximation for maximum weighted k-set packing. In: SODA 2023, pp. 1138–1162. SIAM (2023)
27. Williamson, D.P., Shmoys, D.B.: The Design of Approximation Algorithms. Cambridge University Press, Cambridge (2011)
28. Zhao, J., Xiao, M.: An improved approximation algorithm for metric triangle packing. CoRR abs/2402.08216 (2024)
29. van Zuylen, A.: Deterministic approximation algorithms for the maximum traveling salesman and maximum triangle packing problems. Discrete Appl. Math. **161**(13–14), 2142–2157 (2013)

An Optimal and Practical Algorithm for the Planar 2-Center Problem

Xuehou Tan[(✉)]

Tokai University, 4-1-1 Kitakaname, Hiratsuka 259-1292, Japan
`xtan@tsc.u-tokai.ac.jp`

Abstract. The *2-center* problem for a set S of n points in the plane asks for two congruent circular disks of the minimum radius r^*, whose union covers all points of S. We present an optimal algorithm for computing r^*, with $O(n \log n)$ running time and $O(n)$ space. Our result improves upon the previously known $O(n \log^2 n)$ time algorithm, and solves a long-standing (near-thirty years) open problem. Also, we present $O(n \log n)$ time and $O(n)$ space algorithms for its two variants: The first is to cover a set of points in convex position, and the second is to cover a convex polygon P, whose goal is to find two centers inside P such that the maximum distance from any point of polygon P to its closest center is minimized. Except for efficiency, the other novelty of our algorithms is simplicity: they are built on the standard algorithms for computing the Delaunay triangulation and furthest-site Voronoi diagram of a point set.

1 Introduction

Let S denote a set of n points in the plane. The planar *p-center problem* asks for p congruent closed disks of the minimum radius, whose union covers S. The problem is NP-complete, if p is part of input. However, for small values of p, efficient algorithms are known. The 1-center problem, known as the *smallest enclosing disk* problem, is widely studied and can be solved in $O(n)$ time.

For the planar 2-center problem, Sharir [6] was the first to present a near-linear algorithm with running time $O(n \log^9 n)$, using the powerful parametric search paradigm. Later, Eppstein [5] presented a randomized algorithm with $O(n \log^2 n)$ expected time, and Chan [2] proposed an $O(n \log^2 n \log \log^2 n)$ time deterministic algorithm.

A general method is to divide the 2-center problem into the *well separated* and *nearby* subproblems, depending on whether the distance between the centers of two optimal disks of radius r^* is ϵr^* or $(2 - \epsilon')r^*$, for some constants $\epsilon, \epsilon' > 0$ [2,5]. For the well separated 2-center problem, Eppstein [5] has given an $O(n \log^2 n)$ time deterministic algorithm.

Recently, Wang [7] has given an $O(n \log n \log \log n)$ time solution to the nearby 2-center problem, which was later improved to $O(n \log n)$ by Choi and Ahn [3]. This leads to an $O(n \log^2 n)$ time solution to the 2-center problem. Whether the planar 2-center problem can be solved in $O(n \log n)$ time is a long-standing (near thirty years) open problem in computational geometry.

X. Chen and B. Li (Eds.): TAMC 2024, LNCS 14637, pp. 63–74, 2024.
https://doi.org/10.1007/978-981-97-2340-9_6

A set of points is said to be in *convex position*, if it is the set of vertices of a convex polygon. Wang [7] has also given an $O(n \log n \log \log n)$ time algorithm for a set of points in convex position, which was improved to $O(n \log n)$ by Choi and Ahn [3]. Eppstein, Chan, Wang, Choi and Ahn all made some refinements of Sharir's work [6], and relied on parametric search. For a convex polygon P, Choi, et al. [4] have given an $O(n \log n)$ time algorithm to compute two *centers* inside P such that the maximum distance from any point of P to its closest center in P is minimized. Again, their algorithm relied on parametric search.

Although parametric search is very useful in finding improved asymptotic bounds, it is considered harmful and impractical, because it is highly complicated (consisting of parallel computations) and difficult to implement, see [5, Sect. 2].

1.1   Our Results

We describe a new, optimal approach to the planar 2-center problem. By exploiting the geometric properties of considered 2-centers, we have succeeded in developing an $O(n \log n)$ time and $O(n)$ space algorithm for the planar 2-center problem. A process of enlarging two disks of equal size in the same rate is proposed to solve the 2-center problems for a convex polygon P, and for a set S of points in convex and general position. No parametric searches are needed.

Since the lower time bound on the planar 2-center problem is $\Omega(n \log n)$ [5], our time and space bounds are optimal in the worst case. Our result improves upon the previously known $O(n \log^2 n)$ time algorithm [5,7], and solves a long-standing open problem in computational geometry.

Except for efficiency of our algorithms, the other novelty is their simplicity: Our algorithms are built on the standard ones for computing the Delaunay triangulation and furthest-site Voronoi diagram of a point set. Thus, our result gives the first optimal and practical solution to the planar 2-center problem.

2   Preliminaries

Let S be a set of n planar points. Assume that no four (three) points of S are on a circle (line). Denote by $CH(S)$ the convex hull of S, which can be found in $O(n \log n)$ time. For a disk D (polygon P), denote its boundary by ∂D (∂P), and $D(S)$ the set of the points covered by D. For two points p and q, denote by pq the segment connecting p and q, and $|pq|$ the length of pq. The *bisector* of p and q, denoted by $B_{p,q}$, is the perpendicular line through the midpoint of pq.

The *Voronoi diagram* for S, is a partition of the plane into regions, each containing exactly one point in S, such that for each point $p \in S$, every point within its corresponding region is closer to p than any other point of S. The *Delaunay triangulation* $DT(S)$ of S is the straight-line dual of the Voronoi diagram for S. The furthest-site Voronoi diagram $FV(S)$ of S is a partition of the plane into m ($\leq n$) regions $FV(p_1), \ldots, FV(p_m)$, where $p_1, \ldots, p_m$ are on the convex hull of n sites (points), such that a point z belongs to $FV(p_i)$ if and only if the Euclidean distance of z to p_i is larger than that of z to any other site. In the convex case, both $DT(S)$ and $FV(S)$ can be computed in $O(n)$ time and space [1].

Lemma 1. *Let K be the smallest disk enclosing three points. Any semi-disk of K cannot cover all points, except that two points form a diameter of K.*[1]

2.1 Assumptions on Planar 2-Center Problems in the Convex Case

Suppose that P is a convex polygon and S is the set of vertices of P. Denote by s_1, s_2 and s_3 (if it exists), in clockwise order, the points on the smallest circle enclosing S. Assume that s_1 is to the left of both s_2 and s_3. (Point s_3 is set to s_2, if it does not exist.) Suppose that the disk pair (D_1, D_2) gives a solution to some considered 2-center problem. From the pigeonhole principle, one disk D_1 covers one of s_1, s_2 and s_3, and the other D_2 covers two others.

Let $r^*(s_i)$ $(R^*(s_i))$ be the optimal solution (i.e., radius) to the 2-center problem for set S (polygon P), under the assumption that s_i is covered by D_1 and both s_j and s_k are covered by D_2, where $i, j, k \in \{1, 2, 3\}$. The *overall* optimum solution for S (P) is then the minimum among $r^*(s_1)$, $r^*(s_2)$ and $r^*(s_3)$ $(R^*(s_1)$, $R^*(s_2)$ and $R^*(s_3))$.

We will discuss on how to compute only $r^*(s_1)$ and $R^*(s_1)$. For ease of presentation, denote $r^*(s_1)$ $(R^*(s_1))$ by r^* (R^*), see Sects. 2.2, 3 and 4.

2.2 Two Basic Results on Sets of Points in Convex Position

Denote by P_1 the boundary chain of a convex polygon P from s_1 to s_2, and P_2 the polygonal chain from s_3 to s_1 clockwise. For two vertices u and v of P (resp. P_1, or P_2), denote by $P[u, v]$ (resp. $P_1[u, v]$, or $P_2[u, v]$) the *polygonal chain* of P (resp. P_1, or P_2) from u to v *clockwise*. Also, denote by $P(u, v)$ the *open* polygonal chain of $P[u, v]$, in which u and v are excluded.

Note that a polygonal chain of P is covered by D_1, and the other is by D_2 [7]. Denote the vertices of P_1 by $p_1, p_2, \ldots, p_k$ clockwise, and the vertices of P_2 by $p_{k+1}, p_{k+2}, \ldots, p_{k+h}$. Denote by $r[j+1, i]$ $(r[i+1, j])$ the radius of the smallest disk enclosing $P[p_{j+1}, p_i]$ $(P[p_{i+1}, p_j])$, $1 \le i \le k-1$ $(k+1 \le j \le k+h-1)$.

Let $r_{i,j} = \max\{r[j+1, i], r[i+1, j]\}$. For a point p_i $(1 \le i \le k-1)$, let $r_{p_i}^* = \min\{r_{i,k+1}, r_{i,k+2}, \ldots, r_{i,k+h-1}\}$. We refer to $r_{p_i}^*$ as a *locally optimal* solution to the 2-center problem for S. Then, $r^* = \min_{1 \le i \le k-1}\{r_{p_i}^*\}$. Analogously, for a point p_j $(k+1 \le i \le k+h-1)$, $r_{p_j}^* = \min\{r_{1,j}, r_{2,j}, \ldots, r_{k-1,j}\}$ and $r^* = \min_{k+1 \le j \le k+h-1}\{r_{p_j}^*\}$. In Fig. 1, $r^* = r_{p_2}^*$ $(p_2 \in P_1)$, and $r^* = r_{p_7}^*$ $(p_7 \in P_2)$.

Suppose that ab is an edge of P_1 and cd is an edge of P_2 (see Fig. 1). Also, denote by r_a^* (resp. r_c^*) the local optimal solution for vertex $a \in P_1$ (resp. $c \in P_2$).

Lemma 2. *Let ab $(\in P_1)$ and cd $(\in P_2)$ be two edges of P. Suppose that two disks D_1 and D_2 are the smallest ones enclosing $P[d, a]$ and $P[b, c]$ respectively, and the radii of D_1 and D_2 are at most r^*. Then, D_1 does not cover b nor c, D_2 does not cover a nor d, and $r^* = r_a^* = r_c^*$.*

Corollary 1. *The edge pair (ab, cd) described in Lemma 2 is unique. The pair (ab, cd) leading to the overall solution to the 2-center problem for S is also unique.*

---

[1] Due to space limit, the proofs of all lemmas are omitted in this extended abstract.

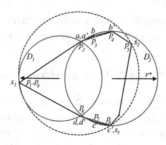

Fig. 1. D_1 does not cover b nor c, and D_2 does not a nor d.

The following tells that r^* can be found by binary search, if a or c is known.

Lemma 3. *Suppose that the vertices a and c specified in Lemma 2 are represented by p_i ($1 \le i \le k - 1$) and p_j ($k + 1 \le j \le k + h - 1$) respectively. Then, $r^* = r_{i,j}$, $r_{i,j} \le r_{i,j-1} \le \cdots \le r_{i,k+1}$ and $r_{i,j} \le r_{i,j+1} \le \cdots \le r_{i,k+h-1}$. (Also, $r_{i,j} \le r_{i-1,j} \le \cdots \le r_{1,j}$ and $r_{i,j} \le r_{i+1,j} \le \cdots \le r_{k-1,j}$.)*

3   The 2-Center of a Convex Polygon in the Plane

Suppose that P is a convex polygon, and the solution r^* (radius) to the 2-center problem for the set S of vertices of P has been obtained. Assume that D_1 and D_2 are the disks of radius at most r^*, which are outputted by our algorithm presented in Sect. 4. Let r_1 (r_2) be the radius of D_1 (D_2). Assume that $r_1 \le r_2$ ($= r^*$). Let s and t be the intersection points of ∂P with ∂D_2, respectively.

A Preprocessing for Checking Whether $r^* < R^*$. Compute (in linear time) the smallest disk enclosing the polygonal chain $P[t, s]$. If r_2 is larger or equal to the radius of found disk, then $R^* = r^*$; otherwise, $r^* < R^*$.

Lemma 4. *Suppose that $r^* < R^*$. Disks K_1 and K_2 of equal radius give the solution R^* to the 2-center problem for P if and only if ∂K_1, ∂K_2 and ∂P share common points i and j, the radius of K_1 and K_2 is strictly larger than $|ij|$, and disks K_1 and K_2 are the smallest ones enclosing $P[j, i]$ and $P[i, j]$ respectively.*

Assume below that $r^* < R^*$. Our algorithm for computing r^* consists of the following two phases. **Phase 1** finds a pair of disks of radius r^*. To be exact, it asks for a disk D_1' of radius r_2 ($= r^*$) such that all points of $D_1(S)$ are covered by D_1', and point s or t, if covered by D_1', is on $\partial D_1'$. **Phase 2** performs a process of enlarging two disks of equal radius, starting from D_1' and D_2, to find two disks E_1 and E_2 of radius R^* such that $P \subseteq E_1 \cup E_2$.

The furthest-site Voronoi diagrams of $D_1(S)$ and $D_2(S)$ can be used to show how this disk-enlarging process is done in linear time and space. Lemma 4 is employed in **Phase 2**. Due to space limit, the detail of our algorithm is omitted.

Lemma 5. *Suppose that the solution r^* to the planar 2-center problem for S is known. Then, one can compute two congruent disks of the smallest radius R^* in $O(n)$ time and space such that the union of them covers polygon P.*

From Lemma 5 and Theorem 2 (Sect. 4), we obtain the following result.

Theorem 1. *For a convex polygon P with n vertices, the 2-center problem for polygon P can be solved in $O(n \log n)$ time and $O(n)$ space.*

The first of following results follows from the definition of r^* and R^*, and the second helps the algorithm given in Sect. 4.2.

Corollary 2. *Let D and D' be the disks of same radius such that $D \cup D'$ covers S. If D is disjoint from D' or the intersection points between ∂D and $\partial D'$ lie in the interior of P, then $r^* < R^*$.*

Lemma 6. *Suppose that $r^* < R^*$. For any pair of disks E_1 and E_2 of radius R^* such that $P \subseteq E_1 \cup E_2$, two common points i and j among ∂E_1, ∂E_2 and ∂P are contained in edges ab and cd (stated in Lemma 2), respectively.*

4   The 2-Center of a Set of Points in Convex Position

Denote by $DT(P[s_3, s_2])$ the Delaunay triangulation of all vertices of $P[s_3, s_2]$.

Lemma 7. *Suppose that ab and cd are the edges described in Lemma 2. If all points a, b, c and d contribute to a path T in the dual of $DT(P[s_3, s_2])$, whose starting and ending nodes represent a triangle with vertex s_1 and the triangle with edge $s_2 s_3$ respectively, then one of ac, bc, ad and bd is an edge of $DT(P[s_3, s_2])$.*

The 2-center problem for S can be divided into the *Delaunay-path* and *non-Delaunay-path* subproblems, depending on whether points a, b, c and d satisfy the condition stated in Lemma 7 or not. In either case, an $O(n \log n)$ time and $O(n)$ space algorithm for computing r^* can be developed (Lemmas 8 and 9).

Our second result can be thus summarized in the following theorem.

Theorem 2. *The planar 2-center problem for a set of n points in convex position can be solved in $O(n \log n)$ time and $O(n)$ space.*

4.1   Algorithm for the Delaunay-Path 2-Center Problem

Consider the paths in the dual of $DT(P[s_3, s_2])$, whose starting and ending nodes represent a triangle with vertex s_1 and the triangle with edge $s_2 s_3$ respectively. There are at most four such *longest* paths. These paths, denoted by T_j ($j = 1$, 2, 3 or 4), can be found in linear time from $DT(P[s_3, s_2])$. We will guess that an edge of T_j is one of ac, ad, bc and bd. By trying all possible situations, the solution of the Deluany-path 2-center subproblem (if exists) can be found.

Let T be one of paths T_j ($j = 1, 2, 3, 4$). Let E_k be a Delaunay edge that is common to two triangles of T. Denote by $E_1, E_2, \ldots, E_m$ the sequence of

Delaunay edges in $\mathcal{T}$. Particularly, let E_0 (E_{m+1}) be the edge having vertex s_1 (s_2), which together with E_1 (E_m) forms a Delaunay triangle. See Fig. 2.

Denote by R the convex polygon formed by all vertices of $E_0, E_1, \ldots, E_{m+1}$. For a vertex v of polygon R, denote by $Succ_R(v)$ and $Pred_R(v)$ be the vertices clockwise before and after v on R, respectively. Let us first guess E_i ($0 \le i \le m+1$) as segment ad. Denote by $u(i)$ the vertex of E_i belonging to P_1, and $v(i)$ the other belonging to P_2. If $Succ_R(u(i)) = b$ and $Pred_R(v(i)) = c$, then the guess is true. To this end, let $P_\alpha(E_i)$ be the set of vertices of $P[v(i), u(i)]$, and $Q_\alpha(E_i)$ the set of vertices of $P[Succ_R(u(i)), Pred_R(v(i))]$. Since $P[v(i), u(i)]$ is supposed to be covered by D_1, we have $s_2, s_3 \ni P[v(i), u(i)]$. Let E_l be the edge with the largest index l such that sets $P_\alpha(E_0), P_\alpha(E_1), \ldots, P_\alpha(E_l)$ are defined.

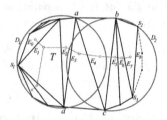

Fig. 2. Illustrating the sequence of Delaunay edges $E_0, E_1, \ldots, E_{m+1}$.

Denote by $r(P_{i,\alpha})$ and $r(Q_{i,\alpha})$ the radii of the smallest disks enclosing $P_\alpha(E_i)$ and $Q_\alpha(E_i)$, respectively. Let $r_{i,\alpha} = \max\{r(P_{i,\alpha}), r(Q_{i,\alpha})\}$, and $r^*(\alpha) = \min\{r_{0,\alpha}, r_{1,\alpha}, \ldots, r_{l,\alpha}\}$. Since $P_\alpha(E_0) \subset \cdots \subset P_\alpha(E_l)$ and $Q_\alpha(E_0) \supset \cdots \supset Q_\alpha(E_l)$, $r(P_{0,\alpha}) \le \cdots \le r(P_{l,\alpha})$ and $r(Q_{0,\alpha}) \ge \cdots \ge r(Q_{l,\alpha})$. Thus, $r^*(\alpha)$ can be found by a binary search on $r_{0,\alpha}, r_{1,\alpha}, \ldots, r_{l,\alpha}$. Since the smallest disk enclosing a point set can be calculated in $O(n)$ time, $r^*(\alpha)$ can be computed in $O(n \log n)$ time.

Analogously, we can define three other solutions, say, $r^*(\beta)$, $r^*(\gamma)$ and $r^*(\delta)$, which deal with the situations in which E_i is guessed as ac, bd or bc. Let $r^*(\mathcal{T}) = \min\{r^*(\alpha), r^*(\beta), r^*(\gamma), r^*(\delta)\}$. So, $r^*(\mathcal{T})$ can be found in $O(n \log n)$ time.

We compute all possible solutions $r^*(\mathcal{T}_j)$, $1 \le j \le 4$. Let $S(j,1)$ and $S(j,2)$ be the sets determining the solution $r^*(\mathcal{T}_j)$. From Corollary 1, the pair $(S(j,1), S(j,2))$ is unique. Since $Succ_R()$ and $Pred_R()$ are defined on R, we have $S(j,1) \cup S(j,2) \subseteq S$; the equality holds only when two corresponding values among $Succ_R()$, $Pred_R()$, $Succ_R()$ and $Pred_R()$ are equal to those defined on polygon P. If $S(j,1) \cup S(j,2) = S$ occurs for some $r^*(\mathcal{T}_j)$, then $r^* = r^*(\mathcal{T}_j)$.

Lemma 8. *In the Delaunay-path case, r^* can be computed in $O(n \log n)$ time and $O(n)$ space.*

4.2 Algorithm for the Non-Delaunay-Path 2-Center Problem

Suppose that the union of two disks of radius at most $r^*(\mathcal{T}_j)$, for any $1 \le j \le 4$, does not cover S. Denote by $P[d', a']$ and $P[b', c']$ two longest chains covered by two disks of radius at most $r^*(\mathcal{T}_1)$ respectively. From Lemma 7 and the algorithm described in Sect. 4.1, $a, b \in P_1[a', b']$ and $c, d \in P_2[c', d']$ (Fig. 3).

Let Q be the convex polygon, whose boundary consists of $P[d', a']$, $a'b'$, $P[b', c']$ and $c'd'$. Denote by r_Q^* (R_Q^*) the solution to the 2-center problem for the set of vertices of Q (the polygon Q). Then, $r_Q^* = r^*(\mathcal{T}_1)$ (Lemma 7), and R_Q^* can be computed. Denote by $E_{1,Q}$ and $E_{2,Q}$ the pair of found disks of radius R_Q^* such that $Q \subseteq E_{1,Q} \cup E_{2,Q}$, and e_1 (resp. e_2) the center of $E_{1,Q}$ (resp. $E_{2,Q}$).

Note that there is a unique path T_1 (T_2) from the node representing a triangle with edge $a'b'$ ($c'd'$) to that with ab (cd) in the dual of $DT(S)$, see Fig. 3. We first claim that T_1 and T_2 can be found dynamically in a process of enlarging two disks of equal size, starting from $E_{1,Q}$ and $E_{2,Q}$.

Denote by $p \in P_1[a', b']$ the vertex such that $\triangle_{a',p,b'}$ is a triangle of $DT(S)$, see Figs. 3(a). Assume that p is not covered by $E_{1,Q}$ and $E_{2,Q}$; otherwise, the triangle adjacent to $\triangle_{a',p,b'}$ in T_1 is considered. Since the pair of disks $E_{1,Q}$ and $E_{2,Q}$ gives the solution R_Q^*, we can assume that $\partial E_{1,Q}$ and $\partial E_{2,Q}$ intersect both $a'p$ and $b'p$. See Fig. 3(b). Connect the intersection point of $\partial E_{1,Q}$ and $a'p$ with that of $\partial E_{2,Q}$ and $b'p$. Also, connect two (possible) intersection points of $P_2[c', d']$ with $\partial E_{1,Q}$ and $\partial E_{2,Q}$. Denote by Q' the resulting polygon. Then, $Q \subset Q'$. Since all vertices of Q' are covered, we can find the 2-center solution for Q'. By considering the polygons like Q' repeatedly, two endpoints of $a'p$ or $b'p$ can be eventually covered. Thus, the triangle adjacent to $\triangle_{a',p,b'}$ in T_1 is found, and its uncovered vertex (if it is) becomes the new target to cover. In this way, we can reach an edge of P_1, whose two endpoints are covered by two present disks respectively. The disk-enlarging process continues until the other edge of P_2 is reached, with the endpoints covered by two disks respectively, before R^* is obtained. From Corollary 2, $r^* < R^*$ holds. Following Lemma 6, these two edges are ab and cd, respectively. Our claim is proved.

Denote by $P_1'[a', b']$ ($P_2'[c', d']$) the convex chain formed by the vertices appeared T_1 (T_2). To give a linear-time process, we perform a disk-enlarging process by moving the centers of enlarging disks towards each other along e_1e_2 such that all points of $E_{1,Q}(S)$ ($E_{2,Q}(S)$) are covered by its enlarged disk. It stops as soon as all vertices of $P_1'[a', b']$ or $P_2'[c', d']$ are covered by two enlarging disks. As discussed above, the lastly uncovered edge of, say, $P_1'[a', b']$ ($P_2'[c', d']$) is ab (cd). (Afterwards, r^* can be computed with binary search using Lemma 3.)

At the time that p is first covered, as the circumcircle of Delaunay triangle $\triangle_{a',p,b'}$ does not contain any other points of S, no points following $\triangle_{a',p,b'}$ in T_1 can be covered. By applying this argument repeatedly, the vertices of $P_1'[a', a]$ are covered in the order of them on $P_1'[a', a]$, and the vertices of $P_1'[b, b']$ are in their counterclockwise order on $P_1'[b, b']$. Also, the vertices of $P_2'[c', c]$ ($P_2'[d, d']$) are covered in their order (counterclockwise order) on $P_2'[c', c]$ ($P_2'[d, d']$).

We can then conclude that any newly covered point in the disk-enlarging process cannot be the neighbor furthest to either moving center, excluding the instant that it appears on the disk's boundary. Thus, $FV(E_{1,Q}(S))$ and $FV(E_{2,Q}(S))$ can be used to maintain the neighbors furthest to moving centers.

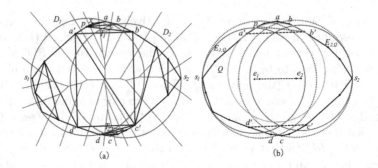

Fig. 3. A situation in which $r_Q^* < R_Q^*$.

Lemma 9. *For the non-Delaunay-path 2-center problem, r^* can be computed in $O(n \log n)$ time and $O(n)$ space.*

5   The 2-Center of a Set of Arbitrarily Given Points

Let S be a set of arbitrarily given points. Still, denote by r^* the solution to the 2-center problem for S. Let H be the set of vertices on polygon $CH(S)$. Denote by r_H^* the *overall* solution to the 2-center problem for H.

Lemma 10. *Suppose that D_1 and D_2 are the disks of equal radius such that $S \subseteq D_1 \cup D_2$, and one of them, say, D_2 is the smallest disk enclosing points $q_1, q_2, q_3 \in S$ (q_3 may not exist). The radius of D_1 and D_2 is the 2-center solution r^* for S if (i) two subsets of H, which are separated by the bisector of the line segment connecting the centers of D_1 and D_2, can be respectively covered by two disks of radius at most r_H^*, and (ii) the radius of every smallest disk enclosing $(S \backslash D_2(S)) \cup \{q_i\}$, $i = 1, 2$ or 3, is at least that of D_2.*

Let $D_{1,H}$ and $D_{2,H}$ be two disks of radius r_H^* such that $H \subseteq D_{1,H} \cup D_{2,H}$. Suppose that disk $D_{1,H}$ is obtained by enlarging the smaller between two disks (if needed), which are outputted by the algorithm of Sect. 4. We first check in linear-time whether the radius of the smallest disk enclosing all the points *not* covered by $D_{2,H}$ is no more than r_H^*. If yes, then $r^* = r_H^*$.

Assume below that $r^* > r_H^*$. Let c_1 and c_2 be the centers of $D_{1,H}$ and $D_{2,H}$, respectively. We will describe a disk-enlarging process, starting from $D_{1,H}$ and $D_{2,H}$, so as to find r^*. It is the key that when the boundary of an enlarging disk touches a new point, we require that the enlarged disks at that moment satisfy the conditions of Lemma 10, with respect to the set of presently covered points.

Let L be the line through c_1 and c_2. Assume that L is not vertical. Let US_1 (US_2) be the set of points, which are vertically below (above) L and uncovered by $D_{1,H} \cup D_{2,H}$. Next, sort in $O(n \log n)$ time the orthogonal projections of points of US_1 onto L into increasing x-order. Denote by $u_1, u_2, \ldots, u_i$ the sequence of sorted points, and C_1 the chain formed by them. Analogously, sort the orthogonal projections of points of US_2 onto L into increasing x-order. Denote by $v_1, v_2, \ldots, v_j$ the sequence of sorted points, and C_2 the chain formed by them.

We will describe how to cover the vertices of a chain C_1 or C_2, ignoring the other, and how the sets of vertices of C_1 and C_2 can be covered one by one.

Covering the Vertices of a Chain C_1 or C_2. Take C_1 as an example. Let $p \in D_{1,H}(S)$ be the point having the largest y-coordinate, among those of the points on both $CH(D_{1,H}(S))$ and $CH(D_{1,H} \cup D_{2,H})$ (it is the convex hull of two disks). Point p can be found from $CH(D_{1,H}(S))$ as well as $CH(D_{1,H} \cup D_{2,H})$. The first step of our algorithm rotates $D_{1,H}$ around p clockwise until a new point $w \in D_{1,H}(S)$ ($\neq p$), which is on the clockwise portion of $CH(S)$ from p to s_2, is encountered (Fig. 4(a)). If w does not exist, then let p, the center of rotated disk and the first uncovered point of C_1 be on a line. Denote by D_1' the resulting disk. Next, the second step enlarges two disks of equal radius, so as to cover the points of C_1, $D_{1,H}(S)$ and $D_{2,H}(S)$ using two disks of the minimum radius.

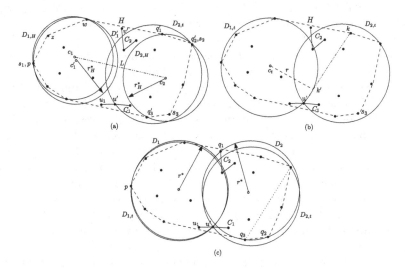

Fig. 4. Illustrating the disk-enlarging process and two polygonal chains C_1 and C_2.

Step 1: Rotating $D_{1,H}$ around p. Denote by u the leftmost uncovered vertex of C_1. Let us rotate the center of $D_{1,H}$ around p clockwise, without changing its coverage of $D_{1,H}(S)$. The rotation stops as soon as a point w of $D_{1,H}(S)$ appears on the boundary of rotated disk (Fig. 4(a)). If u is ever covered in the rotation, then it is updated to the next uncovered vertex on C_1 (if exists).

Consider how to implement the above rotation. Let H' be the set of points, which belong to $D_{1,H}(S)$ and are on the clockwise portion of $CH(S)$ from p to s_2. If H' is empty, then D_1' is obtained by rotating r_H^* until p, the center of D_1' and u are on a line. Otherwise, compute in time $O(n)$ diagram $FV(H')$. Denote by c_1 the center of $D_{1,H}$ as well as the rotating disk. One can then find the *circular direction* F, whose center and radius are p and r_H^* respectively, along which the rotating center moves. Since c_1 is contained in the region $FV(p)$ of $FV(H')$, the circular direction F first hits an edge of $FV(p)$, which is the common edge between $FV(p)$ and $FV(w)$, $w \in D_{1,H}(S)$. Since region $FV(p)$ is convex, the hit point, which is the center of D', can be found in $O(\log n)$ time.

Step 2: Enlarging two disks of equal size. To cover the vertices of C_1, we perform a process of enlarging two disks of equal size, starting from D'_1 and $D_{2,H}$, in the same rate. Since all points of $D'_1(S)$ $(D_{2,H}(S))$ are kept in the disk enlarged from D'_1 $(D_{2,H})$, $FV(D'_1(S))$ and $FV(D_{2,H}(S))$ are initially computed.

Suppose first that point w is on $\partial D'_1$, see Fig. 4(a). The center of D'_1 then moves towards u along $B_{p,w}$. At the time u is touched by the enlarging disk from D'_1, since the center moves *away* from pw, points p, w and u cannot be in any half of present disk. The current disk is thus the smallest one enclosing p, w and u (Lemma 1). Afterwards, u is updated to the next uncovered vertex on C_1.

While the center of enlarged disk from D'_1 moves along $B_{p,w}$, it may reach a Voronoi vertex of $FV(D'_1(S))$. Assume that the met vertex is common to $FV(p)$, $FV(w)$ and $FV(z)$. Here, $z \in D'_1(S)$ has to be a point on the clockwise portion of $CH(S)$ from p to w. The mobile center then changes to move along $B_{p,z}$.

In the special case that w does not exist, the center of enlarging disk from D'_1 moves directly towards u. If u is encountered, then the present disk is the smallest one enclosing p and u, because pu is a diameter of the disk. Afterwards, u is updated accordingly. If a common edge between $FV(p)$ and, say, $FV(p')$ $(p' \in D'_1(S))$, is met, then the center changes to move towards u along $B_{p,p'}$.

Consider how the center of enlarging disk from $D_{2,H}$ moves, particularly, its initial movement. Denote by c_2 the center of $D_{2,H}$ as well as the disk enlarged from $D_{2,H}$. Let u' be the rightmost uncovered vertex of C_1. If two points on $\partial D_{2,H}$ are the endpoints of the disk's diameter, then c_2 moves towards u' along the bisector of them. Otherwise, three points of $D_{2,H}(S)$ are on $\partial D_{2,H}$. If the center moving along the bisector of the segment connecting two of them can shorten its distance to u', then let c_2 move on that (only one) bisector. Otherwise, c_2 moves directly towards u'. In either case, denote by q a neighbor furthest to the moving center c_2, among all points of $D_{2,H}(S)$. When u' is encountered, as discussed above, the meeting disk is the smallest one enclosing the points on its boundary, and u' is then updated to the next uncovered point.

At the time that u or u' is covered, a group of vertices may be covered together. From the definition of C_1, point u or u' can never be the neighbor furthest to either moving center. In this way, all vertices of C_1 can be covered.

Finally, consider how to implement the disk-enlarging process, assuming w exists. (The other situation can be handled analogously.) Denote by G_1 (G_2) the direction along which c'_1 (c_2) moves. Suppose that u is covered when the center of enlarging disk from D'_1 moves to a point l_1 on G_1. Then, (1) l_1 is on G_1, and (2) $|l_1 u| = |l_1 p|$. We can thus compute the x- and y-coordinates of l_1. The radius $(|l_1 p|)$ of obtained disks is termed an *event* of the process. Also, we can determine the point l_2 on G_2, at which u' is covered by the disk enlarged from $D_{2,H}$. After u or u' is updated, the process continues.

As described above, a mobile center may move into some furthest-site Voronoi vertex g_1 (g_2) before l_1 (l_2) is encountered. Vertex g_1 (g_2), if exists, also determines a disk of radius $|g_1 p|$ $(|g_2 q|)$, at the time that the center of enlarging disk from D_1' $(D_{2,H})$ moves to g_1 (g_2). Whenever g_1 (g_2) is met, the direction G_1 (G_2) is updated accordingly, and the disk-enlarging process then continues.

The smallest radius among at most four candidates thus gives the first event occurred in the disk-enlarging process. Also, any point newly covered in the process needn't be maintained as the neighbor furthest to either moving center.

Lemma 11. *The 2-center problem for the set of given points of C_1 (C_2), $D_{1,H}(S)$ and $D_{2,H}(S)$ can be solved in $O(n)$ time and space.*

Proof. The disk-enlarging process may require traversing $O(n)$ edges of $FV(D_1'(S))$ and $FV(D_{2,H}(S))$, which totally takes $O(n)$ time and space. Assume that the last vertex, say, u' of C_1 is covered by the disk enlarged from $D_{2,H}$, at time t. Denote by $D_{2,t}$ $(D_{1,t})$ the disk enlarged from $D_{2,H}$ (D_1') at time t (Fig. 4). Then, $D_{2,t}$ is the smallest one enclosing present (two or three) points on its boundary.

First, since $D_{1,t}(S)$ has at least one point on $CH(D_{1,t} \cup D_{2,t})$, disk $D_{1,t}$ cannot be translated to cover any point on $\partial D_{2,t}$, except that u' happens to be on $\partial D_{1,t}$. Next, we claim that $D_{1,t}$ cannot be rotated to cover any point on $\partial D_{2,t}$, except that u' is also on $\partial D_{1,t}$. Since u' $(\in C_1)$ is covered by $D_{2,t}$, it suffices to consider the rotation of $D_{1,t}$ *upwards*. Note that the points on $\partial D_{2,t}$, excluding u', are also on $CH(D_{2,H}(S))$. Let k be the leftmost point of $CH(D_{2,H}(S))$, which is vertically above L. Thus, k is outside of $D_{1,t}$. See Fig. 4(b). (Point k may happen to be on $CH(D_{2,H})$.) If $\partial D_{1,t}$ does not intersect segment $u'k$, no rotation of $D_{1,t}$ can cover k, without changing its coverage of $D_{1,t}(S)$. Otherwise, let k' be the common point between $u'k$ and $\partial D_{1,t}$, with the larger y-coordinate. Then, $\angle kk'c_t \geq \pi/2$, where c_t denotes the center of $D_{1,t}$ (Fig. 4(b)). Again, $D_{1,t}$ cannot be rotated to cover k. From the definition of point k, our claim is proved.

Therefore, the condition (ii) of Lemma 10 is satisfied by $D_{2,t}$ and $D_{1,t}$, with respect to the set of points of C_1, $D_{1,t}(S)$ and $D_{2,t}(S)$. From the construction of initial disks $D_{1,H}$ and $D_{2,H}$, the condition (i) also holds. The lemma follows. □

Covering the Vertices of C_1 and C_2. Let us perform a disk-enlarging process so as to cover all vertices of C_1, ignoring C_2. Denote by r the radius of $D_{1,t}$ and $D_{2,t}$, which are obtained in the process of covering C_1. Assume that at least one point of C_2 is outside of $D_{1,t} \cup D_{2,t}$. As in the proof of Lemma 11, $D_{1,t}$ is rotated counterclockwise, if needed, so as to close its distance to the leftmost uncovered vertex of C_2, until a new point is encountered. Next, a process of enlarging $D_{2,t}$ and the disk rotated from $D_{1,t}$ is performed, so as to cover all vertices of C_2.

Combining with Theorem 2, the main result of this paper can be obtained.

Theorem 3 *The planar 2-center problem can be solved in $O(n \log n)$ time and $O(n)$ space.*

Proof. Suppose first that the last covered vertex u' of C_1 is touched only by $D_{2,t}$. Two disks of equal radius, starting from $D_{1,t}$ and $D_{2,t}$, are enlarged simultaneously to cover all points of C_2. Denote by D_1 and D_2 the finally stopped disks of equal radius. Assume also that D_2 is the smallest disk enclosing $D_2(S)$.

In the process for covering the points of C_2, diagrams $FV(D_{1,t}(S))$ and $FV(D_{2,t}(S))$ are first computed, in $O(n \log n)$ time. So, a point u of C_1 can be kept as the neighbor furthest to a moving center at some time. It may happen that u is (lately) touched by the other enlarging disk. In this case, we let the disk keeping u as the neighbor furthest to its moving center to leave from u. It can be done by rotating that disk upwards, until a new furthest neighbor (it exists) is met. Since the disk rotations are made upwards, the points of C_1 newly set to be furthest to moving centers of enlarging disks by rotations move *downwards* along $CH(D_{1,t}(S))$ and/or $CH(D_{2,t}(S))$. Thus, the total number of rotations is $O(n)$. As described above, a rotation takes $O(\log n)$ time. The time taken by all rotations is thus $O(n \log n)$. The rest proof is same as that of Lemma 10, i.e., D_1 cannot be rotated nor translated to cover any point on ∂D_2, except that the lastly covered vertex of C_2 happens to be on ∂D_1. Hence, D_1 and D_2 meet the conditions of Lemma 10, i.e., they together give the 2-center solution for S.

Finally, consider the situation in which both $D_{1,t}$ and $D_{2,t}$ touch vertex u' (Fig. 4(c)). Note that this is the only case that a vertex of C_1 can be the neighbor furthest to two moving centers. In this case, we first rotate both $D_{1,t}$ and $D_{2,t}$ upwards, so as to leave from u'. As described above, the disks D_1' and D_2' of equal radius, which together give the two-center solution for $S \backslash \{u'\}$, can be found. The smaller between two radii of smallest enclosing disks of $D_1'(S) \cup \{u'\}$ and $D_2'(S) \cup \{u'\}$ is then the 2-center solution r^* for S (see also Fig. 4(c)). $\square$

References

1. Aggarwa, A., Guibas, L.J., Saxe, J., Shor, P.W.: A linear-time algorithm for computing the Voronoi diagram of a convex polygon. In: DCG, vol. 4, pp. 591–604 (1989)
2. Chan, T.M.: More planar two-center algorithms. Comput. Geom. Theory Appl. **13**, 189–198 (1999)
3. Choi, J., Ahn, H.-K.: Efficient planar two-center algorithms. Comput. Geom. Theory Appl. **97**, 101768:1-101768:10 (2021)
4. Choi, J., Jeong, D., Ahn, H.-K.: Covering convex polygons by two congruent disks. In: Flocchini, P., Moura, L. (eds.) IWOCA 2021. LNCS, vol. 12757, pp. 165–178. Springer, Heidelberg (2021). https://doi.org/10.1007/978-3-030-79987-8_12
5. Eppstein, D.: Faster construction of planar two-centers. In: Proceedings of the 8th ACM-SIAM Symposium on Discrete Algorithms, pp. 131–138 (1997)
6. Sharir, M.: A near-linear time algorithm for the planar 2-center problem. Dis. Comput. Geom. **18**, 125–134 (1997)
7. Wang, H.: On the planar two-center problem and circular hulls. Discrete Comput. Geom. **68**, 1175–1226 (2022). (Its preliminary version appeared in SoCG 2020.)

Endogenous Threshold Selection with Two-Interval Restricted Tests

Zeyu Ren$^{(\boxtimes)}$ and Yan Liu

Renmin University of China, Beijing, China
{zeyuren,liuyan5816}@ruc.edu.cn

Abstract. We continue to study the endogenous threshold selection game. Two firms have interchangeable products with qualities that are drawn i.i.d. from a common prior. A principal wants to minimize the probability of selecting the product of lower quality by offering a set of threshold tests. The threshold test is that given a threshold, the product can be estimated whether its quality exceeds the threshold. Two firms pick the threshold endogenously to maximize the probability of being chosen by the principal.

We consider the scenario where the principal restricts threshold tests to be two-interval. We further characterize properties of Bayes-Nash equilibrium between two firms. Numerically, the minimum probability of incorrect selection is 0.22096 where threshold tests are supported on $[0, 0.19] \cup [0.33, 0.79]$. Moreover, our exploration aids in offering valuable insights into signaling games with one receiver and two senders.

Keywords: Signaling Game · Bayes-Nash Equilibrium · Inversion

1 Introduction

The study of signaling games has been driven by a keen interest. We focus on the endogenous threshold selection game proposed by [2] where a principal wants to choose the better product from two firms. Products are interchangeable and have real-valued qualities that are drawn i.i.d. from a common known distribution. However, the qualities cannot be observed directly. Therefore, the principal offers a set of coarse-grained threshold tests. Then, the two firms each endogenously pick a threshold before knowing the quality of their products. Results that products pass or fail tests are revealed to the principal. Consequently, the principal gets information on whether the products' quality lies above or below the threshold.

Nash equilibrium is a fundamental concept in game theory, helping researchers and analysts model and predict the behavior of participants in strategic interactions. In the endogenous threshold selection game, two firms compete to have their products chosen by the principal with a higher probability. Thus,

Supported by the National Natural Science Foundation of China (Grant No. 62172422).

the interaction gives rise to Bayes-Nash equilibrium. Understanding and identifying Nash equilibrium provides valuable insights into the likely outcomes of strategic interactions.

Signaling game is an important concept in game theory that focuses on how information is communicated between players in strategic interaction. Unlike traditional games where players make decisions without explicitly communicating information, signaling games involve senders and receivers. Senders send signals to receivers to influence their decisions. In the endogenous threshold selection game, two firms act as two senders to send their selected thresholds as signals, and the principal act as the receiver to make better decisions by two receiving signals. The study of endogenous threshold selection game helps us gain insights into how individuals or entities strategically convey information to influence the behavior of others, leading to more informed decision-making in various fields.

We summarize the main two steps of endogenous threshold selection game:

1. Knowing the common prior quality distribution, the principal chooses a restricted threshold set that does not have the full support on the quality distribution. She wants to minimize the probability of choosing the worse product.
2. Two firms endogenously pick thresholds from the restricted set to compete with each other. They both want to maximize the probability of being selected by the principal. Note that the firms themselves do not know which product will be selected for testing. Thus, there exists the Bayes-Nash equilibrium between two firms.

We are interested in the question: which set of tests should the principal restricts to achieve a smaller probability of selecting the worse product? We use I to denote the probability. Previous work [2] has shown that $I(F_{[0,0.79]}) < 0.22975 < 0.23052 < I(F_{[0,1]})$ where $F_{[0,0.79]}$ denotes the unique symmetric Bayes-Nash equilibrium distribution restricted to support on $[0, 0.79]$ and $F_{[0,1]}$ denotes the unique symmetric Bayes-Nash equilibrium distribution with the unrestricted interval. That is, the principal can achieve smaller error by restricting tests. We wonder if the probability can be further smaller if the principal can offer more complicated restricted tests. We move on to two-interval restricted tests and find that the Bayes-Nash equilibrium is still unique and symmetric. The probability exactly get strictly smaller when the principal provides test set that is supported on $[0, 0.19] \cup [0.33, 0.79]$.

1.1   Related Works

The most related work is [2]. They introduced and studied the problem of optimal and endogenous test selection. They explicitly defined the optimal correlated and i.i.d. distributions for the principal, along with the equilibrium distribution when firms have the flexibility to select their own thresholds within an interval $[a, b]$, encompassing the case of the interval $[0, 1]$. They proposed lots of open problems and we choose one of them, which set of tests the principal should provide is optimal, to continue our study.

Besides, there are other related works concerned with signaling games. In the case of multi-sender and a single receiver. There are many results [7,8,10] when senders have commitment power. Please refer to the survey [6,13] for more results.

We also consider the Bayesian persuasion model. [9] examined a model where the information set exhibits Blackwell-connected, signifying that each sender has the ability to independently deviate to any information state where the receiver possesses more information. [11] investigated analogous problems where senders commit to their strategies sequentially, as opposed to simultaneously. [4] studied a model in which each project is either good or bad independently and prior probabilities are known.

In addition, the threshold selection problem discussed bears a close resemblance to multi-sender cheap talk problem. Cheap talk model was first introduced by [5]. Latter, [1] conducted an analysis of multi-sender cheap talk scenarios where state space could be constrained. Then, [3] explored the impact of various modes of communication in a costless information transmission environment with multiple senders. And, [12] investigated how a CEO's sensitivity influences information transmission from multiple unit managers when engaged in strategic communication during decision-making processes.

2 Preliminaries

In the endogenous threshold selection game, a principal aims to choose the better product from two firms. The qualities of products are drawn i.i.d. from a common distribution supported on $[0,1]$. W.l.o.g., we assume that the distribution is uniform.[1] With a little abuse of notation, we refer to X and Y as the firms and their products' qualities interchangeably.

The two firms compete to have their products chosen with a higher probability by endogenously selecting threshold tests. The principal restricts firms from selecting the threshold $\theta \in S$ where $S \subset [0,1]$. A product of quality X passes the test if and only if $X \geq \theta$. Otherwise, it fails the test.

Let θ_X and θ_Y denote the thresholds chosen by firms X and Y respectively. After two test outcomes are revealed, a rational principal's choice follows the principle:

1. If only one product passes its test, then the principal selects the firm with the product.[2]
2. If both firms' products pass or fail their tests, then the principal selects X when $\theta_X > \theta_Y$; the principal selects Y when $\theta_X < \theta_Y$; the principal selects one of the firms uniformly at random when $\theta_X = \theta_Y$.

---

[1] For any quality distribution, its quantile is distributed uniformly in $[0,1]$. Once we get the results of quantile functions, we can derive the results of the original functions. Thus, it allows us to concentrate on the uniform distribution.

[2] We have $\mathbb{E}[X|X < \theta] \leq \mathbb{E}[X] \leq \mathbb{E}[X|X \geq \theta]$ It indicates that the expected quality of the product passing a test will not be lower than the failed one.

Then, the two firms choose threshold tests themselves. The solution falls into a Bayes-Nash equilibrium. We use a pair of distributions (F_X, F_Y) supported on a subset of S to denote the Bayes-Nash equilibrium. That is, if one firm chooses thresholds according to $F_X(F_Y)$, then selecting thresholds based on $F_Y(F_X)$ is the best response for another firm. Note that no matter what threshold in the support of F_X that X chooses, the product of X must be selected with probability $\frac{1}{2}$ at equilibrium. The same goes for Y.

Supposing that Y's test is drawn from F, we define $G(\theta) = \int_0^\theta F(t)dt$. Let $\phi_F = 1 - G(1)$ be the failure probability under F. We denote the point mass at θ as $\delta_\theta = F(\theta) - \lim_{t \uparrow \theta} F(t)$. Using these notations, we will demonstrate that the probability of X being selected when choosing θ can be expressed as:

$$(1 - \theta)\phi_F + (2\theta^2 - 2\theta + 1)\left(F(\theta) - \frac{\delta_\theta}{2}\right) + (1 - 2\theta)G(\theta).$$

We denote $w_{F,\theta}^+$ as the probability that X is selected when X picks threshold θ and passes the test. The probability consists of three parts:

1. Y fails its test and the probability is ϕ_F.
2. Y picks threshold θ and passes the test, but X wins the coin flip. The probability is $\frac{1}{2}(1 - \theta)\delta_\theta$.
3. Y picks the threshold that is smaller than θ and passes the test. The probability is $(1 - \theta)(F(\theta) - \delta_\theta) + G(\theta)$.

Similarly, we denote $w_{F,\theta}^-$ as the probability that X is selected when X picks threshold θ and fails the test. The probability consists of two parts:

1. Y picks threshold θ and fails the test, and X wins the coin flip. The probability is $\frac{1}{2}\theta\delta_\theta$.
2. Y picks the threshold that is smaller than θ and fails the test. The probability is $\theta(F(\theta) - \delta_\theta) - G(\theta)$.

To summarize, we derive that the probability of X being selected is

$$(1 - \theta)w_{F,\theta}^+ + \theta w_{F,\theta}^- = (1 - \theta)\phi_F + (2\theta^2 - 2\theta + 1)\left(F(\theta) - \frac{\delta_\theta}{2}\right) + (1 - 2\theta)G(\theta).$$

We focus on the probability of the principal making incorrect choices, i.e. selecting the product of lower quality. We refer to this probability an inversion. And, we use $I(F)$ to denote it in the scenario where both firms choose tests according to distribution F. Next, we will show that

$$I(F) = \int_0^1 \int_0^x (1 - F(x) + F(y))^2 dy dx.$$

In the beginning, we assume that the qualities of two firms follow $X > Y$. The two firms pick thresholds θ_X and θ_Y. Let $\theta_{\min} = \min\{\theta_X, \theta_Y\}$ and $\theta_{\max} = \max\{\theta_X, \theta_Y\}$. Reconsidering the rational principal's choice principle, we find that inversion happens when X and Y fall into the following two categories:

1. If $X, Y \in [0, \theta_{\min})$ or $X, Y \in [\theta_{\max}, 1]$, then both firms fail or pass. According to the principle, the principal chooses the firm that picks the higher threshold.
2. If $X, Y \in [\theta_{\min}, \theta_{\max})$, then one firm fails while another passes. According to the principle, the principal chooses the firm that passes the test.

That is, the inversion happens if and only if X and Y fall into the same interval. Because the assignment of $\theta_{\min}$ and $\theta_{\max}$ to the two firms is uniformly random, the inversion is created with probability $\frac{1}{2}$. For the first category, it is equivalent to $\theta_X, \theta_Y > X$ or $\theta_X, \theta_Y < Y$. Therefore, the probability is $(1 - F(x))^2 + F(y)^2$. For the second category, it is equivalent to $\theta_X \leq Y < X < \theta_Y$ or $\theta_Y \leq Y < X < \theta_X$. Thus, the probability is $2F(y)(1 - F(x))$. Therefore, for each $X > Y$, the inversion happens with probability $\frac{1}{2}((1 - F(x))^2 + F(y)^2 + 2F(y)(1 - F(x))) = \frac{1}{2}(1 - F(x) + F(y))^2$. Finally, we have

$$I(F) = \int_0^1 \int_0^x (1 - F(x) + F(y))^2 \mathrm{d}y \mathrm{d}x.$$

3 Endogenous Test Selection with Restricted Tests

In the section, we consider the non-empty set $S = [a, b] \cup [c, d] \subset [0, 1]$ where $a < b < c < d$.[3] Both firms are restricted to choose thresholds from S. We first characterize the properties of Bayes-Nash equilibrium distribution. The equilibrium is unique and symmetric. Furthermore, the distribution has no point masses other than possibly at b and d. Then, we give the closed form of the equilibrium distribution. Finally, we implement a Python program and get numerical results of inversion.

3.1   The Properties of Bayes-Nash Equilibrium

Lemma 1. *If both firms are restricted to pick a threshold θ from a non-empty set $S = [a, b] \cup [c, d] \subset [0, 1]$. The Bayes-Nash equilibrium has no point masses other than possibly at b and d.*

Proof. We first show that the distribution F_X and F_Y have no point masses other than b or d. For each $\theta < d$, at most one firm has point mass at θ. Supposing that firm Y has point mass $\delta_\theta > 0$, we have the winning probability of firm X with threshold $\theta + \epsilon$ is

$$(1 - \theta - \epsilon)w^+_{F,\theta+\epsilon} + (\theta + \epsilon)w^-_{F,\theta+\epsilon}$$
$$\geq (1 - \theta - \epsilon)w^+_{F,\theta-\epsilon} + (\theta - \epsilon)w^-_{F,\theta-\epsilon} + \delta_\theta(2\theta^2 - 2\theta + 1) - 2\epsilon.$$

---

[3] For the reason why we do not consider $a \leq b \leq c \leq d$, if $a = b$ or $c = d$, then a point appears in the restricted set. The Bayes-Nash equilibrium may not be unique and symmetric. Because we know that every pair supported on S forms an equilibrium when $S = \{1 - \frac{\sqrt{2}}{2}, \frac{\sqrt{2}}{2}\}$. And if $b = c$, the case is equal to one interval and is well studied.

If $\epsilon > 0$ and $\epsilon \to 0$, the term $\delta_\theta(2\theta^2 - 2\theta + 1) - 2\epsilon$ is strictly positive. Thus, neither θ nor $\theta - \epsilon$ can be the best response for firm X. That is, if firm Y has a point mass at θ, the probability of X selecting a threshold in $[\theta - \epsilon, \theta]$ is 0. By contradiction, we assume that F_Y has a point mass at (a, b) or (c, d). Take $\epsilon \in (0, \theta - a)$ as an example. We have $F_X(\theta - \epsilon) = F_X(\theta)$, i.e. no test in the interval $(\theta - \epsilon, \theta + \epsilon')$ is a best response to the firm X for some $\epsilon' > 0$. Consequently, we get that the distribution F_X and F_Y have no point masses other than b and d.   □

Lemma 2. *If both firms are restricted to pick a threshold θ from a non-empty set $S = [a, b] \cup [c, d] \subset [0, 1]$. The Bayes-Nash equilibrium is unique and symmetric.*

Proof. We know that the Bayes-Nash equilibrium has no point masses other than possibly at b and d. In addition, both firms' distributions obey the formula

$$(1 - \theta)\phi_F + (2\theta^2 - 2\theta + 1)\left(F(\theta) - \frac{\delta_\theta}{2}\right) + (1 - 2\theta)G(\theta) = \frac{1}{2}.$$

Besides, we should always determine the value of point mass d at first, and then move on to check the point mass of b. It indicates that distributions have the same structure. And, the value of F always maintain consistency. Therefore, the only Bayes-Nash equilibrium is symmetric.   □

3.2   The Closed Form of Bayes-Nash Equilibrium Distribution

We know that the Bayes-Nash equilibrium is unique and symmetric. And the distribution has no point masses other than possibly at b and d. Moreover, we figure out the structure of distribution. First, we get the close form if the distribution is continuous.

Lemma 3. *If both firms are restricted to pick a threshold θ from a non-empty set $S = [a, b] \cup [c, d] \subset [0, 1]$. The closed form of the unique symmetric Bayes-Nash equilibrium distribution when it is continuous can be formulated by $\frac{1}{2(1-a)} \cdot$*
$$\left(1 - 2a + \sqrt{2a^2 - 2a + 1} \cdot \frac{2\theta - 1}{\sqrt{2\theta^2 - 2\theta + 1}}\right).$$

Proof. The core of the proof is solving the equality

$$(1 - \theta)\phi_F + (2\theta^2 - 2\theta + 1)F(\theta) + (1 - 2\theta)G(\theta) = \frac{1}{2}. \tag{1}$$

let $\theta = a$, and we know that $F(a) = 0$ and $G(a) = 0$. Thus, we have $\phi_F = \frac{1}{2(1-a)}$. Then, we divide every term by $(2\theta^2 - 2\theta + 1)^{\frac{3}{2}}$ and obtain

$$\frac{G(\theta)'}{(2\theta^2 - 2\theta + 1)^{\frac{1}{2}}} + \frac{(1 - 2\theta)G(\theta)}{(2\theta^2 - 2\theta + 1)^{\frac{3}{2}}} = \frac{1}{2(1 - a)} \cdot \frac{\theta - a}{(2\theta^2 - 2\theta + 1)^{\frac{3}{2}}}.$$

Notice that the LHS of the above formula is the derivative of $\frac{G(\theta)}{(2\theta^2 - 2\theta + 1)^{\frac{1}{2}}}$. That is,

$$\left(\frac{G(\theta)}{(2\theta^2 - 2\theta + 1)^{\frac{1}{2}}}\right)' = \frac{1}{2(1 - a)} \cdot \frac{\theta - a}{(2\theta^2 - 2\theta + 1)^{\frac{3}{2}}}.$$

Next, we consider the integral and get

$$
\begin{aligned}
G(\theta) &= \frac{(2\theta^2 - 2\theta + 1)^{\frac{1}{2}}}{2(1-a)} \cdot \int_a^\theta \frac{t-a}{(2t^2 - 2t + 1)^{\frac{3}{2}}} dt \\
&= \frac{(2\theta^2 - 2\theta + 1)^{\frac{1}{2}}}{2(1-a)} \cdot \frac{(1-2a)t + a - 1}{(2t^2 - 2t + 1)^{\frac{1}{2}}} \Big|_a^\theta \\
&= \frac{(2\theta^2 - 2\theta + 1)^{\frac{1}{2}}}{2(1-a)} \cdot \left(\frac{(1-2a)\theta + a - 1}{(2\theta^2 - 2\theta + 1)^{\frac{1}{2}}} + \sqrt{2a^2 - 2a + 1} \right) \\
&= \frac{1}{2(1-a)} \cdot \left((1-2a)\theta + a - 1 + \sqrt{2a^2 - 2a + 1} \cdot (2\theta^2 - 2\theta + 1)^{\frac{1}{2}} \right).
\end{aligned}
$$

Calculating the derivative of $G(\theta)$, we finally get

$$
F(\theta) = \frac{1}{2(1-a)} \cdot \left(1 - 2a + \sqrt{2a^2 - 2a + 1} \cdot \frac{2\theta - 1}{\sqrt{2\theta^2 - 2\theta + 1}} \right).
$$

□

For convenience, let $H(\theta) := \frac{1}{2(1-a)} \cdot \left(1 - 2a + \sqrt{2a^2 - 2a + 1} \cdot \frac{2\theta-1}{\sqrt{2\theta^2 - 2\theta + 1}} \right)$.
Next, we only need to figure out probabilities of point masses and intervals where the distribution is continuous. Finally, we get the closed form of distribution.

Theorem 1. *If both firms are restricted to choose thresholds from a non-empty set $S = [a, b] \cup [c, d] \subset [0, 1]$, the closed form of the unique symmetric Bayes-Nash equilibrium distribution F_{eq} is:*

– *If $(1-a)d \leq \frac{1}{2}$, then F_{eq} is a step function at d.*
– *Otherwise, let $\delta_d := \frac{1 - a(1-d) - d(1-a)}{(1-a)(2d^2 - 2d + 1)}$ and $\gamma_1 := \frac{1 - a - 2d + 4ad - 2ad^2}{1 - 4(1-a) + 2(1-a)d^2}$.*
  • *If $\gamma_1 \leq b$, then*

$$
F_{eq}(\theta) = \begin{cases} H(\theta) & a \leq \theta < \gamma_1, \\ 1 - \delta_d & \gamma_1 \leq \theta \leq b, c \leq \theta < d, \\ 1 & \theta = d. \end{cases}
$$

  • *If $b < \gamma_1 \leq c$, let δ_b be the solution of*

$$
\frac{1-b}{2(1-a)} + (2b^2 - 2b + 1)\left(1 - \delta_d - \frac{\delta_b}{2}\right) + (1 - 2b)G(b) = \frac{1}{2},
$$

  where $G(b) = d - \frac{1}{2(1-a)} - (d-b)(1 - \delta_d)$.
    * *If $\delta_b + \delta_d \geq 1$, then F_{eq} is a step function at b and d, i.e., both firms choose b with $1 - \delta_d$ probability and choose d with δ_d probability.*
    * *Otherwise, let γ_2 be the solution of*

$$
\frac{1 - \gamma_2}{2(1-a)} + (2\gamma_2^2 - 2\gamma_2 + 1)(1 - \delta_b - \delta_d) + (1 - 2\gamma_2)G(\gamma_2) = \frac{1}{2},
$$

where $G(\gamma_2) = G(b) - (b - \gamma_2)(1 - \delta_b - \delta_d)$. Then

$$F_{eq}(\theta) = \begin{cases} H(\theta) & a \le \theta < \gamma_2, \\ 1 - \delta_b - \delta_d & \gamma_2 \le \theta < b, \\ 1 - \delta_d & \theta = b, c \le \theta < d, \\ 1 & \theta = d. \end{cases}$$

- *If $c < \gamma_1 < d$, let δ_b be the solution of*

$$\frac{1 - b}{2(1 - a)} + (2b^2 - 2b + 1)\left(H(c) - \frac{\delta_b}{2}\right) + (1 - 2b)G(b) = \frac{1}{2},$$

where $G(b) = d - \frac{1}{2(1-a)} - (d - \gamma_1)(1 - \delta_d) - \int_c^{\gamma_1} H(t)dt - (c - b)H(c)$.
  * *If $H(c) - \delta_b \le 0$, then*

$$F_{eq}(\theta) = \begin{cases} 0 & a \le \theta < b, \\ H(c) & \theta = b, \\ H(\theta) & c \le \theta < \gamma_1, \\ 1 - \delta_d & \gamma_1 \le \theta < d, \\ 1 & \theta = d. \end{cases}$$

  * *Otherwise, let γ_2 be the solution of*

$$\frac{1 - \gamma_2}{2(1 - a)} + (2\gamma_2^2 - 2\gamma_2 + 1)(H(c) - \delta_b) + (1 - 2\gamma_2)G(\gamma_2) = \frac{1}{2},$$

where $G(\gamma_2) = G(b) - (b - \gamma_2)(H(c) - \delta_b)$. then

$$F_{eq}(\theta) = \begin{cases} H(\theta) & a \le \theta < \gamma_2, \\ H(\gamma_2) & \gamma_2 \le \theta < b, \\ H(c) & \theta = b, \\ H(\theta) & c \le \theta < \gamma_1, \\ 1 - \delta_d & \gamma_1 \le \theta < d, \\ 1 & \theta = d. \end{cases}$$

Proof. First, we consider the Nash equilibrium distribution when the restricted set $S' = [a, d]$. We know that $\phi_F = \frac{1}{2(1-a)}$. If the failure probability $\phi_F \ge d$, that is, we have $1 - G(1) \ge d$, and the only way to satisfy the condition is that both firms choose d deterministically. Thus, we have if $(1 - a)d \le \frac{1}{2}$, then F_{eq} is a step function at d. Otherwise, we consider the point mass at d. We have

$$(1 - d)\phi_F + (2d^2 - 2d + 1)\left(F(d) - \frac{\delta_d}{2}\right) + (1 - 2d)G(d) = \frac{1}{2},$$

where $F(d) = 1$ and $G(d) = d - \phi_F$. Therefore, the probability of the point mass $\delta_d = \frac{1 - a(1-d) - d(1-a)}{(1-a)(2d^2 - 2d + 1)}$. Notice that if the form of tests changes from one-interval to two-interval, when a and d remain unchanged, regardless of how b and c are adjusted, the probability δ_d does not change.

Then, we figure out the interval when the distribution is continuous. We have

$$(1 - \gamma_1)\phi_F + (2\gamma_1^2 - 2\gamma_1 + 1)F(\gamma_1) + (1 - 2\gamma_1)G(\gamma_1) = \frac{1}{2},$$

where $F(\gamma_1) = 1 - \delta_d$ and $G(\gamma_1) = d - \phi_F - (d - \gamma_1)F(\gamma_1)$. Consequently, we get $\gamma_1 = \frac{1-a-2d+4ad-2ad^2}{1-4(1-a)+2(1-a)d^2}$.

When the interval (b, c) no longer is the support of the Nash equilibrium distribution, the change depends on the relation between γ_1 and the interval (b, c). If $\gamma_1 \leq b$, it indicates that the elimination of the support does not affect the structure of the distribution F_{eq}.

If $b < \gamma_1 \leq c$, the probability of the interval (b, γ_1) should not exist. In order to maintain the equilibrium, there must be a point mass b. Thus, let δ_b be the solution of

$$\frac{1 - b}{2(1 - a)} + (2b^2 - 2b + 1)\left(1 - \delta_d - \frac{\delta_b}{2}\right) + (1 - 2b)G(b) = \frac{1}{2},$$

where $G(b) = d - \frac{1}{2(1-a)} - (d - b)(1 - \delta_d)$. If $\delta_b + \delta_d \geq 1$, it means that there does not exist an interval which is continuous, then F_{eq} is a step function at b and d, i.e., both firms choose b with $1 - \delta_d$ probability and choose d with δ_d probability. Otherwise, we figure out γ_2 to determine the continuous distribution interval. We solve the following equation

$$\frac{1 - \gamma_2}{2(1 - a)} + (2\gamma_2^2 - 2\gamma_2 + 1)(1 - \delta_b - \delta_d) + (1 - 2\gamma_2)G(\gamma_2) = \frac{1}{2},$$

where $G(\gamma_2) = G(b) - (b - \gamma_2)(1 - \delta_b - \delta_d)$.

Finally, we move on to the condition $c < \gamma_1 < d$. Similarly, there must be a point mass b. Let δ_b be the solution of

$$\frac{1 - b}{2(1 - a)} + (2b^2 - 2b + 1)\left(H(c) - \frac{\delta_b}{2}\right) + (1 - 2b)G(b) = \frac{1}{2},$$

where $G(b) = d - \frac{1}{2(1-a)} - (d - \gamma_1)(1 - \delta_d) - \int_c^{\gamma_1} H(t)dt - (c - b)H(c)$. If $H(c) - \delta_b \leq 0$, it indicates that $F_{eq} = 0$ when $\theta < b$. Otherwise, we figure out γ_2 to determine the continuous distribution interval. We calculate γ_2 by the equation

$$\frac{1 - \gamma_2}{2(1 - a)} + (2\gamma_2^2 - 2\gamma_2 + 1)(H(c) - \delta_b) + (1 - 2\gamma_2)G(\gamma_2) = \frac{1}{2},$$

where $G(\gamma_2) = G(b) - (b - \gamma_2)(H(c) - \delta_b)$. Finally, we tackle all conditions and finish the proof. □

To summarize the main proof idea, we know that the distribution has no point masses other than possibly at b and d. In addition, when the distribution is continuous, it satisfies the structure of $H(\theta)$. Therefore, we start with the one-interval restricted tests. Next, we calculate the value of the point mass δ and figure out γ step by step. If necessary, we divide the problem into several cases. Finally, we get the closed form of the Nash equilibrium distribution.

3.3 Numerical Results of Inversion

The closed form characterization of Bayes-Nash equilibrium allows numerical evaluation. Recall that the inversion can be calculated by $I(F) = \int_0^1 \int_0^x (1 - F(x) + F(y))^2 dy dx$. Thus, by Theorem 1, we calculate the inversion.

Theorem 2. *Compared to that both firms are restricted to choose thresholds from a set $S = [a, b] \subset [0, 1]$, the principal can achieve a strictly smaller inversion by offering a threshold test from a set $S = [a, b] \cup [c, d] \subset [0, 1]$.*

Proof. We use Python 3.9 to implement programs. First, we have $I(F_{[0,1]}) \approx 0.23053$ and $I(F_{[0,0.79]}) \approx 0.22977$ in our environment. Then, we search $a, b, c, d \in [0, 1]$ with 0.01 granularity to achieve the optimum. Numerically, the minimum of inversion is achieved at $a = 0, b = 0.19, c = 0.33, d = 0.79$ where $I(F_{[0,0.19] \cup [0.33,0.79]}) \approx 0.22906$. That is, by offering the two-interval set of tests, the principal can get a strictly smaller inversion. The code can be found at our github repository (https://github.com/ruc-renzy/Endogenous-Threshold-Selection-with-Two-Interval-Restricted-Tests). □

4 Conclusion

In this paper, we studied the endogenous threshold selection game with restricted tests. In the setting, a principal aims to select the product of higher probability between two firms. She offers a set of threshold tests that are restricted to a two-interval form. Two firms choose their own thresholds to maximize the probability of being selected by the principal. We characterized Bayes-Nash equilibrium explicitly and gave the minimum inversion numerically. Our study identified that offering two-interval tests can get smaller inversion than an interval $[a, b]$.

There are still several unresolved questions that merit further exploration. Firstly, by which set of tests can the principal achieve the smallest inversion? If we concentrate on the set of n intervals, we conjecture that Bayes-Nash equilibrium is still unique and symmetric. Furthermore, we put forth a conjecture that as the number of intervals increases, the inversion becomes smaller. Besides, the smallest inversion is achieved when n approaches infinity. Secondly, a compelling question arises: when the model extends to more than two firms, how to characterize Bayes-Nash equilibrium? How to calculate inversion? These inquiries may need new and deeper insights.

References

1. Ambrus, A., Takahashi, S.: Multi-sender cheap talk with restricted state spaces. Theor. Econ. **3**(1), 1–27 (2008)
2. Banerjee, S., Kempe, D., Kleinberg, R.: Threshold tests as quality signals: optimal strategies, equilibria, and price of anarchy. In: Feldman, M., Fu, H., Talgam-Cohen, I. (eds.) WINE 2021. LNCS, vol. 13112, pp. 299–316. Springer, Cham (2022). https://doi.org/10.1007/978-3-030-94676-0_17

3. Bayindir, E.E., Gurdal, M.Y., Ozdogan, A., Saglam, I.: Cheap talk games with two-senders and different modes of communication. Games **11**(2), 18 (2020)
4. Boleslavsky, R., Cotton, C.: Limited capacity in project selection: competition through evidence production. Econ. Theor. **65**, 385–421 (2018)
5. Crawford, V.P., Sobel, J.: Strategic information transmission. Econom. J. Econom. Soc. **50**(6), 1431–1451 (1982)
6. Dughmi, S.: Algorithmic information structure design: a survey. ACM SIGecom Exchanges **15**(2), 2–24 (2017)
7. Dughmi, S., Kempe, D., Qiang, R.: Persuasion with limited communication. In: Proceedings of the 2016 ACM Conference on Economics and Computation, pp. 663–680 (2016)
8. Dughmi, S., Xu, H.: Algorithmic Bayesian persuasion. In: Proceedings of the Forty-Eighth Annual ACM Symposium on Theory of Computing, pp. 412–425 (2016)
9. Gentzkow, M., Kamenica, E.: Bayesian persuasion with multiple senders and rich signal spaces. Games Econ. Behav. **104**, 411–429 (2017)
10. Kamenica, E., Gentzkow, M.: Bayesian persuasion. Am. Econ. Rev. **101**(6), 2590–2615 (2011)
11. Li, F., Norman, P.: Sequential persuasion. Theor. Econ. **16**(2), 639–675 (2021)
12. Ogawa, H.: Receiver's sensitivity and strategic information transmission in multi-sender cheap talk. Int. J. Game Theory **50**(1), 215–239 (2021)
13. Sobel, J.: Giving and receiving advice. Adv. Econ. Econom. **1**, 305–341 (2013)

An Improved Kernel and Parameterized Algorithm for Almost Induced Matching

Yuxi Liu and Mingyu Xiao[(✉)] [iD]

University of Electronic Science and Technology of China, Chengdu, China
202211081321@std.uestc.edu.cn, myxiao@uestc.edu.cn

Abstract. An induced subgraph is called an induced matching if each vertex is a degree-1 vertex in the subgraph. The ALMOST INDUCED MATCHING problem asks whether we can delete at most k vertices from the input graph such that the remaining graph is an induced matching. This paper studies parameterized algorithms for this problem by taking the size k of the deletion set as the parameter. First, we prove a $6k$-vertex kernel for this problem, improving the previous result of $7k$. Second, we give an $O^*(1.6765^k)$-time and polynomial-space algorithm, improving the previous running-time bound of $O^*(1.7485^k)$.

1 Introduction

An *induced matching* is an induced regular graph of degree 1. The problem of finding an induced matching of maximum size is known as MAXIMUM INDUCED MATCHING (MIM), which is crucial in algorithmic graph theory. In many graph classes such as trees [6], chordal graphs [2], circular-arc graphs [5], and interval graphs [6], a maximum induced matching can be found in polynomial time. However, MAXIMUM INDUCED MATCHING is NP-hard in planar 3-regular graphs or planar bipartite graphs with degree-2 vertices in one part and degree-3 vertices in the other part [4,10,18]. Kobler and Robotics [11] proved the NP-hardness of this problem in Hamiltonian graphs, claw-free graphs, chair-free graphs, line graphs, and regular graphs. The applications of induced matchings are diverse and include secure communication channels, VLSI design, and network flow problems, as demonstrated by Golumbic and Lewenstein [6].

In terms of exact algorithms, Gupta, Raman, and Saurabh [7] demonstrated that MAXIMUM INDUCED MATCHING can be solved in $O^*(1.6957^n)$ time. This result was later improved to $O^*(1.3752^n)$ by Xiao and Tan [21]. For subcubic graphs (i.e., such graphs with maximum degree 3), Hoi, Sabili and Stephan [8] showed that MAXIMUM INDUCED MATCHING can be solved in $O^*(1.2630^n)$ time and polynomial space. In terms of parameterized algorithms, where the parameter is the solution size k', MAXIMUM INDUCED MATCHING is W[1]-hard in general graphs [17], and is not expected to have a polynomial kernel. However, Moser and Sidkar [16] showed that the problem becomes fixed-parameter tractable (FPT) when the graph is a planar graph, by providing a linear-size problem kernel. The kernel size was improved to $40k'$ by Kanj et al. [9]. In this

© The Author(s), under exclusive license to Springer Nature Singapore Pte Ltd. 2024
X. Chen and B. Li (Eds.): TAMC 2024, LNCS 14637, pp. 86–98, 2024.
https://doi.org/10.1007/978-981-97-2340-9_8

paper, we study parameterized algorithms for MAXIMUM INDUCED MATCHING with the parameter being the number k of vertices not in the induced matching. The problem is formally defined as follows:

ALMOST INDUCED MATCHING

Instance: A graph $G = (V, E)$ and an integer k.

Question: Is there a vertex subset $S \subseteq V$ of size at most k whose deletion makes the graph an induced matching?

ALMOST INDUCED MATCHING becomes FPT when we take the size k of the deletion set as the parameter. Xiao and Kou [19,20] showed that ALMOST INDUCED MATCHING can be solved in $O^*(1.7485^k)$. The same running time bound was also achieved in [13] recently. In terms of kernelization, Moser and Thilikos [17] provided a kernel of $O(k^3)$ vertices. Then, Mathieson and Szeider [15] improved the result to $O(k^2)$. Last, Xiao and Kou [19,20] obtained the first linear-vertex kernel of $7k$ vertices for this problem. In this paper, we first improve the kernel size to $6k$ vertices, and then give an $O^*(1.6765^k)$-time and polynomial-space algorithm for ALMOST INDUCED MATCHING.

For kernelization, the main technique in this paper is a variant of the crown decomposition. We find a maximal 3-path packing $\mathcal{P}$ in the graph, partition the vertex set into two parts $V(\mathcal{P})$ and $Q = V \setminus V(\mathcal{P})$, and reduce the number of the size-1 connected components of $G[Q]$ to at most k. The size-2 components in $G[Q]$ can be reduced to at most k by using the new "AIM crown decomposition" technique. Note that each connected component of $G[Q]$ has a size of at most 2. The size of $\mathcal{P}$ is at most k since at least one vertex must be deleted from each 3-path. In the worst case scenario, when there are k 3-paths in $\mathcal{P}$, k size-2 components, and k size-1 components in $G[Q]$, the graph has at most $3k+2k+k = 6k$ vertices. Our parameterized algorithm is a branch-and-search algorithm. We first handle vertices with degrees of 1 and 2, followed by those with degree at least 5. Next, we only need to deal with degree-3 and degree-4 vertices. The different part from previous algorithms is as follows. We use refined rules to deal with degree-2, degree-3, and degree-4 vertices by carefully checking the local structures. Therefore, we can avoid previous bottlenecks.

The proofs of lemmas and theorems marked ♣ are omitted here due to space constraints, which can be found in the full version of this paper [14].

2 Preliminaries

In this paper, we only consider simple and undirected graphs. Let $G = (V, E)$ be a graph with $n = |V|$ vertices and $m = |E|$ edges. A singleton $\{v\}$ may be denoted as v. We use $V(G')$ and $E(G')$ to denote the vertex set and edge set of a graph G', respectively. A vertex v is called a *neighbor* of a vertex u if there is an edge $\{u, v\} \in E$. Let $N(v)$ denote the set of neighbors of v. For a vertex subset X, let $N(X) = \cup_{v \in X} N(v) \setminus X$ and $N[X] = N(X) \cup X$. We use $d(v) = |N(v)|$ to denote the degree of a vertex v in G. A vertex of degree d is called a degree-d vertex. For a vertex subset $X \subseteq V$, the subgraph induced by X is denoted by $G[X]$, and $G[V \setminus X]$ is also written as $G \setminus X$ or $G - X$. A vertex in a vertex subset X is called an X-vertex. Two vertex-disjoint subgraphs X_1 and X_2 are *adjacent* if there is an edge $\{u, v\} \in E$ with $u \in X_1$ and $v \in X_2$. A graph is called an *induced matching* if the size of each connected component in it is two. A vertex subset S is called an *AIM-deletion set* of G if $G \setminus S$ is an induced matching.

A 3-*path* $P_3 = \{u_1, u_2, u_3\}$ is a path with two edges $\{u_1, u_2\}$ and $\{u_2, u_3\}$. Two 3-paths L_1 and L_2 are *vertex-disjoint* if $V(L_1) \cap V(L_2) = \emptyset$. A set of 3-paths $\mathcal{P} = \{L_1, L_2, ..., L_t\}$ is called a P_3-*packing* if any two 3-paths in it are vertex-disjoint. A P_3-packing is *maximal* if there is no P_3-packing $\mathcal{P}'$ such that $|\mathcal{P}| < |\mathcal{P}'|$ and $\mathcal{P} \subset \mathcal{P}'$. We also use $V(\mathcal{P})$ to denote the set of vertices in 3-paths in a P_3-packing $\mathcal{P}$.

We will use a variant of the classic *VC crown decomposition*. Now, we give the definition of VC crown decomposition [1,3] for the ease of reference.

Definition 1. *A VC crown decomposition of a graph $G = (V, E)$ is a partition (C, H, R) of the vertex set V satisfying the following properties.*

1. *There is no edge between C and R.*
2. *C is an independent set.*
3. *There is an injective mapping (matching) $M : H \to C$ such that $\{x, M(x)\} \in E$ holds for all $x \in H$.*

Lemma 1 ([3]). *If graph $G = (V, E)$ has an independent set $I \subseteq V$ with $|I| > |N(I)|$, then a VC crown decomposition (C, H, R) with $\emptyset \neq C \subseteq I$ and $H \subseteq N(I)$ can be found in linear time.*

3   Kernelization

In this section, we show that ALMOST INDUCED MATCHING allows a kernel of $6k$ vertices. The main idea of our algorithm is as follows. The first step of this algorithm is to find a maximal P_3-packing $\mathcal{P}$ in $G = (V, E)$ by using a greedy method. For a **yes**-instance, we will have $|\mathcal{P}| \leq k$. We partition the vertex set V into two parts $P = V(\mathcal{P})$ and $Q = V \setminus P$. Each connected component in $G[Q]$ is of size at most 2 by the maximality of $\mathcal{P}$. We will bound the number of components in $G[Q]$ to bound the size of Q.

Let Q_0 denote the set of degree-0 vertices in $G[Q]$, and Q_1 denote the set of degree-1 vertices in $G[Q]$. Use Q_1-*edge* to denote the components with size 2 in $G[Q]$. For each $L_i \in \mathcal{P}$, let $Q(L_i)$ denote the set of Q-vertices in the components of $G[Q]$ adjacent to L_i. Let V_i denote $Q(L_i) \cup V(L_i)$. It should be noted that a vertex in $Q(L_i)$ might not be adjacent to any vertex in L_i.

A 3-path $L_i \in \mathcal{P}$ is *good* if at most one vertex in L_i is adjacent to Q-vertices. A 3-path $L_i \in \mathcal{P}$ is *bad* if at least two vertices in L_i are adjacent to Q-vertices. A maximal P_3-packing is *proper* if it holds that $|V_i| \leq 6$ for any bad 3-path L_i in it.

In our kernelization algorithm, we will first find an arbitrary maximal P_3-packing $\mathcal{P}$. After this, we will use two rules in [20] to update $\mathcal{P}$.

Rule 1. *If there is a 3-path $L_i \in \mathcal{P}$ such that $G[V_i]$ contains at least two vertex-disjoint 3-paths, then replace L_i by these 3-paths in $\mathcal{P}$ to increase the size of $\mathcal{P}$ by at least one.*

Lemma 2 ([20]). *Assume that Rule 1 can not be applied on the current instance. For any $L_i \in \mathcal{P}$ with $|V_i| \geq 7$, there is a 3-path L_i' in $G[V_i]$ such that L_i' is a good 3-path after replacing L_i with L_i' in $\mathcal{P}$. Furthermore, the 3-path L_i' can be found in constant time.*

We call the 3-path L_i' in Lemma 2 a *quasi-good 3-path*.

Rule 2. *For any 3-path $L_i \in \mathcal{P}$ with $|V_i| \geq 7$, if it is not good, then replace L_i with a quasi-good 3-path L_i' in $\mathcal{P}$.*

Lemma 3 ([20]). *For any initially maximal P_3-packing $\mathcal{P}$ in G, we can apply Rules 1 and 2 in $O(n^2)$ time to change $\mathcal{P}$ to a proper P_3-packing.*

This lemma implies that the number of Q-vertices in the components of $G[Q]$ adjacent to bad 3-paths in $\mathcal{P}$ is small. So we only need to bound the number of Q_0-vertices and Q_1-edges adjacent to good 3-paths in $\mathcal{P}$. For any proper P_3-packing $\mathcal{P}$ in G, we can bound the number of Q_0-vertices adjacent to good 3-paths by the following Lemma 4 from [20].

Lemma 4 ([20]). *Let $\mathcal{P}$ be a proper P_3-packing in graph G, where the number of bad 3-paths is x. If (G, k) is a **yes**-instance, then the number of Q_0-vertices only adjacent to good 3-paths (not adjacent to bad 3-paths) in $\mathcal{P}$ is at most $k - x$.*

The main contribution in this paper is to bound the number of Q_1-edges adjacent to good 3-paths. We need to use the following technique called *AIM crown decomposition*.

Definition 2 (AIM crown decomposition). *An AIM crown decomposition of a graph $G = (V, E)$ is a decomposition (C, H, R) of the vertex set V such that*

1. *there is no edge between C and R;*
2. *the induced subgraph $G[C]$ is an induced matching;*
3. *there is an injective mapping (matching from vertices to edges) $M : H \to E(C)$ such that for all $v \in H$, there exists $u \in M(v)$ such that $(u, v) \in E$.*

Figure 1 illustrates an AIM crown decomposition. We have the following Lemma 5 for AIM crown decomposition, which allows us to find parts of the solution based on an AIM crown decomposition.

Lemma 5 (♣). *Let (C, H, R) be an AIM crown decomposition of a graph G. There is a minimum AIM-deletion set S such that $H \subseteq S$.*

Once given an AIM crown decomposition (C, H, R) of the graph, we can reduce the instance by including H to the deletion set and removing $H \cup C$ from the graph by Lemma 5. Next, we show that when the number of Q_1-edges is large, we can always find an AIM crown decomposition in polynomial time.

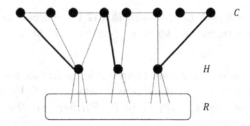

Fig. 1. An AIM crown decomposition (C, H, R), where the three bold edges form an injective matching from H to C.

Lemma 6 (♣). *Let $G = (V, E)$ be a graph with each connected component containing more than two vertices, and A and $B \subseteq V$ be two disjoint vertex sets such that*
(i) no vertex in A is adjacent to a vertex in $V \setminus (A \cup B)$;
(ii) the induced subgraph $G[A]$ is an induced matching.
If $|A| > 2|B|$, then the graph allows an AIM crown decomposition (C, H, R) with $\emptyset \neq C \subseteq A$ and $H \subseteq B$, and the AIM crown decomposition (C, H, R) can be found in polynomial time.

If the number of Q_1-edges only adjacent to vertices of good 3-paths in $\mathcal{P}$ is large, we use Lemma 5 and Lemma 6 to reduce the instance. Our algorithm first finds two vertex-disjoint sets of vertices, A and B, which satisfy the condition in Lemma 6 based on a proper P_3-packing $\mathcal{P}$. Let A be the set of Q_1-vertices that are only adjacent to good 3-paths in $\mathcal{P}$. Let B be the set of vertices in good 3-paths that are adjacent to some vertices in A. If $|A| > 2|B|$, we can find an AIM crown decomposition (C, H, R) by Lemma 6, and we can reduce the instance by including C to the deletion set and removing $C \cup H$ from the graph.

Our algorithm, denoted by $\texttt{Reduce}(G, k)$, is described in Fig. 2.

Lemma 7. *Algorithm $\texttt{Reduce}(G, k)$ runs in polynomial time, and returns either an equivalent instance (G', k') with $|V(G')| \leq 6k'$ and $k' \leq k$ or **no** to indicate that the instance is a **no**-instance.*

Proof. First, let us consider the correctness of each step. Steps 1–6 are trivial cases. Step 7 is based on Lemma 4 and Step 9 is based on Lemma 5. Step 8 is also trivial. Next, we consider Step 10. In this step, $\mathcal{P}$ is a proper P_3-packing and then the number of vertices in P is at most $3k$. Assume that the number of bad 3-paths in $\mathcal{P}$ is x and the number of good 3-paths in $\mathcal{P}$ is y. By the definition of proper P_3-packing, we know that the number of vertices in bad 3-paths and in Q-components adjacent to some bad 3-paths is at most $6x$. The number of vertices in good 3-paths is $3y$, the number of Q_0-vertices only adjacent to good 3-paths is at most $k - x$ by Lemma 4, and the number of Q_1-vertices only adjacent to good 3-paths is at most $2y$ by Lemma 5 and Lemma 6. In total, the number of vertices in the graph is at most $6x + 3y + (k - x) + 2y = k + 5x + 5y \leq 6k$. Thus, we can get that $|V| \leq 6k$ in Step 10.

Input: An undirected graph $G = (V, E)$ and an integer k.
Output: An equivalent instance (G', k') with $|V(G')| \leq 6k'$ and $k' \leq k$ or **no**.

1. **If** there is a connected component C of two vertices, return $\texttt{Reduce}(G \setminus C, k)$.
2. **If** there is a connected component C of one vertice, return $\texttt{Reduce}(G \setminus C, k - 1)$.
3. Find an arbitrary maximal P_3-packing $\mathcal{P}$ in G by using greedy algorithm.
4. Iteratively apply Rules 1 and 2 to update $\mathcal{P}$ until none of them can be applied anymore.
5. **If** $|\mathcal{P}| > k$, **then** return 'no'.
6. Let $P = V(\mathcal{P})$ and $Q = V \setminus P$.
7. **If** the number of Q_0-vertices only adjacent to good 3-paths is more than $k - x$, where x is the number of bad 3-paths in $\mathcal{P}$, **then** return 'no'.
8. Let A be the set of Q_1-vertices that are only adjacent to good 3-paths in $\mathcal{P}$. Let B be the set of vertices in good 3-paths that are adjacent to some vertices in A.
9. **If** $|A| > 2|B|$, find an AIM crown decomposition (C, H, R) by Lemma 6, return $\texttt{Reduce}(G \setminus (C \cup H), k - |H|)$.
10. return (G, k).

Fig. 2. Algorithm $\texttt{Reduce}(G, k)$

Each step in $\texttt{Reduce}(G, k)$ runs in polynomial time. Since each recursive call of $\texttt{Reduce}(G, k)$ decreases $|V|$ by at least 1, $\texttt{Reduce}(G, k)$ will be called at most $|V|$ times. Thus, $\texttt{Reduce}(G, k)$ runs in polynomial time.   □

The following Theorem 1 directly follows from Lemma 7.

Theorem 1. ALMOST INDUCED MATCHING *admits a kernel with* $6k$ *vertices.*

4   A Parameterized Algorithm

In this section, we design a parameterized algorithm for ALMOST INDUCED MATCHING, which is a branch-and-search algorithm. We use a parameter k to measure the instance, and $T(k)$ to denote the maximum size of the search tree generated by the algorithm when running on an instance with the parameter no greater than k. Assume that a branching operation generates l branches and the measure k in the i-th instance decreases by at least c_i. This operation generates a recurrence relation

$$T(k) \leq T(k - c_1) + T(k - c_2) + \dots + T(k - c_l) + 1.$$

The largest root of the function $f(x) = 1 - \sum_{i=1}^{l} x^{-c_i}$ is called the *branching factor* of the recurrence. Let γ denote the maximum branching factor among all branching factors in the search tree. The running time of the algorithm is bounded by $O^*(\gamma^k)$ [12].

4.1 Branching Rules

We have several branching rules that will be applied in different steps.

Branching-Rule (B1). *Branch on v to generate $|N[v]|$ branches by either (i) deleting v from the graph and including it in the deletion set, or (ii) for each neighbor u of v, deleting $N[\{u, v\}]$ from the graph and including $N(\{u, v\})$ in the deletion set.*

When dealing with certain graph structures, we can use a more effective branching rule. A vertex v *dominates* its neighbor u if $N[u] \subseteq N[v]$. A vertex v is called a *dominating* vertex if it dominates at least one vertex. The following property of dominating vertices has been used in [7, 21].

Lemma 8. *Let v be a vertex that dominates a vertex u. If there is a maximum induced matching M of G such that $v \in V(M)$, then there is a maximum induced matching M' of G such that edge $vu \in M'$.*

We can use Lemma 8 to design an effective branching rule.

Branching-Rule (B2). *Assume that vertex v dominates vertex u. Branch on v to generate two instances by either (i) deleting v from the graph and including it in the deletion set, or (ii) deleting $N[\{u, v\}]$ from the graph and including $N(\{u, v\})$ in the deletion set.*

If there is a degree-2 vertex v in a triangle, a more effective branching rule can be applied. The following lemma has been used in [21].

Lemma 9. *If there is a degree-2 vertex v in a triangle vu_1u_2, then there is a maximum induced matching either containing one edge in $\{vu_1, vu_2\}$ or containing no edge incident on a vertex in $\{v, u_1, u_2\}$. Especially, if at least one vertex in $\{u_1, u_2\}$ is of degree at least 3, then there is a maximum induced matching containing one edge in $\{vu_1, vu_2\}$.*

We can use Lemma 9 to design an effective branching rule to deal with degree-2 vertices in triangles.

Branching-Rule (B3). *Branch on a degree-2 vertex v in a triangle with two neighbors u_1 and u_2 as follows*
(i) if $d(u_1) \leq 3$ or $d(u_2) \leq 3$, then generate two instances by either (a) deleting $N[\{v, u_1\}]$ from the graph and including $N(\{v, u_1\})$ in the deletion set, or (b) deleting $N[\{v, u_2\}]$ from the graph and including $N(\{v, u_2\})$ in the deletion set;
(ii) if $d(u_1) \geq 4$ and $d(u_2) \geq 4$, then generate three instances by either (a) deleting $\{v, u_1, u_2\}$ from the graph and including them in the deletion set, (b) deleting $N[\{v, u_1\}]$ from the graph and including $N(\{v, u_1\})$ in the deletion set, or (c) deleting $N[\{v, u_2\}]$ from the graph and including $N(\{v, u_2\})$ in the deletion set.

4.2 The Algorithm

We will use $\mathtt{aim}(G, k)$ to denote our parameterized algorithm. Before executing the main branching steps, the algorithm will first apply some reduction rules to simplify the instance. First of all, we call the kernel algorithm $\mathtt{Reduce}(G, k)$ to reduce the instance, which can be considered as Reduction-Rule 0. We also have three more reduction rules.

Reduction-Rule 1. *If there is a connected component of the graph such that each vertex in it is a degree-2 vertex, then select an arbitrary vertex v in this component and return* $\mathtt{aim}(G \setminus \{v\}, k - 1)$.

Reduction-Rule 2. *If there is a degree-1 vertex v with a degree-2 neighbor u, then return* $\mathtt{aim}(G \setminus N[\{v, u\}], k - 1)$.

Reduction-Rule 1 is trivial. Reduction-Rule 2's correctness is based on the observation: there is always a maximum induced matching containing edge $\{v, u\}$.

A cycle $u_0 u_1 u_2 u_3$ of four vertices is called a *short cycle* if the two vertices u_0 and u_3 are of degree at least 2 and the two vertices u_1 and u_2 are of degree 2.

Lemma 10 (♣). *If a graph G has a short cycle $u_0 u_1 u_2 u_3$, then there is a maximum induced matching of G containing the edge $u_1 u_2$.*

Reduction-Rule 3. *If there is a short cycle $u_0 u_1 u_2 u_3$, then return* $\mathtt{aim}(G \setminus \{u_0, u_1, u_2, u_3\}, k - 2)$.

Now, we are ready to introduce the main branching steps of the algorithm. The algorithm contains nine steps to handle different local structures of the graph. Dominating vertices are processed in Step 1, while Steps 2 to 5 are dedicated to handling degree-2 vertices. Vertices with at least five neighbors are handled in Step 6, and the last three steps focus on graphs with only degree-3/4 vertices. When executing a step, we assume that all previous steps are not applicable to the current graph. We mention that the analysis of the first six steps are omitted here due to space constraints, which can be found in the full version of this paper.

Step 1 (Dominating vertices of degree at least 3). If there is a vertex v of degree at least 3 that dominates a vertex u, then branch on v with Rule (B2).

After Reduction-Rule 2, degree-1 vertices can only be adjacent to vertices of degree at least 3. These vertices will be handled in Step 1. So after Step 1, there are no degree-1 vertices in G.

Next, we consider degree-2 vertices. A path $u_0 u_1 u_2 u_3 u_4$ of five vertices is called a *chain* if the first vertex u_0 is of degree at least 3 and the three middle vertices are of degree 2, where we allow $u_4 = u_0$. A path $u_0 u_1 u_2 u_3$ of four vertices is called a *short chain* if the first vertex u_0 and last vertex u_3 are of degree at least 3 and the two middle vertices are of degree 2, where we allow $u_3 = u_0$.

A chain or a short chain can be found in linear time if it exists. A short chain $u_0u_1u_2u_3$ is called a *good short chain* if $u_3 \neq u_0$ and there is no edge between u_3 and u_0.

Lemma 11 (♣). *If a graph G has a good short chain $u_0u_1u_2u_3$, then there is a maximum induced matching of G containing at least one vertex of u_1 and u_2.*

Step 2 (Chains). If there is a chain $u_0u_1u_2u_3u_4$, then branch on u_1 with Rule (B1). In the branch where u_1 is deleted and included in the deletion set, we get a tail u_2 and then further deal with the tail as we do in Reduction-Rule 2.

After Step 1, there is no short chain $u_0u_1u_2u_3$ with $u_0 = u_3$, since for any short chain $u_0u_1u_2u_3$ with $u_0 = u_3$, vertex u_0 is a dominating vertex, and Step 1 would be applied. After Reduction-Rule 3, no short chain $u_0u_1u_2u_3$ exists with an edge u_0u_3 in G. Therefore, we claim that every short chain in G is a good short chain after Step 1.

Step 3 (Short chains). If there is a short chain $u_0u_1u_2u_3$, then branch on u_1 with Rule (B1). Additionally, by Lemma 11, in the branch of deleting u_1, we can delete $N[\{u_2, u_3\}]$ from the graph and include $N(\{u_2, u_3\})$ in the deletion set.

After Step 3, each degree-2 vertex in the graph has two neighbors of degree at least 3. Since there is no dominating vertex after Step 1, the two neighbors of any degree-2 vertex are not adjacent to each other.

Step 4 (Degree-2 vertices adjacent to a vertex of degree at least 4). If there is a degree-2 vertex v adjacent to a vertex u_1 of degree at least 4, then branch on v with Rule (B1).

After Step 4, if there is a degree-2 vertex v adjacent to two vertices u_1 and u_2, then $d(u_1) \geq 3$ and $d(u_2) \geq 3$. Vertex u_1 is not adjacent to u_2 since there is no dominated vertex. Let $N(u_1) = \{v, w_1, w_2\}$ and $N(u_2) = \{v, w_3, w_4\}$, where it is possible that $\{w_1, w_2\} \cap \{w_3, w_4\} \neq \emptyset$. In Step 5, we are going to deal with such degree vertices v.

Step 5 (Degree-2 vertices with two nonadjacent degree-3 neighbors). We deal with such degree-2 vertices v by considering two different cases.

Case 1. $|\{w_1, w_2\} \cap \{w_3, w_4\}| \geq 1$: We branch on w_1 with Rule (B1) to generate $d(w_1) + 1$ branches. For this case, we assume without loss of generality $w_2 = w_4$. In the branch where w_1 is deleted and included in the deletion set, we get a short cycle $w_2u_1vu_2$ and then further deal with the short cycle as we do in Reduction-Rule 3.

Case 2. $|\{w_1, w_2\} \cap \{w_3, w_4\}| = 0$: Firstly, if $N(\{v, u_1, u_2, w_1, w_2, w_3, w_4\}) = \emptyset$, we delete $\{v, u_1, u_2, w_1, w_2, w_3, w_4\}$ from the graph and include $\{v, w_2, w_4\}$ in the deletion set. The correctness of this operation can be easily verified. Next, we assume that $N(\{v, u_1, u_2, w_1, w_2, w_3, w_4\}) \neq \emptyset$. Without loss of generality, let x^* be a vertex adjacent to w_1.

Case (a). Both of u_1 and u_2 are deleted and included in the deletion set: We delete $\{v, u_1, u_2\}$ from the graph and include them in the deletion set.

Case (b). Only one of u_1 and u_2 is deleted and included in the deletion set: We use u_i ($i \in \{1, 2\}$) to denote the other vertex: We can see that there is a maximum induced matching containing edge $u_i v$. Then we delete $N[\{u_i, v\}]$ from the graph and including $N(\{u_i, v\})$ in the deletion set.

Case (c). Neither u_1 nor u_2 is included in the deletion set: Since u_1 and u_2 are not adjacent, they must be in two different edges in the maximum induced matching M. We first generate two branches by deleting $N[\{u_1, w_i\}]$ for $i = 1$ and 2 and including $N(\{u_1, w_i\})$ in the deletion set. The resulting graph is denoted as G_i. In G_i, since u_2 appears in M, vertex u_2 is not deleted or becomes a degree-0 vertex. If u_2 has only one neighbor v^* or dominates one of its neighbors v^*, we further delete $N[\{u_2, v^*\}]$ and include $N(\{u_2, v^*\})$ in the deletion set. Otherwise we further branch into two branches by deleting $N[\{u_2, w_j\}]$ from the graph and including $N(\{u_2, w_j\})$ in the deletion set for $j = 3$ and 4.

Step 6 (Vertices of degree at least 5). If there is a vertex v of $d(v) \geq 5$, then branch on v with Rule (B1).

Lemma 12 (♣). *If one of Steps 1 to 6 is applied, then we can get a branching factor not greater than 1.6765.*

After the first six steps, the graph contains only vertices with degree 3 and 4, and there is no dominating vertex. The next three steps are used to deal with degree-3/4 vertices in the graph. For any degree-3 vertex v, there is at most one edge between its neighbors, otherwise there must be a dominating vertex in the neighbors of v, and Step 1 would be applied. We first consider a degree-3 vertex v not contained in any triangle.

Step 7 (Degree-3 vertices not in any triangle). If there is a degree-3 vertex v such that there is no edge between its neighbors, then branch on v with Rule (B1) to generate $d(v) + 1$ branches

$$\mathtt{aim}(G \setminus \{v\}, k-1) \quad \text{and} \quad \mathtt{aim}(G \setminus N[\{v, u\}], k - |N(\{v, u\})|) \text{ for each } u \in N(v).$$

Let u_1, u_2 and u_3 denote the three neighbors of v. We have that u_1, u_2, u_3 are nonadjacent vertices of degree at least 3. The branching operation will generate three branches. Since u_1, u_2 and u_3 are not adjacent, we can see that $|N(\{v, u_i\})| = d(u_i) + d(v) - 2 = d(u_i) + 1 \geq 4 (i = 1, 2, 3)$. This leads to a recurrence

$$T(k) \leq T(k-1) + T(k - (d(u_1)+1)) + T(k - (d(u_2)+1)) + T(k - (d(u_3)+1)) + 1,$$

where $d(u_1), d(u_2), d(u_3) \geq 3$. For the worst case that $d(u_1) = d(u_2) = d(u_3) = 3$, the branching factor is 1.6581.

Next we consider degree-3 vertices in triangles that are also adjacent to some degree-4 vertex.

Table 1. The branching factors of each of the first eight steps

Steps	Step 1	Step 2	Step 3	Step 4	Step 5	Step 6	Step 7	Step 8
branching factors	1.6181	1.6181	1.5214	1.6181	**1.6765**	1.6595	1.6581	1.6430

Step 8 (Degree-3 vertices adjacent to some degree-4 vertex). Assume there is a degree-3 vertex v adjacent to at least one degree-4 vertex. After Step 7, each degree-3 vertex is in exactly one triangle since such degree-3 vertex can not be dominated. Let u_1, u_2 and u_3 be the three neighbors of v, where we assume without loss of generality that there is an edge between u_1 and u_2. First, we branch on u_3 with Rule (B1). In the branch of deleting u_3, we further branch on the vertex v with Rule (B3).

Note that the degree-4 neighbor of v can be any one of u_1, u_2 and u_3.

Lemma 13 (♣). *The branching factor of Step 8 is at most 1.6430.*

The worst branching factors in the above eight steps are listed in Table 1. After Step 8, there are no degree-3 vertices adjacent to degree-4 vertices, and each connected component is either 3-regular or 4-regular.

Step 9 (3/4-regular graphs). Pick up an arbitrary vertex v in the 3/4-regular graph and branch on it with Rule (B1).

Step 9 will not exponentially increase the running time bound of the algorithm. We can prove the following theorem.

Theorem 2 (♣). ALMOST INDUCED MATCHING *can be solved in $O^*(1.6765^k)$ time and polynomial space.*

5   Conclusion

In this paper, we study ALMOST INDUCED MATCHING from the prespective of parameterized algorithms, where the parameter k represents the size of the deletion set.

In the context of kernelization, we introduce an enhanced structure called AIM crown decomposition, which effectively yields a $6k$-vertex kernel. For further improvements, we may need to explore new structural properties and employ different techniques. Note that the number of vertices in the kernel is already small. It would also be interesting to achieve some nontrivial lower bounds for the kernel size.

In our parameterized algorithm, by using new methods to deal with degree-3 vertices adjacent to degree-4 vertices in the graph, we successfully avoid the bottlenecks in previous papers. Table 2 shows the new bottleneck case in our algorithm generated by Step 5, which is to deal degree-2 vertices adjacent to two degree-3 vertices without an edge between them.

Acknowledgements. This work was supported by the National Natural Science Foundation of China (Grant No. 62372095 and 62172077) and the Sichuan Natural Science Foundation (Grant No. 2023NSFSC0059).

References

1. Abu-Khzam, F.N., Collins, R.L., Fellows, M.R., Langston, M.A., Suters, W.H., Symons, C.T.: Kernelization algorithms for the vertex cover problem: theory and experiments. In: Arge, L., Italiano, G.F., Sedgewick, R. (eds.) Proceedings of the Sixth Workshop on Algorithm Engineering and Experiments and the First Workshop on Analytic Algorithmics and Combinatorics, New Orleans, LA, USA, January 10, 2004, pp. 62–69. SIAM (2004)
2. Cameron, K.: Induced matchings. Discret. Appl. Math. **24**(1–3), 97–102 (1989)
3. Chor, B., Fellows, M., Juedes, D.: Linear kernels in linear time, or how to save k colors in $o(n^2)$ steps. In: Hromkovic, J., Nagl, M., Westfechtel, B. (eds.) Graph-Theoretic Concepts in Computer Science. Lecture Notes in Computer Science, vol. 3353, pp. 257–269. Springer, Berlin (2004). https://doi.org/10.1007/978-3-540-30559-0_22
4. Duckworth, W., Manlove, D.F., Zito, M.: On the approximability of the maximum induced matching problem. J. Discret. Algorithms **3**(1), 79–91 (2005)
5. Golumbic, M.C., Laskar, R.C.: Irredundancy in circular arc graphs. Discret. Appl. Math. **44**(1–3), 79–89 (1993)
6. Golumbic, M.C., Lewenstein, M.: New results on induced matchings. Discret. Appl. Math. **101**(1–3), 157–165 (2000)
7. Gupta, S., Raman, V., Saurabh, S.: Maximum r-regular induced subgraph problem: fast exponential algorithms and combinatorial bounds. SIAM J. Discret. Math. **26**(4), 1758–1780 (2012)
8. Hoi, G., Sabili, A.F., Stephan, F.: An exact algorithm for finding maximum induced matching in subcubic graphs (2022). arXiv:2201.03220
9. Kanj, I., Pelsmajer, M.J., Schaefer, M., Xia, G.: On the induced matching problem. J. Comput. Syst. Sci. **77**(6), 1058–1070 (2011)
10. Ko, C., Shepherd, F.B.: Bipartite domination and simultaneous matroid covers. SIAM J. Discret. Math. **16**(4), 517–523 (2003)
11. Kobler, D., Rotics, U.: Finding maximum induced matchings in subclasses of claw-free and P 5-free graphs, and in graphs with matching and induced matching of equal maximum size. Algorithmica **37**(4), 327–346 (2003)
12. Kratsch, D., Fomin, F.: Exact Exponential Algorithms. Springer, Cham (2010)
13. Kumar, A., Kumar, M.: Deletion to induced matching (2020). arXiv:2008.09660
14. Liu, Y., Xiao, M.: An improved kernel and parameterized algorithm for almost induced matching. arXiv preprint: arXiv:2308.14116 (2023)
15. Mathieson, L., Szeider, S.: Editing graphs to satisfy degree constraints: a parameterized approach. J. Comput. Syst. Sci. **78**(1), 179–191 (2012)
16. Moser, H., Sikdar, S.: The parameterized complexity of the induced matching problem. Discret. Appl. Math. **157**(4), 715–727 (2009)
17. Moser, H., Thilikos, D.M.: Parameterized complexity of finding regular induced subgraphs. J. Discret. Algorithms **7**(2), 181–190 (2009)
18. Stockmeyer, L.J., Vazirani, V.V.: NP-completeness of some generalizations of the maximum matching problem. Inf. Process. Lett. **15**(1), 14–19 (1982)

19. Xiao, M., Kou, S.: Almost induced matching: linear kernels and parameterized algorithms. In: Heggernes, P. (ed.) Graph-Theoretic Concepts in Computer Science. Lecture Notes in Computer Science(), vol. 9941, pp. 220–232. Springer, Berlin (2016). https://doi.org/10.1007/978-3-662-53536-3_19
20. Xiao, M., Kou, S.: Parameterized algorithms and kernels for almost induced matching. Theoret. Comput. Sci. **846**, 103–113 (2020)
21. Xiao, M., Tan, H.: Exact algorithms for maximum induced matching. Inf. Comput. **256**, 196–211 (2017)

A Tight Threshold Bound for Search Trees with 2-Way Comparisons

Sunny Atalig and Marek Chrobak[(✉)]

University of California at Riverside, Riverside, USA
marek@cs.ucr.edu

Abstract. We study search trees with 2-way comparisons (2WCST's), which involve separate less-than and equal-to tests in their nodes, each test having two possible outcomes, yes and no. These trees have a much subtler structure than standard search trees with 3-way comparisons (3WCST's) and are still not well understood, hampering progress towards designing an efficient algorithm for computing minimum-cost trees. One question that attracted attention in the past is whether there is an easy way to determine which type of comparison should be applied at any step of the search. Anderson, Kannan, Karloff and Ladner studied this in terms of the ratio between the maximum and total key weight, and defined two threshold values: λ^- is the largest ratio that forces the less-than test, and λ^+ is the smallest ratio that forces the equal-to test. They determined that $\lambda^- = \frac{1}{4}$, but for the higher threshold they only showed that $\lambda^+ \in [\frac{3}{7}, \frac{4}{9}]$. We give the tight bound for the higher threshold, by proving that in fact $\lambda^+ = \frac{3}{7}$.

1 Introduction

Search trees are decision-tree based data structures used for identifying a query value within some specified set $\mathcal{K}$ of keys, by applying some simple tests on the query. When $\mathcal{K}$ is a linearly ordered set, these tests can be comparisons between the query and a key from $\mathcal{K}$. In the classical model of 3-way comparison trees (3WCST's), each comparison has three outcomes: the query can be smaller, equal, or greater than the key associated with this node. In the less studied model of search trees with 2-way comparisons (2WCST's), proposed by Knuth [7, §6.2.2, Example 33], we have separate less-than and equal-to tests, with each test having two outcomes, yes or no. In both models, the search starts at the root node of the tree and proceeds down the tree, following the branches corresponding to these outcomes, until the query ends up in the leaf representing the key equal to the query value[1].

---

[1] Here we assume the scenario when the query is in $\mathcal{K}$, often called the *successful-query* model. If arbitrary queries are allowed, the tree also needs to have leaves representing inter-key intervals. Algorithms for the successful-query model typically extend naturally to this general mode without increasing their running time.

[2] There is of course vast amount of research on the dynamic case, when the goal is to have the tree to adapt to the input sequence, but it is not relevant to this paper.

Research supported by NSF grant CCF-2153723.

X. Chen and B. Li (Eds.): TAMC 2024, LNCS 14637, pp. 99–110, 2024.
https://doi.org/10.1007/978-981-97-2340-9_9

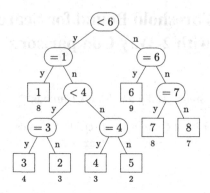

Fig. 1. An example of a 2WCST. This tree handles keys $1, 2, 3, 4, 5, 6, 7, 8$ with respective weights $8, 3, 4, 3, 2, 9, 8, 7$. Computing the cost in terms of leaves, we get $8 \cdot 2 + 3 \cdot 4 + 4 \cdot 4 + 3 \cdot 4 + 2 \cdot 4 + 9 \cdot 2 + 8 \cdot 3 + 7 \cdot 3 = 127$.

The focus of this paper is on the static scenario, when the key set $\mathcal{K}$ does not change over time[2]. Each key k is assigned a non-negative weight w_k, representing the frequency of k, or the probability of it appearing on input. Given these weights, the objective is to compute a tree T that minimizes its cost, defined by $\text{cost}(T) = \sum_{k \in \mathcal{K}} w_k \cdot \text{depth}(k)$, where $\text{depth}(k)$ is the depth of the leaf representing key k in T. This concept naturally captures the expected cost of a random query (Fig. 1).

Using the now standard dynamic programming algorithm, optimal 3WCST's can be computed in time $O(n^3)$, where $n = |\mathcal{K}|$ denotes the number of keys. As shown already by Knuth [6] in 1971 and later by Yao [10] in 1980, using a more general approach, the running time can be improved to $O(n^2)$. This improvement leverages the property of 3WCST's called the quadrangle inequality (which is essentially equivalent to the so-called Monge property or total monotonicity—see [3]). In contrast, the first (correct[3]) polynomial-time algorithm for finding optimal 2WCST's was developed only in 2002 by Anderson, Kannan, Karloff and Ladner [1]. Its running time is $O(n^4)$. A simpler and slightly more general $O(n^4)$-time algorithm was recently given in [4].

The reason for this disparity in the running times lies in the internal structure of 2WCST's, which is much more intricate than that of 3WCST's. Roughly, while the optimal cost function of 3WCST's has a dynamic-programming formulation where all sub-problems are represented by key intervals, this is not the case for 2WCST's. For 2WCST's, a similar approach produces intervals with holes (corresponding to earlier failed equal-to tests), leading to exponentially many sub-problems. As shown in [1] (and conjectured earlier by Spuler [8,9]), this can be reduced to $O(n^3)$ sub-problems using the so-called heaviest-first key property.

One other challenge in designing a faster algorithm is that, for any sub-problem, it is not known a priori whether the root should use the less-than test or the equal-to test. The intuition is that when some key is sufficiently heavy then the optimum tree must start with the equal-to test (to this key) at the root. On the other hand, if all weights are

---

[3] Anderson et al. [1] reference an earlier $O(n^5)$-time algorithm by Spuler [8,9]. However, as shown in [5], Spuler's proof is not correct.

roughly equal (and there are sufficiently many keys), then the tree should start with a less-than test, to break the instance into two parts with roughly the same total weight. Addressing this, Anderson et al. [1] introduced two threshold values for the maximum key weight. Denoting by W the total weight of the keys in the instance, these values are defined as follows:

- λ^- is the largest λ such that if all key weights are smaller than λW then there is no optimal tree with an equal-to test at the root.
- λ^+ is the smallest λ such that if any key has weight at least λW then there is an optimal tree with an equal-to test at the root.

In their paper[4], they proved that $\lambda^- = \frac{1}{4}$ and $\lambda^+ \in [\frac{3}{7}, \frac{4}{9}]$. These thresholds played a role in their $O(n^4)$-time algorithm for computing optimal 2WCST's. The more recent $O(n^4)$-time algorithm in [4] uses a somewhat different approach and does not rely on any threshold bounds on key weights.

Nevertheless, breaking the $O(n^4)$ barrier will require deeper understanding of the structure of optimal 2WCST's; in particular, more accurate criteria for determining which of the two tests should be applied first are likely to be useful. Even if not improving the asymptotic complexity, such criteria reduce computational overhead by limiting the number of keys to be considered for less-than tests.

With these motivations in mind, in this work we give a tight bound on the higher weight threshold, by proving that the lower bound of $\frac{3}{7}$ for λ^+ in [1] is in fact tight.

Theorem 1. *For all $n \geq 2$, if an instance of n keys has total weight W and a maximum key weight at least $\frac{3}{7}W$ then there exists an optimal 2WCST rooted at an equal-to test. In other words, $\lambda^+ \leq \frac{3}{7}$.*

The proof is given in Sect. 3, after we introduce the necessary definitions and notation, and review fundamental properties of 2WCST's, in Sect. 2.

Note: For interested readers, other structural properties of 2WCST's were recently studied in the companion paper [2]. In particular, that paper provides other types of threshold bounds, including one that involves two heaviest keys, as well as examples showing that the speed-up techniques for dynamic programming, including the quadrangle inequality, do not work for 2WCST's.

2 Preliminaries

Notation. Without any loss of generality we can assume that set of keys is $\mathcal{K} = \{1, 2, ..., n\}$, and their corresponding weights are denoted $w_1, w_2, \ldots, w_n$. Throughout the paper, we will typically use letters $i, j, k, \ldots$ to represent keys. The total weight of the instance is denoted by $W = \sum_{k \in \mathcal{K}} w_k$. For a tree T by $w(T)$ we denote the total weight of keys in its leaves, calling it the *weight of T*. For a node v in T, $w(v)$ denotes the weight of the sub-tree of T rooted at v.

---

[4] In [1] the notation for the threshold values λ^- and λ^+ was, respectively, λ and μ. We changed the notation to make it more intuitive and consistent with [2].

Each internal node is either an equal-to test to a key k, denoted by $\langle = k \rangle$, or a less-than test to k, denoted by $\langle < k \rangle$. Conventionally, the left and right branches of the tree at a node are labelled by "yes" and "no" outcomes, but in our proof we will often depart from this notation and use relation symbols "=", "$\neq$", "<", etc., instead. For some nodes only the comparison key k will be specified, but not the test type. Such nodes will be denoted $\langle * k \rangle$, and their outcome branches will be labeled "$=/\geq$" and "$\neq/<$". The interpretation of these is natural: If $\langle * k \rangle$ is $\langle = k \rangle$ then the first branch is taken on they "yes" answer, otherwise the second branch is taken. If $\langle * k \rangle$ is $\langle < k \rangle$ then the second branch is taken on the "yes" answer and otherwise the first branch is taken. This convention will be very useful in reducing the case complexity.

A branch of a node is called *redundant* if there is no query value $q \in \mathcal{K}$ that traverses this branch during search. A 2WCST T is called *irreducible* if it does not contain any redundant branches. Each tree can be converted into an irreducible tree by "splicing out" redundant branches (linking the sibling branch directly to the grandparent). This does not increase cost. So throughout the paper we tacitly assume that any given tree is irreducible. In particular, note that any key k appearing in a node $\langle * k \rangle$ of an irreducible tree must satisfy all outcomes of the tests on the path from the root to $\langle * k \rangle$. (The only non-trivial case is when $\langle * k \rangle$ is $\langle < k \rangle$ and there is a node $\langle = k \rangle$ along this path. In this case, $\langle < k \rangle$ can be replaced by $\langle < k + 1 \rangle$, as in this case k cannot be n if T is irreducible.)

Side Weights. We use some concepts and auxiliary lemmas developed by Anderson et al. [1]. In particular, the concept of side-weights is useful. The *side-weight* of a node v in a 2WCST T is defined by

$$sw(v) = \begin{cases} 0 & \text{if } v \text{ is a leaf} \\ w_k & \text{if } v \text{ is an equal-to test } \langle = k \rangle \\ \min\{w(L), w(R)\} & \text{if } v \text{ is a less-than test with sub-trees } L, R \end{cases}$$

Lemma 1. [1] *Let T be an optimal 2WCST. Then $sw(u) \geq sw(v)$ if u is a parent of v.*

The lemma, while far from obvious, can be proved by applying so-called "rotations" to the tree, which are local rearrangements that swap some adjacent nodes. As a simple example, if $u = \langle = k \rangle$ is a child of $v = \langle = l \rangle$ and $sw(u) > sw(v)$ then exchanging these two comparisons would reduce the tree cost, contradicting optimality. For the complete proof, see [1].

Lemma 1 implies immediately the following:

Lemma 2. [1] *For $n > 2$, if an optimal 2WCST for an instance of n keys is rooted at an equal-to test on i, then i is a key of maximum-weight.*

We remark that in case of ties between maximum-weight keys a subtlety arises that led to some complications in the algorithm in [1]. It was shown later in [4] that this issue can be circumvented. This issue does not arise in our paper, and the above statement of Lemma 2 is sufficient for our argument.

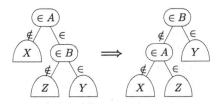

Fig. 2. A tree rotation using general queries.

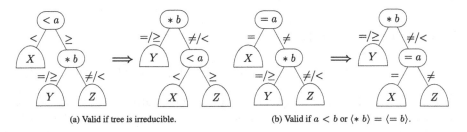

(a) Valid if tree is irreducible.   (b) Valid if $a < b$ or $\langle * \, b \rangle = \langle = b \rangle$.

Fig. 3. Rotating a node with double outcomes.

More About Rotations. We will extend the concept of rotations to involve nodes of the form $\langle * \, k \rangle$, with unspecified tests. This makes the rotations somewhat non-trivial, since such nodes represent two different possible subtrees, and we need to justify that the tree obtained from the rotation is correct in both cases.

To give a simple condition for ensuring valid rotations, we generalize 2WCSTs by considering arbitrary types of tests, not merely equal-to or less-than tests. Any binary test can be identified by the set of keys that satisfy the "yes" outcome. We can thus represent such a test using the set element relation, denoted $\langle \in A \rangle$, with branches labelled "$\in$" and "$\notin$". An equal-to test $\langle = k \rangle$ can be identified by the singleton set $\{k\}$ or its complement $\mathcal{K} - \{k\}$, and a less-than test $\langle < k \rangle$ can be identified by $(-\infty, k)$ or $[k, \infty)$.

Consider the rotation in Fig. 2, where we denote the sub-trees and the keys they contain by X, Y, and Z. Comparing the trees on the left and the right, we must have $Y = A \cap B = B$, which implies that for the rotation to be correct we need $B \subseteq A$. (If the tree on the left is irreducible, the containment must in fact be strict.) On the other hand, if $B \subseteq A$ then in both trees we have $Z = A \setminus B$ and $X = \mathcal{K} - A$. This gives us the following property:

Containment Property: The rotation shown in Fig. 2 is valid if and only if $B \subseteq A$.

We remark that other rotations, such as when $\langle \in B \rangle$ is in the $\notin$-branch of $\langle \in A \rangle$, can be accounted for by the above containment property, by replacing $\langle \in A \rangle$ with $\langle \in \bar{A} \rangle$, where $\bar{A}$ is the complement of A.

Consider Fig. 3a, with tests $\langle < a \rangle$ and $\langle * \, b \rangle$. Assuming the original tree is irreducible, we have $b \geq a$, with strict inequality if $\langle * \, b \rangle$ is a less-than test. We identify $\langle < a \rangle$ with set $A = [a, \infty)$ and $\langle * \, b \rangle$ with $B = \{b\}$ or $[b, \infty)$ (second option corresponds to $\langle < b \rangle$). Then by our inequalities, it is clear that $B \subset A$ and the rotation is valid. By a similar reasoning, the rotation shown in Fig. 3b is also valid, assuming either $a < b$ or $\langle * \; b \rangle$ is an equal-to test. For proving Theorem 1, we need only consider these two rotations (or the corresponding reverse rotations).

Not all tree modifications in our proof are rotations. Some modifications also *insert* a new comparison test into the tree. This modification has the effect of making a tree reducible, though it can be converted into a irreducible tree, as explained earlier in this section. Insertions are used by Anderson et al. [1] in proving the tight bound for λ^- and will also be used in proving Theorem 1.

3   Proof of Theorem 1

In this section we prove Theorem 1, namely that the lower bound of $\frac{3}{7}$ on λ^+ in [1] is tight.

Before proceeding with the proof, we remind the reader that $\langle * \; k \rangle$ denotes an unspecified comparison test on key k, and that its outcomes are specified with labels "$=$ /$\geq$" and "$\neq$/$<$".

The proof is by induction on the number of nodes n. The cases $n = 2, 3$ are trivial so we'll move on to the inductive step. Assume that $n \geq 4$ and that Theorem 1 holds for all instances where the number of keys is in the range $\{2, \ldots, n-1\}$.

To show that the theorem holds for any n-key instance, consider a tree T for an instance with n keys and whose maximum-weight key m satisfies $w_m \geq \frac{3}{7} w(T)$, and suppose that the root of T is a less-than test $\langle < r \rangle$. We show that we can then find another tree T' that is rooted at an equal-to test node and has cost not larger than T.

If $r \in \{2, n\}$, then one of the children of $\langle < r \rangle$ is a leaf, and we can simply replace $\langle < r \rangle$ by an appropriate equal-to test. So we can assume that $2 < r \leq n - 1$, in which case each sub-tree of $\langle < r \rangle$ must have between 2 and $n - 1$ nodes. By symmetry, we also can assume that a heaviest-weight key m is in the left sub-tree L of $\langle < r \rangle$. Since $w_m \geq \frac{3}{7} w(T) \geq \frac{3}{7} w(L)$, we can replace L with a tree rooted at $\langle = m \rangle$ without increasing cost, by our inductive assumption. (A careful reader might notice that, if a tie arises, the inductive assumption only guarantees that the root of this left subtree will be an equal-to test to a key of the same weight as m. For simplicity, we assume that this key is m, for otherwise we can just use the other key in the rest of the proof.) Let T_1 be the $\neq$-branch of $\langle = m \rangle$ and T_2 be the $\geq$-branch of $\langle < r \rangle$.

By applying Lemma 1 to node $\langle = m \rangle$ and its parent $\langle < r \rangle$, we have that $w(T_2) \geq w_m \geq \frac{3}{7} w(T)$, which in turn implies that $w(T_1) \leq \frac{1}{7} w(T)$. Since T_2 is not a leaf, let its root be $\langle * \; i \rangle$ with sub-trees T_3 and T_4, where T_3 is the $=$/$\geq$-branch. The structure of T is illustrated in Fig. 4a.

To modify the tree, we break into two cases based on T_3's weight:

Case 1: $w(T_3) \geq \frac{1}{3} w(T_2)$ To obtain T', rotate $\langle = m \rangle$ to the root of the tree, then rotate $\langle * \; i \rangle$ so that it is the $\neq$-branch of $\langle = m \rangle$. (The irreducibility of T implies that the Containment Property in Sect. 2 holds for both rotations.) Node m goes up by 1 in depth and subtrees T_1 and T_4 both go down by 1 (see Fig. 4b). So the total change in cost is

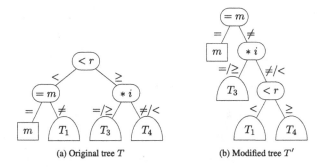

(a) Original tree T (b) Modified tree T'

Fig. 4. On the left, tree T in the proof of Theorem 1. On the right, the modification for Case 1. Notice that T_3 is either the leaf i or contains all keys greater than or equal to i, depending on the comparison test $\langle * i \rangle$.

$$
\begin{aligned}
\text{cost}(T') - \text{cost}(T) &= w(T_1) + w(T_4) - w_m \\
&= w(T) - 2w_m - w(T_3) \quad & w_m + w(T_1) + w(T_3) + w(T_4) = w(T) \\
&\leq w(T) - 2w_m - \tfrac{1}{3}w_m \quad & w(T_3) \geq \tfrac{1}{3}w(T_2), w(T_2) \geq w_m \\
&= w(T) - \tfrac{7}{3}w_m \\
&\leq 0 \quad & w_m \geq \tfrac{3}{7}w(T)
\end{aligned}
$$

Case 2: $w(T_3) < \tfrac{1}{3}w(T_2)$ Note that in this case, $w(T_4) > \tfrac{2}{3}w(T_2)$. We'll first handle some trivial cases. If T_4 is a leaf j, we can replace $\langle * i \rangle$ with $\langle = j \rangle$ and do the same rotations as in Case 1 (swapping the roles of T_4 and T_3). If T_4 contains only two leaves, say k and j, applying Lemma 1 we obtain that $w_k, w_j \leq w(T_3)$, which in turn implies that $w(T_2) \leq 3w(T_3) < w(T_2)$—a contradiction.

Therefore, we can assume that T_4 contains at least 3 leaves and at least 2 comparison nodes. Let $\langle * j \rangle$ be the test in the root of T_4. If $\langle * j \rangle$ is a less-than test and any of its branches is a leaf, we can replace $\langle * j \rangle$ with an equal-to test. If $\langle * j \rangle$ is an equal-to test than its $\neq$-branch is not a leaf (because T_4 has at least 3 leaves). Thus we can assume that the $\neq/<$-branch of $\langle * j \rangle$ is not a leaf, and let $\langle * k \rangle$ be the root of this branch. This means that both $\langle * j \rangle$ and $\langle * k \rangle$ follow from an $\neq/<$-outcome. Let T_5 be the $=/\geq$-branch of $\langle * j \rangle$ and T_6 and T_7 be the branches of $\langle * k \rangle$. The structure of T is shown in Fig. 6a.

This idea behind the remaining argument is this: We now have that T_5, T_6 and T_7 together are relatively heavy (because T_4 is), while T_1 and T_3, which are at lower depth, are light. If this weight difference is sufficiently large, it should be thus possible to rebalance the tree and reduce the cost. We will accomplish this rebalancing by using keys i, j and k, although it may require introducing a new less-than test to one of these keys.

For convenience, re-label the keys $\{i, j, k\} = \{b_1, b_2, b_3\}$ such that $b_1 \leq b_2 \leq b_3$. The goal is to use $\langle < b_2 \rangle$ as a "central" cut-point to divide the tree, with $\langle < r \rangle$ and $\langle * b_1 \rangle$ in its $<$-branch. The $\geq$-branch will contain $\langle * b_3 \rangle$ and, if $\langle * b_2 \rangle$ is an equal-to test, also $\langle * b_2 \rangle$.

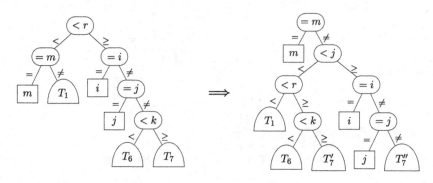

Fig. 5. An example conversion of T (on the left) into T' (on the right) in Case 2, where $k < j < i$. T_7' and T_7'' are copies of T_7.

The idea is illustrated by the example in Fig. 5. In tree T' we have two copies of T_7, denoted T_7' and T_7''. (Because of this T' is not irreducible. As explained in Sect. 2, T' can be then converted into an irreducible tree by splicing out redundant branches.) These copies are needed because the range of T_7 is partitioned by the $\langle < j \rangle$ test into two subsets, with query keys smaller than j following the $<$-branch and the other following the $\geq$-branch. We refer to this in text later as "fracturing" of T_7. However, since these subsets form a disjoint partition of the range of T_7, the total contribution of T_7' and T_7'' to the cost of T' is the same as the contribution of T_7 to the cost of T. More generally, such fracturing does not affect the cost as long as the fractured copies of a subtree are at the same depth as the original subtree.

Ultimately, using the case assumption that $w(T_3) < \frac{1}{3}w(T_2)$, we want to show the following:

Claim 1: There exists a tree T' such that

$$\text{cost}(T') - \text{cost}(T) \leq w(T_1) - w_m + 2w(T_3).$$

In other words, we want to show that in the worst case scenario, we have a modified tree whose cost is at worst equivalent to moving key m up by one, T_1 down by one, and T_3 down by 2. This suffices to prove Case 2 as then

$$
\begin{aligned}
\text{cost}(T') - \text{cost}(T) &\leq w(T_1) - w_m + 2w(T_3) \\
&= w(T) - 2w_m + w(T_3) - w(T_4) && w_m + w(T_1) + w(T_3) + w(T_4) = w(T) \\
&\leq w(T) - 2w_m + \tfrac{1}{3}w(T_2) - \tfrac{2}{3}w(T_2) && w(T_3) < \tfrac{1}{3}w(T_2), w(T_4) > \tfrac{2}{3}w(T_2) \\
&= w(T) - 2w_m - \tfrac{1}{3}w(T_2) \\
&\leq w(T) - 2w_m - \tfrac{1}{3}w_m && w(T_2) \geq w_m \\
&\leq 0
\end{aligned}
$$

Before describing the construction of T', we'll first establish Claim 2 below.

Claim 2: $r < b_2$ (so that $\langle < r \rangle$ and $\langle < b_2 \rangle$ are distinct tests).

Because keys i, j, and k are in the $\geq$-branch of $\langle < r \rangle$, we have that $r \leq b_1, b_2, b_3$. As any irreducible tree can perform at most two tests on the same key, if $b_i = r$ then b_i is distinct from the other two keys. Then $b_i = b_1$ must hold to preserve order, and thus $b_2 > b_1 = r$. This proves Claim 2.

The modified tree T' then has the following form: $\langle = m \rangle$ is at the root, $\langle < b_2 \rangle$ is at the $\neq$-branch of $\langle = m \rangle$, and $\langle < r \rangle$ is at the $<$-branch of $\langle < b_2 \rangle$. Notably, T_1 will still be in the $<$-branch of $\langle < r \rangle$ in T', which implies that m moves up by 1 and T_1 moves down by 1, thus matching the $w(T_1) - w_m$ terms in Claim 1. The right branch of $\langle < r \rangle$ leads to $\langle * b_1 \rangle$. The rest of T' will be designed so that all subtrees T_3, T_5, T_6 and T_7 will have roots at depth at most 4, so in particular T_6 and T_7 will never move down.

Then it suffices to show that the new depths of T_3 and T_5 imply a cost increase no greater than $2w(T_3)$. More precisely, we will show that T' has one of the following properties: Compared to T, in T'

(j1) T_3 moves down by at most 2 and T_5's depth doesn't change, *or*
(j2) If $\langle * j \rangle$ is an equal-to test then T_5 and T_3 move down at most by 1 each. (This suffices because $w(T_5) = w_j \leq w(T_3)$ if $\langle * j \rangle$ is an equal-to test, by applying Lemma 1 to T.)

To describe the rest of our modification, we break into two sub-cases, depending on whether $\langle * b_2 \rangle$ is and equal-to test or less-than test.

Case 2.1: $\langle * b_2 \rangle = \langle < b_2 \rangle$ In this case, $\langle * b_3 \rangle$ is at the $\geq$-branch of $\langle < b_2 \rangle$ in T'. We do not introduce any new comparison tests, obtaining T' shown in Fig. 6b. We now break into further cases based on which key b_2 is.

First, we observe that $b_2 \neq i$, as for $b_2 = i$ the structure of T would imply that $j, k < i$, meaning that i would be the largest key, instead of the middle one. Thus, $b_2 \in \{j, k\}$. This gives us two sub-cases.

Case 2.1.1: $b_2 = j$. Then we have $k < j$ by the structure of T, implying $k = b_1$, and $i \geq j$ since i must now be b_3. Then, in T', $\langle * b_1 \rangle$ has branches T_6 and T_7, while $\langle * b_3 \rangle$ has branches T_3 ($=/\geq$-branch) and T_5 ($\neq/<$-branch). In which case, only T_3 moves down by 1, satisfying (j1).

Case 2.1.2: $b_2 = k$. Then the structure of T implies that $k \neq \{i, j\}$, so $b_1 < k$, which in turn implies that $\langle * b_1 \rangle$ is an equal-to test. We now have two further sub-cases. If $(i, j) = (b_1, b_3)$ then, in T', T_5 is in the $=/\geq$-branch of $\langle * b_3 \rangle$ and leaf $T_3 = i$ is in the $=/\geq$-branch of $\langle * b_1 \rangle$, implying T_5's depth doesn't change and T_3 moves down by 2, satisfying (j1). If $j = b_1$, then leaf $T_5 = j$ is below $\langle * b_1 \rangle$ and T_3 is below $\langle * b_3 \rangle$, both moving down by 1, satisfying (j2).

Case 2.2: $\langle * b_2 \rangle = \langle = b_2 \rangle$ and both tests $\langle * b_1 \rangle$, $\langle * b_3 \rangle$ are different from $\langle < b_2 \rangle$. In this case, $\langle < b_2 \rangle$ (the $\neq$-child of $\langle = m \rangle$ in T') is a newly introduced test. In T', we will have $\langle = b_2 \rangle$ and $\langle * b_3 \rangle$ in the $\geq$-branch of $\langle < b_2 \rangle$. The order in which we perform $\langle = b_2 \rangle$ and $\langle * b_3 \rangle$ will be determined later.

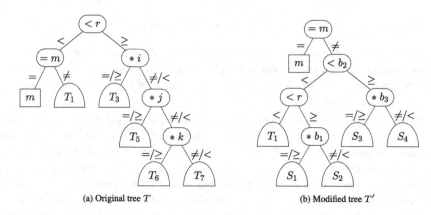

(a) Original tree T (b) Modified tree T'

Fig. 6. Original and modified tree for Case 2.1. S_1, S_2, S_3, S_4 is simply some permutation of T_3, T_5, T_6, T_7.

We will say $\langle = b_2 \rangle$ is "performed first" if it is the root of the $\geq$-branch of $\langle < b_2 \rangle$, in which case $\langle * b_3 \rangle$ is rooted at the $\neq$-branch of $\langle = b_2 \rangle$, or "performed second". Likewise if $\langle * b_3 \rangle$ is performed first, then $\langle = b_2 \rangle$ is performed second, rooted at the $\neq/<$-branch of $\langle * b_3 \rangle$. Figure 7 illustrates these two possible configurations. In most cases, $\langle = b_2 \rangle$ and $\langle * b_3 \rangle$ are performed in the same order as in the original tree T (i.e. $\langle * i \rangle$ comes before $\langle * j \rangle$, which comes before $\langle * k \rangle$), though one case (Case 2.2.2) requires going out-of-order, implying a rotation.

Because $\langle < b_2 \rangle$ is a new comparison, one of the sub-trees T_3, T_5, T_6, T_7 in T may fracture, meaning that some of its keys may satisfy this comparison and other may not. We will in fact show that only T_6 or T_7 can fracture, but their depths do not increase. So, as explained earlier, this fracturing will not increase cost. (As also explained before, the redundancies created by this fracturing, and other that can occur as a result of the conversion, can be eliminated by post-processing T' that iteratively splices out redundant branches.) We again break into further cases based on which key b_2 is.

Case 2.2.1: $b_2 = i$ and $\langle * j \rangle$ is an equal-to test. Then $T_3 = i$ and $T_5 = j$ are both leaves and can't fracture. Perform $\langle = b_2 \rangle$ first and $\langle * b_3 \rangle$ second (since $i = b_2$, this follows order of comparisons in the original tree). Then T_3 and T_5 both move down by 1, satisfying (j2), with $T_3 = i = b_2$ in the =-branch of $\langle = b_2 \rangle$ and $T_5 = j$ in the =-branch of either $\langle * b_1 \rangle$ or $\langle * b_3 \rangle$ (depending on whether j is b_1 or b_3).

Case 2.2.2: $b_2 = i$ and $\langle * j \rangle$ is a less-than test. Then $k < j$, so $j = b_3$ and $k = b_1$. By the case assumption, we have that $j > i$. Then in the $\geq$-branch of $\langle < b_2 \rangle$, we perform $\langle < j \rangle$ *before* $\langle = i \rangle$ (going out-of-order compared to the original tree), which will be in the <-branch of $\langle < j \rangle$. Then $T_3 = i$ will be below $\langle = i \rangle$ moving down twice and T_5 will be in the $\geq$-branch of $\langle < j \rangle$ staying at the same depth, satisfying (j1).

Case 2.2.3: $b_2 = j$. Then we may simply perform $\langle = b_2 \rangle$ and $\langle * b_3 \rangle$ in the same order as in the original tree. If $\langle * i \rangle$ is an equal-to test, then T_3 and T_5 are both leaves, and either both will go down by 1 (if $i = b_3$ and $k = b_1$), satisfying (j2), or only T_3 goes down by 2 (if $i = b_1$ and $k = b_3$), satisfying (j1). If $\langle * i \rangle$ is a less-than test,

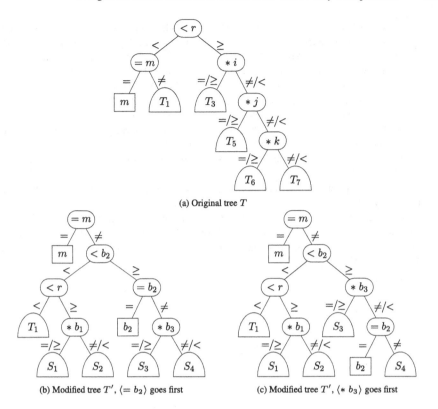

(a) Original tree T

(b) Modified tree T', $\langle = b_2 \rangle$ goes first

(c) Modified tree T', $\langle * b_3 \rangle$ goes first

Fig. 7. Original and modified tree for Case 2.2. In this case, $T_5 = b_2$, and S_1, S_2, S_3, S_4 is some permutation of T_3, T_6, T_7, T'', where T'' is a copy of T_6 or T_7.

then $j, k < i$, so $k = b_1$ and $i = b_3$. $\langle < i \rangle$ will be performed first in the $\geq$-branch of $\langle < b_2 \rangle$ and T_3 will be in the $\geq$-branch of $\langle < i \rangle$, implying T_3 and T_5 both go down by 1, satisfying (j2).

Case 2.2.4: $b_2 = k$. The analysis is similar to Case 2.1.2. By the structure of T, $k \neq \{i, j\}$, so $b_1 < k$, implying $\langle * b_1 \rangle$ is an equal-to test. Then perform $\langle = b_2 \rangle$ and $\langle * b_3 \rangle$ in the same order as the original tree (that is, $\langle = k \rangle$ is always performed second). Thus $\langle * b_1 \rangle$ is in the $\geq$-branch of $\langle < r \rangle$ and $\langle * b_3 \rangle$ is in the $\geq$-branch of $\langle < b_2 \rangle$. If $(i, j) = (b_1, b_3)$, then T_3 (which is a leaf i) is at the $=/\geq$-branch of $\langle * b_1 \rangle$ and T_5 is at the $=/\geq$-branch of $\langle * b_3 \rangle$, implying T_5's depth stays the same and T_3 moves down by 2, satisfying (j1). If $(i, j) = (b_3, b_1)$, then leaf $T_5 = j$ is below $\langle * b_1 \rangle$ and T_3 is below $\langle * b_3 \rangle$, implying both trees move down by 1, satisfying (j2).

References

1. Anderson, R., Kannan, S., Karloff, H., Ladner, R.E.: Thresholds and optimal binary comparison search trees. J. Algorithms **44**, 338–358 (2002)
2. Atalig, S., Chrobak, M., Mousavian, E., Sgall, J., Vesely, P.: Structural properties of search trees with 2-way comparisons. CoRR, abs/2311.02224 (2023)

3. Bein, W., Golin, M.J., Larmore, L.L., Zhang, Y.: The Knuth-Yao quadrangle-inequality speedup is a consequence of total monotonicity. ACM Trans. Algorithms **6**(1), 1–17 (2009)
4. Chrobak, M., Golin, M., Munro, J.I., Young, N.E.: A simple algorithm for optimal search trees with two-way comparisons. ACM Trans. Algorithms **18**(1), 1–11 (2021)
5. Chrobak, M., Golin, M., Munro, J.I., Young, N.E.: On Huang and Wong's algorithm for generalized binary split trees. Acta Informatica **59**(6), 687–708 (2022)
6. Knuth, D.E.: Optimum binary search trees. Acta Informatica **1**, 14–25 (1971)
7. Knuth, D.E.: The Art of Computer Programming, Volume 3: Sorting and Searching, 2nd edn. Addison-Wesley Publishing Company, Redwood City (1998)
8. Spuler, D.: Optimal search trees using two-way key comparisons. Acta Informatica **31**(8), 729–740 (1994)
9. Spuler, D.A.: Optimal search trees using two-way key comparisons. PhD thesis, James Cook University (994)
10. Yao, F.F.: Efficient dynamic programming using quadrangle inequalities. In: Miller, R.E., Ginsburg, S., Burkhard, W.A., Lipton, R.J. (eds.) Proceedings of the 12th Annual ACM Symposium on Theory of Computing, 28–30 April 1980, Los Angeles, California, USA, pp. 429–435. ACM (1980)

Kleene Theorems for Lasso Languages and ω-Languages

Mike Cruchten$^{(\boxtimes)}$ (ORCID)

The University of Sheffield, Sheffield, England
m.cruchten@sheffield.ac.uk

Abstract. Automata operating on pairs of words were introduced as an alternative way of capturing acceptance of regular ω-languages. Families of DFAs and lasso automata operating on such pairs were defined subsequently, giving rise to minimisation algorithms, a Myhill-Nerode theorem and language learning algorithms. Yet Kleene theorems for these well-studied classes are still missing. We introduce rational lasso languages and expressions, show a Kleene theorem for lasso languages and explore the connection between rational lasso and ω-expressions, which yields a Kleene theorem for ω-languages and saturated lasso automata. For one direction of the Kleene theorems, we also provide a Brzozowski construction for lasso automata from rational lasso expressions.

Keywords: Lasso Languages · ω-Languages · Kleene Theorem

1 Introduction

Lassos occur naturally in the study of ω-automata, where they manifest themselves in nondeterministic Büchi automata as paths consisting of a prefix leading from the initial state to some state q, and a period, which leads from q back to itself, traversing some accepting state infinitely often. These infinite paths correspond to infinite words of the shape uv^{ω}, called ultimately periodic words. Such words play an important role, as two regular ω-languages are equal if and only if they contain the same ultimately periodic words. By representing an ultimately periodic word uv^{ω} as a string $u\$v$, Calbrix et al. show that for any regular ω-language L, the set $L_\$ = \{u\$v \mid uv^{\omega} \in L\}$ is regular [4]. Their work shows that certain DFAs over an extended alphabet, called $L_\$$-automata, act as acceptors of regular ω-languages. These results came with the hope of improving existing algorithms for deciding emptiness and language inclusion of regular ω-languages, which are prominently used in software verification and model checking.

Angluin et al. introduced families of DFAs (FDFAs) operating on pairs (u, v) that represent uv^{ω} [2]. They combined work by Calbrix et al. [4] and by Maler et al. [12] to produce different variations of FDFAs. From these, they devise language learning algorithms for regular ω-languages and show that learning can be done in polynomial time in the size of the FDFA. Moreover, they investigate

© The Author(s), under exclusive license to Springer Nature Singapore Pte Ltd. 2024
X. Chen and B. Li (Eds.): TAMC 2024, LNCS 14637, pp. 111–123, 2024.
https://doi.org/10.1007/978-981-97-2340-9_10

the complexity of certain operations and decision procedures on FDFAs, including deciding emptiness and language inclusion, and the performance of Boolean operations [1]. These can be performed in nondeterministic logarithmic space, validating the hopes of Calbrix et al.

An equivalent automaton to the FDFA is the lasso automaton defined by Ciancia and Venema [5]. Lasso automata also operate on pairs (u, v), which they call lassos (we follow this naming convention). They give a Myhill-Nerode theorem and show that lasso automata can be minimised using partition refinement.

Although automata operating on lassos are well-established in many regards, they still lack a Kleene theorem. Our main goal is to establish a Kleene theorem for lasso languages with respect to lasso automata, and to show how rational lasso and ω-expressions relate. This paves a way towards a Kleene theorem for ω-languages with respect to saturated lasso automata (Definition 3).

Our contributions are drawn as dashed arrows in the figure to the right. We define rational lasso languages as those lasso languages which can be obtained from rational languages using rational lasso operations. Our first contribution is a Kleene theorem for lasso lan-

$$\text{rational } \omega\text{-lang.} \overset{8}{\dashrightarrow} \text{rational lasso lang.}$$

$$[4]\Big(\ \ \Big\downarrow 4 \qquad\qquad 1 \Big\uparrow\ \ 3\ \Big\downarrow 5,2$$

$$\text{regular } \omega\text{-lang.} \xrightarrow[[4]]{} \text{regular lasso lang.}$$

guages: we show that a lasso language is rational if and only if it is accepted by a finite lasso automaton (Theorem 3). For one direction, we provide a Brzozowski construction, which turns a rational lasso expression into a finite lasso automaton accepting the corresponding rational lasso language (Theorem 1). For the other direction, we dissect a finite lasso automaton into several DFAs and prove that the lasso language accepted by the lasso automaton can be obtained from the rational languages corresponding to the DFAs by using the rational lasso operations (Proposition 5 and Corollary 2), following ideas by [4].

Secondly, we study the relationship between rational lasso and ω-expressions. We introduce the notion of a rational lasso expression *representing* a rational ω-expression (Definition 9), which expresses that the language semantics of either expression determines that of the other. We show that for any given rational ω-expression, we can syntactically construct a representing rational lasso expression using two additional operations on rational expressions (cf. Proposition 8).

Our two contributions together with a result by [4] allow us to re-establish Kleene's theorem for ω-languages with respect to saturated lasso automata. Given a rational ω-language, we can turn it into a rational lasso expression (Proposition 8) and apply our Brzozowski construction (Theorem 1) to obtain the desired finite saturated lasso automaton, hence every rational ω-language is regular (Theorem 4). The other direction is given by [4], showing that every ω-language accepted by a finite saturated lasso automaton is rational.

In this extended abstract we can only outline our constructions and present theorems without proofs. Details can be found in an extended version [7].

2   Preliminaries

We use Σ to denote a finite alphabet. The free monoid over Σ is $(\Sigma^*, \cdot, \varepsilon)$, which consists of finite words written u, v, w with concatenation $u \cdot v = uv$ and the empty word ε. We write Σ^ω for the set of infinite words and Σ^{up} for the set of ultimately periodic words, i.e. those of the form uv^ω. A *lasso* is a pair $(u, v) \in \Sigma^* \times \Sigma^+$ (which we abbreviate Σ^{*+}), with u the *spoke* and $v \neq \varepsilon$ the *loop*. We think of the lasso (u, v) as a representative for uv^ω.

We write U, V, W for languages of words and L, K for languages of infinite words (ω-languages) or of lassos, depending on context. The rational languages and ω-languages are defined as usual, so are their operations. A language (resp. ω-language) is regular if it is accepted by a DFA (resp. finite nondeterministic Büchi automaton). For an ω-language L, $\mathrm{UP}(L) = L \cap \Sigma^{\mathrm{up}}$ is its *ultimately periodic fragment*. We use $t, r, s \in \mathrm{Exp}$ for rational expressions and $[\![-]\!] : \mathrm{Exp} \to 2^{\Sigma^*}$ maps a rational expression to its corresponding rational language in the usual way. We write $t \in N$ if $\varepsilon \in [\![t]\!]$, i.e. t has the *empty word property*. We use $T \in \mathrm{Exp}_\omega$ to denote rational ω-expressions, and $[\![-]\!]_\omega : \mathrm{Exp}_\omega \to 2^{\Sigma^\omega}$ is the usual language map for rational ω-expressions.

The set of lassos can be equipped with two rewrite rules

$$\frac{a \in \Sigma \quad (ua, va)}{(u, av)} (\gamma_1) \qquad \text{and} \qquad \frac{(u, v^k) \quad (k > 1)}{(u, v)} (\gamma_2).$$

We write $(u, v) \to_{\gamma_i} (u', v')$ if (u, v) γ_i-*reduces* to (u', v') in one step (equiv. (u', v') γ_i-*expands* to (u, v)). Moreover, let $\to_\gamma = \to_{\gamma_1} \cup \to_{\gamma_2}$ and $\sim_\gamma$ be the least equivalence relation including $\to_\gamma$. If $(u, v) \sim_\gamma (u', v')$, we call the lassos γ-*equivalent*. The relation $\to_\gamma$ is confluent and strongly normalising, so each lasso (u, v) has a *normal form*. The relation $\sim_\gamma$ captures when two lassos represent the same ultimately periodic word: $(u, v) \sim_\gamma (u', v') \iff uv^\omega = u'(v')^\omega$ [6]. This shows that $(-)^\omega$ is captured by $(-)^\circ$ together with the rotation law $(uv)^\omega = u(vu)^\omega$ and the pumping law $u^\omega = (u^k)^\omega$.

Definition 1 ([5]). A *lasso automaton* is a structure $\mathcal{A} = (X, Y, \overline{x}, \delta_1, \delta_2, \delta_3, F)$ where $\delta_1 : X \to X^\Sigma$, $\delta_2 : X \to Y^\Sigma$, $\delta_3 : Y \to Y^\Sigma$ and $F \subseteq Y$. We call X and Y the sets of *spoke* and *loop states*. The maps δ_1, δ_2 and δ_3 are called the *spoke*, *switch* and *loop transition maps* of $\mathcal{A}$. The set F denotes the *accepting states* and $\overline{x} \in X$ is the *initial state* of $\mathcal{A}$.

For convenience, we assume $X \cap Y = \emptyset$ and define the map $(\delta_2 : \delta_3) : X \uplus Y \to Y^\Sigma$, which is equal to δ_2 on X and equal to δ_3 on Y. The maps $\delta_1, (\delta_2 : \delta_3)$ and δ_3 can be extended from symbols to finite words in the usual way.

Definition 2 ([5]). A lasso $(u, v) \in \Sigma^{*+}$ is *accepted* by $\mathcal{A}$, if $(\delta_2 : \delta_3)(\delta_1(\overline{x}, u), v) \in F$. The set of lassos accepted by $\mathcal{A}$ is denoted $L_\circ(\mathcal{A})$. A lasso language L is *regular* if it is accepted by a finite lasso automaton.

In [5], the authors define an Ω-automaton as a lasso automaton with special structural properties. We give an equivalent definition using saturation.

Definition 3. *A lasso automaton* $\mathcal{A}$ *is* saturated *if for any two* γ-*equivalent lassos* $(u_1, v_1), (u_2, v_2) \in \Sigma^{*+} : (u_1, v_1) \in L_\circ(\mathcal{A}) \iff (u_2, v_2) \in L_\circ(\mathcal{A})$. *An* Ω-*automaton is a saturated lasso automaton.*

Finite Ω-automata act as acceptors of regular ω-languages. A finite Ω-automaton $\mathcal{A}$ accepts the regular ω-language L if $L_\circ(\mathcal{A}) = \{(u, v) \mid uv^\omega \in L\}$ [5]. Note that $L_\circ(\mathcal{A})$ is always $\sim_\gamma$-saturated for an Ω-automaton $\mathcal{A}$, and that $\{(u, v) \mid uv^\omega \in L\} = \{(u, v) \mid uv^\omega \in K\}$ implies $L = K$ for regular ω-languages K, L [4]. The *regular* ω-*language accepted by a finite* Ω-*automaton* $\mathcal{A}$ is denoted $L_\omega(\mathcal{A})$.

Example 1. Consider the Ω-automaton $\mathcal{A}$ (left) and lasso automaton $\mathcal{B}$ (right). $L_\circ(\mathcal{A}) = \{(u, a^k) \mid u \in \Sigma^*, k \geq 1\}$, $L_\omega(\mathcal{A}) = \{ua^\omega \mid u \in \Sigma^*\}$ and $L_\circ(\mathcal{B}) = \{(a^k, ba^{k'}) \mid k, k' \in \mathbb{N}\}$. Note that $(\varepsilon, b) \sim_\gamma (b, b)$ but $(\varepsilon, b) \in L_\circ(\mathcal{B})$ and $(b, b) \notin L_\circ(\mathcal{B})$, so $\mathcal{B}$ is not saturated.

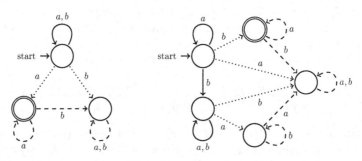

Spoke, switch and loop transitions are drawn as solid, dotted and dashed lines.

3   Rational Lasso Expressions

In this section we introduce rational lasso expressions, languages and an algebra thereof, which we show to be sound. A lasso language is *rational* if it is obtained from rational languages using the operations

$$U^\circ = \{(\varepsilon, u) \mid u \in U\}, \quad U \cdot K = \{(uv, w) \mid u \in U, (v, w) \in K\}, \quad K_1 \cup K_2,$$

with U a rational language and K, K_1, K_2 rational lasso languages. In contrast to U^ω, where we construct an infinite word by choosing infinitely many words from U, constructing a lasso in U° requires the choice of one single word in U.

From here on, we assume RA to be an arbitrary but fixed algebra of rational expressions of signature $(0, 1, +, \cdot, {}^*)$ (e.g. KA, the theory of Kleene Algebra [10]). We write $\vdash t = r$ if $t = r$ is deducible in RA and drop the turnstyle when it is clear from context. We write $t \leq r$ for $\vdash t + r = r$, as the $+$-reduct of an RA-algebra is a join-semilattice. Finally, for a formula φ, the *Iverson bracket* of φ is defined as $[\varphi] = 1$ if φ is true and $[\varphi] = 0$ otherwise.

Definition 4. *Let* $t, r \in Exp$ *with* $r \notin N$. *The set* $Exp_\circ$ *of* rational lasso expressions *is defined by the following grammar*

$$\rho, \sigma ::= 0 \mid t \cdot \rho \mid \rho + \sigma \mid r^\circ.$$

Definition 5. *The* language semantics *for rational lasso expressions is given by the map* $[\![-]\!]_\circ : Exp_\circ \to 2^{\Sigma^{*+}}$ *defined recursively on rational lasso expressions:*

$$[\![0]\!]_\circ = \emptyset, \qquad [\![t \cdot \rho]\!]_\circ = [\![t]\!] \cdot [\![\rho]\!]_\circ, \qquad [\![\rho + \sigma]\!]_\circ = [\![\rho]\!]_\circ \cup [\![\sigma]\!]_\circ, \qquad [\![r^\circ]\!]_\circ = [\![r]\!]^\circ.$$

Note how the lasso language semantics $[\![-]\!]_\circ$ *extends the language semantics of rational expressions* $[\![-]\!]$.

Example 2. Let $\Sigma = \{a, b\}$ and $\rho = b(a^*b^\circ)$. On the left we compute the associated rational lasso language $[\![\rho]\!]_\circ$. On the right we give a finite lasso automaton for $[\![\rho]\!]_\circ$ in particular showing that this rational lasso language is also regular.

$$
\begin{aligned}
[\![\rho]\!]_\circ &= [\![b(a^*b^\circ)]\!]_\circ \\
&= [\![b]\!][\![(a^*b^\circ)]\!]_\circ \\
&= \{b\}([\![a^*]\!][\![b^\circ]\!]_\circ) \\
&= \{b\}(\{a^k \mid k \geq 0\}[\![b]\!]^\circ) \\
&= \{b\}(\{a^k \mid k \geq 0\}\{(\varepsilon, b)\}) \\
&= \{(ba^k, b) \mid k \geq 0\}.
\end{aligned}
$$

Next, we introduce a theory to reason about regular lasso expressions. This theory is sound with respect to the lasso language semantics, which we require for the construction of a Brzozowski lasso automaton.

Definition 6. *The* two-sorted theory LA of lasso algebras *extends the theory RA by the following axioms:*

$$
\begin{array}{lll}
1 \cdot \rho = \rho & 0 \cdot \rho = 0 & \rho + \sigma = \sigma + \rho \\
(t + r) \cdot \rho = t \cdot \rho + r \cdot \rho & 0^\circ = 0 & (\rho + \sigma) + \tau = \rho + (\sigma + \tau) \\
t \cdot (\rho + \sigma) = t \cdot \rho + t \cdot \sigma & 0 + \rho = \rho & \rho + \rho = \rho \\
t \cdot (r \cdot \rho) = (t \cdot r) \cdot \rho & t \cdot 0 = 0 & (t + r)^\circ = t^\circ + r^\circ \ (t, r \notin N)
\end{array}
$$

with $t, r \in Exp$ *and* $\rho, \sigma, \tau \in Exp_\circ$. *The axioms together with the laws for equality and the substitution of provably equivalent rational and rational lasso expressions gives us the deductive system* **LA**. *We write* $\vdash_{\mathbf{LA}} \rho = \sigma$ *when the equation* $\rho = \sigma$ *is deducible in LA. Whenever it is clear from context, we drop the turnstile* $\vdash_{\mathbf{LA}}$.

Remark 1. We briefly highlight the differences between lasso and Wagner algebras [13]. A Wagner algebra is a two-sorted algebra similar to the lasso algebra but having an operation $(-)^\omega$ instead of $(-)^\circ$. They are used to reason about rational ω-expressions and Wagner showed completeness of his axiomatisation with respect to the language semantics for rational ω-expressions. Wagner's axiomatisation looks very similar to that of a lasso algebra. However, the unary operations $(-)^\circ$ and $(-)^\omega$ satisfy different laws: the rotation and pumping laws for $(-)^\omega$ do not hold for $(-)^\circ$; conversely, $(t + r)^\circ = t^\circ + r^\circ$ but $(t + r)^\omega = (t^* \cdot r)^\omega + (t^* \cdot r)^* \cdot t^\omega \neq t^\omega + r^\omega$. Other than this, two more subtle differences can be pointed out:

1. from the $(-)^\omega$-axioms, one can deduce that $0^\omega = 0$, this is not the case for lasso algebras (i.e. we need the axiom $0^\circ = 0$),
2. Wagner has an additional derivation rule, which allows to solve equations of a particular type, such a rule is not given for lasso algebras.

The theory of lasso algebras is sound with respect to the language semantics for rational lasso expressions. We make no claim about its completeness.

Proposition 1 (Soundness). *For all $\rho, \sigma \in Exp_\circ$, $\vdash_{\mathbf{LA}} \rho = \sigma \Rightarrow [\![\rho]\!]_\circ = [\![\sigma]\!]_\circ$.*

Analogously to the situation for rational expressions and ω-expressions, each rational lasso expression is provably equivalent to one of the form $\sum_{i=1}^n t_i \cdot r_i^\circ$. Such a form is called a *disjunctive form* (and it is not unique).

We use (t, r) as a shorthand for $t \cdot r^\circ$, giving us a more direct correspondence to the language semantics as $[\![(t, r)]\!]_\circ = \{(u, v) \in \Sigma^{*+} \mid u \in [\![t]\!], v \in [\![r]\!]\}$.

4   Rational Lasso Languages are Regular

In this section, given a rational lasso expression ρ, our aim is to build a finite lasso automaton $\mathcal{A}$, which accepts $[\![\rho]\!]_\circ$. This shows one direction of Kleene's Theorem, namely that every rational lasso language is regular.

We assume that the reader is familiar with the Brzozowski construction for deterministic finite automata (c.f. [7]). We write d for the *Brzozowski derivative*. The automaton $\mathcal{B} = (\mathrm{Exp}, d, N)$ is the *Brzozowski automaton*, a deterministic automaton, which has the property that for $t \in \mathrm{Exp}$, $L(\mathcal{B}, t) = [\![t]\!]$ ([3]).

Example 3

Let $\rho = (b(ab)^*, ab^*)$, then $(u, v) \in [\![\rho]\!]_\circ$ if and only if $u \in [\![b(ab)^*]\!]$ and $v \in [\![ab^*]\!]$. So the idea is to build one DFA for $b(ab)^*$ and one DFA for ab^*, which correspond to the spoke and loop part of the lasso automaton, and then link them. The construction of the DFAs is done using Brzozowski derivatives. Note that we are allowed to transition from the first to the second DFA only after

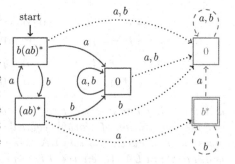

reading u, that is, only once we have reached an accepting state in the spoke DFA. From any other state in the spoke DFA, attempting to transition just leads to a dead state. The initial state of the second DFA is omitted as it is not reachable.

Definition 7. *Define the* spoke *and* switch *Brzozowski derivatives* $d_1\colon Exp_\circ \to Exp_\circ^\Sigma$ *and* $d_2\colon Exp_\circ \to Exp^\Sigma$ *recursively on the structure of* $Exp_\circ$:

$$d_1(0,a) = 0 \qquad\qquad d_2(0,a) = 0$$
$$d_1(t^\circ,a) = 0 \qquad\qquad d_2(t^\circ,a) = d(t,a)$$
$$d_1(\rho+\sigma,a) = d_1(\rho,a)+d_1(\sigma,a) \qquad d_2(\rho+\sigma,a) = d_2(\rho,a)+d_2(\sigma,a)$$
$$d_1(r\cdot\rho,a) = d(r,a)\cdot\rho + [r\in N]\cdot d_1(\rho,a) \qquad d_2(r\cdot\rho,a) = [r\in N]\cdot d_2(\rho,a)$$

The next two propositions give an understanding of the relationship between the derivatives we just defined, and the theory of lasso algebras.

Proposition 2. *For* $\rho,\sigma \in Exp_\circ$ *with* $\vdash_{\mathbf{LA}} \rho = \sigma$ *we have* $\vdash d_2(\rho,a) = d_2(\sigma,a)$ *and* $\vdash_{\mathbf{LA}} d_1(\rho,a) = d_1(\sigma,a)$ *for all* $a \in \Sigma$.

Proposition 3 (Fundamental Theorem). *Let* $\rho \in Exp_\circ$. *Then*

$$\vdash_{\mathbf{LA}} \rho = \left(\sum_{a\in\Sigma} a\cdot d_1(\rho,a)\right) + \left(\sum_{a\in\Sigma} a\cdot d_2(\rho,a)\right)^\circ.$$

Using the Fundamental Theorem and Soundness for rational lasso expressions, we obtain our first Brzozowski lasso automaton.

Proposition 4. *We call* $\mathcal{C} = (Exp_\circ, Exp, d_1, d_2, d, N)$ *the* Brzozowski lasso automaton. *If* $\rho \in Exp_\circ$, *then* $L_\circ(\mathcal{C},\rho) = [\![\rho]\!]_\circ$.

The Brzozowski lasso automaton is not necessarily finite. The easiest fix to this is to assume that our rational lasso expressions are disjunctive forms. To accommodate this assumption, we slightly adapt the definition of d_1 so that $d_1(\rho,a)$ is again a disjunctive form for $a \in \Sigma$. This is done by defining

$$d_1(t\cdot s^\circ,a) := d(t,a)\cdot s^\circ.$$

It is important to point out that this does not jeopardise our earlier result (Proposition 4). Indeed without this change we get the provably equivalent

$$d_1(t\cdot s^\circ,a) = d(t,a)\cdot s^\circ + [t\in N]\cdot d(s^\circ,a) = d(t,a)\cdot s^\circ + [t\in N]\cdot 0.$$

With this in mind, we see that d_1 only acts on the spoke expressions (where it acts just like a normal Brzozowski derivative), d_2 acts on the loop expression to give a rational expression (provided that the spoke expression has the empty word property) and finally d acts on the obtained rational expression. In this sense, we are really constructing two DAs using Brzozowski derivatives and linking them appropriately with d_2, as in Example 3. It is well-known that Brzozowski's construction yields a DFA when quotienting it by a suitable equivalence relation $\sim_B$ (c.f. [7]). The same can also be accomplished in our case when dealing with disjunctive forms, leading to the next definition and result.

Definition 8. *We define* $\sim_C \subseteq Exp_\circ^2$ *to be the least equivalence such that*

$$\sum_{i=1}^{n} t_i \cdot r_i^\circ \sim_C \sum_{i=1}^{n} t_i' \cdot (r_i')^\circ \iff \forall\, 1 \le i \le n : t_i \sim_B t_i' \text{ and } r_i = r_i'.$$

We define $\widehat{d_1}, \widehat{d_2}, \widehat{d}$ and $\widehat{N}$ on equivalence classes in the obvious way.

Theorem 1. *The lasso automaton* $\widehat{\mathcal{C}} = (Exp_\circ/\!\sim_C, Exp/\!\sim_B, \widehat{d_1}, \widehat{d_2}, \widehat{d}, \widehat{N})$ *satisfies for all* $\rho \in Exp_\circ$ *(ρ a disjunctive form)*:

1. $L_\circ(\widehat{\mathcal{C}}, [\rho]_{\sim_C}) = [\![\rho]\!]_\circ$ *and*
2. *the set of states reachable from* $[\rho]_{\sim_C}$ *is finite.*

5   Regular Lasso Languages are Rational

The previous section gives us one direction of Kleene's theorem for lasso languages. We accomplish the other direction by slightly modifying a result by [4], which is interesting in its own right. The authors show how to build the rational ω-language accepted by a finite Ω-automaton $\mathcal{A}$ from some rational languages defined on the basis of $\mathcal{A}$. This shows that every regular ω-language is rational.

We first state the results (Lemma 1 and Theorem 2) by [4] and then follow it up with a similar result for lasso languages, which shows that every regular lasso language is rational. The next lemma is of a technical nature.

Lemma 1 ([4]). *Given regular languages* U, V *with* $\varepsilon \notin V$, $UV^* = U$ *and* $V^+ = V$, *then* $\forall uv^\omega \in UV^\omega, \exists u' \in U, v' \in V : uv^\omega = u'v'^\omega$.

Theorem 2 ([4]). *Let* $\mathcal{A} = (X, Y, \overline{x}, \delta_1, \delta_2, \delta_3, F)$ *be a finite* Ω-*automaton. Then* $L_\omega(\mathcal{A}) = \bigcup_{x \in X} \bigcup_{y \in F} S_x \cdot R_{x,y}^\omega$, *where* $(x \in X, y \in F)$:

$$S_x = \{u \in \Sigma^* \mid \delta_1(\overline{x}, u) = x\},$$
$$R_{x,y} = \{u \in \Sigma^+ \mid \delta_1(x, u) = x \text{ and } (\delta_2 : \delta_3)(x, u) = \delta_3(y, u) = y\}.$$

The rational language S_x consists of all words that lead from the initial state $\overline{x}$ to x. As $R_{x,y}$ contains those words which at the same time bring us from x back to x via δ_1, from x to y via $(\delta_2 : \delta_3)$ and from y back to itself via δ_3, concatenating infinitely many such words traces in some sense paths which intersect a final state infinitely often. As such, a saturated lasso automaton can be seen as a nondeterministic Büchi automaton ([5, Remark 21]).

As S_x and $R_{x,y}$ are both rational languages, we can build rational expressions from them and the next corollary follows immediately.

Corollary 1. *Given a finite* Ω-*automaton* $\mathcal{A}$, *one can construct a rational* ω-*expression* T *such that* $L_\omega(\mathcal{A}) = [\![T]\!]_\omega$.

In order to obtain a similar result for lasso languages, we modify the definition of $R_{x,y}$ from Theorem 2. This shows that every regular lasso language is rational.

Proposition 5. *Let $\mathcal{A} = (X, Y, \overline{x}, \delta_1, \delta_2, \delta_3, F)$ be a finite lasso automaton. For $x \in X$ and $y \in F$, let $R_{x,y} = \{u \in \Sigma^+ \mid (\delta_2 : \delta_3)(x, u) = y\}$ and S_x be defined as in Theorem 2. Then $L_\circ(\mathcal{A}) = \bigcup_{x \in X} \bigcup_{y \in F} S_x \cdot R_{x,y}^\circ$.*

Corollary 2. *Given a finite lasso automaton $\mathcal{A}$, one can construct a rational lasso expression ρ such that $L_\circ(\mathcal{A}) = [\![\rho]\!]_\circ$.*

Corollary 2 and Theorem 1 give us a Kleene theorem for lasso languages.

Theorem 3. *A lasso language is regular if and only if it is rational.*

6 Rational Lasso and ω-Expressions

In this section we explore the connection between rational lasso and ω-expressions, and define what it means for a rational lasso expression τ to represent a rational ω-expression T. Intuitively, this notion expresses that $[\![\tau]\!]_\circ$ and $[\![T]\!]_\omega$ completely determine each other. We show that for each T, we can construct τ in a syntactic way from T with the help of two additional operations on rational expressions. If we pair this with the Brzozowski construction, this shows that every rational ω-language is accepted by a finite Ω-automaton. By [4], we know that the converse also holds. This re-establishes the Kleene theorem for ω-languages with respect to Ω-automata. Additionally, we also obtain a more direct method for constructing Ω-automata from rational ω-expressions, which has not been done before to the best of our knowledge.

Definition 9. *Let $\varphi \colon \Sigma^{*+} \to \Sigma^{up}$ be given by $(u, v) \mapsto uv^\omega$. For a rational lasso expression $\tau \in Exp_\circ$ and a rational ω-expression $T \in Exp_\omega$, we say that*

1. *τ weakly represents T if $\varphi([\![\tau]\!]_\circ) = UP([\![T]\!]_\omega)$ and*
2. *τ represents T if $[\![\tau]\!]_\circ = \varphi^{-1}(UP([\![T]\!]_\omega))$.*

Example 4. Let $T = (a + b)^* a^\omega$. Then $((a + b)^*, a)$ constitutes a weak representation but not a representation of T. This is seen from the following facts:

1. $\varphi([\![((a + b)^*, a)]\!]_\circ) = \varphi(\{(u, a) \mid u \in \Sigma^*\}) = \{ua^\omega \mid u \in \Sigma^*\} = UP([\![T]\!]_\omega)$,
2. $\varphi(\varepsilon, aa) = a^\omega \in UP([\![T]\!]_\omega)$ but $(\varepsilon, aa) \notin [\![((a + b)^*, a)]\!]_\circ$.

Remark 2. If τ represents T it also weakly represents T as φ is surjective. Conversely, if τ weakly represents T, then it represents T if and only if $[\![\tau]\!]_\circ = \varphi^{-1}(\varphi([\![\tau]\!]_\circ))$, that is if and only if $[\![\tau]\!]_\circ$ is $\sim_\gamma$-saturated. For the previous example, $((a + b)^*, a^+)$ is a weak representation of T and moreover its semantics is $\sim_\gamma$-saturated, hence it is a representation of T.

According to the last remark, to obtain a representation of T, we can start with a weak representation and then syntactically modify it such that its semantics is $\sim_\gamma$-saturated, while making sure that it stays a weak representation. We recall the rotation and pumping law, which play an important role.

$$u(vu)^\omega = (uv)^\omega \qquad (1) \qquad (u^k)^\omega = u^\omega \quad (k \geq 1) \qquad (2)$$

In order to accomplish our goal we have to mimic these properties on rational lasso expressions. This is easy for 2. For a rational language U, the root of U is $\sqrt{U} = \{u \mid u^k \in U, k \geq 1\}$ ([11]). As $\sqrt{-}$ preserves rationality, we assume that we have an operation $\sqrt{-}$ on rational expressions. Let $t \in \text{Exp}$. Then we can show that $(\varepsilon, u) \in [\![(\sqrt{t^+})^\circ]\!]_\circ \iff (\varepsilon, u^k) \in [\![(\sqrt{t^+})^\circ]\!]_\circ$. In other words, we can take care of 2 by introducing the transitive closure and the root underneath $(-)^\circ$. We remark that the order is important as $(-)^+$ and $\sqrt{-}$ do not commute. Property 1 is technically harder to deal with. In Eq. 1, from right to left, if $(\varepsilon, uv) \in [\![t^\circ]\!]_\circ$, the idea is to split t into two rational lasso expressions t_1, t_2 such that $u \in [\![t_1]\!]$ and $v \in [\![t_2]\!]$. Then $(u, vu) \in [\![t_1(t_2 t_1)^\circ]\!]_\circ$ and so we should include $t_1(t_2 t_1)^\circ$ in our final expression. For the other direction, if $(u, vu) \in [\![t \cdot r^\circ]\!]_\circ$, we split r into suitable r_1, r_2 such that $u \in [\![t \cap r_2]\!]$ and $v \in r_1$, so that $(\varepsilon, uv) \in [\![((t \cap r_2) r_1)^\circ]\!]_\circ$. So we should include $((t \cap r_2) r_1)^\circ$ in our final expression. This second idea requires us to define $\cap$ on rational expressions, so we assume this on top of $\sqrt{-}$. We now formally define the idea of (sequential) splits.

Definition 10 ([8,9]). Define $\nabla \colon \text{Exp} \to 2^{\text{Exp} \times \text{Exp}}$ as follows:

$$\nabla(0) = \emptyset \qquad \nabla(1) = \{(1,1)\} \qquad \nabla(a) = \{(1,a),(a,1)\}$$

$$\nabla(t+r) = \nabla(t) \cup \nabla(r) \quad \nabla(t^*) = \{(t^* \cdot t_0, t_1 \cdot t^*) \mid (t_0, t_1) \in \nabla(t)\} \cup \{(1,1),(t^* \cdot t, 1)\}$$

$$\nabla(t \cdot r) = \{(t_0, t_1 \cdot r) \mid (t_0, t_1) \in \nabla(t)\} \cup \{(t \cdot r_0, r_1) \mid (r_0, r_1) \in \nabla(r)\}$$

We write ∇_t for $\nabla(t)$ and call ∇_t the *sequential splitting relation of t*.

The following lemma establishes some properties of the splitting relation, of which points 1. and 2. are taken from [9].

Lemma 2. *Let $t \in \text{Exp}$. The sequential splitting relation ∇_t satisfies:*

1. $|\nabla_t|$ *is finite and* $\forall (t_0, t_1) \in \nabla_t \colon t_0 \cdot t_1 \leq t$,
2. *if* $u \cdot v \in [\![t]\!]$, *then there is a split* $(t_0, t_1) \in \nabla_t$ *such that* $u \in [\![t_0]\!]$ *and* $v \in [\![t_1]\!]$,
3. *if* $(t_0, t_1) \in \nabla_t$ *and* $(r_0, r_1) \in \nabla_{t_1}$, *then there exists* $(s_0, s_1) \in \nabla_t$ *such that* $t_0 \cdot r_0 \leq s_0$ *and* $r_1 \leq s_1$.

Next we define a function h, which we show to map a rational ω-expression to a weak representation. The most challenging part of the definition is that of t^ω. In its definition we use the sequential splitting relation to simulate one direction of Property 1, and also apply the transitive closure (by using Kleene stars) to obtain one direction of Property 2.

Definition 11. *Let* $h \colon \text{Exp}_\omega \to \text{Exp}_\circ$ *be defined as*

$$h(0) = 0 \qquad\qquad h(t^\omega) = \sum_{(t_0, t_1) \in \nabla_t} (t^* \cdot t_0, t_1 \cdot t^* \cdot t_0)$$

$$h(T_1 + T_2) = h(T_1) + h(T_2) \qquad h(t \cdot T) = t \cdot h(T)$$

This function gives us not only a weak representation, the language semantics of the rational lasso expression we obtain is also closed under γ-expansion. Indeed, the direction of both properties we included expand words.

Proposition 6. *Let $T \in Exp_\omega$. Then $h(T)$ weakly represents T and $[\![h(T)]\!]_\circ$ is closed under γ-expansion.*

The next definition incorporates the remaining ingredients. We now apply the operations $\cap$ and $\sqrt{-}$ to get the other direction of both properties. Afterwards, we follow up with a proposition which relates the semantics of τ to that of $\Gamma(\tau)$.

Definition 12. *Let $\tau = \sum_{i=1}^{n} t_i \cdot r_i^\circ \in Exp_\circ$. We define the map Γ as*

$$\Gamma(\tau) = \sum_{i=1}^{n} \sum_{\substack{(t_0', t_1') \in \nabla_{t_i} \\ (s_0', s_1') \in \nabla_{s_i}}} t_0' \cdot \left(\sqrt{(t_1' \cap s_1') \cdot s_0'} \right)^\circ.$$

Proposition 7. *Let $\tau = \sum_{i=1}^{n} t_i \cdot r_i^\circ \in Exp_\circ$. Then $(u, v) \in [\![\Gamma(\tau)]\!]_\circ$ if and only if $\exists k_1, k_2 \geq 0, \exists v_1, v_2 \in \Sigma^*: v = v_1 v_2$ and $(uv^{k_1}v_1, v_2 v^{k_2+k_1}v_1) \in [\![\tau]\!]_\circ$.*

Corollary 3. *Let $\tau \in Exp_\circ$. Then*

1. $\forall (u, v) \in [\![\Gamma(\tau)]\!]_\circ, \exists (u', v') \in [\![\tau]\!]_\circ : (u', v') \rightarrow^*_\gamma (u, v)$,
2. $\{uv^\omega \mid (u, v) \in [\![\tau]\!]_\circ\} = \{uv^\omega \mid (u, v) \in [\![\Gamma(\tau)]\!]_\circ\}$ *and*
3. $[\![\tau]\!]_\circ \subseteq [\![\Gamma(\tau)]\!]_\circ$.

With the use of the previous corollaries, we show that if the language semantics of τ is closed under γ-expansion, then $[\![\Gamma(\tau)]\!]_\circ$ is $\sim_\gamma$-saturated. The requirement that $[\![\tau]\!]_\circ$ be closed under γ-expansion is not redundant. While property 1 does not have any restrictions on the length of words we can shift from the prefix underneath the $(-)^\omega$, $\cap$ does as it relies on the expression under the $(-)^\circ$.

Lemma 3. *Let $\tau \in Exp_\circ$. If $[\![\tau]\!]_\circ$ is closed under γ-expansion, then $[\![\Gamma(\tau)]\!]_\circ$ is $\sim_\gamma$-saturated.*

By Remark 2, Proposition 6, and Lemma 3, we can obtain a representation for any rational ω-expression by making use of the operations $\sqrt{-}$ and $\cap$.

Proposition 8. *Let $T \in Exp_\omega$. Then $\Gamma(h(T))$ represents T.*

The last proposition together with the Brzozowski construction for lasso automata give us the main result of this section.

Theorem 4. *Every rational ω-language is accepted by a finite Ω-automaton.*

7 Conclusion

We have introduced rational lasso expressions and languages and shown a Kleene Theorem for lasso languages and ω-languages. In order to obtain these results we gave a Brzozowski construction for lasso automata. Moreover, we introduced the notion of representation and showed how to construct a representing rational lasso expression from a rational ω-expression. As a consequence, we obtained a construction method for converting rational ω-expressions to Ω-automata.

Our results present interesting directions for future work. Angluin et al. introduce syntactic and recurring FDFAs and show that they can be up to exponentially smaller than periodic FDFAs [2]. This raises the question whether, from a given rational ω-expression, one can construct a rational lasso expression such that the Brzozowski construction yields a syntactic or recurring FDFA. Another line of work is the exploration of our results in a categorical setting. Finally, there is the question of complexity of our constructions, and investigating applications of our Brzozowski construction for Ω-automata. Further directions can be found in the extended version [7].

Acknowledgments. The author would like to thank Yde Venema for suggesting this topic as an MSc project and also both Tobias Kappé and Yde Venema for valuable discussions. Furthermore, the author would like to thank Harsh Beohar and Georg Struth for valuable discussions and for reading earlier drafts of this paper.

References

1. Angluin, D., Boker, U., Fisman, D.: Families of DFAs as acceptors of omega-regular languages. In: Faliszewski, P., Muscholl, A., Niedermeier, R. (eds.) 41st International Symposium on Mathematical Foundations of Computer Science, MFCS 2016, August 22-26, 2016 - Kraków, Poland. LIPIcs, vol. 58, pp. 1–14. Schloss Dagstuhl - Leibniz-Zentrum für Informatik (2016)
2. Angluin, D., Fisman, D.: Learning regular omega languages. Theor. Comput. Sci. **650**, 57–72 (2016)
3. Brzozowski, J.A.: Derivatives of regular expressions. J. ACM **11**(4), 481–494 (1964)
4. Calbrix, H., Nivat, M., Podelski, A.: Ultimately periodic words of rational ω-languages. In: Brookes, S.D., Main, M.G., Melton, A., Mislove, M.W., Schmidt, D.A. (eds.) Mathematical Foundations of Programming Semantics, 9th International Conference, New Orleans, LA, USA, April 7–10, 1993, Proceedings. Lecture Notes in Computer Science, vol. 802, pp. 554–566. Springer, Cham (1993). https://doi.org/10.1007/3-540-58027-1_27
5. Ciancia, V., Venema, Y.: Omega-automata: a coalgebraic perspective on regular omega-languages. In: Roggenbach, M., Sokolova, A. (eds.) 8th Conference on Algebra and Coalgebra in Computer Science (CALCO). LIPIcs, vol. 139, pp. 5:1–5:18. Schloss Dagstuhl - Leibniz-Zentrum für Informatik (2019)
6. Cruchten, M.: Topics in Ω-Automata – A Journey through Lassos, Algebra, Coalgebra and Expressions. Master's thesis, The University of Amsterdam (2022)
7. Cruchten, M.: Kleene theorems for lasso languages and ω-languages **abs/2402.13085** (2024). https://arxiv.org/abs/2402.13085

8. Foster, N., Kozen, D., Milano, M., Silva, A., Thompson, L.: A coalgebraic decision procedure for NetKAT. In: Rajamani, S.K., Walker, D. (eds.) Proceedings of the 42nd Annual ACM SIGPLAN-SIGACT Symposium on Principles of Programming Languages, pp. 343–355. ACM (2015)

9. Kappé, T., Brunet, P., Silva, A., Zanasi, F.: Concurrent Kleene algebra: free model and completeness. In: Ahmed, A. (ed.) Programming Languages and Systems. Lecture Notes in Computer Science(), vol. 10801, pp. 856–882. Springer, Cham (2018). https://doi.org/10.1007/978-3-319-89884-1_30

10. Kozen, D.: A completeness theorem for Kleene algebras and the algebra of regular events. Inf. Comput. **110**(2), 366–390 (1994)

11. Krawetz, B., Lawrence, J., Shallit, J.O.: State complexity and the monoid of transformations of a finite set. Int. J. Found. Comput. Sci. **16**(3), 547–563 (2005)

12. Maler, O., Staiger, L.: On syntactic congruences for omega-languages. Theor. Comput. Sci. **183**(1), 93–112 (1997)

13. Wagner, K.W.: Eine Axiomatisierung der Theorie der regulären Folgenmengen. J. Inf. Process. Cybern. **12**(7), 337–354 (1976)

Tight Double Exponential Lower Bounds

Ivan Bliznets[1]([✉])[ID] and Markus Hecher[2][ID]

[1] University of Groningen, Groningen, The Netherlands
i.bliznets@rug.nl
[2] Massachusetts Institute of Technology, Cambridge, USA
hecher@mit.edu

Abstract. The majority of established algorithms have polynomial, exponential, or factorial runtime complexities. Examples of problems that admit tight double or even triple exponential bounds on computational complexity are relatively rare. The Choosability problem is one such example. For this problem, Marx and Mitsou [ICALP 2016] presented a quite sophisticated proof that shows there is no $\mathcal{O}(2^{2^{o(tw)}})n^{O(1)}$ time algorithm parameterized by treewidth tw, assuming ETH. In our paper, we show how we almost immediately come to the same conclusion knowing the reduction from ∀∃-TQBF to the Choosability problem. Besides, in some sense, we provide a factory that produces problems with tight double exponential lower bounds not only in terms of treewidth but also pathwidth, cutwidth, and bandwidth. It was suspected that the Π_2^P-complete or Σ_2^P-complete problems require a double exponential time algorithm in terms of treewidth. However, in our paper, we provide a counterexample to this statement.

Keywords: Choosability · double exponential · lower bounds

1 Introduction

Computational complexity [20] focuses on classifying computational problems and algorithms according to resource usage. One of the most studied resources is running time, which makes the study on required runtime to solve a problem, an important part of complexity theory. Unfortunately, the most interesting problems are probably not solvable in polynomial time, assuming P ≠ NP. However, fixed-parameter tractability (FPT) [6,7,10] oftentimes still enables efficient runtimes, assuming a specific parameter of interest is bounded. One of the most prominent and widely applied parameters is *treewidth*, which was originally introduced for graph problems [2,21]. However, treewidth is by far not

This research was funded by the Austrian Science Fund (FWF), grants J 4656 and P 32830, the Society for Research Funding in Lower Austria (GFF) grant ExzF-0004, the Vienna Science and Technology Fund (WWTF) grant ICT19-065, as well as by the project CRACKNP that has received funding from the European Research Council (ERC) under the European Union's Horizon 2020 research and innovation programme (grant agreement No. 853234).

X. Chen and B. Li (Eds.): TAMC 2024, LNCS 14637, pp. 124–136, 2024.
https://doi.org/10.1007/978-981-97-2340-9_11

restricted to graph problems, as it renders a large variety of NP-hard problems tractable [3,5]. Among these applications on more general input structures are, e.g., the well-known Boolean satisfiability (SAT) problem.

A way to establish tight lower bounds in parameterized complexity theory is to assume the *exponential time hypothesis (ETH)* [13] and construct reductions. ETH is a widely accepted standard hypothesis for exact and parameterized algorithms, stating that there is some $s > 0$ such that we cannot decide satisfiability of a given 3-CNF formula with n variables in time $2^{s \cdot n} \cdot n^{\mathcal{O}(1)}$ [6, Ch.14]. While the thereby obtained lower bounds are expected to be superexponential, natural problems with tight lower bounds beyond the factorial function are rare. For example in a recent preprint [11] Foucaud et al. state

> *These types of lower bounds, **which show that at least double-exponential factors in the running time** are necessary, exhibit the extraordinary level of computational hardness for such problems, and **are rare in the current literature: there are only a handful of such lower bounds** (for treewidth and vertex cover parameterizations) and all of them are for problems that are either Σ_2^P-, Σ_3^P- or #NP-complete.*
> — FOUCAUD ET. AL.   [11]

Among those problems are k-CHOOSABILITY and k-CHOOSABILITY DELETION problems, which are located on the second and third level of the polynomial hierarchy, respectively. Naturally, under ETH these problems require a double- and triple-exponential lower bound in the treewidth [18]. Further, also projected model counting is expected to be double exponential. Another prominent problem is deciding the validity (TQBF) of a quantified Boolean formula [4]. Despite TQBF not being FPT for treewidth [1,19], in case the number ℓ of alternating quantifier blocks is fixed, under ETH, we can not avoid an ℓ-fold exponential runtime in the treewidth of the Boolean formula [9,15].

In general, it has been suspected that implicit quantifier alternations of a problem are the underlying reasons for causing unusually large dependence on the treewidth [18]. We address this expectation and provide a counter example. In more details, in this paper we present the following contributions.

- We demonstrate a natural problem that — despite being Σ_2^P-complete — admits a simple single-exponential runtime for treewidth. This is not entirely obvious, as it is expected that being Σ_2^P-complete already gives sufficient explanation why double-exponential dependence on treewidth is needed [18]. We thereby provide a counterexample and demonstrate that quantifier alternations in the problem definition are indeed not necessarily the underlying reason for requiring large dependence on the treewidth.
- We give an alternative proof for double-exponentiality of $(2,3)$-CHOOSABILITY that is not only inherently simpler than the existing proof [18], we also provide stronger results for the weaker parameters cutwidth, and bandwidth. The resulting bounds are simply obtained after generalizing the lower bound of $\forall\exists$-TQBF for parameters cutwidth and bandwidth, followed by a reduction to $(2,3)$-CHOOSABILITY.

Due to space constraints we omit proofs of observations, lemmas, theorems marked by $\star$.

2 Preliminaries

List Coloring Problems. Let $\mathcal{C}$ be a given set of elements, called *colors*, let $G = (V, E)$ be a graph and let $L : V \rightarrow 2^{\mathcal{C}}$, called *list function*, map every vertex $v \in V$ to a set $L(v) \subseteq \mathcal{C}$ of allowed colors. A *(proper) list coloring* $c : V \rightarrow \mathcal{C}$ maps every vertex $v \in V$ to a color contained in $L(v)$ such that c never assigns two adjacent vertices in G the same color. Graph G is *k-choosable* [8] if for any set $\mathcal{C}$ of colors of size at least k there exists a proper list coloring for any list function assigning k colors to every vertex. In the k-CHOOSABILITY problem, one is given graph G and is asked whether the graph is k-Choosable.

More generally, for a function $f : V \rightarrow \mathbb{N}^+$ assigning every vertex a positive integer, G is *f-choosable* if there is a proper list coloring for any list function assigning $f(v)$ colors to every vertex $v \in V$. The problem (a, b)-CHOOSABILITY decides for given $a, b \in \mathbb{N}^+$ with $a \leq b$ and a given function $f : V \rightarrow \{a, \dots, b\}$, whether G is f-choosable. Obviously, (a, a)-CHOOSABILITY corresponds to a-CHOOSABILITY. The BIPARTITE $(2, 3)$-CHOOSABILITY problem is a special case of the $(2, 3)$-CHOOSABILITY problem where our inputs are restricted to bipartite graphs only. Finally, (a, b)-CHOOSABILITY DELETION [18] asks for given $a, b \in \mathbb{N}^+$ with $a \leq b$ and a given function $f : V \rightarrow \{a, \dots, b\}$ and a number $r \in \mathbb{N}$, whether there is a set $S \subseteq V$ with $|S| = r$, such that deleting S from G makes $G-S$ f-choosable. We also consider the 2-CLIQUE COLORING problem, where we are given a graph G and we need to check if it is possible to color vertices in two colors such that every maximal clique in G contains vertices of both colors.

Quantified Boolean Formulas (QBFs). Boolean formulas are defined in the usual way [14], where *literals* are variables or their negations. For a Boolean formula F, we denote by $\text{var}(F)$ the set of variables of F. A *term* or *clause* is a conjunction or disjunction of literals, respectively, which is interpreted as a set S of literals. A Boolean formula F is in *conjunctive normal form (CNF)* if F is a conjunction of clauses and F is in *disjunctive normal form (DNF)* if F is a disjunction of terms. In both cases, we identify F by its set of clauses or terms, respectively. Formula F is in *d-CNF* or *d-DNF* if each set in F consists of at most d literals.

Let $\ell \geq 0$ be integer. A *quantified Boolean formula* Q is of the form $\mathcal{Q}.F$ for *prefix* $\mathcal{Q} = Q_1 V_1.Q_2 V_2. \cdots Q_\ell V_\ell$, where *quantifier* $Q_i \in \{\forall, \exists\}$ for $1 \leq i \leq \ell$ and $Q_j \neq Q_{j+1}$ for $1 \leq j \leq \ell - 1$; and where V_i are disjoint, non-empty sets of Boolean variables with $\text{var}(F) = \bigcup_{i=1}^{\ell} V_i$; and F is a Boolean formula. We call ℓ *quantifier depth* of Q and let $\text{matr}(Q) = F$.

An *assignment* is a mapping $\alpha : X \rightarrow \{0, 1\}$ from a set X of variables.

Let F be a Boolean formula in CNF and let α be a partial assignment for $\text{var } F$, then by $F[\alpha]$ we denote a Boolean formula obtained by removing every $c \in F$ with $x \in c$ and $\neg x \in c$ if $\alpha(x) = 1$ and $\alpha(x) = 0$, respectively, and by removing from every remaining clause $c \in F$ literals x and $\neg x$ with $\alpha(x) = 0$ and $\alpha(x) = 1$, respectively. Analogously, for F in DNF values 0 and 1 are swapped. For a given QBF Q and an assignment α, $Q[\alpha]$ is a QBF obtained from Q, where variables x mapped by α are removed from preceding quantifiers accordingly, and $\mathsf{matr}(Q[\alpha]) = (\mathsf{matr}(Q))[\alpha]$. A Boolean formula F *evaluates to true* (or *is satisfied*) if there exists an assignment α for $\mathrm{var}(F)$ such that $F[\alpha] = \emptyset$ if F is in CNF or $F[\alpha] = \{\emptyset\}$ if F is in DNF. We say that then α *satisfies* F or α is a *satisfying assignment* of F. A QBF Q *evaluates to true (or is valid)* if $\ell = 0$ and $\mathsf{matr}(Q)$ evaluates to true under the empty assignment. Otherwise, i.e., if $\ell \neq 0$, we distinguish according to Q_1. If $Q_1 = \exists$, then Q evaluates to true if and only if there exists an assignment $\alpha : V_1 \to \{0,1\}$ such that $Q[\alpha]$ evaluates to true. If $Q_1 = \forall$, then Q evaluates to true if for any assignment $\alpha : V_1 \to \{0,1\}$, we have that $Q[\alpha]$ evaluates to true. We say two QBFs are *equivalent* if one evaluates to true whenever the other does. Given a QBF Q, the *evaluation problem* TQBF asks whether Q evaluates to true. TQBF_ℓ refers to TQBF restricted to quantifier depth ℓ and $\mathcal{Q}$-TQBF restricts the problem to prefix $\mathcal{Q}$. In general, TQBF is PSPACE-complete [20,23].

In order to apply parameters, we define graph representations. For a Boolean formula F in CNF or DNF we define the *primal graph* $P_F = (\mathrm{var}(F), E)$ over variables of F, where two variables are adjoined by an edge, whenever they appear together in at least one clause or term of F, i.e., $E = \{\{x,y\} \mid f \in F, \{x,y\} \subseteq \mathrm{var}(f), x \neq y\}$. The *incidence graph* $I_F = (\mathrm{var}(F) \cup F, E')$ of F is over variables and clauses (or terms) of F and $E' = \{\{f,x\} \mid f \in F, x \in \mathrm{var}(f)\}$. For a QBF Q, $P_Q = P_{\mathsf{matr}(Q)}$ and $I_Q = I_{\mathsf{matr}(Q)}$.

Treewidth and Pathwidth. For a rooted tree T and a node t in T, we let $\mathrm{children}(t)$ be the set of all *child nodes* of t in T. Let $G = (V,E)$ be a graph. A *tree decomposition (TD)* [21] of graph G is a pair $\mathcal{T} = (T, \chi)$ where T is a tree, and χ is a mapping that assigns to each node t of T a set $\chi(t) \subseteq V$, called a *bag*, such that the following conditions hold: (i) $V = \bigcup_{t \text{ of } T} \chi(t)$ and $E \subseteq \bigcup_{t \text{ of } T}\{\{u,v\} \mid u, v \in \chi(t)\}$; and (ii) for each q, s, t, such that s lies on the path from q to t, we have $\chi(q) \cap \chi(t) \subseteq \chi(s)$. Then, $\mathrm{width}(\mathcal{T}) = \max_{t \text{ of } T} |\chi(t)| - 1$. The *treewidth* $\mathrm{tw}(G)$ of G is the minimum $\mathrm{width}(\mathcal{T})$ over all TDs $\mathcal{T}$ of G.

Analogously, we define *pathwidth* $\mathrm{pw}(G)$ as the minimum $\mathrm{width}(\mathcal{T})$ over all TDs of G that are paths. The *2-local pathwidth* of G is the minimum $\mathrm{width}(\mathcal{T})$ over all TDs of G that are paths, such that every vertex of G appears in at most two different bags of $\mathcal{T}$. We denote 2-local pathwidth of a graph G by $\mathrm{pw}_2^{\mathrm{loc}}(G)$.

We define function $\mathrm{tow}(\cdot, \cdot)$ in the following way: $\mathrm{tow}(1, k) = 2^k$, $\mathrm{tow}(i + 1, k) = 2^{\mathrm{tow}(i,k)}$. For TQBF, the following lower bound result is known.

Theorem 1 [9]. *Assume we are given an arbitrary QBF of the form $Q = Q_1 V_1 Q_2 V_2 Q_3 V_3 \ldots Q_\ell V_\ell F$, where $\ell \geq 1$ and F in 3-CNF (if $Q_\ell = \exists$) or 3-DNF (if $Q_\ell = \forall$). Then, unless ETH fails, Q cannot be solved in time $\mathrm{tow}(\ell, o(k)) \, \mathrm{poly}(|\mathrm{var}(F)|)$, where k is the pathwidth of graph P_Q.*

Bandwidth and Cutwidth. Let $G = (V, E)$ be a given graph and $f : V \to \{1, \ldots, |V|\}$ be a bijective (one-to-one) *ordering* that uniquely assigns a vertex to an integer. Then, the *dilation* of G and f is defined as $\max_{\{u,v\} \in E} |f(u) - f(v)|$. The *bandwidth* $\mathsf{bw}(G)$ of G is the minimum dilation among every ordering f for G. The *cutwidth* $\mathsf{cw}(G)$ is the minimum $\max_{1 \leq i \leq |V|} |\{\{x, y\} \in E \mid f(x) \leq i, f(y) > i\}|$ among every ordering f for G.

Lemma 1 ($\star$). *Let* $G = (V, E)$ *be a graph then* $\mathsf{bw}(G) \leq 2\mathsf{pw}_2^{loc}(G)$.

Observation 1 ($\star$). *For any graph* $G = (V, E)$, *we have that* $\mathsf{cw}(G) \leq \Delta(G) \cdot \mathsf{bw}(G)$, *where* $\Delta(G)$ *is the maximum degree of* G.

3   A Σ_2^P-Complete Problem – Single-Exponential for Treewidth

One might expect that if a problem is hard for the second level of the polynomial hierarchy, then this automatically implies a double exponential lower bound in terms of treewidth, pathwidth, or cutwidth. However, this is not always the case. We show that 2-CLIQUE COLORING admits a single-exponential time algorithm parameterized by treewidth, even though 2-CLIQUE COLORING is Σ_2^P-complete [17].

Theorem 2 ($\star$). *The problem* 2-CLIQUE COLORING *on any graph* $G = (V, E)$ *can be solved in time* $2^{\mathcal{O}(\mathsf{tw}(G))} \cdot \mathrm{poly}(|V|)$.

4   Simpler Lower Bound Proofs

The following lemma is a key ingredient of our paper, as it provides lower bounds for the most standard problems from different levels of the polynomial hierarchy.

Lemma 2. *Assume we are given an arbitrary QBF of the form* $Q = Q_1V_1.Q_2V_2.Q_3V_3 \cdots Q_\ell V_\ell.F$, *where* $\ell \geq 1$ *and* F *is in 3-CNF (if* $Q_\ell = \exists$) *or 3-DNF (if* $Q_\ell = \forall$), *and a path decomposition of the primal graph* P_F *of width* w. *Then in polynomial time we can construct a formula* F' *and QBF* $Q' = Q_1V_1'.Q_2V_2'.Q_3V_3' \cdots Q_{\ell-1}V_{\ell-1}'Q_\ell V_\ell'.F'$ *such that* Q *and* Q' *are equivalent, each variable in* F' *appears at most in 5 clauses (if* $Q_\ell = \exists$) *or 5 terms (if* $Q_\ell = \forall$) *and has at most 4 neighbors in the primal graph* $P_{F'}$. *Moreover,* $\mathsf{pw}_2^{loc}(P_{F'}) \leq 2w+1$, $\mathsf{pw}_2^{loc}(I_{F'}) \leq 4w+4$, $\mathsf{bw}(P_{F'}) \leq 4w + 2$, $\mathsf{bw}(I_{F'}) \leq 8w + 8$, $\mathsf{cw}(P_{F'}) \leq 16w + 8$, $\mathsf{cw}(I_{F'}) \leq 40w + 40$.

We assume that $Q_\ell = \exists$, since the case $Q_\ell = \forall$ works analogously, but inverts Q first. In other words, if $Q_\ell = \forall$ we proceed on the inverse QBF $\overline{Q} = \overline{Q_1}V_1.\overline{Q_2}V_2.\overline{Q_3}V_3 \cdots \overline{Q_\ell}V_\ell.\overline{F}$, where $\overline{F}$ is the negation of F in 3-DNF and for i with $1 \leq i \leq \ell$, $\overline{Q_i} = \exists$ if $Q_i = \forall$ and $\overline{Q_i} = \forall$ if $Q_i = \exists$. From now on, assume $Q_\ell = \exists$.

Let $F = C_1 \wedge \ldots \wedge C_m$, where F depends on variables $x_1, x_2, \ldots x_n$. Let $\mathcal{T} = (T, \chi)$ be a given path decomposition of P_F. For each i, there is a node t_i, such that all variables of clause C_i are contained in the bag of t_i. In Fig. 1 for a formula $F_1 = C_1 \wedge C_2 \wedge C_3 \wedge C_4 \wedge C_5$ with $C_1 = x_1 \vee x_2 \vee \neg x_3$, $C_2 = \neg x_1 \vee v_2 \vee x_4$, $C_3 =$

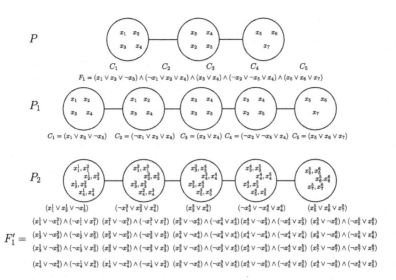

Fig. 1. A modification of a 3-CNF formula F_1 with a path decomposition P of the primal graph P_{F_1}, thereby constructing a formula F_1' with a path decomposition P_2.

$x_3 \vee x_4, C_4 = \neg x_2 \vee \neg x_5 \vee x_4, C_5 = x_5 \vee x_6 \vee x_7$ we have that bag t_1 contains variables of clauses C_1, C_2, C_3, bag t_2 contains variables of clauses C_3, C_4. Bag t_3 contains variables of the clause C_5. Having a path decomposition of some formula, we can construct a new path decomposition of the same width with the property that there is a bijection f between clauses and bags, such that for each clause C bag $f(C)$ contains all variables from the clause C. In order to do this, we simply delete bags to which no clauses were assigned and repeat bags that were assigned to ℓ clauses ℓ times. In Fig. 1, we show the modification of a path decomposition P for the formula F_1 considered before.

By renumbering clauses, we can assume that we have a path decomposition P_1 such that variables of clause C_i are contained in the i-th bag of the path decomposition. Now, we create a new path decomposition P_2 based on P_1. If variable x_j is contained in the i-th bag of path decomposition P_1, we create variables x_j^i, x_j^{i+1} and place them into i-th bag of the path decomposition P_2, see Fig. 1. We also create a new formula F' based on F and the path decomposition P_1. Let C_i be a clause in F with three literals $\ell_a \vee \ell_b \vee \ell_c$. Then F' contains the clause $\ell_a^i \vee \ell_b^i \vee \ell_c^i$, which is assigned to the i-th bag of the path decomposition P_2 (recall that C_i is assigned to the i-th bag of the path decomposition P_1). We proceed similarly for every clause in F with less then 3 literals. Moreover, if i-th bag of the path decomposition P_2 contains variables x_j^i, x_j^{i+1}, then we add clauses $x_j^i \vee \neg x_j^{i+1}, \neg x_j^i \vee x_j^{i+1}$. We call these clauses *equality clauses*, as they are true if and only if $x_j^i = x_j^{i+1}$. An example of the construction of formula F' and its path decomposition P_2 is depicted in Fig. 1 for $F = F_1$.

In order finish the definition of Q', it remains to specify sets $V_1', V_2', \ldots, V_\ell'$. For each $j \in \{1, 2, \ldots n\}$, let i_j be the smallest index such that bag i_j contains a variable x_j and $x_j \in V_{p_j}$. Then we put variable $x_j^{i_j}$ into the set V'_{p_j}, and put variables $x_j^{i_j+1}, x_j^{i_j+2}, x_j^{i_j+3}, \ldots$ into the set V'_ℓ.

If for the formula F_1 we consider the following QBF $\forall x_1, x_2, x_5, x_7 \exists x_3,$ $x_4, x_6 F_1$ and path decomposition P, then our transformation outputs the following QBF $Q' = \forall x_1^1, x_2^1, x_5^3, x_7^5 \; \exists x_1^2, x_2^2, x_3^3, x_2^4, \; x_3^1, x_3^2, x_3^3, x_3^4, x_3^5, x_4^1,$ $x_4^2, x_4^3, x_4^4, x_4^5, x_5^4, x_5^5, x_6^5, x_6^6, x_7^6 F_1'$, where F_1' contains all clauses shown in Fig. 1 below the path decomposition P_2.

It is easy to see that Q' is equivalent to Q, as essentially variable $x_j^{i_j}$ in Q' plays the same role as variable x_j in Q and due to equality clauses and the fact that variables $x_j^{i_j+1}, x_j^{i_j+2}, \ldots$ are placed into innermost quantifier ($\exists$), we have $x_j^{i_j} = x_j^{i_j+1} = x_j^{i_j+2} = \ldots$.

Note that in the path decomposition P_2, each variable appears at most in two bags and if P has width w, then the width of P_2 is $2w + 1$. Therefore, the 2-local pathwidth of $P_{F'}$ is at most $2w + 1$. Moreover, note that each variable appears in at most in 5 clauses and has at most 4 neighbors.

For the constructed path decomposition P_2 it is easy to convert it to a path decomposition of the incidence graph of the formula F'. Indeed, we simply add a vertex corresponding to the clause C_i, to the i-th bag, and add vertices corresponding to equality clauses between variables x_j^i, x_j^{i+1} to the i-th bag as well. In this case, the bag with $2q$ vertices/variables additionally acquires $1 + 2q$ vertices. Hence, overall each bag of the constructed decomposition of $I_{F'}$ contains at most $4w + 5$ vertices, and each vertex appears in at most two bags. Therefore, the 2-local pathwidth of the graph $I_{F'}$ is at most $4w + 4$.

By Lemma 1, we have $\mathsf{bw}(P_{F'}) \leq 4w + 2$, $\mathsf{bw}(I_{F'}) \leq 8w + 8$). Moreover, by Observation 1, $\mathsf{cw}(P_{F'}) \leq 4\mathsf{bw}(P_{F'}) \leq 16w + 8$, $\mathsf{cw}(I_{F'}) \leq 5\mathsf{bw}(I_{F'}) \leq 40w + 40$.

In some sense, the next theorem provides a factory that produces double- or triple-exponential lower bounds. Now, in order to design a problem with a double-exponential lower bound, one needs to take a Σ_2^P-complete or Π_2^P-complete (one can find such problems in the excellent survey [22]) and design a reduction that proves completeness of the problem and satisfies the properties described in the following lemma. As we will see later, $(2,3)$-CHOOSABILITY is one of these problems that satisfies the criteria.

Theorem 3 ($\star$). *Assuming ETH, TQBF on formulas Q with quantifier depth ℓ, clause length at most 3 and occurrences of each variable at most 5 does not admit $\mathsf{tow}(\ell, o(k)) \mathsf{poly}(|\mathsf{var}(\mathsf{matr}(Q))|)$ algorithm where k is one of the following parameters: 2-local pathwidth, bandwidth, cutwidth of primal graph P_Q or incidence graph I_Q.*

Lemma 3. *Let P be some graph problem. Assume that P is Σ_ℓ^P-hard or Π_ℓ^P-hard for some $\ell > 1$. Further, let the proof of hardness be based on a reduction R that takes a QBF Q with $F = \mathsf{matr}(Q)$ and constructs a graph $G = (V, E)$, such that the following properties hold: (i) each variable is replaced with a constant size subgraph/gadget; (ii) each clause is replaced with a constant size subgraph/gadget; (iii) a gadget for a variable x can be connected with a gadget of a clause C only* *by a constant-size gadget and only if $x \in C$. We note that our connector-gadget have common vertices with variable-gadget and clause-gadget that it connects. However, variable-gadgets and clause-gadgets do not share common vertices.*

Then, (i) P does not admit an algorithm running in time $\text{tow}(\ell, o(\text{cw}(G))) \cdot \text{poly}(|V|)$, unless ETH is false; (ii) P does not admit a runtime $\text{tow}(\ell, o(\text{bw}(G))) \cdot \text{poly}(|V|)$, unless ETH is false. Hence, P also does not admit $\text{tow}(\ell, o(\text{tw}(G))) \cdot \text{poly}(|V|)$ or $\text{tow}(\ell, o(\text{pw}(G))) \cdot \text{poly}(|V|)$ runtimes, unless ETH is false.

Proof. For convenience, we denote the variable-gadget corresponding to a variable x by $G[x]$, the clause-gadget of a clause C by $G[C]$, the connector-gadget that corresponds to an edge between variable x and a clause C by $G[x, C]$. The largest number of vertices in a variable, clause, or connector-gadget is denoted by V_v, V_c, or V_{con}, respectively. Similarly, the largest number of edges in these gadgets are denoted by E_v, E_c, or E_{con}, respectively. Vertices from the set $G[x, C] \setminus G[x] \setminus G[C]$ we call internal vertices of the gadget $GF[x, C]$.

First we prove (i). We know by Theorem 3 that a QBF Q with quantifier rank ℓ, does not admit an algorithm running in time $\text{tow}(\ell, o(\text{cw}(I_Q))) \cdot \text{poly}(|\text{var}(\text{matr}(Q))|)$, even on formulas where each variable appears at most 5 times. Let G' be the graph obtained after the reduction is applied to a QBF Q. So, it is enough to show that G' has cutwidth at most $c \cdot \text{cw}(I_Q)$ for some constant c depending only on a variable-, clause-, and connector-gadget. Having an arrangement π of the graph I_Q with cutwidth $\text{cw}(I_Q)$, we construct an arrangement, i.e., vertex ordering, of vertices of G' with cutwidth at most $c \cdot \text{cw}(I_Q)$. In order to do this, we replace each vertex in π by a corresponding variable-gadget or clause-gadget. All vertices from one gadget are placed successively. In graph G', we also might have some additional vertices that are internal vertices of some connector gadgets. We place such internal vertices successively somewhere between vertices corresponding to variable-gadget and clause-gadget. The operation is performed in such way that finally all vertices of any variable-gadget and any clause-gadget are placed successively, as well as internal vertices of any connector-gadgets. Now, in order to show that the arrangement has at most cutwidth $c \cdot \text{cw}(I_Q)$ for some c, we perform some changes to the graph G'. The modifications can only increase the cutwidth in the considered arrangement.

For each connector-gadget $G[x, C]$, we delete all internal vertices of the gadget, however, we connect each vertex from gadget $G[x]$ with each vertex of gadget $G[C]$ by $|E(G[x, C])|$ parallel edges. It is easy to see that the cutwidth in the induced arrangement can only increase, as we delete at most $|E(G[x, C])|$ edges and add at least $|E(G[x, C])|$ edges going across, were the deletion occurs. After replacement of internal vertices for each connector gadget, we are ready to bound the cutwidth. Note that the cut size in the position between two gadgets is at most $E_{con} \cdot V_v \cdot V_c$ times larger than the cut size in the same position in ordering π of graph I_Q. Now, we bound the cut size in the middle of some variable or clause gadget. The number of edges over the gadget can increase at most by a factor $E_{con} \cdot V_v \cdot V_c$ as in the previous case. However, we should also add edges that are contained in the gadget, as well as edges that were entering the corresponding vertex from left and from right. Overall, it shows that the cut size is at most $\max\{E_c, E_v\} + 3E_{con} \cdot V_v \cdot V_c \cdot \mathsf{cw}(I_Q)$. So we proved $\mathsf{cw}(G') \leq c \cdot \mathsf{cw}(I_Q)$.

Now we prove a lower bound in terms of bandwidth. As before, we can assume that we are given a QBF Q, whose incidence graph I_Q has degree at most 5. We assume that $R(Q)$ is a graph G'. Similarly as for cutwidth, we consider an arrangement π of vertices of the graph I_Q with the smallest bandwidth. Additionally we orient all edges from left to right. Now we construct an arrangement of graph G' with bandwidth at most $c' \cdot \mathsf{bw}(I_Q)$. We replace each vertex from graph I_Q in the arrangement with a sequence of vertices of the corresponding gadget. Recall that all our edges in I_Q are directed. For all edges out-going from vertex u, we put all internal vertices of all corresponding connector-gadgets immediately after the right-most vertex from the gadget corresponding to vertex u. Note that vertices of the same gadget are placed consecutively in the ordering. Let us estimate how the edge length can change after such an operation. Recall that in the incidence graph all edges connect variables with clauses. Let $x \in C$ and assume that x is to the left of C in the optimal arrangement (case when x is to the right of C is identical). After the first operation (replacing variables and clauses by gadgets), the worst case is when the leftmost vertex of gadget $G[x]$ is connected to the rightmost vertex of a gadget $G[C]$, so the edge distance is bounded by $(\pi(C) - \pi(x)) \cdot \max\{V_c, V_v\} + V_c + V_v$. After that, between $G[x]$ and $G[C]$ we add internal vertices of connector gadgets. Note that between gadgets $G[x]$ and $G[C]$, we have at most $(\pi(C) - \pi(x))$ variable or clause gadgets. Hence, at most $5(\pi(C) - \pi(x))V_{con}$ vertices are be added between gadgets $G[x]$ and $G[C]$. So, the bandwidth of G' is at most $\mathsf{bw}(I_Q) \cdot \max\{V_c, V_v\} + V_c + V_v + 5\mathsf{bw}(I_Q)V_{con}$, which leads to the desired lower bound, assuming ETH.

Corollary 1. *Problem* $(2,3)$-CHOOSABILITY *on any graph* $G = (V, E)$ *does not admit algorithms running in time* $2^{2^{o(\mathsf{cw}(G))}} \cdot \mathrm{poly}(|V|)$, $2^{2^{o(\mathsf{bw}(G))}} \cdot \mathrm{poly}(|V|)$ *unless ETH is false. Hence,* $(2,3)$-CHOOSABILITY *also does not admit algorithms running in time* $2^{2^{o(\mathsf{pw}(G))}} \cdot \mathrm{poly}(|V|)$, $2^{2^{o(\mathsf{tw}(G))}} \cdot \mathrm{poly}(|V|)$ *unless ETH is false, since* $\mathsf{cw}(G) \geq \mathsf{pw}(G), \mathsf{tw}(G)$.

Proof. The reduction in [8] establishes Π_2^P-completeness of BIPARTITE $(2,3)$-CHOOSABILITY, using constant size connector-gadgets (single edge) and clause-gadgets (one vertex). However, the variable gadget, which is a union of $\exists$-graph or $\forall$-graph gadgets (depicted in the Fig. 2) with a chain of propagators (Fig. 3) (see details in [8]), is of size proportional to m (number of clauses in the formula). If a variable is quantified by an $\exists$ quantifier, its gadget contains the $\exists$-graph gadget, otherwise it contains the $\forall$-graph gadget. The only reason why the variable-gadget has m propagators, is the fact that the reduction from [8] reduces from an arbitrary 3-CNF formula and in their input formula, a variable can appear up to m times in the whole formula. However, we know that TQBF_2 over matrix F does not admit an algorithm running in time $2^{2^{o(\mathsf{cw}(I_F))}} \cdot \mathrm{poly}(|\mathrm{var}(F)|)$, even on instances, where each variable appears at most 5 times, assuming ETH. Therefore, we can use almost identical gadgets as in [8]. In our case, the variable

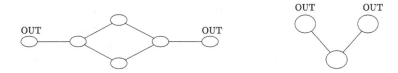

Fig. 2. ∀-graph (left) and ∃-graph (right). All shown vertices have list length 2 in the $(2, 3)$-CHOOSABILITY problem.

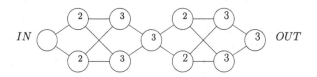

Fig. 3. Propagator gadget.

gadget has constant size. Hence, by Lemma 3, BIPARTITE $(2, 3)$-CHOOSABILITY does not admit an algorithm running in time $2^{2^{o(\mathsf{cw}(G))}} \cdot \mathrm{poly}(|V|)$, assuming ETH.

We note that even though we depicted the ∀-graph, ∃-graph (Fig. 2), and propagator graph (Fig. 3), we never use them, since it is only relevant that variable-gadgets, clause-gadgets and connector gadgets have constant size. We emphasize that our proof is simple by knowing the reduction that proves Π_2^P-completeness of $(2, 3)$-CHOOSABILITY.

As mentioned in [18], Gutner and Tarsi [12] gave a reduction from $(2, 3)$-CHOOSABILITY to k-CHOOSABILITY, where the pathwidth of the constructed graph G' is $\mathcal{O}(\mathsf{pw}(G))$ which implies a lower bound $2^{2^{o(\mathsf{pw}(G))}} \cdot \mathrm{poly}(|V|)$, $2^{2^{o(\mathsf{tw}(G))}} \cdot \mathrm{poly}(|V|)$ for k-CHOOSABILITY. However, the reduction does not allow to rule out $2^{2^{o(\mathsf{cw}(G))}} \cdot \mathrm{poly}(|V|)$ algorithm for k-CHOOSABILITY. One can also show modifications of the construction so that it works for cutwidth, however, this is not the essence of the current submission, so we do not provide proof details here.

Corollary 2. *Assuming ETH, k-CHOOSABILITY for any $k \geq 3$ on any input graph $G = (V, E)$ cannot be decided in time $2^{2^{o(\mathsf{pw}(G))}} \cdot \mathrm{poly}(|V|)$.*

5   Discussion and Conclusion

Prior to our work there was no systematic way of proving double or triple-exponential lower bounds that have matching upper bounds. Indeed, proving such lower bounds required significant creativity and a lot of work. For example, the proof by Mitsou and Marx that $(2, 3)$-CHOOSABILITY requires double exponential running time in the pathwidth used new tricky gadgets, employed reductions from EDGE 3-COLORING, and did not rely on the Π_2^P-completeness proof of k-CHOOSABILITY in any way. They even wrote:

> *The reader may feel that the Π_2^P or Σ_3^P-completeness of these problems already give sufficient explanation why double- or triple-exponential dependence on treewidth is needed. This is true in some sense: the quantifier alternations in the problem definitions are the common underlying reasons for being in the higher levels of the polynomial hierarchy and for requiring unusually large dependence on treewidth.* **But let us point out that these two types of complexity results require very different proof structures.** *In Π_2^P or Σ_3^P-completeness proofs, we start with a canonical Π_2^P or Σ_3^P-complete quantified satisfiability problem and we use the alternations inherent in the definitions of k-CHOOSABILITY or k-CHOOSABILITY DELETION to express the alternations in quantified satisfiability. On the other hand, in the proofs of Theorems 2(2) and 3(2), we start with problems in NP and use the alternations inherent in k-CHOOSABILITY or k-CHOOSABILITY DELETION for compression: we want to express the original instance by a graph having treewidth only $\mathcal{O}(\log n)$ or $\mathcal{O}(\log \log n)$. Thus the main theme of our proofs is trading alternation for compression: we want to use alternation to allow the succinct encoding and verification of information.*
>
> — MARX AND MITSOU   [18]

We stress that the full version (private communication) of the paper [18] contains two additional results, namely the triple-exponential lower bound for k-CHOOSABILITY DELETION with $k \geq 4$ and Σ_3^P-completeness of k-CHOOSABILITY DELETION where $k \geq 3$. The first proof is significantly more complicated than the second one, as it uses a larger amount of non-trivial gadgets and ideas. It even takes *15 pages*, while the second one takes **less than 2 pages**. Moreover, the first reduction does not work for $k = 3$ while the second does. However, with our construction we are convinced that their proof of Σ_3^P-completeness fits the requirements for the reduction R from Lemma 3 and hence *almost immediately* implies a $2^{2^{2^{o(pw)}}} n^{\mathcal{O}(1)}$ lower bound for k-CHOOSABILITY DELETION even if $k = 3$.

Beyond that, the construction from Lemma 2 can be enhanced using standard gadgets for converting SAT into PLANAR SAT, see [16], so that the incidence graph of the formula F' is planar. This immediately implies double exponential lower bounds even for planar variants of some problems.

We expect that this work opens several interesting research questions. First, we are certain that the quest of finding other interesting problems with higher dependence on structural parameters like treewidth, will be rewarding and insightful. On top, reducing from TQBF enables simple(r) conditional lower bounds for further interesting problems. In the spirit of *the book*[1] and since existing lower bound proofs are oftentimes more involved, we should aim for easier proofs. Therefore the focus on future developments should be further simplifications and improvements in making complexity easier to grasp.

---

[1] For proofs from the BOOK, see en.wikipedia.org/wiki/Proofs_from_THE_BOOK.

Acknowledgments. We would like to thank anonymous reviewers for their feedback that improved the presentation of our results. Ivan Bliznets is grateful to Valia Mitsou for useful discussions in a very early stage.

References

1. Atserias, A., Oliva, S.: Bounded-width QBF is PSPACE-complete. J. Comput. Syst. Sci. **80**(7), 1415–1429 (2014)
2. Bertelè, U., Brioschi, F.: Contribution to nonserial dynamic programming. J. Math. Anal. Appl. **28**(2), 313–325 (1969)
3. Bodlaender, H.L., Koster, A.M.: Combinatorial optimization on graphs of bounded treewidth. Comput. J. **51**(3), 255–269 (2008). https://doi.org/10.1093/comjnl/bxm037
4. Chen, H.: Quantified constraint satisfaction and bounded treewidth. In: ECAI'04, pp. 161–165. IOS Press (2004)
5. Chimani, M., Mutzel, P., Zey, B.: Improved Steiner tree algorithms for bounded treewidth. J. Discret. Algorithms **16**, 67–78 (2012). https://doi.org/10.1016/j.jda.2012.04.016
6. Cygan, M., et al.: Parameterized Algorithms. Springer, Cham (2015). https://doi.org/10.1007/978-3-319-21275-3
7. Downey, R.G., Fellows, M.R.: Fundamentals of Parameterized Complexity. TCS, Springer, London (2013). https://doi.org/10.1007/978-1-4471-5559-1
8. Erdos, P., Rubin, A.L., Taylor, H.: Choosability in graphs. Congr. Numer. **26**(4), 125–157 (1979)
9. Fichte, J.K., Hecher, M., Pfandler, A.: Lower bounds for QBFs of bounded treewidth. In: LICS'20, pp. 410–424. ACM (2020)
10. Flum, J., Grohe, M.: Parameterized Complexity Theory. TTCSAES, Springer, Heidelberg (2006). https://doi.org/10.1007/3-540-29953-X
11. Foucaud, F., et al.: Tight (Double) Exponential Bounds for NP-Complete Problems: Treewidth and Vertex Cover Parameterizations. CoRR **abs/2307.08149** (2023)
12. Gutner, S., Tarsi, M.: Some results on (a: b)-choosability. Discret. Math. **309**(8), 2260–2270 (2009)
13. Impagliazzo, R., Paturi, R., Zane, F.: Which problems have strongly exponential complexity? J. Comput. Syst. Sci. **63**(4), 512–530 (2001). https://doi.org/10.1006/jcss.2001.1774
14. Kleine Büning, H., Lettman, T.: Propositional Logic: Deduction and Algorithms. Cambridge University Press, Cambridge (1999)
15. Lampis, M., Mitsou, V.: Treewidth with a quantifier alternation revisited. In: IPEC'17, vol. 89, pp. 26:1–26:12. Dagstuhl Publishing (2017)
16. Lichtenstein, D.: Planar formulae and their uses. SIAM J. Comput. **11**(2), 329–343 (1982)
17. Marx, D.: Complexity of clique coloring and related problems. Theor. Comput. Sci. **412**(29), 3487–3500 (2011)
18. Marx, D., Mitsou, V.: Double-exponential and triple-exponential bounds for choosability problems parameterized by treewidth. In: ICALP'16. LIPIcs, vol. 55, pp. 28:1–28:15. Schloss Dagstuhl - Leibniz-Zentrum für Informatik (2016)
19. Pan, G., Vardi, M.Y.: Fixed-parameter hierarchies inside PSPACE. In: LICS'06, pp. 27–36. IEEE Computer Society (2006). https://doi.org/10.1109/LICS.2006.25

20. Papadimitriou, C.H.: Computational Complexity. Addison-Wesley, Boston (1994)
21. Robertson, N., Seymour, P.D.: Graph minors. I. Excluding a forest. J. Comb. Theory Ser. B **35**(1), 39–61 (1983). https://doi.org/10.1016/0095-8956(83)90079-5
22. Schaefer, M., Umans, C.: Completeness in the polynomial-time hierarchy: a compendium. SIGACT News **33**(3), 32–49 (2002)
23. Stockmeyer, L.J., Meyer, A.R.: Word problems requiring exponential time. In: STOC'73, pp. 1–9. ACM (1973). https://doi.org/10.1145/800125.804029

Source-Oblivious Broadcast

Pierre Fraigniaud[1]([✉]) and Hovhannes A. Harutyunyan[2]

[1] Institut de Recherche en Informatique Fondamentale, CNRS and Université Paris Cité, Paris, France
`pierre.fraigniaud@irif.fr`

[2] Department of Computer Science and Software Engineering, Concordia University, Montréal, Canada

Abstract. This paper revisits the study of (minimum) broadcast graphs, i.e., graphs enabling fast information dissemination from every source node to all the other nodes (and having minimum number of edges for this property). This study is performed in the framework of compact distributed data structures, that is, when the broadcast protocols are bounded to be encoded at each node as an ordered list of neighbors specifying, upon reception of a message, in which order this message must be passed to these neighbors. We show that this constraint does not limit the power of broadcast protocols, as far as the design of (minimum) broadcast graphs is concerned. Specifically, we show that, for every n, there are n-node graphs for which it is possible to design protocols encoded by lists yet enabling broadcast in $\lceil \log_2 n \rceil$ rounds from every source, which is optimal even for general (i.e., non space-constrained) broadcast protocols. Moreover, we show that, for every n, there exist such graphs with the additional property that they are asymptotically as sparse as the sparsest graphs for which $\lceil \log_2 n \rceil$-round broadcast protocols exist, up to a constant multiplicative factor. Concretely, these graphs have $O(n \cdot L(n))$ edges, where $L(n)$ is the number of leading 1s in the binary representation of $n - 1$, and general minimum broadcast graphs are known to have $\Omega(n \cdot L(n))$ edges.

Keywords: Broadcast graphs · Minimum broadcast graphs · Network design · Information dissemination · Distributed data structures

1 Introduction

1.1 Context

The *broadcast* problem is the information dissemination problem consisting of passing a piece of information (i.e., an atomic message) from a node of a connected graph to all the nodes of the graph. A broadcast protocol proceeds in synchronous *rounds*. Initially, a single node is informed, called the *source* node.

P. Fraigniaud—Additional support from ANR project DUCAT (ref. ANR-20-CE48-0006).

At each round, every informed node (i.e., every node possessing the information) can transmit the information to at most one of its neighbors. It follows that the number of informed nodes can at most double at each round, and thus, for every source node s, the minimum number of rounds required to broadcast from s in an n-node graph G, denoted by $b(G, s)$, is at least $\lceil \log_2 n \rceil$.

Broadcast Graphs. For every n-node connected graph G, let

$$b(G) = \max_{s \in V(G)} b(G, s)$$

be the *broadcast time* of G. Since $b(G, s) \geq \lceil \log_2 n \rceil$ for every source node s, we have $b(G) \geq \lceil \log_2 n \rceil$. For every $n \geq 1$, any graphs G with n nodes and satisfying $b(G) = \lceil \log_2 n \rceil$ is called *broadcast graph* [2]. Observe that broadcast graphs do exist. In particular, for every $n \geq 1$, the complete graph K_n is a broadcast graph. Indeed, in the complete graph, it is always possible to construct a matching from the set of informed nodes to the set of non-informed nodes, which saturates one of the two sets, and therefore $b(K_n) = \lceil \log_2 n \rceil$.

Minimum Broadcast Graphs. Motivated by the design of networks supporting efficient communication protocols but consuming few resources, the construction of sparse broadcast graphs has attracted lot of attention (cf. Sect. 1.4). For every $n \geq 1$, let $B(n)$ denotes the minimum number of edges of broadcast graphs with n nodes. As all complete graphs are broadcast graphs, $B(n)$ is well defined for all n. However, there are broadcast graphs much sparser than complete graphs. For instance, for every $d \geq 0$, the d-dimensional hypercube Q_d, with $n = 2^d$ nodes, is a broadcast graph as $b(Q_d) = d = \lceil \log_2 n \rceil$ (an optimal broadcast protocol merely consists to transmit the information sequentially through edges of increasing dimension). As a consequence, for n is a power of 2, we have $B(n) \leq \frac{1}{2} n \log_2 n$. In fact, it is known [6] that

$$B(n) = \Theta(n \cdot L(n)), \tag{1}$$

where, for every positive integer x, $L(x)$ denotes the number of consecutive leading 1 s in the binary representation of $x - 1$. For instance, $L(12) = 1$ as $11 = (1011)_2$. In particular, if $n = 2^d$ for $d \geq 1$, then $B(n) = \Theta(n \log n)$ as $n - 1 = (11 \ldots 11)_2$, and thus $L(n) = d = \log_2 n$. In fact, it is easy to see (see [2]) that all hypercubes Q_d, $d \geq 1$, are *minimum* broadcast graphs, that is, broadcast graphs with the smallest number of edges — this is simply because, for $n = 2^d$, every source node must be active at each round $r = 1, \ldots, d$ for insuring that all nodes receive the information after d rounds, and hence every (source) node must have degree d. Therefore, for n power of 2, we have $B(n) = \frac{1}{2} n \log_2 n$.

1.2   Objective

We are interested in the *encoding* of broadcast protocols at each node of a graph. For any source node s of a graph G, every node v of G receiving a piece of information originated from s must inform its neighbors (non necessarily all) in a right order for insuring that the information is broadcast fast, ideally in $\lceil \log_2 n \rceil$ rounds whenever G is a broadcast graph. If the local encoding of the protocol is done in a brute force manner, every node v stores a table T_v with n entries, one for each source s, such that

$$T_v[s] = (u_{s,1}, \ldots, u_{s,k_s})$$

provides v with an ordered list of neighbors that v must sequentially inform upon reception of a piece of information broadcast from s. That is, upon receiving a message broadcast from s, node v forwards that message to $u_{s,1}$ first, then to $u_{s,2}$, and so on, up to u_{s,k_s}, in k_s successive rounds. In the worst case, this encoding may consume up to $O(n \log d!)$ bits to be stored at a degree-d node v, which can be almost as high as storing the entire graph G at v whenever $d = \Theta(n)$.

Source-Oblivious Broadcast. With the objective of limiting the space complexity of encoding broadcast protocols locally at each node, we consider *source-oblivious* broadcast protocols, as previously considered in, e.g., [1,11,12]. Any such protocol can be encoded at each node v by a unique ordered list

$$\ell_v = (u_1, \ldots, u_k) \tag{2}$$

of k distinct neighbors of v, where $k \leq \deg(v)$, hence consuming only $O(d \log d)$ bits at degree-d nodes. That is, upon receiving a message, node v forwards that message to u_1 first, then to u_2, and so on, up to u_k, in k successive rounds, no matter the source of the information is.

Fully-Adaptive Source-Oblivious Broadcast. We actually focus on the variant of source-oblivious broadcast introduced in [4,5], called *fully-adaptive*. Specifically, we assume that, upon reception of a piece of information broadcast from a source node s, every node v acknowledges reception by sending a signal message to all its neighbors. Note that the signal messages are short in comparison to the broadcast messages, which could be arbitrarily large. It follows that, at the end of each round, every node v is aware of which of its neighbors have received the information, and this holds even if the node v has not yet received that information.

Broadcast thus performs as follows in the fully-adaptive source-oblivious model. Upon receiving a piece of information originated from any source s, every node v initiates a series of *calls* to its neighbors during subsequent rounds. Let $\ell_v = (u_1, \ldots, u_k)$ be the list of node v, and assume that v received the broadcast message at round r. Then, for $i = 1, \ldots, k$, node v aims at forwarding the message to node u_i at round $r + i$. However, if node u_i has already received the information at the round when v is supposed to send the message to u_i, then u_i is skipped, and the message is transmitted to u_{i+1} instead, unless u_{i+1} is also already informed, in which case u_{i+2} is considered, etc. More generally, at a given round, node v forwards the broadcast message to the next node in its list ℓ_v that has not already received that message. It stops when the list is exhausted.

1.3   Our Results

In a nutshell, we show that constraining broadcast protocols by bounding them to be source-oblivious does not limit their power, as far as the design of broadcast graphs and minimum broadcast graphs is concerned. Specifically, we first establish the following.

Theorem 1. *In the fully-adaptive source-oblivious model, there are n-node broadcast graphs for every $n \geq 1$, i.e., n-node graphs for which there exists a collection of lists $(\ell_v)_{v \in V(G)}$ achieving broadcast in $\lceil \log_2 n \rceil$ rounds from any source node. In particular, for every $n \geq 1$, the broadcast time of the clique K_n is $\lceil \log_2 n \rceil$ in the fully-adaptive source-oblivious model.*

Note that this result contrasts with the current knowledge about weaker variants of the source-oblivious model, like those defined in [1]. In particular, in the *adaptive* variant (where nodes are not aware whether their neighbors received the broadcast message, apart from the neighbors from which they actually received the message), and in the *non-adaptive* variant (where the nodes forward the message blindly, by following the orders specified by their lists, and ignoring the fact that there is no need to send the message to neighbors from which they actually received that message), the best known upper bound on the minimum number of rounds required to broadcast in the complete graph K_n is $\log_2 n + O(\log \log n)$ [10].

Next, we focus on the construction of minimum broadcast graphs, and establish the following.

Theorem 2. *In the fully-adaptive source-oblivious model, for every $n \geq 1$, there are n-node broadcast graphs with $O(n \cdot L(n))$ edges.*

It follows from this result combined with from Eq. (1) that fully-adaptive source-oblivious broadcast protocols enable the design of broadcast graphs as sparse as what can be achieved with general broadcast protocols (i.e., protocols taking into account the source of the broadcast), while drastically reducing the space complexity of the broadcast table to be stored at each node, from $O(n \log d!)$ to $O(d \log d)$ bits at degree-d nodes in n-node graphs.

Remark. The graphs whose existence is guaranteed by Theorem 2 are broadcast graphs, and therefore Theorem 1 can be viewed as a corollary of Theorem 2. However, the graphs exhibited in the proof of Theorem 1 have maximum degree $O(\log n)$, whereas the graphs used in the proof of Theorem 2 have maximum degree as large as $\Omega(n)$, which may be an issue from a practical perspective, as far as network design is concerned. On the other hand, the graphs in the proof of

Theorem 1 have $O(n \log n)$ edges, while the graphs in the proof of Theorem 2 are sparser, and actually have the smallest possible number of edges. In particular, for every $k \geq 0$, and every n satisfying $2^k < n \leq 2^k + 2^{k-1}$, the n-node broadcast graphs in Theorem 2 have a linear number of edges, as $L(n) = 1$ for n in this range.

1.4   Related Work

For more about the broadcast problem in general, we refer to the many surveys on the topic, such as [3,8,9]. We provide below a quick survey of the literature dealing with broadcast under *universal lists*, i.e., source-oblivious broadcast.

The broadcast problem under universal lists was first discussed indirectly by Slater et al. [12]. The first formal definition of the problem of broadcasting with universal lists was given by Rosenthal and Scheuermann [11], who described an algorithm for constructing optimal broadcast schemes for trees under the adaptive model. Later, Diks and Pelc [1] distinguished between non-adaptive and adaptive models with universal lists, and formally defined them. They designed optimal broadcast schemes for paths, cycles, and grids under both models. They also gave tight upper bounds for tori and complete graphs, for adaptive and non-adaptive models. Diks and Pelc also described an infinite family of graphs for which the adaptive broadcast time is strictly larger than the broadcast time. Later, Kim and Chwa [10] designed non-adaptive broadcast schemes for paths and grids. They also came up with upper bounds for hypercubes, and improved the upper bound from [1] under non-adaptive model. More recently, the lower and upper bounds for trees under the non-adaptive model were tightened by Harutyunyan et al. [7], as well as the upper bounds on general graphs. Harutyunyan et al. also presented a polynomial-time dynamic programming algorithm for finding the non-adaptive broadcast time of any tree. The most recent papers on the matter, by Gholami and Harutyunyan [4,5] defined the fully adaptive model with universal lists. Under this model they computed the broadcast time of grids, tori, hypercubes, and Cube Connected Cycles (CCC).

2   Preliminaries

Recall that, according to the definition of fully-adaptive source-oblivious broadcast introduced in [4], once a vertex v is informed, say at round t, it will follow its list of neighbors ℓ_v (cf. Eq. (2)), and pass the message to the first vertex on the list which is not already informed before round $t + 1$. In other words, not only node v skips all its neighbors from which it received the message, but v also skip all other informed neighbors. Given a set $L = (\ell_v)_{v \in V(G)}$ of lists, the number of rounds for broadcasting a message from $s \in V(G)$ to all the other nodes of G using lists L under the fully adaptive model is denoted by $b_{\mathsf{fa}}(G, s, L)$. The broadcast time of a graph G under the fully adaptive model is then defined as

$$b_{\mathsf{fa}}(G) = \min_{L} \max_{s \in V(G)} b_{\mathsf{fa}}(G, s, L).$$

The following results illustrate the above definition, and may serve as a warm up for the rest of the paper. The first proposition shows that the broadcast time using protocols encoded with lists cannot be more than twice the broadcast time using general protocols.

Proposition 1. *For every graph $G = (V, E)$, the broadcast complexity of G under the fully-adaptive source-oblivious model is at most $2 \min_{s \in V} b(G, s)$, that is,*

$$b_{\mathsf{fa}}(G) \leq 2 \min_{s \in V} b(G, s).$$

Proof. Let $s \in V$ with minimum broadcast time among all nodes in G, i.e., $b(G, s) = \min_{s' \in V} b(G, s')$. Let T be a broadcast tree rooted at s, enabling broadcast from s in $b(G, s)$ rounds. (Such a tree is a spanning tree of G rooted at s in which the children of any node v are ordered, specifying the order in which these children must be informed by v upon reception of a message from s.) Let $v \neq s$ be a node of T, let w be the parent of v in T, and let $u_1, \ldots, u_d$ be the d children of v in T enumerated in the order in which they are called in an optimal broadcast protocol from s in T. The list assigned to v is

$$\ell_v = (w, u_1, \ldots, u_d).$$

Similarly, the list assigned to s is

$$\ell_s = (u_1, \ldots, u_d)$$

where $u_1, \ldots, u_d$ are the d children of s in T enumerated in the order in which they are called in an optimal broadcast protocol from s in T.

To show that this set of lists enable fully adaptive broadcast to perform in at most $2b(G, s)$ rounds, let u be a source node. Following the assigned universal lists every vertex makes the first call to its parent once gets informed. Thus, a message broadcast from u reaches the root s of T after at most $\text{depth}(T)$ rounds. Since $b(G, s) \geq \text{depth}(T)$, the message reaches s after at most $b(G, s)$ rounds. Once at s, the message is broadcast down the tree in at most $b(G, s)$ rounds since, for every node $v \neq s$, the parent w of v in the list ℓ_v is skipped in the fully adaptive model. The upward and downward phases amount for a total of $2\, b(G, s)$ rounds.   □

The following is a direct consequence of Proposition 1.

Corollary 1. *For every graph G, the broadcast complexity of G under the fully-adaptive source-oblivious model is at most $2\, b(G)$, i.e.,*

$$b_{\mathsf{fa}}(G) \leq 2\, b(G).$$

Note that Proposition 1 also holds in the adaptive source-oblivious model (in this model every node v just skips the neighbors in ℓ_v from which it received the message). The second proposition shows that there is a price to pay for using broadcast protocols encoded with lists, in the sense that the broadcast time using such protocols may be larger than the broadcast time using general protocols.

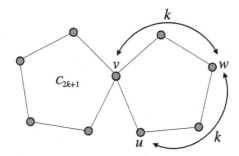

Fig. 1. Graph in the proof of Proposition 2, for $k = 2$.

Proposition 2. *There is an infinite family of graphs $\mathcal{F}$ such that, for every $G \in \mathcal{F}$, the broadcast complexity of G in the fully-adaptive source-oblivious model is larger than its broadcast time, that is $b_{\mathsf{fa}}(G) > b(G)$.*

Proof. A basic example of such a family of graphs is obtained from two cycles C_{2k+1} of length $2k + 1$, by merging two of their vertices into one (see Fig. 1). For every $k \geq 1$, the resulting graph G_k has $4k + 1$ vertices, with a cut vertex v at the intersection of the two cycles. It is known [8] that $b(C_{2k+1}) = k + 1$, from which it follows that $b(G_k) = 2k + 1$.

In the fully-adaptive source-oblivious model, let ℓ_v be the list of node v, and let u be the neighbor of v occurring first in this list. Let w be the vertex antipodal to the edge $\{u, v\}$ in the cycle containing both u and v. If w does not call first its neighbor on the shortest path from w to v in G_k, then broadcast from w will take at least $2k + 2$ rounds. On the other hand, if w calls first its neighbor on the shortest path from w to v in G_k, then v will receive the information in round k. Since u has not yet received the information at the end of round k, v will proceed according to its list ℓ_v, and call u at round $k + 1$. As a result, v will start broadcasting in the other cycle (the one not containing u and w), no sooner than round $k + 2$, and thus the whole protocol will not complete before $2k + 2$ rounds. $\qquad\square$

3   Broadcast Graphs

This section is devoted to a sketch of proof for Theorem 1. Recall that this theorem states that, in the fully-adaptive source-oblivious model, there are n-node broadcast graphs for every $n \geq 1$, i.e., n-node graphs for which there exists a collection of lists $(\ell_v)_{v \in V(G)}$ achieving broadcast in $\lceil \log_2 n \rceil$ rounds from any source node. Due to the space constraint the proof of Theorem 1 is omitted, but Fig. 2 provides the intuition of the construction in the proof.

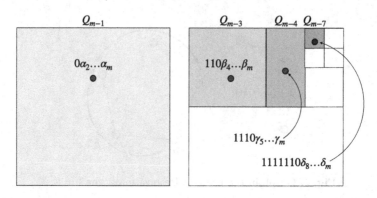

Fig. 2. Construction in the proof of Theorem 1. For every i, $\alpha_i, \beta_i, \gamma_i, \delta_i \in \{0, 1\}$.

4   Minimum Broadcast Graphs

This section is entirely devoted to the proof of Theorem 2. Recall that this theorem states that, in the fully-adaptive source-oblivious model, for every $n \geq 1$, there are n-node broadcast graphs with $O(n \cdot L(n))$ edges.

Proof of Theorem 2. Recall that the d-dimensional hypercubes are minimum broadcast graphs for the fully adaptive model, i.e., for $n = 2^d$, $B_{\mathsf{fa}}(n) = \frac{1}{2}n \log n$ (see [5]). For any n not a power of 2, we will construct a graph on n vertices and $O(nL(n))$ edges which is a broadcast graph under the fully adaptive model. More precisely, our graph has $n(L(n) + 1)$ edges. Its construction is directly inspired from the construction in [6]. Any n not a power of 2 can be presented as

$$n = 2^m - 2^k - r,$$

where $0 \leq k \leq m - 2$ and $0 \leq r \leq 2^k - 1$.

   We begin the construction of our graph $G = (V, E)$ using $m-k$ binomial trees $T_{m-1}, T_{m-2}, \ldots, T_k$ of sizes $2^{m-1}, 2^{m-2}, \ldots, 2^k$, respectively rooted at vertices $v_{m-1}, v_{m-2}, \ldots, v_k$. Recall that the binomial tree of size 1 consists of a single node (which is the root of the tree), and, for $d > 0$, the binomial tree of size 2^d is obtained by connecting the two roots of two copies of a binomial tree of size 2^{d-1} by an edge, and selecting one of these two roots as the root of the resulting tree. The union of the trees $T_{m-1}, T_{m-2}, \ldots, T_k$ contains $2^m - 2^k$ vertices and $2^m - 2^k - (m - k)$ edges.

   Next, we delete r vertices (and r edges) from T_k by repeating r times the removal of a leaf that is furthest away from the root of T_k. In order to simplify the notation, we will abuse notation, and still call the resulting tree T_k. Note that since $r \leq 2^k - 1$, T_k is not empty.

   This union of the trees $T_{m-1}, \ldots, T_k$ now contains $n = 2^m - 2^k - r$ vertices, and $2^m - 2^k - r - (m - k) = n - (m - k)$ edges. To complete the construction of our graph G, we connect every vertex of $V = \cup_{i=k}^{m-1} V(T_i)$ to all the roots $v_{m-1}, \ldots, v_k$ of the $m - k$ binomial trees. Thus, the graph $G = (V, E)$ will have

$$|V| = n = 2^m - 2^k - r$$

vertices, and

$$|E| = n - (m - k) + (m - k)(n - 1) = (m - k + 1)n - 2(m - k)$$

edges.

To show that $b_{\mathrm{fa}}(s) = \lceil \log_2 n \rceil$ for any originator $s \in V$, we first assign the lists of all vertices in V. Observe that each vertex u of G is actually the root of some binomial tree T_{m-p} for some $1 \le p \le m$ — by definition of binomial trees. Let us denote $u = \mathrm{root}(T_{m-p})$. ALso, for the root w of a binomial tree of dimension at least $m - (p+1)$, let $c_{m-p}(w)$ be the child of w that is the root of a binomial tree of dimension $m - p$. The list ℓ_u assigned to a vertex $u = c_{m-p}(w)$ distinct from $v_{m-1}, \dots, v_k$ is

$$\ell_u = v_{m-1}, v_{m-2}, \dots, v_{m-p}, c_{m-p-1}(u), c_{m-p-2}(u), \dots, c_0(u). \qquad (3)$$

For a root vertex $u = v_{m-i}$ for some $1 \le i \le m - k$, we set

$$\ell_{v_{m-i}} = v_{m-1}, v_{m-2}, \dots, v_{m-i}, c_{m-i-1}(v_{m-i}), c_{m-i-2}(v_{m-i}), \dots, c_0(v_{m-i}). \qquad (4)$$

These lists have a desirable elementary property:

Remark 1. Any vertex w that is the root of binomial tree of dimension d will inform all the vertices of its binomial tree during the last d rounds of broadcast. Indeed, each of w's children $c_{l-1}(w), c_{l-2}(w), \dots, c_0(w)$ will receive the message from w at round $m - l + 1, m - l + 2, \dots, m$, respectively, and, by following their own lists, each will complete broadcast within its binomial trees of respective dimensions $d - 1, d - 2, \dots, 0$. As a result, all the vertices in the binomial tree of dimension d rooted at w will be informed by round m.

More generally, let us consider a vertex $u \in V$ as the source of broadcast, and let us assume that u is the root of a binomial tree of dimension $m - p$ (see Fig. 3). Also, as u belongs to one of the trees $T_{m-1}, \dots, T_k$, let us assume that it belongs to tree T_{m-r} rooted at vertex v_{m-r}, where $m - r > m - p$. Following its list defined in Eq. (3), vertex u first informs the root vertices $v_{m-1}, \dots, v_{m-p}$ during the first p rounds, and then it informs all of its $m - p$ children in its binomial tree of dimension $m - p$ during rounds $p + 1, p + 2, \dots, p + (m - p)$, completing after m rounds.

All root vertices $v_{m-1}, v_{m-2}, \dots, v_{m-p}$ will receive the information from u at rounds $1, 2, \dots, p$, respectively, and will act according to their lists defined in Eq. (4). In particular, following its list $\ell_{v_{m-1}}$, node v_{m-1} will skip v_{m-1}, and will inform all of its children within the binomial tree T_{m-1} during the remaining $m - 1$ rounds. Thus, all vertices of tree T_{m-1} will be informed by round m thanks to Remark 1. In general, any root vertex v_{m-q} for $q = 1, 2, \dots, p$, except the root vertex v_{m-r}, will receive the message from u at round q, and will follow its list $\ell_{v_{m-q}}$ specified in Eq. (4). Since by round q all root vertices $v_{m-1}, v_{m-2}, \dots, v_{m-q-1}$ are already informed by u, the root v_{m-q} will skip the

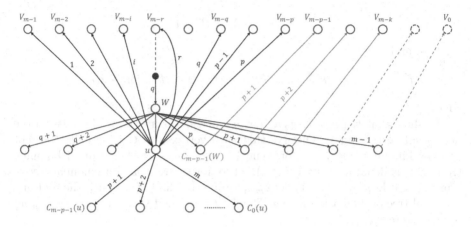

Fig. 3. Broadcast from source node u

vertices $v_{m-1}, \ldots, v_{m-q}$ in its list, and, starting from round q, it will inform all its $m - q$ children $c_{m-q-1}(v_{m-q}), c_{m-q-2}(v_{m-q}), \ldots, c_0(v_{m-q})$. Again, all vertices of the binomial tree T_{m-q} will be informed by time unit m thanks to Remark 1.

Next, let us describe how broadcast proceeds in the binomial tree T_{m-r}, as well as in the binomial trees $T_{m-p-1}, T_{m-p-2}, \ldots, T_k$. Since $m - r > m - p$, the root v_{m-r} of tree T_{m-r} will receive the information from u at round r, and, as mentioned above, will inform all of its children in its binomial tree, starting at time unit $r + 1$. Let us assume that the parent w of u in T_{m-r} receives the information at round s, which means that w is the root of a binomial tree of dimension $m - s$. We have $m - p < m - s \leq m - r$. Vertex w will follow its list, and will inform all its children, starting from round $m - s + 1$. Following its list, node w had to inform vertex u at round p. However, since u is already informed, w will skip u, and will inform its children $c_{m-p-1}(w), c_{m-p-2}(w), \ldots, c_0(w)$ one round earlier, at rounds $p, p + 1, \ldots, m - 1$, respectively. Now, following its list $\ell_{c_{m-p-1}(w)} = v_{m-1}, v_{m-2}, \ldots, v_{m-p}, v_{m-p-1}, c_{m-p-2}(u), \ldots, c_0(u)$, vertex $c_{m-p-1}(w)$ will skip all informed vertices $v_{m-1}, v_{m-2}, \ldots, v_{m-p}$ (they were all informed from u), and will send the information to the root vertex v_{m-p-1} at round $p + 1$. Starting from round $p + 2$, vertex w will inform all of its children in its binomial tree of dimension $m - p - 1$, and will complete broadcast by round m.

Similarly, each of w's children, i.e., nodes $c_{m-p-t}(w)$ for $t = 1, 2, \ldots, m - p - k$, will skip all informed vertices $v_{m-1}, v_{m-2}, \ldots, v_{m-p}, \ldots, v_{m-p-t}$, since they are informed earlier, either by u, or by the children $c_{m-p-1}(w), \ldots, c_{m-p-t}$ of w. Thus, w's child $c_{m-p-t}(w)$ will send the information to the root v_{m-p-t} at round $p + t$. Starting from round $p + t + 1$, vertex w will then inform all of its children in its binomial tree of dimension $m - p - 1$, and will complete broadcast by round m. Thanks to Remark 1, the Binomial tree T_{m-p-t} will be

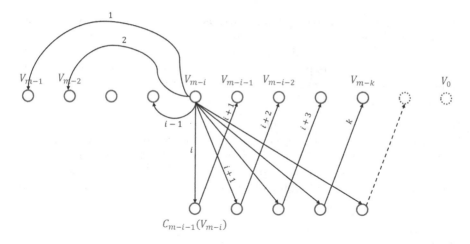

Fig. 4. Broadcast from source node v_{m-i}

fully informed by round m once its root vertex v_{m-p-t} receives the message at round $p + t$ from $c_{m-p-t}(w)$ (see Fig. 3).

It remains to prove that all root vertices $v_{m-p-1}, \ldots, v_k$ will complete broadcast within their respective binomial trees. This directly follows from Remark 1, as each root v_{m-l}, $p + 1 \le l \le m - k$, receives the information from $c_{m-l}(w)$ at round l, and completes broadcast in its Binomial tree T_{m-l} during the remaining $m - l$ rounds, for all $l = p + 1, \ldots, m - k$.

Finally, if the source of broadcast is one of the root vertices $v_{m-1}, \ldots, v_k$, then the process is even simpler. The details of broadcast from a root vertex source is displayed in Fig. 4. □

5   Conclusion

We have shown that, as far as the design of minimum broadcast graphs is concerned, the power of broadcast protocols is not limited by bounding them to be encoded by a single ordered list of neighbors at each node, which has the profitable feature of drastically reducing the space-complexity of the local encoding of the protocols.

Our results hold under the assumption that every node can signal its neighbors to let them know that it has received the broadcast information. The cost of signaling the neighbors is negligible compared to the cost of transmitting a potentially long message, but getting rid of this assumption may be desirable, by focusing, e.g., on adaptive or even non-adaptive source-oblivious protocols. The analysis of adaptive and non-adaptive protocols however appears to be quite challenging. In fact, it is not even clear whether n-node broadcast graphs exists for all n under these constraints. Indeed, as already mentioned, the best known upper bound on the broadcast time of cliques is $\log_2 n + O(\log \log n)$ [10]. On the other hand, a systematic study of the minimum broadcast graphs with small number of nodes show that optimal broadcast protocols for these graphs (i.e., protocols performing in $\lceil \log_2 n \rceil$ rounds) can be implemented by lists in the adaptive source-oblivious model. So, it may actually be the case that the aforementioned signaling assumption can be removed, while essentially preserving the good properties of the protocols. This is however not clear, and we state that issue as an open problem.

Open Problem. Is there an infinite family of n-node graphs $(G_n)_{n\geq 1}$ such that, for every $n \geq 1$, the broadcast time of G_n in the adaptive (or even non-adaptive) source-oblivious model is $\lceil \log_2 n \rceil$? And, independently from whether the answer to the previous question is positive or not, what is the minimum number of edges of n-node graphs with optimal broadcast time in the adaptive (or non-adaptive) source-oblivious model?

References

1. Diks, K., Pelc, A.: Broadcasting with universal lists. Networks **27**(3), 183–196 (1996)
2. Farley, A.M., Hedetniemi, S.T., Mitchell, S.M., Proskurowski, A.: Minimum broadcast graphs. Discret. Math. **25**(2), 189–193 (1979)
3. Fraigniaud, P., Lazard, E.: Methods and problems of communication in usual networks. Discret. Appl. Math. **53**(1–3), 79–133 (1994)
4. Gholami, M.S., Harutyunyan, H.A.: Broadcast graphs with nodes of limited memory. In: 13th International Conference on Complex Networks (CompleNet), pp. 29–42 (2022)
5. Gholami, M.S., Harutyunyan, H.A.: Fully-adaptive model for broadcasting with universal lists. In: 24th International Symposium on Symbolic and Numeric Algorithms for Scientific Computing (SYNASC), pp. 92–99 (2022)
6. Grigni, M., Peleg, D.: Tight bounds on minimum broadcast networks. SIAM J. Discret. Math. **4**(2), 207–222 (1991)
7. Harutyunyan, H.A., Liestman, A.L., Makino, K., Shermer, T.C.: Nonadaptive broadcasting in trees. Networks **57**(2), 157–168 (2011)
8. Hedetniemi, S.M., Hedetniemi, S.T., Liestman, A.L.: A survey of gossiping and broadcasting in communication networks. Networks **18**(4), 319–349 (1988)
9. Hromkovic, J., Klasing, R., Pelc, A., Ruzicka, P., Unger, W.: Dissemination of Information in Communication Networks - Broadcasting, Gossiping, Leader Election, and Fault-Tolerance. Texts in Theoretical Computer Science. An EATCS Series. Springer, Berlin, Heidelberg (2005). https://doi.org/10.1007/b137871
10. Kim, J., Chwa, K.: Optimal broadcasting with universal lists based on competitive analysis. Networks **45**(4), 224–231 (2005)
11. Rosenthal, A., Scheuermann, P.: Universal rankings for broadcasting in tree networks. In: 25th Allerton Conference on Communication, Control and Computing, pp. 641–649 (1987)
12. Slater, P.J., Cockayne, E.J., Hedetniemi, S.T.: Information dissemination in trees. SIAM J. Comput. **10**(4), 692–701 (1981)

On the 3-Tree Core of Plane Graphs

Debajyoti Mondal[1]([✉])[ID] and Md. Saidur Rahman[2][ID]

[1] Department of Computer Science, University of Saskatchewan, Saskatoon, Canada
dmondal@cs.usask.ca
[2] Graph Drawing and Information Visualization Laboratory,
Department of Computer Science and Engineering (CSE), Bangladesh University
of Engineering and Technology (BUET), Dhaka, Bangladesh
saidurrahman@cse.buet.ac.bd

Abstract. Plane 3-trees have been well studied in graph drawing literature. For many graph drawing styles, the aesthetic qualities that have been achieved for plane 3-trees are much better than the ones known for general plane graphs. This motivates us to investigate whether one can find a large plane 3-tree type structure in a general plane graph, and if so, whether it can be leveraged to obtain a better drawing for the graph. We thus introduce the concept of a 3-tree core H of a 3-connected plane graph G. Here, H is an edge-labeled plane 3-tree that represents G, and the distance d between H and G is the number of vertices of G that are missing in H. As an application of this concept, we consider the planar ortho-path visibility drawing, where each vertex is drawn as an orthogonal polygonal chain on an integer grid and each edge is drawn as an orthogonal line segment between the paths corresponding to its end vertices. We show that if H has a flat visibility drawing (i.e., each ortho-path is a horizontal line segment) with height k, then G has an ortho-path visibility drawing with height $O(k2^d)$. In particular, if G is a planar triangulation with $d = O(1)$, then G can be drawn with height $4n/9 + O(1)$ by choosing an appropriate planar embedding. This bound is significantly smaller than the $2n/3 + O(1)$ lower bound for the ortho-path visibility drawing when one must respect the input embedding.

1 Introduction

A *straight-line grid drawing* of a planar graph is a planar drawing where every vertex is drawn as an integer grid point in $\mathbb{R}^2$ and every edge is drawn as a straight line segment between its corresponding end vertices. In the *fixed embedding setting*, the drawing algorithm must respect the given embedding of the input graph, and in the *variable embedding setting*, the drawing algorithm can choose any embedding to draw the input graph. The *area (height)* of a drawing is the area (height) of its axis-aligned bounding box. We distinguish between the

The work of Debajyoti Mondal is supported in part by NSERC. The work of Md. Saidur Rahman is done under RISE Internal Research Grant 2021-01-06 and partly supported by "Basic Research Grant" of BUET.

X. Chen and B. Li (Eds.): TAMC 2024, LNCS 14637, pp. 149–160, 2024.
https://doi.org/10.1007/978-981-97-2340-9_13

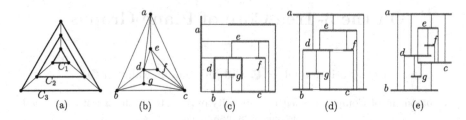

Fig. 1. (a) A nested triangles graph. (b) A plane 3-tree G, where the vertices are inserted in the following order: d, g, e, f. (c) An ortho-path visibility drawing of G. (c) A flat-visibility drawing of G. (e) A visibility drawing of G.

terms 'planar graph' and 'plane graph'. By a *plane graph*, we denote a planar graph with a fixed combinatorial embedding, where both the clockwise ordering of the neighbors of each vertex and the outerface are specified.

In the fixed embedding setting, every n-vertex plane graph admits a straight-line grid drawing with $O(n^2)$ area [4,16]. A rich body of research studies minimizing one dimension of the straight-line drawing (i.e., the height or, equivalently width), where the other dimension could be unbounded [3,9,11,13]. The best-known upper bound on the height of straight-line drawings of plane graphs is $2n/3 + O(1)$, where the width can be unbounded [6]. A lower bound of $\lfloor 2(n-1)/3 \rfloor$ is determined by the *nested triangles graph*, which is a plane graph formed by a sequence of $n/3$ vertex disjoint cycles, $C_1, C_2, \ldots, C_{n/3}$, where for each $i \in \{2, \ldots, n/3\}$, cycle C_i contains the cycles $C_1, \ldots, C_{i-1}$ in its interior, and the jth vertex of C_i is adjacent to the jth vertex of C_{i-1} (Fig. 1(a)). The lower bound can be improved slightly by some additive constant by using a subclass of plane 3-trees [14]. Here, a *plane 3-tree* is a triangulated plane graph that can be constructed by starting with a triangle, and then repeatedly adding a vertex to some inner face of the current graph and triangulating that face (Fig. 1(b)).

In this paper, our primary focus is on examining plane 3-tree type structures in general planar graphs. For many graph drawing styles, a plane 3-tree can be drawn with better aesthetics compared to a general planar triangulation [7,10,12,14]. For example, in the variable embedding setting, the upper bound on height for general n-vertex planar triangulations is the same as that of the fixed-embedding setting, which is $2n/3 + O(1)$. However, plane 3-trees can be drawn with a height of $4n/9 + O(1)$ by choosing an appropriate planar embedding [8]. Due to such successes with plane 3-trees, it is natural to expect that planar graphs that are structurally similar to plane 3-trees would also admit good drawings. This motivates us to examine whether one can find a plane 3-tree type structure H in a general planar graph G, and if such a structure exists, then to leverage the drawing of H to obtain a better drawing for G. We formalize this concept of H as a *3-tree core*, and explore 3-tree cores in the context of ortho-path visibility drawing.

An *ortho-path visibility drawing* of a plane graph is a planar drawing that represents each vertex as an axis-aligned polygonal chain (i.e., *ortho-path*) on an integer grid and each edge as a horizontal or a vertical segment between the ortho-paths corresponding to the end vertices (Fig. 1(c)). This is a special case of *ortho-polygon visibility drawing* (introduced in [5]), where instead of ortho-paths one can use orthogonal polygons. A *flat visibility drawing* is a special case of an ortho-path visibility drawing where each ortho-path is drawn as a horizontal segment (Fig. 1(d)). A *visibility drawing* is a special case of a flat visibility drawing where each edge is drawn as a vertical line segment (Fig. 1(e)). A rich body of literature studies visibility drawings and their variants motivated by their theoretical appeal and application in VLSI layout [2,5,15,17,18]. Every plane graph admits a visibility drawing with height $2n/3 + O(1)$ [20], which is also the lower bound in the fixed embedding setting [19]. This lower bound can also be realized by nested triangles graph for both flat visibility drawing and ortho-path visibility drawing. Therefore, it is natural to ask whether one can achieve a smaller upper bound in the variable embedding setting.

Our Contribution: The main contribution of this paper is to formalize the concept of the distance from a plane graph to a plane 3-tree type structure. We show that every 3-connected plane graph G can be represented using a *3-tree core H*, which is an edge-labeled plane 3-tree (Sect. 3). The distance d between G and H is the number of vertices of G that are missing in H. We show how a flat-visibility drawing of H can be leveraged to construct an ortho-path visibility drawing for G with a small height (Sect. 4). In particular, if H has a flat visibility drawing with height k, then we can construct an ortho-path visibility drawing for G with height $O(k2^d)$. For $k = o(n)$ and $d = O(1)$, this provides drawings with sublinear height for a nontrivial class of planar graphs.

  If G is a planar triangulation and not too distant from a 3-tree core H, i.e., $d = O(1)$, then we can construct an ortho-path visibility drawing for G with height $k + O(1)$ (Sect. 6). Every planar 3-tree and hence H admits a flat visibility drawing with height at most $4n/9 + O(1)$ in the variable embedding setting [8]. Hence we obtain a drawing for G with height $4n/9+O(1)$. This bound is interesting as it is significantly smaller than the lower bound of $2n/3+O(1)$ in the fixed embedding setting. We also show that every plane triangulation with $n \geq 4$ vertices has a 3-tree core which is 2-outerplanar and contains at least $\lceil (n + 1)/2 \rceil$ vertices (Sect. 5).

2   Preliminaries

A *planar graph* is a graph that admits a planar drawing in $\mathbb{R}^2$, i.e., each vertex is mapped to a distinct point, each edge is mapped to a simple curve between its corresponding end vertices and no two curves cross except possibly at their common endpoint. A planar graph delimits the plane into some connected regions which are called *faces*. The unbounded region is called the *outerface* and the other regions are called the *inner faces*. The vertices (edges) on the outerface are *outer vertices* (*outer edges*) and the remaining vertices are *inner vertices* (*inner edges*). A planar graph $G = (V, E)$ is called *maximal* if addition of any more edge to G results in a nonplanar graph. Let D be a straight-line drawing of

G. Then every face in D corresponds to a triangle in $\mathbb{R}^2$. Therefore, these graphs are also known as *planar triangulations*. A planar graph is called *internally triangulated* if each of its inner faces is a cycle of length three. A planar graph G is called *outerplanar* if it has an embedding where all the vertices are on its outerface. A planar graph is *2-outerplanar* if the deletion of all its outer vertices yields an outerplanar graph. A planar graph is *3-connected* if removal of two or fewer vertices does not disconnect the graph. A plane graph G is *internally 3-connected* if it can be extended to a 3-connected graph by adding a vertex to the outerface and connecting it to all vertices on the outerface. The *inner dual* of G is a graph G' that contains one vertex for each inner face of G, and there is an edge between two vertices of G' if and only if the corresponding faces in G have at least one edge in common.

3   Construction of a 3-Tree Core

Given a plane 3-connected graph G, a *3-tree core* of G is an edge-labeled plane 3-tree H, which is constructed as follows.

(a) Construction of the outerface: The outerface of H is constructed as an edge-labeled 3-cycle (a, b, c, a), where a, b and c are three vertices on the outerface of G in clockwise order. The edge (a, b) is labeled by the list of vertices that appear between a and b while walking from a to b on the outerface of G. The edge labels of (b, c) and (c, a) are defined similarly. Figure 2(a) illustrates an example. The edge labels are shown in purple.

(b) Insertion of a vertex: If G contains an inner vertex v, then by Menger's theorem, v has three vertex-disjoint (except at v) paths to the outer vertices a, b, c. We choose each vertex-disjoint path to be *minimal* or 'without short-cut edges', i.e., if a pair of vertices are non-adjacent in the path, they must also be non-adjacent in G. One can make each vertex-disjoint path minimal by repeatedly replacing a subpath of three or more vertices with an edge connecting the end vertices of the subpath (if exists). Figures 2(c)–(d) illustrate such an example. We insert v into the face (a, b, c, a) of H and triangulate the face by adding the edges $(v, a), (v, b), (v, c)$. The labels of these edges consist of the list of vertices that appear on the corresponding vertex-disjoint paths from v to a, b, c.

By construction, a face (q, r, s, q) in the resulting 3-tree corresponds to three paths P_{qr}, P_{rs}, P_{sq} in G. Furthermore, the union of P_{qr}, P_{rs}, P_{sq} (merging parallel edges into a single edge) contains exactly one cycle C that includes at least one edge from each path (Fig. 2(e)). We will refer to this property as the *cycle property* and show that this is an invariant of the construction.

(c) Repeated insertion: We now repeat the process that we followed in Step (b) for the subgraphs that remain in the connected regions delimited by the vertex-disjoint paths considered so far. Each time we find an inner vertex v in G that did not appear in any previous vertex-disjoint paths, we can repeat this process of finding minimal vertex-disjoint paths from v and inserting v into the resulting plane 3-tree, as follows. Let (q, r, s, q) be a face in the current plane 3-tree and let P_{qr}, P_{rs}, P_{sq} be the paths in G corresponding to $(q, r), (r, s), (s, q)$,

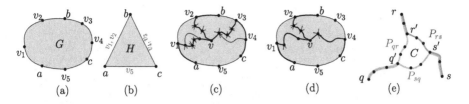

Fig. 2. (a)–(b) Construction of the outerface of H. (c)–(d) Construction of minimal vertex-disjoint paths from v to a, b, c. (e) Illustration for the cycle property.

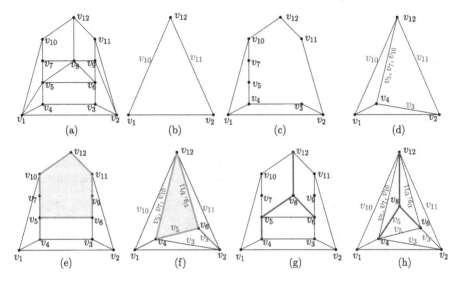

Fig. 3. Construction of a 3-tree core. (a)–(b) show the construction of the outerface, and (c)–(d), (e)–(f), and (g)–(h) correspond to vertex insertions. (Color figure online)

respectively. By the cycle invariant, the union of P_{qr}, P_{rs}, P_{sq} contains exactly one cycle C that includes at least one edge from each path. Let r' be the unique vertex on C that belongs to both P_{qr} and P_{rs}. Similarly, define q' and s'.

Let v be a vertex of G that is interior to C and has not been considered in any vertex-disjoint paths so far. Since G is 3-connected, the subgraph of G inside C is internally 3-connected. Therefore, by Menger's theorem, there are three minimal vertex-disjoint paths from v to $q', r's'$, and we can extend these paths to q, r, s maintaining the vertex disjointness. We now insert v into the face (q, r, s, q) of the plane 3-tree and use the vertex-disjoint paths from v to q, r, s to compute the edge labels for $(v, q), (v, r), (v, s)$. Figure 3 illustrates the construction of a 3-tree core and Figs. 3(e)–(h) illustrate the process of inserting a vertex $v = v_8$ into a face (v_4, v_{12}, v_6, v_4).

The insertion of v does not destroy the cycle invariant of any of the old faces, and by leveraging the structure of C, it is straightforward to observe that the cycle invariant holds for each new face in the plane 3-tree. We thus have the following theorem.

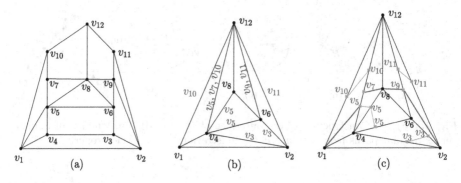

Fig. 4. (a) A 3-connected plane graph G, (b) its 3-tree core H, and (c) a pseudopath augmentated 3-tree core $\mathcal{H}$. The pseudopaths are shown in orange. (Color figure online)

Theorem 1. *Every 3-connected plane graph admits a 3-tree core.*

A 3-tree core H of a 3-connected plane graph G is not necessarily unique as it depends on the choice of vertices being picked at each step. We use the following property of a 3-tree core, whose proof is omitted due to space constraints.

Lemma 1 (Non-edge property). *Let G be a 3-connected plane graph and let H be a 3-tree core of G. Let (a, b) be an edge in H whose label is an ordered list L of three or more vertices. Let v_i and v_k be two vertices in L that are not consecutive in L. Then v_i cannot be adjacent to v_k in G.*

3.1   Sequence of Triangles

A *sequence of triangles* in a triangulated plane graph is either a single edge or an ordered set of faces $f_1, f_2, \ldots, f_j$ such that for each i, where $1 \le i < j$, f_i has exactly one edge in common with f_{i+1} and the inner dual of the plane graph determined by these faces is a path. The following lemma states that the edges of H that contain a common vertex in their labels, correspond to a sequence of triangles in H. The proof of the lemma is omitted due to space constraints.

Lemma 2 (Sequence of Triangles Property). *Let G be a 3-connected plane graph with a 3-tree core H. Let w be a vertex of G. Then the edges that contain w in their labels correspond to a sequence of triangles in H.*

3.2   Pseudopath Augmented 3-Tree Core

Let G be a 3-connected plane graph and let H be a 3-tree core of G (Fig. 4(a)–(b)). Let D be a straight-line drawing of H. For each edge (a, b) with a label consisting of t vertices, replace the edge with a path $a, v_1, \ldots, v_t, b$ of $(t + 2)$ vertices. The vertices between a and b are called the *division* vertices and they are labeled respecting the order of the vertices in the label of (a, b). For each face, if there are two division vertices with a common label, then we join them by a straight-line edge. By the sequence of triangles property (Lemma 2), the division vertices with the same label w now lie on a simple path. We call this a *pseudopath of w*. Figure 4(c) represents the pseudopaths in orange. Let the resulting drawing be D' and let the corresponding graph be H'. Note that for each edge (a, b) in H, the corresponding path in H' respects the vertex ordering specified in the label of (a, b). Therefore, no two pseudopaths intersect in D'. Hence H' is also a planar graph. For each face F in D', we now draw the edges that are in G but missing in H' with straight line segments. Figure 4(c) represents these edges in blue. By Lemma 1, such an edge must connect two different sides of the polygon that represents F. Therefore, the resulting drawing D'' is a planar straight-line drawing. We refer to the corresponding plane graph as the pseudopath augmented 3-tree core $\mathcal{H}$. Later, we will use $\mathcal{H}$ to construct the ortho-path visibility drawing of G. We will use the following property of $\mathcal{H}$.

Remark 1. Let G be a 3-connected plane graph and let $\mathcal{H}$ be a pseudopath augmented 3-tree core of G. Then $\mathcal{H}$ is planar and for each vertex w in G, the cyclic order of its neighbors in G is the same as that of in $\mathcal{H}$.

4   Ortho-Path Visibility Drawing for Planar Graphs

We now construct ortho-path visibility drawings for 3-connected planar graphs.

Theorem 2. *Let G be a 3-connected plane graph and let H be a 3-tree core of G. Assume that the distance between G and H is d. If H admits a flat visibility drawing D with height k, then G admits an ortho-path visibility drawing with height $O(k2^d)$. Here the embedding of D may differ from the embedding of H.*

Proof. Let H' be the plane graph corresponding to D and let $\mathcal{H}$ be the pseudopath augmented 3-tree core corresponding to H'. Since each face F of D is a triangle, the drawing of F corresponds to a polygon $\texttt{poly}(F)$, which is either rectangular or L-shaped. Figures 5(a)–(c) give a few examples. If the outerface of D is rectangular, then assume without loss of generality that the bottommost layer contains two outer vertices. We then add an extra layer to modify the outerface to have an L shape (Fig. 5(d)), i.e., we drag one outer vertex to the new layer and extend it to have vertical visibility to the other outer vertex.

For every inner face, we now can find a vertical slab $\texttt{slab}(F)$ that lies interior to F and touches all horizontal segments on the topmost and bottommost grid lines, as shown in gray (Figs. 5(a)–(c)). We also define $\texttt{slab}(F)$ for the outerface F. Here the slab touches a topmost and a bottommost horizontal segment, and the remaining horizontal segment is routed following an orthogonal path to obtain a horizontal visibility to the slab (Fig. 5(e)).

Drawing the Pseudopaths: For each vertex w of G that is missing in D, we have a pseudopath P_w in $\mathcal{H}$. We identify the sequence of faces σ of D that are split by P_w. By Remark 1, such a sequence exists even if the embedding of D is

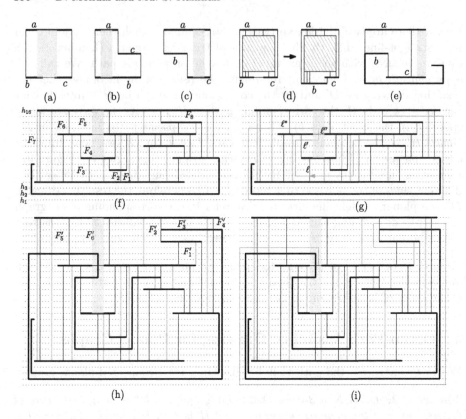

Fig. 5. (a)–(e) Illustration for slab(F), and outerface modification. (f)–(g) A pseudopath (in orange) insertion; $\sigma = (F_1, \ldots, F_8)$. (h)–(i) A subsequent insertion. Each identified face corresponds to a face in D (e.g., F_3' corresponds to F_8). (Color figure online)

different from H. We then draw P_w using an orthogonal path.

Pseudopaths with Two or More Vertices: We scale up the height of D by a factor of two. Let $h_{2k}, h_{2k-1}, \ldots, h_1$ be the horizontal grid lines from top to bottom in D. Then the grid lines h_{2i}, where $1 \leq i \leq k$, must be empty of any horizontal segment of the drawing, as shown in Fig. 5(f). We will refer to these as even grid lines and use them to draw P_w. We split the faces σ of D maintaining the order they appear in σ. We process a face F, as follows.

Assume that the polyline L corresponding to P_w has reached a vertical side ℓ of poly(F) along a horizontal grid line h, and it needs to exit through another vertical side ℓ' (Fig. 5(g)). If h intersects ℓ', then we route L along h crossing slab(F) and then ℓ'. Otherwise, we route L vertically along slab(F) (without intersecting slab(F)) to reach an even grid line h' that intersects ℓ', and then route L along h' to split F.

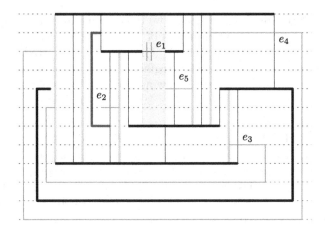

Fig. 6. Drawing of the pseudopaths that contain a single vertex are shown in orange. A previously drawn pseudopath of two vertices is shown in blue. (Color figure online)

Consider now the case when L reaches a vertical side ℓ' and needs to exit through a horizontal side ℓ''. By the definition of $\texttt{slab}(F)$, ℓ' is contained in $\texttt{slab}(F)$. Here we route L horizontally so that it enters $\texttt{slab}(F)$ and then route L vertically to intersect ℓ''.

If L reaches a horizontal side, then the only option is to exit through a vertical side ℓ^*. We thus route L vertically inside the slab to reach an even grid line h that intersects ℓ^*, and then route L along h to split F.

Pseudopaths with One Vertex: Each time we draw a pseudopath, we scale up the drawing by a factor of two. However, for simplicity, Fig. 6 illustrates various situations simultaneously, and hence the pseudopaths may appear to share even grid lines. Assume that P_w consists of a single vertex w. We split its corresponding edge e using a line segment L orthogonal to e.

If e is drawn as a vertical segment, then we place L at an even grid line that intersects e. If e is an inner edge, then we extend L horizontally so that it touches the slabs of the faces adjacent to e (if visibility to the slabs exists). The edge e_2 in Fig. 6 illustrates a scenario when L can be extended to reach one of the slabs, whereas for e_5, L can reach both slabs. If e is an outer edge, then we extend L orthogonally on the outerface F using an even grid line to reach a vertical side of $\texttt{slab}(F)$ whenever visibility exists (e.g., e_3 and e_4 in Fig. 6).

If e is drawn as a horizontal segment, then we place L anywhere interior to e without intersecting any other segment. If e is an inner edge, then L already belongs to the slabs of the faces adjacent to e (e.g., edge e_1 in Fig. 6). Note that e cannot be an outer edge of D as we modified the outerface to be L-shaped.

Drawing the Missing Edges: Let D' be the drawing after inserting all pseudopaths. We insert the remaining edges of $\mathcal{H}$ (e.g., blue edges in Fig. 4(c)) at each face F of D by leveraging $\mathtt{slab}(F)$. The details are omitted.

Finally, we remove unnecessary line segments, i.e., keep only one line segment to represent an edge, and stretch the drawing horizontally to place every vertical segment on a vertical grid line (see [2] for such transformations).

Bounding the Height: We may need two additional grid lines to transform the outerface of D (Figs. 5(d)–(e)). Each time we insert a pseudopath into D, we scale up the drawing by a factor of two. Finally, we scale up by a factor of two to remove degeneracies that appear while drawing the missing edges. Hence the height of the drawing is at most $(k+2) \times 2^{d+1}$, where d is the number of pseudopaths to be inserted.   □

5   Finding a Large 3-Tree Core in a Planar Triangulation

In this section we show that every n-vertex plane triangulation G has a 3-tree core with at least $\lceil (n+1)/2 \rceil$ vertices. The key is to use the Schnyder realizer of a planar graph, which is a well-known decomposition of its inner edges into three trees T_r, T_b and T_g (e.g., Fig. 7(a)). Let $\mathtt{leaf}(T_r)$, $\mathtt{leaf}(T_b)$ and $\mathtt{leaf}(T_g)$ be the number of leaves in T_r, T_b and T_g, respectively, where leaves in each tree are counted including the two outer vertices. Then the following property holds.

Lemma 3 (Zhang and He [18]). *Every n-vertex plane triangulation admits a Schnyder realizer with a tree in the realizer having at least $\lceil (n+1)/2 \rceil$ leaves.*

The following lemma shows how to construct a 3-tree core H of G leveraging a tree of the Schnyder realizer such that all the leaves become inner vertices of H (e.g., Fig. 7(b)). The proof is omitted due to space constraints.

Lemma 4. *Let G be an n-vertex plane triangulation, and let T be a tree in its Schnyder realizer. Then G admits a 3-tree core H with $\mathtt{leaf}(T)$ vertices of G.*

By Lemma 3 and Lemma 4, H contains $\lceil (n+1)/2 \rceil$ vertices. In the proof of Lemma 4, we make the inner vertices of H adjacent to an outer vertex of H. Therefore, H is 2-outerplanar, and hence we obtain the following theorem.

Theorem 3. *Every n-vertex plane triangulation, where $n \geq 4$, has a 3-tree core that is 2-outerplanar and contains at least $\lceil (n+1)/2 \rceil$ vertices.*

6   Ortho-Path Visibility Drawing for Triangulations

Theorem 4. *Let G be a plane triangulation and let H be a 3-tree core of G with distance d. If H admits a flat visibility drawing D with height k, then G admits an ortho-path visibility drawing with height $k + O(9^d)$. Note that the embedding of D may differ from the embedding of H.*

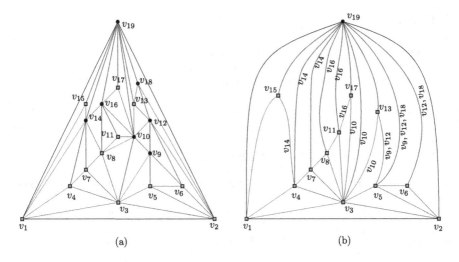

Fig. 7. (a) A plane triangulation G where the trees in a Schnyder realizer (except for the edges that are on the outerface on G) are drawn in red, blue, and green. (b) An edge-labeled plane 3-tree of G. (Color figure online)

Proof (sketch). We insert the pseudopaths in the order they appear in a tree of the Schnyder realizer. Each time we insert a pseudopath, we insert new horizontal grid lines only around three ortho-paths. We show that the insertion of the ith pseudopath creates a polygonal chain with at most $O(9^i)$ horizontal segments. Our construction ensures that the number of grid lines inserted at each step is within a constant factor of the number of horizontal segments in the inserted psedopath. Thus the number of horizontal lines inserted is at most $O(9^d)$.   □

Corollary 1. *Let G be an n-vertex plane triangulation with a 3-tree core at distance $O(1)$. Then G has an ortho-path visibility drawing with height $4n/9 + O(1)$.*

7   Conclusion

Our work opens up some interesting questions. Does every n-vertex planar triangulation admit a 3-tree core with larger than $n/2 + O(1)$ vertices? We examined ortho-path visibility drawing style, which appears to be applicable also for constructing 'pixel drawings'. Since 2-outerplanar graphs have 'pixel drawings' in $O(n)$ area [1], our technique may provide $O(n)$-area pixel drawings for triangulations with constant 3-tree core distance. Can we prove similar results for other drawing styles? Can we design fixed-parameter tractable algorithms for graph-theoretic problems parameterized by the 3-tree core distance?

References

1. Alam, M.J., Bläsius, T., Rutter, I., Ueckerdt, T., Wolff, A.: Pixel and voxel representations of graphs. In: Di Giacomo, E., Lubiw, A. (eds.) GD 2015. LNCS, vol. 9411, pp. 472–486. Springer, Cham (2015). https://doi.org/10.1007/978-3-319-27261-0_39
2. Biedl, T.: Height-preserving transformations of planar graph drawings. In: Duncan, C., Symvonis, A. (eds.) GD 2014. LNCS, vol. 8871, pp. 380–391. Springer, Heidelberg (2014). https://doi.org/10.1007/978-3-662-45803-7_32
3. Chrobak, M., Nakano, S.I.: Minimum-width grid drawings of plane graphs. Comput. Geom. Theory Appl. **11**(1), 29–54 (1998)
4. de Fraysseix, H., Pach, J., Pollack, R.: How to draw a planar graph on a grid. Combinatorica **10**(1), 41–51 (1990)
5. Di Giacomo, E., et al.: Ortho-polygon visibility representations of embedded graphs. Algorithmica **80**(8), 2345–2383 (2018)
6. Dolev, D., Leighton, T., Trickey, H.: Planar embedding of planar graphs. Adv. Comput. Res. **2**, 147–161 (1984)
7. Dujmović, V., Suderman, M., Wood, D.R.: Really straight graph drawings. In: Pach, J. (ed.) GD 2004. LNCS, vol. 3383, pp. 122–132. Springer, Heidelberg (2005). https://doi.org/10.1007/978-3-540-31843-9_14
8. Durocher, S., Mondal, D.: Drawing planar graphs with reduced height. J. Graph Algorithms Appl. **21**(4), 433–453 (2017)
9. Durocher, S., Mondal, D.: Relating graph thickness to planar layers and bend complexity. SIAM J. Discret. Math. **32**(4), 2703–2719 (2018)
10. Felsner, S., Knauer, K.B., Mertzios, G.B., Ueckerdt, T.: Intersection graphs of L-shapes and segments in the plane. Discret. Appl. Math. **206**, 48–55 (2016)
11. Hossain, M.I., Mondal, D., Rahman, M.S., Salma, S.A.: Universal line-sets for drawing planar 3-trees. J. Graph Algorithms Appl. **17**(2), 59–79 (2013)
12. Jelínek, V., Jelínková, E., Kratochvíl, J., Lidický, B., Tesar, M., Vyskocil, T.: The planar slope number of planar partial 3-trees of bounded degree. Graphs Comb. **29**(4), 981–1005 (2013)
13. Mondal, D., Alam, M.J., Rahman, M.S.: Minimum-layer drawings of trees. In: Katoh, N., Kumar, A. (eds.) WALCOM 2011. LNCS, vol. 6552, pp. 221–232. Springer, Heidelberg (2011). https://doi.org/10.1007/978-3-642-19094-0_23
14. Mondal, D., Nishat, R.I., Rahman, M.S., Alam, M.J.: Minimum-area drawings of plane 3-trees. J. Graph Algorithms Appl. **15**(2), 177–204 (2011)
15. Nishat, R.I., Mondal, D., Rahman, M.S.: Visibility drawings of plane 3-trees with minimum area. Math. Comput. Sci. **5**(1), 119–132 (2011)
16. Schnyder, W.: Embedding planar graphs on the grid. In: Proceedings of the 1st Annual ACM-SIAM Symposium on Discrete Algorithms (SODA), San Francisco, California, USA, pp. 138–148. ACM (1990)
17. Tamassia, R., Tollis, I.G.: A unified approach to visibility representations of planar graphs. Discret. Comput. Geom. **1**, 321–341 (1986)
18. Zhang, H., He, X.: Canonical ordering trees and their applications in graph drawing. Discret. Comput. Geom. **33**(2), 321–344 (2005)
19. Zhang, H., He, X.: Visibility representation of plane graphs via canonical ordering tree. Inf. Process. Lett. **96**(2), 41–48 (2005)
20. Zhang, H., He, X.: Optimal st-orientations for plane triangulations. J. Comb. Optim. **17**(4), 367–377 (2009)

A Coq-Based Infrastructure for Quantum Programming, Verification and Simulation

Wenxuan Tao and Gang Chen[✉]

College of Computer Science and Technology, Nanjing University of Aeronautics
and Astronautics, Nanjing 211106, China
gangchensh@qq.com

Abstract. Quantum programming presents a significant departure from
traditional programming due to its non-intuitive algorithm design and
reliance on intricate linear algebraic derivations. Therefore, formal verification using theorem provers is essential and highly suitable for quantum programming. The mathematical foundation of quantum computing
lies in matrix and vector computation. Hence, establishing a proper formal theory of matrices and vectors within theorem provers becomes of
utmost significance. This paper expands and refines a Coq-based type
system tailored for quantum computing, focusing specifically on powers-
of-2 matrices and vectors. This enables precise descriptions of quantum
states and operations while facilitating the formal verification of quantum programs. By utilizing this type system, we offer an environment for
writing and verifying quantum programs. The verified programs can be
extracted as equivalent OCaml files and executable simulators. Our work
provides an effective infrastructure for quantum programming, verification and simulation, thereby laying a foundation for future developments.

Keywords: Quantum programming · Formal verification · Coq ·
Quantum simulation · OCaml

1 Introduction

Quantum computing has emerged as a transformative paradigm with the potential to revolutionize multiple domains [1–3]. Unlike traditional computing, quantum computing utilizes the principles of quantum mechanics to perform computations on qubits, which can exist in superposition states and exhibit entan-
glement [4]. This unique behavior of qubits allows quantum computers to solve
certain problems exponentially faster than classical computers.

To harness the power of quantum computing, effective programming languages and tools are necessary. Quantum programming relies heavily on linear
algebraic derivations, making existing quantum programs, while not lengthy,
often challenging to understand. Therefore, ensuring the correctness of the
derivation process is crucial for enhancing the reliability of quantum programming. Quantum programming languages and tools that support linear algebraic
derivations are more suitable for the development of quantum algorithms. The
use of theorem proving techniques that support formal verification of linear

X. Chen and B. Li (Eds.): TAMC 2024, LNCS 14637, pp. 161–172, 2024.
https://doi.org/10.1007/978-981-97-2340-9_14

algebraic derivations is an important approach [5]. Given that quantum programs mainly consist of matrix and vector operations, leveraging these types as the foundation for the formal verification of quantum program correctness is essential. There have been several quantum programming languages based on theorem provers [6–9], but their matrix and vector type systems are either not strict enough or too complicated to master.

In previous work [10], we introduced a type system for powers-of-2 matrices and their related operations. It provided a rigorous and concise type system for matrices of quantum states and operations. Based on it we developed a quantum intermediate representation called PQIR, serving as an effective tool for quantum programming. Because the type system encompassed neither the definition of quantum state vectors nor the properties of matrices and vectors, this representation is insufficient for the verification of quantum programs.

In this paper, we expand and refine the formalization system for powers-of-2 matrices and vectors, capturing the characteristics of quantum states and operations accurately. By implementing our type system in Coq [11], we complete the definitions and properties of numerous operations. This refinement facilitates the formal verification of quantum programs, establishing an infrastructure for quantum programming, verification and simulation. To showcase its expressive capability, we implement and formally verify the quantum Fourier transform(QFT) [12]. An interesting outcome of this work is the development of quantum program simulators through code extraction for Coq to OCaml [13]. This enables the execution of verified quantum programs by an efficient functional programming language. For instance, the verified QFT program becomes a reliable executable in OCaml.

2   Related Work

Existing formalization approaches for matrices and vectors in quantum computing can be divided into two categories, each with its own advantages and limitations.

The first approach, exemplified by SQIR [7], treats matrices and vectors as functions from natural numbers to complex numbers. This model offers simplicity and ease of understanding, but lacks the inclusion of matrix or vector size in the type description. As a result, additional proofs are required to establish the size constraints, which can complicate the verification process.

The second approach, exemplified by CoqQ [9], represents matrices and vectors as functions from a finite set of natural numbers $\{k|0 \leq k < n\}$ to complex numbers [14]. While this model provides better rigor, it involves complicated formalization theories that may pose challenges for non-expert programmers in terms of understanding and mastery.

In the context of qubit systems, matrices and vectors used are of sizes that are equivalent to powers of 2. The above approaches do not explicitly exclude the possibility of users specifying matrices and vectors of other sizes. We previously addressed this limitation by constructing powers-of-2 matrices and defining the related matrix operations, yielding a viable way of quantum programming [10].

However, we did not define the vector type to describe pure states, nor did we prove any properties of powers-of-2 matrices and vectors, preventing the verification of quantum programs.

This paper refines the type system, including the definition and operations of powers-of-2 vectors as well as properties of these matrices and vectors. This provides a more precise and accessible formalization system for quantum computing, enhancing both reliability and usability in quantum programming, verification and simulation.

3   Preliminaries

Quantum states of qubits can be divided into pure states and mixed states. Any pure state of a single qubit can be expressed as a 2^1 vector like Eq. (3.1), where a and b are both complex numbers, and satisfy $|a|^2 + |b|^2 = 1$.

$$|\psi\rangle = \begin{bmatrix} a \\ b \end{bmatrix}. \tag{3.1}$$

$|0\rangle$ and $|1\rangle$ are a commonly used set of basis states.

$$|0\rangle = \begin{bmatrix} 1 \\ 0 \end{bmatrix}, |1\rangle = \begin{bmatrix} 0 \\ 1 \end{bmatrix}. \tag{3.2}$$

A system of n qubits with the ith qubit in the $|\psi_i\rangle$ state is in the $|\psi_1\rangle \otimes |\psi_2\rangle \otimes \cdots \otimes |\psi_n\rangle$ state. The operator $\otimes$ represents the Kronecker product, where each element of the first operand is multiplied separately by the second operand.

If a system is in the $|\psi_i\rangle$ state with probability $0 < p_i < 1$, $\sum_i p_i = 1$, then it's in a mixed state, which is often described by a density matrix. The density matrix of a pure state $|\psi_i\rangle$ is $\rho_i = |\psi_i\rangle\langle\psi_i|$, and that of a mixed state is the probability distribution of p_i:

$$\rho = \sum_i p_i |\psi_i\rangle\langle\psi_i|. \tag{3.3}$$

Quantum unitary transformations are basic operations to control and change the state of qubits. The transformation acting on n qubits is a $2^n \times 2^n$ unitary matrix. Applying a transformation A to $|\psi\rangle$ will get final state $|\psi'\rangle = A|\psi\rangle$. Applying A to Eq. (3.3) will get $\rho' = A\rho A^\dagger$. If $\{A_1, A_2, \cdots, A_k\}$ are applied on n qubits in turn, then the complete transformation is $A_k \cdots A_2 A_1$. If A_i is applied on the ith of n qubits at the same time, then the complete transformation is $A_1 \otimes A_2 \otimes \cdots \otimes A_n$.

Quantum measurement is another basic operation, which is used to detect state information. Quantum measurement realizes a non-unitary transformation that is probabilistic and irreversible. All possible results and the probabilities of their occurrences can be described by a set of measurement operators $\{M_m\}$ satisfying $\sum_m M_m^\dagger M_m = I$ with the subscript m denoting the possible result. Equation (3.3) will become $\rho' = \sum_m M_m \rho M_m^\dagger$ after measurement.

4   Powers-of-2 Matrix and Vector

Quantum computing in qubit systems only involve $2^n \times 2^n$ matrices and 2^n vectors in complex space. So the question of modeling a quantum computing system is to establish a type system encompassing such matrices and vectors.

4.1   Type and Operation Definition

In previous work [10], we defined a type for $2^n \times 2^n$ matrices, denoted as PMatrix, and completed the associated operations. This type represents the block matrix structure A_{2^n} with elements of the complex type C.

$$A_{2^n} = \begin{cases} C, & n = 0, \\ \begin{bmatrix} A_{2^{n-1}} & A_{2^{n-1}} \\ A_{2^{n-1}} & A_{2^{n-1}} \end{bmatrix}, & n \geq 1. \end{cases} \quad (4.1)$$

While advantageous for representing both pure and mixed states, utilizing PMatrix becomes somewhat redundant when exclusively dealing with pure states. Consequently, we introduce a novel vector type named PVector, which embodies the structure V_{2^n} with elements of type C.

$$V_{2^n} = \begin{cases} C, & n = 0, \\ \begin{bmatrix} V_{2^{n-1}} \\ V_{2^{n-1}} \end{bmatrix}, & n \geq 1. \end{cases} \quad (4.2)$$

Algorithm 1. Powers-of-2 vector type

Input: a natural number n
Output: a 2^n complex vector
1: Fixpoint PVector (n : nat) : Type :=
2: match n with
3: | O => C
4: | S n' => (PVector n', PVector n')
5: end.

With the introduction of PVector, we can now depict quantum states as state vectors instead of being limited to density matrices. This approach is more convenient and less computationally intensive, particularly when dealing exclusively with pure states.

As both PMatrix and PVector employ recursive structures, most operations can be implemented recursively, and there is a resemblance in their definitions. The following is an example of the Kronecker product of powers-of-2 matrices. The ellipsis refers to a recursive application of the Kronecker product.

$$A \otimes B = \begin{bmatrix} A_1 & A_2 \\ A_3 & A_4 \end{bmatrix} \otimes B = \begin{bmatrix} A_1 \otimes B & A_2 \otimes B \\ A_3 \otimes B & A_4 \otimes B \end{bmatrix} = \cdots \quad (4.3)$$

Similar to the above process, we implement operations for powers-of-2 matrices and vectors essential for quantum computing. These include operations such as the inner product $\langle \psi_1 | \psi_2 \rangle$ and outer product $| \psi_1 \rangle \langle \psi_2 |$, multiplication between matrices and vectors and so on. Subsequently, various operations between quantum transformations and states can be executed within our type system.

4.2 Property Verification

As PMatrix and PVector are recursive types, inductive proof emerges as a common approach to establishing the validity of a property for both basic and recursive cases. Consequently, many properties are proven, leading to a notable similarity in the proof steps. The following exemplifies the proof of the distributive law for the Kronecker product of powers-of-2 matrices.

Theorem 1. *For any $2^n \times 2^n$ matrices x and y and $2^m \times 2^m$ matrix z, $(x+y) \otimes z = x \otimes z + y \otimes z$.*

Proof. An inductive analysis of n is performed:

- Basis: If $n = 0$, x and y are single complex numbers. The goal is: for any complex numbers x and y and $2^m \times 2^m$ matrix z, $(x + y)z = xz + yz$. We apply the distributive law for the scalar multiplication to complete the proof;
- Induction step: If $n \geq 1$, the goal is: for any $2^{n+1} \times 2^{n+1}$ matrices x and y and $2^m \times 2^m$ matrix z, $(x + y) \otimes z = x \otimes z + y \otimes z$.

$$
\begin{aligned}
(x + y) \otimes z &= \left(\begin{bmatrix} x_1 & x_2 \\ x_3 & x_4 \end{bmatrix} + \begin{bmatrix} y_1 & y_2 \\ y_3 & y_4 \end{bmatrix} \right) \otimes z \\
&= \begin{bmatrix} x_1 + y_1 & x_2 + y_2 \\ x_3 + y_3 & x_4 + y_4 \end{bmatrix} \otimes z \\
&= \begin{bmatrix} (x_1 + y_1) \otimes z & (x_2 + y_2) \otimes z \\ (x_3 + y_3) \otimes z & (x_4 + y_4) \otimes z \end{bmatrix} \\
&= \begin{bmatrix} x_1 \otimes z + y_1 \otimes z & x_2 \otimes z + y_2 \otimes z \\ x_3 \otimes z + y_3 \otimes z & x_4 \otimes z + y_4 \otimes z \end{bmatrix} \quad (4.4) \\
&= \begin{bmatrix} x_1 \otimes z & x_2 \otimes z \\ x_3 \otimes z & x_4 \otimes z \end{bmatrix} + \begin{bmatrix} y_1 \otimes z & y_2 \otimes z \\ y_3 \otimes z & y_4 \otimes z \end{bmatrix} \\
&= \begin{bmatrix} x_1 & x_2 \\ x_3 & x_4 \end{bmatrix} \otimes z + \begin{bmatrix} y_1 & y_2 \\ y_3 & y_4 \end{bmatrix} \otimes z \\
&= (x \otimes z) + (y \otimes z)
\end{aligned}
$$

The overall proof steps can be extracted into an automated proof tactic called *prove_PMV*, utilizing the Ltac mechanism in Coq. Most properties within the PMatrix and PVector system can be automatically established using *prove_PMV*. For instance, Theorem 1 can be rapidly proved by *prove_PMV* (Fig. 1).

Theorem PMkron_plus_1 : forall (m n : nat) (x y : PMatrix m) (z : PMatrix n),
(x + y) ⊗ z = (x ⊗ z) + (y ⊗ z).
Proof. prove_PMV. apply PMscale_Cplus. **Qed.**

Fig. 1. Finish the proof by *prove_PMV*

In a manner analogous to the aforementioned process, we verified over 300 commonly used properties related to powers-of-2 matrices and vectors. All the details can be found in https://gitee.com/biexiangtao/Chalcis. This comprehensive verification provides a feasible environment for subsequent quantum program verification.

5 PQIR

In previous work [10], we implemented a quantum intermediate representation based on the PMatrix type system(PQIR), offering a more convenient method for quantum programming. The current work builds upon PQIR, so this section briefly introduces its syntax and semantics.

Basic quantum operations can be divided into unitary and non-unitary commands. The unitary commands, defined as *ucom*, consist of four types.

$$ucom := \ | \ app \ i \ U(\alpha, \beta, \gamma) \ | \ CNOT \ i \ j \ | \ u_1; u_2 \ | \ skip. \tag{5.1}$$

app i U signifies the application of gate U with three real parameters to the ith qubit. $CNOT \ i \ j$ denotes the application of a $CNOT$ gate with the ith qubit as the control and the jth qubit as the target. $u_1; u_2$ represents the sequence of unitary commands u_1 and u_2. *skip* indicates a move to the next command.

In a system comprising n qubits, the semantics of *ucom* are outlined as follows.

1. *app i U* will be translated into $I^{\otimes i} \otimes U \otimes I^{\otimes n-i-1}$.
2. $CNOT \ i \ j$ will be translated into $(I^{\otimes j} \otimes X \otimes I^{n-j-1})(I^{\otimes i} \otimes |1\rangle\langle 1| \otimes I^{n-i-1}) + I^{\otimes i} \otimes |0\rangle\langle 0| \otimes I^{n-i-1}$.
3. The semantics of $u_1; u_2$ is the product of the semantics of u_2 and u_1.
4. The semantics of *skip* is a $2^n \times 2^n$ identity matrix.

The full syntax defined as *com* contains three commands.

$$com := \ | \ ucom \ | \ measure \ i \ | \ c_1; c_2. \tag{5.2}$$

All unitary commands are encompassed within the full syntax. *measure i* represents the measurement of the ith qubit. $c_1; c_2$ generates a sequence of commands, which can be either unitary or non-unitary.

While a system comprising n qubits is in the ρ state, the semantic translation rules for these commands are outlined as follows.

1. For *ucom* that can be translated into a powers-of-2 matrix A, the final state is $\rho' = A\rho A^\dagger$.
2. For *measure i*, the final state is $\rho' = M_0'\rho M_0' + M_1'\rho M_1'$ with $M_0' = I^{\otimes i} \otimes |0\rangle\langle 0| \otimes I^{n-i-1}$ and $M_1' = I^{\otimes i} \otimes |1\rangle\langle 1| \otimes I^{n-i-1}$.
3. For $c_1; c_2$, we translate c_1 and then c_2 in turn.

6   Quantum Program Verification and Simulation

Utilizing the PMatrix and PVector type system built in Coq, along with PQIR, we can write quantum programs and formally verify their correctness. The verified quantum programs can serve as secure interfaces for programming calls, a challenging goal for mainstream quantum programming languages.

Furthermore, these quantum programs can be extracted into equivalent OCaml files through the extraction mechanism in Coq. The advantage lies in the ability to generate quantum program simulators based on OCaml, enhancing computational efficiency and potentially facilitating parallel simulation.

There are countless quantum programs that can be written and verified within our framework. This section will provide an overview of the general method, using the representative QFT algorithm as an illustrative example.

6.1   QFT Algorithm

The quantum Fourier transform(QFT) algorithm serves as the implementation of the Fourier transform within quantum computing and stands as a crucial element in various other quantum algorithms. For any basis state $|j_0, j_1, \cdots, j_{n-1}\rangle$ over n qubits with $j_i \in \{0, 1\}$, the circuit illustrating QFT is shown below.

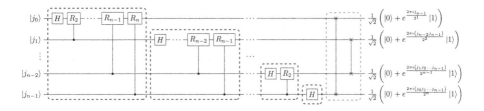

Fig. 2. QFT circuit

It consists of two primary components: the QFT_iter circuit(all blue dashed boxes) and the $reverse\_qubits$ circuit(the yellow dashed box). The QFT_iter is employed to generate superposition states and phase e-exponentials in the final states, while the $reverse\_qubits$ is used to reverse the order of the qubits.

Upon traversing the QFT_iter circuit, the initial state $|j_0, j_1, \cdots, j_{n-1}\rangle$ undergoes the following evolution.

$$|j_0, j_1, \cdots, j_{n-1}\rangle \rightarrow \frac{1}{\sqrt{2^n}} \otimes_{k=0}^{n-1} \left[|0\rangle + e^{2\pi i (j_k j_{k+1} \cdots j_{n-1})/2^{n-k}} |1\rangle \right] \qquad (6.1)$$

After the $reverse\_qubits$ circuit, the intermediate state will evolve as follows.

$$|j_0, j_1, \cdots, j_{n-1}\rangle \rightarrow \frac{1}{\sqrt{2^n}} \otimes_{k=0}^{n-1} \left[|0\rangle + e^{2\pi i (j_k j_{k+1} \cdots j_{n-1})/2^{n-k}} |1\rangle \right]$$

$$\rightarrow \frac{1}{\sqrt{2^n}} \otimes_{k=0}^{n-1} \left[|0\rangle + e^{2\pi i (j_{n-k-1} j_{n-k} \cdots j_{n-1})/2^{k+1}} |1\rangle \right] \qquad (6.2)$$

The Kronecker product of the quantum states in Eq. (6.2) can be transformed into a summation form. Consequently, the effect of QFT can be expressed through the following equation.

$$|j\rangle \rightarrow \frac{1}{\sqrt{2^n}} \sum_{k=0}^{2^n-1} e^{2\pi i jk/2^n} |k\rangle \tag{6.3}$$

6.2 QFT Program and Verification

According to Fig. 2, we can formulate the corresponding QFT program to operate on any quantity of qubits with PQIR.

In order to implement single-controlled R_k gates in the circuit, our initial step involves defining the single-controlled unitary gate CU. Given that the combination of U and $CNOT$ gate is a set of universal quantum gates that can execute arbitrary quantum gates [15], we can decompose the CU gate [16] (Fig. 3).

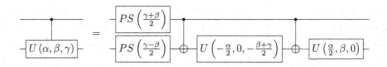

Fig. 3. Decomposition of CU gate

The $PS(\theta)$ gate is equivalent to the $U(0,0,\theta)$ gate. So we can define the CU command accordingly:

$$CU\ i\ j\ U(\alpha,\beta,\gamma) :=app\ i\ PS(\frac{\gamma+\beta}{2});\ app\ j\ PS(\frac{\gamma-\beta}{2});\ CNOT\ i\ j;$$

$$app\ j\ U(-\frac{\alpha}{2},0,-\frac{\beta+\gamma}{2});\ CNOT\ i\ j; \tag{6.4}$$

$$app\ j\ U(\frac{\alpha}{2},\beta,0).$$

Hence, we have the single-controlled R_k command denoted as $CU\ i\ j\ R_k$ with $R_k = U(0,0,\frac{2\pi}{2^k})$. We assemble a sequence of single-controlled R_k gates other than H gate in any one blue dashed box, termed as the QFT_ctrl circuit.

Algorithm 2. QFT_ctrl circuit

Input: a natural number n
Output: a sequence of single-controlled R_k gates contains $\{R_2, R_3, \cdots, R_n\}$
 1: Fixpoint QFT_ctrl (n : nat) : ucom :=
 2: match n with
 3: | S (S O) => CU 1 0 (Rk_gate 2)
 4: | S (S _ as n') => QFT_ctrl n'; CU n' 0 (Rk_gate n)
 5: | _ => skip
 6: end.

A blue dashed box is constituted by an H gate and a QFT_ctrl circuit, and an incremental assembly of n such blue dashed boxes constructs the QFT_iter circuit. So it can be realized using the QFT_ctrl circuit.

Algorithm 3. QFT_iter circuit

Input: a natural number n
Output: QFT_iter circuit on n qubits
 1: Fixpoint QFT_iter (n : nat) : ucom :=
 2: match n with
 3: | O => skip
 4: | S O => H 0
 5: | S n' => H 0 ; QFT_ctrl n ; ucom_map S (QFT_iter n')
 6: end.

$ucom\_map$ is a function offered by PQIR to reconstruct the qubit sequence numbers, enabling the realization of multiple ladder descent modules.

To construct the $reverse\_qubits$ circuit, we need to define the $SWAP$ gate first. It can be decomposed into a combination of $CNOT$ gates, as shown below (Fig. 4).

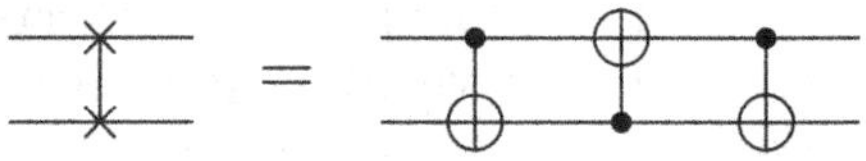

Fig. 4. SWAP decomposition

Then we define the $SWAP$ command as follows.

$$SWAP\ i\ j := CNOT\ i\ j;\ CNOT\ j\ i;\ CNOT\ i\ j. \tag{6.5}$$

Therefore, the $reverse\_qubits$ circuit in the yellow dashed box can be defined to perform a series of $SWAP$ gates, reversing the order of the qubits.

Algorithm 4. $reverse\_qubits$ circuit

Input: a natural number n
Output: reverse the states of n qubits from top to bottom
 1: Fixpoint reverse_qubits' (n i : nat) : ucom :=
 2: match i with
 3: | O => skip
 4: | S O => SWAP 0 (n - 1)
 5: | S i' => reverse_qubits' n i' ; SWAP i' (n - i' - 1)
 6: end.
 7: Definition reverse_qubits (n : nat) : ucom := reverse_qubits' n (n/2).

Finally, the complete QFT program can be obtained by integrating the circuits QFT_iter and $reverse\_qubits$. The subsequent theorem verifies the correctness of the QFT program based on our PMatrix and PVector system. Specifically, it confirms that the program effectively implements the transformation outlined in Eq. (6.3) (Fig. 5).

```
Theorem QFT_on_basis_state : forall (n k : nat),
(0 < n /\ k < 2 ^ n)%nat ->
ucom_eval n (QFT n) × (basis_state n k) = ((1/sqrt(2^n))%R ×' state_sum (2 ^ n)
(fun (i : nat) => Cexp(2 * PI * INR k / (2^n) * INR i) ×' (basis_state n i)))%PV.
Proof.
intros. generalize (dec_to_binlist n k). intros. destruct H. apply H0 in H1.
do 2 destruct H1. rewrite H2. apply QFT_on_basis_state_binlist. all : easy.
Qed.
```

Fig. 5. Verification of the QFT program

This theorem articulates that the QFT program we just wrote undergoes translation by the $ucom\_eval$ function of PQIR, resulting in the corresponding $2^n \times 2^n$ transformation matrix. Then the matrix is applied to the kth basis state of n qubits using the multiplication function between a $2^n \times 2^n$ matrix and a 2^n vector, yielding the final state. The final state aligns precisely with the outcome of QFT evolution specified by Eq. (6.3).

It's important to highlight that our proposition revolves around presenting a framework and methodology for quantum program verification rather than delving into the exhaustive steps for verifying the QFT program. As a result, the intricate details of the verification process are not explicitly delineated here. In addition to the QFT algorithm, we also verified the Greenberger-Horne-Zeilinger(GHZ) state preparation algorithm [17] and the quantum phase estimation(QPE) algorithm [18] based on our PMatrix and PVector type system. All the codes are shown in https://gitee.com/biexiangtao/Chalcis.

6.3   QFT Simulation

Although Coq provides robust verification functions, its execution performance tends to be slow. Fortunately, Coq offers an extraction mechanism that converts Coq code into OCaml code. OCaml is an efficient functional programming language with better code execution performance. Leveraging the OCaml compiler to generate quantum program simulators allows us to strike a better balance between verifiability and executability.

Firstly, we define a QFT program simulator in Coq.

Algorithm 5. QFT simulator in Coq

Input: two natural numbers n and k
Output: the effect of QFT on the kth basis state on n qubits
 1: Definition QFT_simulator (n k : nat) : PVector n :=
 2: ucom_eval n (QFT n) × (basis_state n k).

Then we extract it into an OCaml simulator through the extraction mechanism in Coq.

Algorithm 6. QFT simulator in OCaml

Input: two natural numbers n and k
Output: the effect of QFT on the kth basis state on n qubits
1: let qft_simulator n k =
2: pmv n (ucom_eval n (qft n)) (basis_state n k).

In this way, we get an OCaml simulator of the QFT program that has been verified in Coq. This simulator will quickly perform the calculations and return the corresponding results when provided with specific parameters. For example, applying the QFT to the initial state $|001\rangle$ results in the final state Eq. (6.6).

$$\left[\frac{1}{2\sqrt{2}}, \frac{1}{4}(1+i), \frac{1}{2\sqrt{2}}i, \frac{1}{4}(i-1), -\frac{1}{2\sqrt{2}}, -\frac{1}{4}(1+i), -\frac{1}{2\sqrt{2}}i, \frac{1}{4}(1-i)\right] \quad (6.6)$$

Executing this evolutionary process on the simulator reveals that the output aligns with expectations. The results are presented with three decimal places, and complex numbers are expressed as pairs of their real and imaginary components (Fig. 6).

```
$ ./chalcis -qprog qft -qubits 3 -basis 1
[(0.354, 0) (0.250, 0.250) (0, 0.354) (-0.250, 0.250) (-0.354, 0)
(-0.250, -0.250) (0, -0.354) (0.250, -0.250)]
```

Fig. 6. An example of QFT simulator in OCaml

Comparative experiments highlight the superior speed of the QFT simulator in OCaml over Coq by PQIR. Moreover, our QFT simulator in PQIR exhibits enhanced computational performance when contrasted with the relevant work in SQIR. In Table 1, × indicates that there is no result within 5 min.

Table 1. Comparison on QFT simulators between OCaml, PQIR and SQIR.

number of qubits	OCaml/s	PQIR/s	SQIR/s
1	0.095	2.172	65.615
2	0.143	×	×
3	0.175	×	×
4	0.206	×	×
5	0.740	×	×
6	7.133	×	×

7   Conclusion

This paper tackles the challenges inherent in quantum programming and formal verification by expanding and refining our robust type system tailored for powers-of-2 matrices and vectors. Based on this type system, we offer a way to write and verify the correctness of quantum programs. The verified quantum programs can be extracted into executables, yielding efficient quantum program simulators for testing. We hope that this work will pave the way for more reliable and accessible formal verification and simulation of quantum programs, ultimately propelling advancements in the field of quantum computing.

References

1. Mitra, S., Jana, B., Bhattacharya, S., Pal, P., Poray, J.: Quantum cryptography: overview, security issues and future challenges. In: 4th International Conference on Opto-Electronics and Applied Optics, pp. 1–7. IEEE, India (2017)
2. Zhou, L., Wang, S.T., Choi, S., Pichler, H., Lukin, M.D.: Quantum approximate optimization algorithm: performance, mechanism, and implementation on near-term devices. Phys. Rev. X **10**, 021067 (2020)
3. Pyrkov, A., et al.: Quantum computing for near-term applications in generative chemistry and drug discovery. Drug Discov. Today **28**(8), 103675 (2023)
4. Nielsen, M.A., Chuang, I.L.: Quantum Computation and Quantum Information. Cambridge University Press, Cambridge (2010)
5. Lewis, M., Soudjani, S., Zuliani, P.: Formal verification of quantum programs: theory, tools and challenges. arXiv:2110.01320 (2022)
6. Paykin, J., Rand, R., Zdancewic, S.: QWIRE: a core language for quantum circuits. SIGPLAN **52**(1), 846–858 (2017)
7. Hietala, K., Rand, R., Hung, S.H., Li, L., Hicks, M.: Proving quantum programs correct. In: 12th International Conference on Interactive Theorem Proving, pp. 21:1–21:19 (2021)
8. Liu, J., et al.: Formal verification of quantum algorithms using quantum Hoare logic. In: Dillig, I., Tasiran, S. (eds.) CAV 2019. LNCS, vol. 11562, pp. 187–207. Springer, Cham (2019). https://doi.org/10.1007/978-3-030-25543-5_12
9. Zhou, L., Barthe, G., Strub, P.Y., Liu, J., Ying, M.: CoqQ: foundational verification of quantum programs. Proc. ACM Program. Lang. **7**(POPL) (2023)
10. Tao, W., Chen, G.: Quantum intermediate representation and translation based on power-of-two matrices. J. Comput. Appl. (2024)
11. Coq manual. https://coq.inria.fr/refman/index.html
12. Coppersmith, D.: An approximate Fourier transform useful in quantum factoring. arXiv:quant-ph/0201067 (2002)
13. OCaml manual. https://v2.ocaml.org/releases/4.14/htmlman/index.html
14. The Mathematical Components Library. https://math-comp.github.io
15. Kaye, P., Laflamme, R., Mosca, M.: An Introduction to Quantum Computing. Oxford University Press, New York (2006)
16. Qiskit manual. https://qiskit.org/documentation/tutorials.html
17. Greenberger, D.M., Horne, M.A., Shimony, A., Zeilinger, A.: Bell's theorem without inequalities. Am. J. Phys. **58**(12), 1131–1143 (1990)
18. Kitaev, A.Y.: Quantum measurements and the Abelian stabilizer problem (1995). arXiv:quant-ph/9511026

A Local Search Algorithm
for Radius-Constrained k-Median

Gaojie Chi and Longkun Guo[✉]

Fuzhou University, Fuzhou, China
lkguo@fzu.edu.cn

Abstract. Given a set X of n points in a metric space and a radius R, we consider the combining version of k-center and k-median which is with the same objective as k-median, but with the constraint that any x in X is assigned to a center within the range R. In this paper, we propose a bicriteria approximation algorithm achieving a bicriteria approximation ratio of $(3 + \varepsilon, 7)$, by incorporating local search with the technology of finding key balls with consequent center exchanges therein. Compared to the state-of-the-art approximation ratio of $(8, 4)$ that is based on an LP formulation for k-median, we improve the ratio of the total distance from 8 to $3 + \varepsilon$ at the price of compromising the ratio of radius from 4 to 7.

Keywords: local search · k-median · k-center · approximation algorithm

1 Introduction

Clustering is a process of grouping data members that are similar in some way. Clustering is a technique for discovering the intrinsic structure of data. It is often referred to as unsupervised learning. Clustering algorithms are valuable tools for studying data structures and have been applied in various fields, including geography, astronomy, information retrieval, image segmentation, and more.

In typical clustering problems, the relationships between observational points are generally not considered. However, in many practical scenarios, observational points are correlated or accompanied by additional constraints. Constrained k-clustering is an approach that incorporates specific constraint conditions into the clustering process. Commonly encountered constrained clustering problems include lower-bound constrained clustering, capacity-constrained clustering, and chromatic clustering, among others. In our paper, we focus on the k-median problem with radius constraints.

1.1 Related Work

In the k-median problem, Aray [3] was the first to integrate local search with k-median for the analysis of the local search heuristic algorithm applied to both

Supported by the National Science Foundation of China (Nos. 12271098 and 61772005).

the k-median and facility location problems. They demonstrated that the single exchange local search algorithm achieves an approximation ratio of $5 + \varepsilon$. When allowing the simultaneous exchange of p facilities where p is the maximum number of swaps, the approximation ratio is improved to $3 + \varepsilon$. Cohen et al. [9] introduced a new local search algorithm, building upon Aray's results, to achieve an improved approximation ratio of $2.836 + \varepsilon$. In addition to the local search algorithm, several other algorithms have been devised to address the k-median problem. Li et al. [14] presented a pseudo-approximation algorithm with an approximation ratio of $2.732 + \varepsilon$. Byrka et al. [4] introduced preprocessing methods that form the basis for the current best result, which is $2.675 + \varepsilon$. This result was obtained by combining the primal-dual algorithm [10] with a sophisticated rounding technique applied to the two-point solution.

Moreover, a more complex problem is the fairness clustering problem. Jung et al. [11] demonstrated that this problem is NP-Hard and provided a solution achieving a 2-approximation, meaning there exists a feasible solution containing k centers, where for each point x, a center can be obtained within a radius of $2r_x$. However, their solution does not explicitly consider the objective function associated with standard clustering problems, such as k-median or k-means. In recent work by Plesnı[13], Jung et al.'s problem was rebranded as the priority k-center problem. In the research by Mahabadi and Vakilian [15], approximation algorithms were explored and developed for priority k-median and priority k-means problems. Their motivation was to integrate individual fairness requirements with the radius proposed by Jung et al., in addition to traditional clustering objectives. They devised a bicriteria approximation algorithm using local search and provided a 7-approximation for fairness concerning the p-norm objective $(\sum_{x \in X} d^p(x, S))^p$. In their paper, the main focus was on k-median clustering, where they provided a dual approximation result of $(84, 7)$. Addressing this problem, Negahbani and Chakrabarty [16] made significant contributions by providing improved algorithms using linear programming rounding techniques. Their primary contribution lies in enhancing algorithms for this problem through linear programming rounding.

If all r_x are identical (which may not always be the case), the clustering problem is also studied under the guise of the central ordered median problem [1,5,6]. Chakrabarty and Negahbani [16] combining ideas from both these works, designed an $(8,8)$ algorithm for the fair k-median problem, achieving an 8 approximation in both cost and fairness. Vakilian and Yalciner [19] proposed an enhanced $(16^p, 3)$ bicriteria approximation for this problem for any $\varepsilon > 0$. Additionally, they introduced improved cost approximation factors of 7.081 for $p = 1$ (k-median) and 3 for $p = \infty$ (k-center). To achieve these guarantees, they extended the framework introduced by Swamy [18] and designed a $16p$ approximation algorithm for facility locations with p-norm costs under matroid constraints.

A simplified problem related to the fair clustering problem is the standard clustering problem, and a further simplified aspect of the bicriteria approximate fair clustering is to focus solely on the k-center and k-median problems. Alamdari and Shmoys [1] combined the k-median and k-center problems. By considering the minimization of the k-median objective function, resulting in a standard approximation of $(8, 4)$. Their research was partially motivated by the ordered median problem [2,5,17]. Kamiyama [12] explored the extension of this uniform radius requirement to the context of matroid medians for clients and derived a double standard approximation algorithm with a ratio of $(11, 16)$.

1.2   Our Results

The main result of this paper is as below:

Theorem 1. *The radius-constrained k-median clustering problem admits a bicriteria approximation algorithm with ratio $(3 + \varepsilon, 7)$.*

Compared to the state-of-the-art approximation ratio $(8, 4)$ that is achieved by an LP-based method, our algorithm decreases the ratio of the total distance from 8 to $3 + \varepsilon$ while growing the ratio of radius from 4 to 7.

2   Preliminaries

In this section, we start by explaining the symbols and notations that are consistently used in this document, providing a clear definition of the problem we are addressing.

2.1   Notations and Problem Definition

Let's consider X as the collection of points that we aim to group together. The term $d(x, y)$ is used to represent the distance between any two points x and y within X. Furthermore, for a given point x in X, the set $B(x, R) = \{y \in X : d(x, y) \le R\}$ is defined as a ball with x at its center and a radius of R. In the context of a specific point x and a set S, the distance between them is expressed as $d(x, S) = \min\{d(x, s) : s \in S\}$. In cases where the set S is empty, the distance is designated as $d(x, \emptyset) = +\infty$.

In the context of the k-median problem, where k is a given positive value, the objective is to select a subset S consisting of k points from the available space. The aim is to minimize the function $cost(S) = \sum_{x \in X} d(x, S)$, which represents the cumulative distance from each point in X to the set S. Generalizing k-median, our problem can be formally defined as follows:

Definition 1. *Let X be a set of points of size n in a metric space, the objective is to partition the given dataset into k disjoint clusters, minimizing the sum of distances from each data point to some point within its respective cluster. Additionally, it is assumed that each point in X can find its optimal cluster center within a range of R.*

Throughout this paper, we use S and O to represent the center set provided by our algorithm and the optimal center set, respectively. In addition, we consider employing a local search algorithm to obtain a bicriteria approximation algorithm, denoted as (α, β)-approximation algorithm, which implies that the output of the algorithm satisfies the following conditions: (1) $cost(S) \leq \alpha \cdot cost(O)$; (2) For every $x \in X$, we always have $maxd(x, S) \leq \beta \cdot R$.

2.2 Basic Structures

We introduce a key concept used in our algorithm: *key ball*. The radius used to bind the cluster can be formally defined as follows:

Definition 2. *Consider a set of n points X in a metric space along with a set of balls $\mathscr{B}$ with identical radius R. Let C_0 be the set of centers of these balls. Then a ball $B_i \in \mathscr{B}$ can be represented as $B_i = B(c_i, R)$ where c_i in C_0. We say $\mathscr{B}$ is a set of key balls if and only if it satisfies the following two conditions:*

(1) $\forall x \in X$, $d(x, C_0) \leq 6R$ holds;
(2) $\forall c_i, c_j \in C_0$, $d(c_i, c_j) > 6R$ holds.

We say a set of centers S is *feasible* regarding a set of key balls $\mathscr{B}$, if and only if it satisfies the following criterion: for each ball B within the set $\mathscr{B}$, $B \cap S \neq \emptyset$ holds. In other words, a set of centers S is considered *feasible* regarding $\mathscr{B}$ if every key ball $B \in \mathscr{B}$ contains at least one center from S.

Lemma 3. *Let o be an optimal center for ball $B \in \mathscr{B}$. Assume that s_o is the point in S closest to o. Then, s_o will not appear in any other ball of $\mathscr{B} \setminus B$. Moreover, for any pair of optimal center points o_1 and o_2 appearing within distinct key balls, $s_{o_1} \neq s_{o_2}$ always holds.*

Proof. Consider the key ball $B \in \mathscr{B}$ that contains o_1. Since S is a feasible center set regarding $\mathscr{B}$, so $d(o_1, s_{o_1}) \leq 2R$. However the distance of o_1 to any other key ball $B_2 = B(c_2, R)$ is at least $d(c_1, c_2) - R - R > 4R$. Therefore, s_{o_1} cannot be in any ball other than B_1.

If two points $o_1 \in B_1$ and $o_2 \in B_2$ are in different key balls and have the same $s = s_{o_1} = s_{o_2}$, which means that their distance is at most $d(o_1, o_2) \leq d(o_1, s) + d(o_2, s) \leq 2R + 2R = 4R$. However, by the previous argument, their distance is larger than $4R$ which is a contradiction. □

2.3 Description of the Algorithm

Through our algorithm, we assume that the optimal radius R of the problem is known, because we can always find a replacement for the optimal radius with a binary search when it is unknown.

The algorithm commences by establishing a set of initial centers, denoted as C_0, ensuring that each point $x \in X$ lies within a distance of $6R$ from at least one center in C_0. These centers effectively delineate key balls, each centered at a

---

Algorithm 1: Local search algorithm regarding key balls

Input: A set of points X of size n, the optimal radius R, positive integer k and the size bound of swaps t;

Output: The center set S.

// Phase I: Constructing the center of the key balls C_0.

1 Set $P \leftarrow X$, $C_0 \leftarrow \emptyset$

2 **for** each $p \in P$ **do**

3   **if** $d(p, C_0) > 6R$ **then**

4 | Set $C_0 \leftarrow C_0 \cup \{p\}$ and $P \leftarrow P \setminus \{C_0\}$

5   **end**

6 **end**

7 **return** C_0 and the number of points in set C_0 is denoted by m.

// Phase II: Construct the initial center Set C.

8 Set $C \leftarrow C_0$;

9 **for** $i = 1$ to $k - m$ **do**

10   set $z \leftarrow \arg\max_{x \in X \setminus C} \{d(x, C)\}$ and then $C \leftarrow C \cup \{z\}$;

11 **end**

// Phase III: Local Search.

12 $S \leftarrow C$;

13 **for** $i = 1$ to t **do**

14   **for** every pair of $T_1 \subseteq S$ and $T_2 \subset X \setminus S$ with $|T_1| = |T_2| = i$ **do**

15   **if** $(S \cup T_2) \setminus T_1$ is feasible regarding $\mathscr{B}$ **then**

16              $S' \leftarrow (S \cup T_2) \setminus T_1$;

17   **if** $cost(S') - cost(S) > 0$ **then**

18 | **return** S;

19   **end**

20   Set $S := S'$;

21   **end**

22   **end**

23 **end**

---

point within C_0 and with a radius of R. The primary objective in constructing C_0 is to confine the eventual cluster radius.

Subsequently, the algorithm iteratively expands the set of centers from C_0 to C, with the aim of attaining a total of k centers. In the final stage, we employ a slightly modified version of the renowned local search k-median algorithm to rearrange the centers within C, thereby enhancing the overall proximity of data points to C. Crucially, throughout the optimization phase, the algorithm ensures that each key ball contributes at least one center to the final set C.

Our method is similar to that proposed by Mahabadi and Vakilian [16], which is a slight modification of the greedy method in [7,8]. After identifying the key balls, we establish pairwise connections between them, which are associated with corresponding weights.

3   Deal with Radius Constraint and Analyze Local Search Algorithm

First of all, the radius bound of our algorithm can be stated as below:

Lemma 4. *Let* $S = \{s_1, s_2, \ldots, s_k\}$ *represent the set of clustering centers output by Algorithm 1. Then, the radius of the balls according to* S *does not exceed* $7R$, *i.e.* $maxd(x, S) \leq 7R$.

Next, we analyze the distance sum bound of the local search method. Recall that S and O are respectively output of our algorithm solution and the optimal solution. Then, we construct an auxiliary weighted graph G, where $V(G) = S \cup O$ and the details of constructing $E(G)$ are given in Section 4. Let $\mathscr{E} = \{E_1, E_2, \ldots, E_z\}$ be a family of subsets of $E(G)$, where E_i is a subset of $E(G)$. We aim to construct E satisfying the following conditions:

Definition 1 *((t, w)-Weight bounded covering). Let* $\mathscr{B} = \{B_1, B_2, \ldots, B_m\}$ *be the set of key balls in the point set* X. *Assume that* S *and* O *are the output of our algorithm solution and the optimal solution, respectively. There exist* $\mathscr{E} = \{E_1, E_2, \ldots, E_z\}$ *as a covering of all edges in* $E(G)$, *Let* $O(E_i)$ *denote the set of points in* E_i *that belong to* O *and* $S(E_i)$ *denote the set of points in* E_i *that belong to* S. *We say that the covering* $\mathscr{E}$ *is* (t, w)-*weight bounded covering if it satisfies the following properties:*

1) *For each* E_i, $S \setminus S(E_i) \cup O(E_i)$ *is a set of centers that is feasible regarding* $\mathscr{B}$.
2) $\forall o_1, o_2 \in O(E_i)$, *if* $o_1 \neq o_2$, *then* $(s_1, o_1) \neq (s_2, o_2)$ *where* $s_1 \neq s_2$, $o_1 \neq o_2$ *and* $|S(E_i)| = |O(E_i)| \leq t$.
3) $\forall o \in O \setminus O(E_i)$, $s_o \notin S(E_i)$ *always holds.*
4) *The weight* w_i *of* E_i *equals the minimum weight among the weights of the included edge* (s, o) *in* $E(G)$ *where* s *in* S *and* o *in* O. w *is the maximum value of weights for each* E_i *in* $\mathscr{E}$, *that is, any* $w_i \leq w$.

In property 4, the weight of E_i being w_i means that, at this point, the weight of each edge included in E_i becomes w_i. In Sections 4, we prove: Assuming S is the solution obtained by the algorithm and O is the optimal solution, we can find a weighted graph G and a $(4p - 2, 1)$-weighted bounded cover $\mathscr{E}$ where $p \geq 2$. In $\mathscr{E}$, the weight of each point s and point o is at most $1 + \frac{1}{p}$, and the weight of each point o is at least 1. Next, we will show that if such a weighted graph and cover exist, we exchange the point sets in each E_i, calculate the weighted cost, and sum up the changes in the objective function obtained for all E_i as follows:

Lemma 5. *Consider a set of points* X *in a metric space, where* S *is the solution obtained by the algorithm. Suppose there exists a weighted graph* G *and a* (t, w)-*weight bounded covering* $\mathscr{E}$. *We can obtain that* $cost(S) \leq (3 + \varepsilon)cost(O)$, *where the* $cost(O)$ *represents the objective function generated by the optimal solution for the k-median problem within the point set* X.

4   Construction and Analysis of Graph G

In this section, we will provide a detailed explanation of constructing edges and the corresponding weights in the auxiliary weighted graph G.

4.1   Introduce the Concept of Capture

We assume S is the output of our algorithm solution, for each optimal solution o belonging to the set O, there exists a corresponding center s_o in S that is nearest to it. In cases where multiple points in the algorithm solution s are equidistant to the same o, any one of them suffices. This relationship is denoted as a "capture relationship", signifying that s_o captures o. It is noteworthy that each optimal center o is exclusively captured by a singular center $s \in S$. However, a center s from the algorithm solution may capture multiple o or none at all. We categorize a center s that captures at least one optimal center o as a "bad point". Conversely, an algorithm center s that captures no optimal center o is termed a "good point".

We use the structures proposed by Mahabadi et al. in [15] as restated in the following.

Definition 6. *Here we provide definitions for the terms "Safe ball", "Safe point", and "Safe edge".*

Safe Ball: A ball $B \in \mathscr{B}$ is considered safe if it satisfies one of the following conditions: (1) The key ball B contains at least two points from the algorithm solution S; (2) Within the key ball B, there exists an optimal center o such that s_o captures no other optimal center except for o. This implies a one-to-one capture relationship between s_o and o.

Safe Point: A point $s \in S$ is considered safe if it meets one of the following criteria: (1) s is not contained in any of the key balls; (2) s resides in a key ball that is classified as a safe ball.

Safe Edge: An edge (s, o), where $s \in S$ and $o \in O$, is deemed safe if it fulfills one of the following conditions: (1) s is a safe point; (2) Both endpoints s and o of this edge are located within the same key ball.

We classify a key ball, a point, or an edge as unsafe if they do not meet the criteria for being safe. The above definition reveals a fact: an unsafe point among the good points must be inside a key ball.

4.2   Partition of Solution S and Optimal Solution O

We partition S and O as follows. Assume that $Z \subseteq S$ is the set of all the "bad points" and without loss of generality that $Z = \{s_1, s_2, \ldots, s_r\}$, where r is the total number of "bad points" and $r \leq k$. Building on the previously explained "capture" concept, we can infer that each "bad point" in set Z captures at least one point from the optimal centers O.

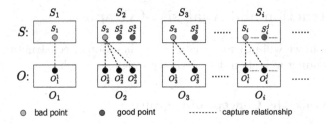

Fig. 1. The algorithm solution S and the optimal solution O are partitioned based on the definition of capture relation. In the diagram, we have $|S_1| = |O_1| = 1$, with $s_{o_1^1} = s_1$; $|S_2| = |O_2| = 3$, with $s_{o_2^1} = s_{o_2^2} = s_{o_2^3} = s_2$ and s_2^2, s_2^3 are "good points"; when $|S_3| = |O_3| = 2$, with $s_{o_3^1} = s_{o_3^2} = s_3$ and s_3^2 is "good points". Similarly, for any s_i in Z, we have that $|S_i| = |O_i| = t_i$ and $s_{o_i^j} = s_i$ for all values of index j from 1 to t_i and $S_i \setminus s_i$ are all "good points".

Then the optimal centers O can be partitioned into r disjoint subsets according to Z, say $\{O_1, O_2, \ldots, O_r\}$, where each subset O_i contains t_i points of the optimal solution, all of which are captured by the "bad point" s_i. In other words, $O_i = \{o \in O | s_o = s_i\}$ for $i = 1, 2, \ldots, r$. For each subset O_i, we construct a corresponding subset $S_i \subseteq S$ with $|S_i| = t_i$. The construction is as below:

Suppose the optimal solution o lies within a key ball B, and this key ball contains a "good point" s that is unsafe. In that case, we place such s into the corresponding S_i, where the remaining points in S_i consist of a "bad point" s_i and any $t_i - 2$ arbitrary "good points". If such s does not exist, then S_i consists of a "bad point" s_i and any $t_i - 1$ arbitrary good points. We ensure $|S_i| = |O_i| = t_i$ and that all subsets S_i are mutually disjoint. Through Lemma 3 on the properties of key balls, we can conclude that for each S_i, there is at most one unsafe point belonging to the "good points", where $i = 1, 2, \ldots, r$.

For simplicitly, we denote each set S_i as consisting of points $\{s_i, s_i^2, \ldots, s_i^{t_i}\}$, where $i = 1, 2, \ldots, r$. Correspondingly, we denote $O_i = \{o_i^1, o_i^2, \ldots, o_i^{t_i}\}$ for each $i = 1, 2, \ldots, r$. It is worth noting that each o_i^j in O_i is captured by s_i, where $j = 1, 2, \ldots t_i$, and in S_i, apart from s_i, all other points are considered "good points" (Fig. 1 illustrates an example of capture, division, and "good points").

4.3   Construction of the Auxiliary Graph

Initially, we define an event F: We prioritize forming edges between the "good points" s belonging to unsafe points in S_i and the points o in O_i that belong to the same key ball as s. If there are multiple such o in O_i, any one of them can be arbitrarily chosen. The weight of edges formed by event F is 1. For ease of notation, we represent the points in the edges formed from event F where one endpoint belongs to the algorithm solution S as S_F and the points where the other endpoint belongs to the optimal centers O as O_F. Similarly, we denote the points in S_i that include points from event F as S_{i_F} and the points in O_i that include points from event F as O_{i_F}.

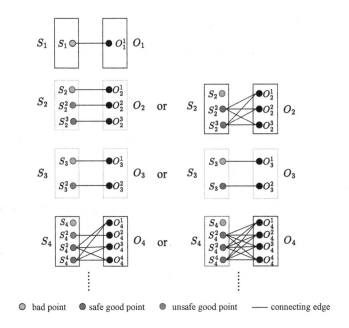

Fig. 2. When $p = 2$, construct the exchange edges. Gray boxes indicate that such S_i and O_i are must in the same E_i.

Assume p is the maximum number of edges exchanged between the pairs of points from S_i and O_i, excluding the event F, and $p \geq 2$. We can deduce that for each point pair of S_i and O_i, where $i = 1, 2, \ldots, r$, the edges from event F either do not exist or exist only once.

After prioritizing the formation of edges for event F, we proceed to create edges (s, o) for other point pairs S_i and O_i as follows (See Fig. 2 for an example of constructing such edges with $p = 2$):

(1) For each point pair S_i and O_i where $|S_i| = |O_i| = t_i$ and $t_i - I_{S_{i_F}} \leq p$, we add $t_i - 1 - I_{S_{i_F}}$ "good points" to S_i to ensure that the edges formed by these point pairs are one-to-one correspondences, with each edge having a weight of 1. Here, $I_{S_{iF}}$ is an indicator function; $I_{S_{i_F}} = 1$ when S_i contains points from S_F and $I_{S_{i_F}} = 0$ when S_i does not contain points from S_F. We simultaneously swap the edges (s, o) composed of s in S_i and o in O_i.

(2) For each point pair where $|S_i| = |O_i| = t_i$ and $t_i - I_{S_{i_F}} > p$, we add $t_i - 1 - I_{S_{i_F}}$ "good points" to S_i from $S_i \setminus (\{s_i\} \cup S_{i_F})$. Each point s in $S_i \setminus (\{s_i\} \cup S_{i_F})$ forms an edge (s, o) with every point o in $O_i \setminus O_{i_F}$, with each edge having a weight of $\frac{1}{t_i - 1 - I_{S_{i_F}}}$. For each such pair of S_i and O_i, $\left(t_i - I_{S_{i_F}}\right) \cdot \left(t_i - 1 - I_{S_{i_F}}\right)$ edges are generated, each with a weight of $\frac{1}{t_i - 1 - I_{S_{i_F}}}$.

4.4   Analysis of Local Search

Next, we prove some important properties of the constructed edges (s, o). In Lemma 8, we prove the value of the upper limit t of the exchange in the algorithm.

Lemma 7. *The edge (s, o) in any $E_i \in \mathscr{E}$ satisfies one of the following properties:*

1. *The edges formed by the event F are denoted as (s, o), all of which are safe with a weight of 1;*
2. *When $|S_i| = |O_i| = t_i$ and $t_i - I_{S_{i_F}} > p$, the edge (s, o) formed by the pair of points S_i and O_i is safe, and the weight of each edge is $\frac{1}{t_i - 1 - I_{S_{i_F}}}$;*
3. *When $|S_i| = |O_i| = t_i$ and $t_i - I_{S_{i_F}} \leq p$, the pairs of points S_i and O_i constitute edges (s, o) in a one-to-one correspondence, where s is in S_i and o is in O_i. For the sake of discussion, let us denote these pairs as $S_i = \{s_i, s_i^2, \ldots, s_i^{t_i}\}$ and $O_i = \{o_i^1, o_i^2, \ldots, o_i^{t_i}\}, i = 1, 2, \ldots, r$. The weight of these edges (s, o) is uniformly 1, and each edge possesses one of the following properties:*
   - *(s, o) is safe;*
   - *(s, o) is unsafe, but it will enter a key ball with a safe exit edge.*

Proof. According to the partition of point pairs, we establish the properties of (s, o) as follows.

For the first property, according to the construction of edges, it is evident that the endpoints of the edges (s, o) created by event F must belong to the same key ball. Consequently, the edges (s, o) formed by event F are classified as safe edges. As per the previous explanation of edge construction, these edges are guaranteed to weigh these edges 1.

For the second, when $|S_i| = |O_i| = t_i$ and $t_i - I_{S_{i_F}} \geq p$, considering the previous edge construction, we know that for each S_i, we need to add $t_i - I_{S_{i_F}} - 1$ "good point". Next, we will prove by contradiction that these edges are all safe.

Assume there exists an unsafe edge (s_1, o_1) where $s_1 \in S_i$ and $o_1 \in O_i$. Based on the earlier edge construction and the fact that the edge (s_1, o_s) is unsafe, we can infer that s_1 is an unsafe point among the "good points", and o_1 is not within the same key ball as s_1. Let's assume $s_1 \in B$, where $B \in \mathscr{B}$. Within the key ball B, there must exist an optimal center, let's denote it as o, and there is no one-to-one relationship between o and s_o. According to the edge construction, we should prioritize forming the edge (s_1, o), which contradicts our assumption of the existence of an unsafe edge (s_1, o_1).

Hence, when $|S_i| = |O_i| = t_i$ and $t_i - I_{S_{i_F}} > p$, the constructed edges are denoted as (s, o) are all safe. Based on the construction of these edges, it follows that each edge appears with a weight of $\frac{1}{t_i - 1 - I_{S_{i_F}}}$.

Lastly, when $|S_i| = |O_i| = t_i$ and $t_i - I_{S_{i_F}} \leq p$, based on the previous edge construction, the edge (s, o) formed by the point pairs S_i and O_i is a one-to-one correspondence. Here $S_i = \{s_i, s_i^2, \ldots, s_i^{t_i}\}$ and $O_i = \{o_i^1, o_i^2, \ldots, o_i^{t_i}\}$ for $i = 1, 2, \ldots, r$ where s_i captures all $o \in O_i$ and points in $S_i \setminus s_i$ are all "good points".

Based on property 2, we can conclude that the added $t_i - 1 - I_{S_{i_F}}$ "good points" are all safe points within the set of "good points". Therefore, in conjunction with property 1, it can be deduced that the edges (s, o) formed by $s_i^t \in S_i$, and $o_i^t \in O_i$ for $t = 2, 3, \ldots, t_i$, denoted as $(s, o) = (s_i^t, o_i^t)$, are all safe.

Next, we analyzing all the two cases of $(s, o) = (s_i, o_i^1)$ as below:

- If s_i is a safe point or s_i and o_i^1 are in the same key ball, then (s_i, o_i^1) is clearly safe.
- Otherwise, i.e., s_i is an unsafe point and o_i^1 and s_i are not in the same key ball. According to the definition of an unsafe point, s_i must be within a key ball, and this key ball contains no other points from S except s_i. Let's assume $B \in \mathscr{B}$, $B \cap S = s_i$, and $o_i^1 \notin B$. For convenience, let's denote one of the optimal centers in the key ball as o'. According to Lemma 3, we know that $s_{o'}$ is not in any other ball except for the key ball B, so (s', o') is safe. Therefore, (s_i, o_i^1) constitutes an unsafe edge, yet its associated key ball has a safe outgoing edge.   □

Since the maximum value of the weight for each edge is 1, the weight w_i of each $E_i \in \mathscr{E}$ is equal to the minimum weight of the edges it contains. Therefore, we can conclude that the weights of the elements in $\mathscr{E}$ do not exceed 1, i.e., $w_i \leq w = 1$.

Lemma 8. *Each edge (s, o) satisfies Lemma 7, leading to the conclusion that the number of edges t contained in each exchange partition E_i in $\mathscr{E}$ does not exceed $4p - 2$ where $p \geq 2$. Furthermore, in $\mathscr{E}$, the weight of each point s and point o is at most $1 + \frac{1}{p}$, and the weight of each point o is at least 1.*

5   Conclusion

This paper addresses a hybridized version of the k-center and k-median problems, utilizing k-median as the objective function while enforcing the constraint that each point x in the set X must be allocated to a center within a radius of R. Employing a local search algorithm, we present a bicriteria approximation ratio of $(3 + \varepsilon, 7)$. The primary contribution of the paper lies in the implementation of this algorithm through local search, the identification of key balls, and subsequent center exchanges. The algorithm enhances the ratio from $(8, 4)$ based on the k-median LP formulation, reducing the total distance ratio from 8 to $3 + \varepsilon$ while increasing the radius ratio from 4 to 7.

Acknowledgments. This work is supported by the Taishan Scholars Young Expert Project of Shandong Province (No. tsqn202211215) and the National Science Foundation of China (Nos. 12271098 and 61772005).

References

1. Alamdari, S., Shmoys, D.: A bicriteria approximation algorithm for the k-center and k-median problems. In: Solis-Oba, R., Fleischer, R. (eds.) WAOA 2017. LNCS,

vol. 10787, pp. 66–75. Springer, Cham (2018). https://doi.org/10.1007/978-3-319-89441-6_6

2. Aouad, A., Segev, D.: The ordered k-median problem: surrogate models and approximation algorithms. Math. Program. **177**(1–2), 55–83 (2019)

3. Arya, V., Garg, N., Khandekar, R., Meyerson, A., Munagala, K., Pandit, V.: Local search heuristic for k-median and facility location problems. In: Proceedings of The Thirty-third Annual ACM Symposium on Theory of Computing, pp. 21–29 (2001)

4. Byrka, J., Pensyl, T., Rybicki, B, et al.: An improved approximation for k-median and positive correlation in budgeted optimization. ACM Transactions on Algorithms (TALG) **13**(2), 1–31 (2017)

5. Byrka, J., Sornat, K., Spoerhase, J.: Constant-factor approximation for ordered k-median. In: Proceedings of The 50th Annual ACM SIGACT Symposium on Theory of Computing, pp. 620–631 (2018)

6. Chakrabarty, D., Swamy, C.: Interpolating between k-median and k-center: approximation algorithms for ordered k-median. In: 45th International Colloquium on Automata, Languages, and Programming (ICALP 2018) (2018)

7. Chan, T.-H.H., Dinitz, M., Gupta, A.: Spanners with slack. In: Azar, Y., Erlebach, T. (eds.) ESA 2006. LNCS, vol. 4168, pp. 196–207. Springer, Heidelberg (2006). https://doi.org/10.1007/11841036_20

8. Charikar, M., Makarychev, K., Makarychev, Y.: Local global tradeoffs in metric embeddings. SIAM J. Comput. **39**(6), 2487–2512 (2010)

9. Cohen-Addad, V., Gupta, A., Hu, L., Oh, H., Saulpic, D.: An improved local search algorithm for k-median. In: Proceedings of The 2022 Annual ACM-SIAM Symposium on Discrete Algorithms (SODA), pp. 1556–1612 (2022)

10. Jain, K., Mahdian, M., Markakis, E., Saberi, A., Vazirani, V.V.: Greedy facility location algorithms analyzed using dual fitting with factor-revealing LP. J. ACM (JACM) **50**(6), 795–824 (2003)

11. Jung, C., Kannan, S., Lutz, N.: A center in your neighborhood: fairness in facility location. arXiv e-prints, pp. arXiv–1908 (2019)

12. Kamiyama, N.: The distance-constrained matroid median problem. Algorithmica **82**(7), 2087–2106 (2020)

13. Khoury, B., Pardalos, P.M.: A heuristic for the Steiner problem in graphs. Comput. Optim. Appl. **6**, 5–14 (1996)

14. Li, S., Svensson, O.: Approximating k-median via pseudo-approximation. SIAM J. Comput. **45**(2), 530–547 (2016)

15. Mahabadi, S., Vakilian, A.: Individual fairness for k-clustering. In: International Conference on Machine Learning, pp. 6586–6596 (2020)

16. Negahbani, M., Chakrabarty, D.: Better algorithms for individually fair k-clustering. Adv. Neural. Inf. Process. Syst. **34**, 13340–13351 (2021)

17. Nickel, S., Puerto, J.: Location Theory: A Unified Approach. Springer, Heidelberg (2006). https://doi.org/10.1007/3-540-27640-8

18. Swamy, C.: Improved approximation algorithms for matroid and knapsack median problems and applications. ACM Trans. Algorithms (TALG) **12**(4), 1–22 (2016)

19. Vakilian, A., Yalciner, M.: Improved approximation algorithms for individually fair clustering. In: International Conference on Artificial Intelligence and Statistics, pp. 8758–8779 (2022)

Energy and Output Patterns in Boolean Circuits

Jayalal Sarma[1] and Kei Uchizawa[2(✉)]

[1] Indian Institute of Technology Madras, Chennai, India
`jayalal@cse.iitm.ac.in`
[2] Graduate School of Science and Engineering, Yamagata University, Yamagata, Japan
`uchizawa@yz.yamagata-u.ac.jp`

Abstract. Consider a Boolean circuit over a basis $\{\wedge, \vee, \neg\}$, where $\wedge$ and $\vee$ are the conjunction of unbounded fan-in and disjunction of unbounded fan-in, respectively, and $\neg$ is the negation. The energy complexity of a circuit is defined as the number of gates outputting ones in the circuit, where the maximum is taken over all the input assignments.

We prove that the number of output patterns (the set of possible outputs of all the internal gates in the circuit) in a circuit of energy e is at most $2^{e \log e + 4e}$ vastly improving the trivial bound of $\binom{s}{e}$ where s is the size of the circuit. Building on this tool, we prove the following:

- Any Boolean circuit C of energy e can be converted to a depth-3 circuit of size $s2^{e \log e + 4e}$. This implies a strong limitation on computational power of energy-bounded circuits even if the circuit is allowed to use unbounded fan-in.
- The problem of counting output patterns of a given circuit is known to be #P-complete in general, but is fixed-parameter tractable when parameterized by energy e (over the basis $\mathcal{B}_2 = \{\wedge_2, \vee_2, \neg\}$). We also show further applications of this algorithm.
- If a function $f : \{0, 1\}^n \times \{0, 1\}^n \to \{0, 1\}$ is computed by a circuit of energy e, then, the unbounded-error communication complexity $U(f)$ is at most $O(e \log e)$. In particular, this implies that any circuit computing IP_n has energy $\exp^{W(\Omega(n))}$, where W denotes the product logarithm. This improves the lower bound of $\Omega(\sqrt{n})$ that follows from the relation between energy complexity and decision tree height of a function.

1 Introduction

Energy complexity of a Boolean circuits was introduced in the 1960s, by Vaintsvaig [14] towards evaluating static power consumed in integrated circuits caused by dissipation of heat while maintaining a given potential at the nodes of the circuit. Let $\mathcal{B}$ be a set of Boolean functions. Let C be a Boolean circuit consisting of gates computing functions chosen from $\mathcal{B}$, and n be the number of input variables of C. For an input assignment $\mathbf{a} \in \{0, 1\}^n$ to C, the energy

© The Author(s), under exclusive license to Springer Nature Singapore Pte Ltd. 2024
X. Chen and B. Li (Eds.): TAMC 2024, LNCS 14637, pp. 185–196, 2024.
https://doi.org/10.1007/978-981-97-2340-9_16

complexity $EC_\mathcal{B}(C, \mathbf{a})$ of C for $\mathbf{a}$ is defined as the number of the gates in C that evaluate to 1 for $\mathbf{a}$. Then, for a Boolean function $f : \{0,1\}^n \to \{0,1\}$, we define $EC_\mathcal{B}(f) = \min_C \max_{\mathbf{a} \in \{0,1\}^n} EC_\mathcal{B}(C, \mathbf{a})$, where the minimum is taken over all the circuits C computing f, and $EC_\mathcal{B}(n) = \max_f EC_\mathcal{B}(f)$, where the maximum is taken over all n-variate Boolean functions. Vaintsvaig [14] showed that for any finite basis $\mathcal{B}$, it holds that $\Omega(n) \le EC_\mathcal{B}(f) \le O(2^n/n)$ (possibly with exponential size). Kasim-zade [4] then refined this result further by showing an interesting trichotomy: for any finite complete basis $\mathcal{B}$, either (i) $EC_\mathcal{B}(n) = \Theta(2^n/n)$, (ii) $\Omega(2^{n/2}) \le EC_\mathcal{B}(n) \le O(\sqrt{n}2^{n/2})$ or (iii) $\Omega(n) \le EC_\mathcal{B}(n) \le O(n^2)$. Thus, $EC_\mathcal{B}(n)$ notably varies according to bases that we employ. Note that the size of the circuits can also be exponential in the above bounds. Lozhkin and Shupletsov [5] improved the energy bound to $3n(1+\epsilon)$ where the circuit size is also bounded by $\frac{2^n}{n}(1+\epsilon)$.

Along with $EC_\mathcal{B}(n)$, $EC_\mathcal{B}(f)$ has been recently investigated for several common bases. For the standard basis $\mathcal{B}_2 = \{\wedge_2, \vee_2, \neg\}$ where $\wedge_2$ and $\vee_2$ are a conjunction and a disjunction of fan-in two, respectively, and $\neg$ is a negation, Dinesh et al. [1] showed that $\mathrm{psens}(f)/3 \le EC_{\mathcal{B}_2}(f) \le O(DT(f)^3)$ holds for all Boolean functions f, where $\mathrm{psens}(f)$, called positive sensitivity, is an asymmetric variant of the sensitivity of f, and $DT(f)$ is the optimal decision tree depth of f. Sun et al. [8] refined the relationship between $EC_{\mathcal{B}_2}(f)$ and $DT(f)$, and proved that $\Omega(\sqrt{DT(f)}) \le EC_{\mathcal{B}_2}(f) \le \min\{DT(f)^2/2 + O(DT(f)), n + 2DT(f) - 2\}$, which implies that $EC_{\mathcal{B}_2}(f)$ and $DT(f)$ are polynomially related.

For a basis $\mathcal{B}_{th}$ consisting of linear threshold functions, it is known that there exist Boolean functions for which any constant-depth threshold circuit needs even exponential size to compute if the energy is bounded, while these are computable by constant-depth and linear-size circuit [6, 10, 11]. For another basis $\mathcal{B}_\ell$ consisting of arbitrary Boolean functions of arity ℓ, Suzuki et al. [9] showed that $\Omega\left(\frac{n-m_f}{\ell}\right) \le EC_{\mathcal{B}_\ell}(f) \le O(n/\ell)$ for any symmetric function f, where m_f is the maximum number of consecutive "0"s or "1"s in the value vector of f.

The energy complexity has been studied also from an algorithmic point of view. Let $\wedge$ and $\vee$ denote conjunction and disjunction of arbitrary arity, respectively. Silva and Souza [7] considered a computational problem MINEC_M^+ asking if, given a monotone circuit C over $\mathcal{B}^+ = \{\wedge, \vee\}$ and an integer e, there exists a satisfying assignment $\mathbf{a}$ of C such that $EC_{\mathcal{B}^+}(C, \mathbf{a}) \le e$. They prove that MINEC_M^+ is NP-hard even if C is planar, and W[1]-hard but in XP when parameterized by the size of the solution.

Our Results: In this paper, we consider the standard basis $\mathcal{B}^* = \{\wedge, \vee, \neg\}$, where $\wedge$ and $\vee$ are the conjunction of unbounded fan-in and disjunction of unbounded fan-in, respectively, and $\neg$ is the negation. We then investigate to what extent a restriction on energy affects the computational power of circuits over $\mathcal{B}^*$.

As our main tool, we prove a structural limitation on the circuit if the energy is limited by e. Let C be a Boolean circuit of size s, and $g_1, g_2, \ldots, g_s$ be the gates in C. For an input assignment $\mathbf{a} \in \{0,1\}^n$, we call a vector $(g_1(\mathbf{a}), g_2(\mathbf{a}), \ldots, g_s(\mathbf{a})) \in \{0,1\}^s$ an output pattern. For a circuit C, let $P(C)$

denote the set of possible output patterns of C. We obtain the following upper bound on $|P(C)|$.

Theorem 1. *If C is a circuit with energy e, $|P(C)| \leq 2^{e \log e + 4e}$.*

The above theorem vastly improves the trivial upper bound of $\binom{s}{e}$ where s is the size of the circuit. We show three applications of this new bound.

Transformation to Depth-3 Circuits: We show a transformation from bounded energy circuits to depth-3 circuits. More formally, we prove that:

Theorem 2. *If a Boolean function f can be computed by a circuit C of size s and energy e, then f can also be computed by a depth-3 circuit C' of size $s2^{e \log e + 4e}$.*

This shows that restriction on energy affects the computational power of Boolean circuits even if unbounded fan-in and depth are allowed. It is important to note that a known bound [8] on energy, $\mathsf{DT}(f) \leq e^2$ implies that any function f can be computed by a depth-2 circuit of size at most 2^{e^2}. Thus, Theorem 2 can also be considered as an improvement of this bound at the expense of extra depth by one.

One potential use of Theorem 2 is that exponential lower bounds on the size against depth-3 circuits translate to lower bounds on energy for any circuit. Indeed, the best known lower bound for depth-three circuits (for explicit functions) are of the form $2^{\Omega(\sqrt{n})}$ [3]. Note that Theorem 1 implies strong lower bounds for circuits of energy $e = \frac{\log n}{\log \log n}$.

FPT Algorithms for Counting the Number of Output Patterns: Theorem 1 also shows the weakness of energy-bounded circuits from an algorithmic aspect. In [13], the computational complexity of the following counting problem #PAT($\mathcal{B}$) is studied: Given a circuit C over $\mathcal{B}$, how many output patterns arise in C? The authors of [13] provide a dichotomy result on this counting problem among various bases. In particular, it was shown that #PAT($\{\wedge_2\}$) and #PAT($\{\vee\}_2$) are #P-complete even when a depth-2 circuit is given. We show that:

Theorem 3. *#PAT($\mathcal{B}_2$) is solvable in polynomial time in n, s and $2^{e \log e}$.*

Thus, #PAT($\mathcal{B}_2$) is fixed parameter tractable when parameterized by e.

By applying the same algorithm, we can also provide a FPT algorithm for MINEC_M^+ if a given circuit has bounded fan-in.

Theorem 4. *MINEC($\mathcal{B}_2$) is solvable in polynomial time in n, s and $2^{e \log e}$.*

Note that our FPT algorithm can solve the problem even if a given circuit is not monotone.

Communication Protocols from Bounded Energy Circuits: Following an argument in [12], we show, as another application of Theorem 1, that construction of energy bounded circuits for a function leads to communication protocols for the same function. For a Boolean function f, let $U(f)$ be the unbounded-error communication complexity of f. Then, we can show that $U(f)$ is bounded by the logarithm of the number of output patterns of C for f. That is, we can obtain the following theorem:

Theorem 5. *If a function $f : \{0,1\}^n \times \{0,1\}^n \to \{0,1\}$ is computed by a circuit of energy e, then $U(f) \leq O(e \log e)$.*

Consequently, lower bounds for unbounded-error communication complexity translates to size-independent lower bounds on energy as well. For example, it is known that $U(\mathrm{IP}_n) = \Omega(n)$, where the inner product function IP_n is defined as $\mathrm{IP}_n(\mathbf{x}, \mathbf{y}) = \bigoplus_{i=1}^{n} x_i \wedge y_i$, and hence we conclude that any circuit computing IP_n has energy $\exp^{W(\Omega(n))}$, where W denote the product logarithm. However, for the special case of IP_n function, we can also obtain a better lower bound without employing Theorem 6 by a variant of the standard elimination argument (see Theorem 10).

2   Preliminaries

For a positive integer n, we denote by $[n]$ the set $\{1, 2, \ldots, n\}$. Let $\mathcal{B}^* = \{\wedge, \vee, \neg\}$, where $\wedge$ is the conjunction of unbounded fan-in, $\vee$ is the disjunction of unbounded fan-in, and $\neg$ is the negation.

A Boolean circuit C over $\mathcal{B}^*$ is expressed by a directed acyclic graph where nodes with in-degree zero are labeled by input variables, and the other nodes by gates computing functions chosen from $\mathcal{B}^*$. We assume that C contains a unique gate with out-degree zero, and regard the output of the gate as the output of C.

We define size of a circuit C as the number of the gates in C, and depth of C as the length of a longest path in the underline graph of C. For $\mathbf{a} \in \{0,1\}^n$, we define $\mathrm{EC}(C, \mathbf{a})$ as the number of $\wedge$-gates, $\vee$-gates and $\neg$-gates outputting ones for $\mathbf{a}$. Then we define

$$\mathrm{EC}_{\mathcal{B}^*}(C) = \max_{\mathbf{a} \in \{0,1\}^n} \mathrm{EC}_{\mathcal{B}^*}(C, \mathbf{a}) \qquad \text{and} \qquad \mathrm{EC}_{\mathcal{B}^*}(f) = \min_{C \in \mathcal{C}_f} \mathrm{EC}_{\mathcal{B}^*}(C)$$

where $\mathcal{C}_f$ is a set of circuits computing f of n variables. Note that if C has energy e and m $\neg$-gates, we have $m \leq e$.

Let C be a circuit with n input variables, and s be the size of C. Let $g_1, g_2, \ldots, g_s$ be the gates in C. Recall that we define $(g_1(\mathbf{a}), g_2(\mathbf{a}), \ldots, g_s(\mathbf{a})) \in \{0,1\}^s$ as an output pattern of C for an input assignment $\mathbf{a}$. We define a set $P(C)$ of the output patterns that arise in C as

$$P(C) = \{(g_1(\mathbf{a}), (g_2(\mathbf{a}), \ldots, (g_s(\mathbf{a})) \mid \mathbf{a} \in \{0,1\}^n\}.$$

The problem $\#\mathrm{PAT}(\mathcal{B})$ asks to compute the value of $|P(C)|$ for a given circuit C over $\mathcal{B}$. It is known that the problem is $\#P$-complete even if a given circuit consists solely of $\wedge_2$-gates (or $\vee_2$-gates), where $\wedge_2$ (resp., $\vee_2$) is a conjunction (resp., disjunction) gate of fan-in two, respectively [13]. We also define

MINEC($\mathcal{B}$) as a decision problem asking if, given a circuit C over $\mathcal{B}$ and an integer e, there exists a satisfying assignment $\mathbf{a}$ of C such that $EC_\mathcal{B}(C, \mathbf{a}) \leq e$. Note that MINEC($\mathcal{B}^+$) is identical to the original problem MINEC$_M^+$ considered in [7].

Let f be a Boolean function of $2n$ variables, and Alice and Bob have unlimited computational power together with private random coins. In the setting of communication complexity, Alice and Bob receive $\mathbf{a} \in \{0,1\}^n$ and $\mathbf{b} \in \{0,1\}^n$, respectively, and wish to evaluate $f(\mathbf{a}, \mathbf{b})$ by communicating bits with each other according to a protocol. We say that a protocol Π computes f if

$$\Pr[\Pi(\mathbf{a}, \mathbf{b}) = f(\mathbf{a}, \mathbf{b})] > \frac{1}{2}$$

for every $\mathbf{a}, \mathbf{b} \in \{0,1\}^n$ where the probability is taken over all the random coins. The complexity of a protocol is defined as the maximum number of bits communicated, where the maximum is taken over $\mathbf{a}, \mathbf{b}$. The unbounded-error communication complexity $U(f)$ is defined as the minimum complexity of the protocols computing f.

3   Bound on the Number of Output Patterns

In this section, we prove the main technical claim of the paper - an upper bound on the number of output patterns for a circuit whose energy is bounded by e. We restate it from the introduction (Theorem 1).

Theorem 6. *If C is a circuit with energy e, $|P(C)| \leq 2^{e \log e + 4e}$.*

Proof. For a positive integer e and a non-negative integer m, we define $P(e, m)$ as the maximum number of the output patterns that any circuit of energy at most e and at most m ¬-gates can possess. Since $m \leq e$, it suffices to prove that

$$P(e, m) \leq 2^{e \log e + 3e + m}. \tag{1}$$

We verify Eq. (1) by induction on e and m.

For the base case, we consider (i) $e < m$, (ii) $e = 1$ and (iii) $m = 0$. Note that the right-hand side of Eq. (1) is at least 8. In the case (i), no circuit simultaneously has energy e and m negation gates, and hence we write $P(e, m) = 0$. In the case (ii), C has at most one of $\wedge$-gate and $\vee$-gate together with a ¬-gate, and hence $P(e, m) \leq 2$. In the case (iii), circuits are monotone, and hence any of them has at most e gates. Thus, $P(e, m) \leq 2^e$.

Suppose for the induction hypothesis that Eq. (1) holds for the case where either energy is at most $e - 1$ or the number of negation gates is at most $m - 1$. We below consider $P(e, m)$. Let C be an arbitrary circuit of energy e and m negation gates.

If there is an input variable x that is directly connected to a negation gate, we have two cases $x = 0$ and $x = 1$. The negation gate outputs one for $x = 0$ and zero for $x = 1$. Thus, it holds that $P(e, m) \leq P(e - 1, m - 1) + P(e, m - 1)$ and hence the induction hypothesis implies that $P(e,m) \leq 2^{(e-1)\log(e-1)+3(e-1)+m-1} + 2^{e\log e+3e+m-1}$ which in turn is bounded by $2^{e\log e+3e+m}$.

Otherwise, all the input variables are connected to AND-gates or OR-gates. We say that a gate g is monotone if every path from any input variable to g contains no negation gates. Let C' be a subcircuit of C consisting of all the monotone gates. Since C has energy e, C' has energy at most e. Thus C' has at most e gates. For each $\mathbf{p} \in P(C')$, we define $C[\mathbf{p}]$ as a circuit obtained from C by fixing the outputs of the gates in C' accordingly to $\mathbf{p}$, where we leave the input variables untouched. Note that $C[\mathbf{p}]$ may have extra output patterns, because $C[\mathbf{p}]$ may receive input assignments that does not cause $\mathbf{p}$ in C'. We denote by $|\mathbf{p}|$ the hamming weight of $\mathbf{p}$. Then, $C[\mathbf{p}]$ has energy at most $e - |\mathbf{p}|$, and at most $m - 1$ negation gates, because C' has at least a gate connected to a negation gate. Thus,

$$P(C) \leq \sum_{\mathbf{p}\in P(C')} P(e-|\mathbf{p}|, m-1) \leq P(e-1, m-1) + \sum_{k=1}^{e} \binom{e}{k} P(e-k, m-1).$$

where the term "$P(e-1, m-1)$" corresponds to the case where $\mathbf{p} = (0, 0, \ldots, 0)$, and at least a $\neg$-gate outputs one. Therefore, the induction hypothesis implies that

$$P(C) \leq 2^{(e-1)\log(e-1)+3(e-1)+m-1} + \sum_{k=1}^{e} \binom{e}{k} 2^{(e-k)\log(e-k)+3(e-k)+m-1}.$$

By applying $\binom{e}{k} < ((c\cdot e)/k)^k \leq 2^{k(\log e - \log k + 2)}$ where $c < 4$, the exponent of the kth component of the second term is written as $k(\log e - \log k + 2) + (e-k)\log(e-k) + 3(e-k) + m - 1$, which can be rewritten as $k\log e - k\log k + e\log(e-k) - k\log(e-k) + 3e - k + m - 1$ by using the fact that $2k + 3(e-k) = 3e - k$. We bound this, analysing the two cases:

Case 1 : $k \leq e/2$. In this case, we have:

$$k\log e - k\log k + e\log(e-k) - k\log(e-k) + 3e - k + m - 1$$
$$= (e\log(e-k) + 3e + m - 1) - k\log k + (k\log e - k\log 2(e-k))$$
$$\leq (e\log e + 3e + m - 1) - k\log k$$
$$\leq (e\log e + 3e + m - 1) - k$$

Case 2: $e/2 \leq k \leq e$

$$k\log e - k\log k + e\log(e-k) - k\log(e-k) + 3e - k + m - 1$$
$$\leq (k\log e - k\log(e/2)) + e\log(e/2) + 3e - k + m - 1$$
$$\leq k + e\log e - e + 3e - k + m - 1$$
$$\leq (e\log e + 3e + m - 1) - e \leq (e\log e + 3e + m - 1) - k.$$

Therefore,

$$P(C) \le 2^{(e-1)\log(e-1)+3(e-1)+m-1} + 2^{e\log e+3e+m-1} \cdot \left(\sum_{k=1}^{e} 2^{-k} \right)$$

$$\le 2 \cdot 2^{e\log e+3e+m-1} \le 2^{e\log e+3e+m},$$

as desired. □

The upper bound given in Theorem 6 is almost tight, as follows.

Proposition 1. *There exists a circuit C of energy e such that $P(C) = 2^{e-1}$.*

Proof. Consider a depth-2 circuit C of $e-1$ $\wedge$-gates $g_1, g_2, \ldots, g_{e-1}$ in the bottom layer together with a single $\vee$-gate in the second layer. Let $x_1, x_2, \ldots, x_n$ be input variables, then the input of g_i is x_i for every i, $1 \le i \le e-1$. C has size e, and hence energy e. Since the outputs of $g_1, \ldots, g_{e-1}$ are independent of each other by construction, we have $P(C) = 2^{e-1}$. □

4   Three Applications of the Bound

We present three applications of the bound on the number of output patterns of an energy bounded circuit.

4.1   From Bounded-Energy Circuits to Depth-3 Circuits

As mentioned in the introduction, we show that any energy bounded circuit can be simulated by a depth-3 circuit with a limited blow up on the size. We prove Theorem 2 from the introduction which we restate below.

Theorem 7. *If a Boolean function f is computable by a circuit of size s and energy e, then f is also computable by a circuit C' of size at most $s2^{e\log e+4e} + 1$ and depth 3.*

Proof. Let C be a circuit with n input variables. We use Theorem 6 to prove the theorem. We define a set P_1 of output patterns in which the output of C is one:

$$P_1 = \{\mathbf{p} \in P(C) \mid \mathbf{p} \text{ is an output pattern for which the output of } C \text{ is one}\}.$$

For each output pattern $\mathbf{p} = (p_1, p_2, \ldots, p_s) \in P_1(C) \subseteq P(C)$, we construct a depth-2 circuit with at most s gates in the first layer, and an $\wedge$-gate at the top that detects the pattern $\mathbf{p}$ arises for $\mathbf{a}$: In the first layer, we put g_j for each $j \in [s]$ modified so that (i) If g_j receives an output of $g_{j'}$ for $j' \in [j-1]$, we feed the value $p_{j'}$ into g_j instead; and (ii) if $p_j = 0$, the output of g_j is connected to the $\wedge$-gate at the top through a $\neg$-gate. Taking a disjunction of the $\wedge$-gates, we complete the construction. Thus, C has at most $s \cdot |P(C)| + 1$ gates, and hence Theorem 6 implies the claim. □

4.2   FPT Algorithm for Counting Output Patterns

As a second application of Theorem 6, we design an FPT algorithm for $\#\mathrm{PAT}(\mathcal{B}_2)$ and $\mathrm{MINEC}(\mathcal{B}_2)$, where $\mathcal{B}_2 = \{\wedge_2, \vee_2, \neg\}$. This proves Theorem 3 from the introduction, which we restate below in equivalent terms.

Theorem 8. $\#\mathrm{PAT}(\mathcal{B}_2)$ *is fixed-parameter tractable when parameterized by the energy* e.

Proof. We prove that if C has size s and $\mathrm{EC}_{\mathcal{B}_2}(C) \le e$, we can obtain $P(C)$ in time polynomial in n, s and $2^{e\log e + 4e}$. Let $g_1, \ldots, g_s$ be the gates in C, where the gates are named in topological order. For each ℓ, $1 \le \ell \le s$, we denote by P_ℓ a projection of $P(C)$ onto the first l components:

$$P_\ell = \{(\alpha_1, \ldots, \alpha_l) \mid (\alpha_1, \ldots, \alpha_s) \in P(C)\}.$$

Note that $P(C) = P_s$. Theorem 6 implies that $|P_\ell| \le 2^{e\log e + 4e}$ for any ℓ. Below, we inductively obtain P_l from $l = 1$ to $\ell = s$.

We can construct P_1 by checking if there is an assignment $\mathbf{a}$ for which g_1 outputs zero (and also one). Suppose we have obtained $P_{\ell-1}$. Consider each $\alpha = (\alpha_1, \ldots, \alpha_{\ell-1}) \in P_{\ell-1}$. For each $z \in \{0, 1\}$, we solve the following 2SAT problem

$$[g_1(\mathbf{x}) = \alpha_1] \wedge \cdots \wedge [g_{\ell-1}(\mathbf{x}) = \alpha_{\ell-1}] \wedge [g_\ell(\mathbf{x}) = z].$$

If it is satisfiable, we add $(\alpha_1, \ldots, \alpha_{\ell-1}, z)$ to P_ℓ. We can solve the 2SAT in time polynomial in n and s, The total number of times we solve 2SAT is bounded by $|P_1| + |P_2| + \cdots + |P_s|$. Thus, Theorem 6 implies that we can obtain $P(C) = P_s$ in time polynomial in n, s and $2^{e\log e + 4e}$. $\square$

Since the above algorithm enumerates all output patterns of the circuit, it can also be used directly to solve the $\mathrm{MINEC}(\mathcal{B}_2)$ problem: We get the following corollary, which proves Theorem 4 from the introduction.

Corollary 1. $\mathrm{MINEC}(\mathcal{B}_2)$ *is fixed-parameter tractable when parameterized by the energy* e.

4.3   Size-Independent Lower Bound on Energy

As a third application of Theorem 6, we show a connection between energy complexity and communication complexity of the function which we restate below. The proof follows an argument given in [12], but we give a proof for completeness noting that the argument generalizes to Boolean circuits as well.

Theorem 9. *If a function* $f : \{0, 1\}^n \times \{0, 1\}^n \to \{0, 1\}$ *is computed by a circuit of energy* e, *then,* $U(f)$ *is at most* $O(e\log e)$.

Proof. Let C be a circuit computing f, and s, d, and e be size, depth, and energy of C. We denote by $g_1, \ldots, g_s$ the gates in C. We prove the theorem by constructing the desired protocol Π.

We first show that C can be converted to a function L such that

$$L(\mathbf{x}, \mathbf{y}) = \sum_{\mathbf{p} \in P(C)} w_{\mathbf{p}} F_{\mathbf{p}}(\mathbf{x}, \mathbf{y})$$

where $F_{\mathbf{p}}$ is a formula of conjunctive normal form (that is, CNF) and $w_{\mathbf{p}}$ is an integer weight for the output of $F_{\mathbf{p}}$.

Fix an arbitrary pattern $\mathbf{p} \in P(C)$. Below we give $F_{\mathbf{p}}$ and $w_{\mathbf{p}}$. Let z be the number of ones in $\mathbf{p}$, and $i_1, \ldots, i_z$ be the corresponding indices of the z gates characterizing $\mathbf{p}$. For each j, $1 \leq j \leq z$, we obtain h_{i_j} from g_{i_j}, as follows: If g_{i_j} receives an output of g_r, we fix the output to one (resp., zero) if the rth element of $vecp$ is one (resp., zero). For example, if g_{i_j} is a $\vee$-gate, h_{i_j} is either a $\vee$-gate receiving only some of the input variables or a constant function no matter which layer the original gate g_{i_j} is in C. Then we define $F_{\mathbf{p}}$ as the conjunction of $h_{i_1}, \ldots, h_{i_z}$. Since $z \leq e$, $F_{\mathbf{p}}$ is a CNF of at most e disjunctions. We now define $\mathbf{k} = (k_1, \ldots, k_d)$ where for each $\ell \in [d]$, k_ℓ is the number of gates among $g_1, \ldots, g_z$ that are in the ℓth layer. Then we define

$$w_{\mathbf{p}} = \begin{cases} a^{q(\mathbf{p})} & \text{if one of } g_{i_1}, \ldots, g_{i_z} \text{ is the top gate;} \\ -a^{q(\mathbf{p})} & \text{otherwise.} \end{cases}$$

where

$$a = |P(C)| + 1$$

and

$$q(\mathbf{p}) = \sum_{\ell=1}^{d} k_\ell \cdot e^{d-\ell}.$$

We now verify the following claim.

Claim. $L(\mathbf{a}, \mathbf{b})$ is positive if $C(\mathbf{a}, \mathbf{b}) = 1$, and $L(\mathbf{a}, \mathbf{b})$ is negative, otherwise.

Proof. $L(\mathbf{a}, \mathbf{b})$ is positive if $C(\mathbf{a}, \mathbf{b}) = 1$, and $L(\mathbf{a}, \mathbf{b})$ is negative, otherwise. Consider an arbitrary input $(\mathbf{a}, \mathbf{b}) \in \{0, 1\}^{2n}$. Let $\mathbf{p}^*$ be the output pattern that arises in C for $(\mathbf{a}, \mathbf{b})$. Note that $F_{\mathbf{p}^*}(\mathbf{a}, \mathbf{b}) = 1$. Thus, it suffices to show that $w_{\mathbf{p}}$ dominates the sum of the other weights $w_{\mathbf{p}}$ such that $\mathbf{p} \neq \mathbf{p}^*$ and $F_{\mathbf{p}}(\mathbf{a}, \mathbf{b}) = 1$. Consider any other output pattern $\mathbf{p}$ than $\mathbf{p}^*$, and $\mathbf{k} = (k_1, \ldots, k_d)$ for $\mathbf{p}$. Since $\mathbf{p} \neq \mathbf{p}^*$, there exists ℓ' such that $k_\ell = k_{\ell'}$ for every $\ell < \ell'$, but $k_{\ell'} \neq k_{\ell'}^*$. If $k_{\ell'} \geq k_{\ell'}^*$, $F_{\mathbf{p}}(\mathbf{a}, \mathbf{b}) = 0$, since one of the at most e gates characterizing $\mathbf{p}$ in the ℓ'th layer do not output one. If $k_{\ell'} < k_{\ell'}^*$, we have

$$q(\mathbf{p}) \leq q(\mathbf{p}^*) - 1$$

Consequently, it holds that

$$\sum_{\mathbf{p} \in P(C) \backslash \{\mathbf{p}^*\}} |w_{\mathbf{p}}| < |P(C)| \cdot a^{q(\mathbf{p}^*)-1} = |P(C)| \cdot \frac{|w_{\mathbf{p}^*}|}{a} < |w_{\mathbf{p}^*}|,$$

as desired. □

By Theorem 6, we have $|P(C)| \leq 2^{O(e \log e)}$, which implies a protocol Π computing f, as follows. Alice randomly chooses a CNF $F_\mathbf{p}$ with probability proportionally to $|w_\mathbf{p}|$, and let Bob know the chosen $\mathbf{p}$ by communicating $O(\log |P(C)| = O(e \log e)$ bits. Besides that, Alice lets Bob know if each of at most e disjunctions in $F_\mathbf{p}$ is satisfied by Alice's inputs by $O(e)$ bits (If the conjunction receives Alice's inputs directly, she also sends the result of the conjunction by additional one bit). Bob then evaluates $F_\mathbf{p}$, and outputs one if $F_\mathbf{p}$ is evaluated to one and $0 < w_\mathbf{p}$, and outputs zero if $F_\mathbf{p}$ is evaluated to one and $w_\mathbf{p} < 0$. If $F_\mathbf{p}$ is evaluated to zero, Bob outputs one with probability $1/2$, and zero with probability $1/2$.

Clearly, the number of bits communicated is $O(e \log e)$. Let ϵ be the probability that Alice chooses $\mathbf{p}$ such that $F_\mathbf{p}(\mathbf{a}, \mathbf{b}) = 1$. Then, the claim implies that Alice can choose $\mathbf{p}$ such that $w_\mathbf{p}$ has the correct sign with probability strictly greater than $\epsilon/2$. Consequently, we have

$$\Pr[\Pi(\mathbf{a}, \mathbf{b}) = f(\mathbf{a}, \mathbf{b})] > \frac{\epsilon}{2} + \frac{(1 - \epsilon)}{2}.$$

where the second term corresponds to the event that Alice chooses $\mathbf{p}$ such that $F_\mathbf{p}(\mathbf{a}, \mathbf{b}) = 0$, but Bob's random guess is correct.   □

Recall the Inner-Product function IP_n of $2n$ variables $\mathbf{x} = (x_1, \ldots x_n)$ and $\mathbf{y} = (y_1, \ldots, y_n)$ as $\mathrm{IP}_n(\mathbf{x}, \mathbf{y}) = \bigoplus_{i=1}^{n} x_i \wedge y_i$

Suppose IP_n can be computed by a circuit C of energy e. Since it is known that the unbounded-error communication complexity of IP_n is $\Omega(n)$ [2], we have the following corollary:

Corollary 2. *Any circuit computing* IP_n *has energy* $\exp^{W(\Omega(n))}$, *where* W *denote the product logarithm.*

However, for the special case of IP_n function, we can also obtain a lower bound without employing Theorem 6 by a variant of the standard elimination argument.

Theorem 10. *Any circuit C computing* IP_n *has energy* $\Omega(n)$.

Proof. We prove the claim by a combination of energy and gate elimination. We below show that we can obtain either the energy reduction or size reduction. Let C be a circuit computing IP_n, and e and d be energy and depth of C, respectively.

Suppose there exists a $\vee$-gate g directly connected to an input variable x_i. In this case, we set x_i to one and y_i to zero. Then g outputs one no matter what we fix the rest of the input variables, and so the resulting circuit computes IP_{n-1} and has energy $e - 1$. Similarly, if there exists a $\neg$-gate g directly connected to an input variable x_i, we set x_i and y_i to zeros. Then g outputs one and the resulting circuit computes IP_{n-1} and has energy $e - 1$.

Consider the other case where every input is directly connected to $\wedge$-gate. The number of these $\wedge$-gates is at most e, since C has energy e. We can thus fix the outputs of the $\wedge$-gates to zeros by setting at most e input variables to zeros. Consequently, the output of C is fixed.

Thus, we can fix the output of C by setting at most $3e$ input variables. Thus, $e = \Omega(n)$.

5 Discussion and Open Problems

In this paper, we show a bound on the number of output patterns of an energy bounded Boolean circuit, and show applications of the same. We showed three applications of this, which improves known bounds in three frontiers of energy bounded circuits - an improved constant depth simulation of energy bounded circuits, an FPT algorithm for counting the number of output patterns, and finally a new lower bound for energy of circuits computing the inner product function.

An interesting special case of energy complexity of circuits is that of monotone circuits. Indeed, the energy of Boolean circuits which does not contain negation gate is at least as large as the size of the circuits, since the substitution of all 1's input will ensure all the gates evaluate to 1. Generalizing this model, in the case of formulas, is that of De Morgan circuits (circuits which use negations only at the leaves). We conjecture that De Mrogan circuits must also have energy linear in the size of the circuit.

References

1. Dinesh, K., Otiv, S., Sarma, J.: New bounds for energy complexity of Boolean functions. Theoret. Comput. Sci. **845**, 59–75 (2020)
2. Forster, J.: A linear lower bound on the unbounded error probabilistic communication complexity. J. Comput. Syst. Sci. **65**(4), 612–625 (2002). Special Issue on Complexity 2001
3. Jukna, S.: Boolean Function Complexity: Advances and Frontiers. Springer, Heidelberg (2012). https://doi.org/10.1007/978-3-642-24508-4
4. Kasim-zade, O.M.: On a measure of active circuits of functional elements (Russian). Math. Probl. Cybern. "Nauka" **4**, 218–228 (1992)
5. Lozhkin, S.A., Shupletsov, M.S.: Switching activity of Boolean circuits and synthesis of Boolean circuits with asymptotically optimal complexity and linear switching activity. Lobachevskii J. Math. **36**(4), 450–460 (2015)
6. Maniwa, H., Oki, T., Suzuki, A., Uchizawa, K., Zhou, X.: Computational power of threshold circuits of energy at most two. IEICE Trans. Fundam. Electron. Commun. Comput. Sci. **E101.A**(9), 1431–1439 (2018)
7. Silva, J.C.N., Souza, U.S.: Computing the best-case energy complexity of satisfying assignments in monotone circuits. Theoret. Comput. Sci. **932**, 41–55 (2022)
8. Sun, X., Sun, Y., Wu, K., Xia, Z.: On the relationship between energy complexity and other Boolean function measures. J. Comb. Optim. **43**, 1470–1492 (2022)
9. Suzuki, A., Uchizawa, K., Zhou, X.: Energy and fan-in of logic circuits computing symmetric Boolean functions. Theoret. Comput. Sci. **505**, 74–80 (2013)
10. Uchizawa, K., Douglas, R.J., Maass, W.: On the computational power of threshold circuits with sparse activity. Neural Comput. **18**(12), 2994–3008 (2008)

11. Uchizawa, K.: Size, depth and energy of threshold circuits computing parity function. In: Proceedings of 31st International Symposium on Algorithms and Computation (ISAAC 2020), vol. 181, pp. 54:1–54:13 (2020)
12. Uchizawa, K., Takimoto, E.: Exponential lower bounds on the size of constant-depth threshold circuits with small energy complexity. Theoret. Comput. Sci. **407**(1–3), 474–487 (2008)
13. Uchizawa, K., Wang, Z., Morizumi, H., Zhou, X.: Complexity of counting output patterns of logic circuits. In: Proceedings of the Nineteenth Computing: The Australasian Theory Symposium, vol. 141, pp. 37–41 (2013)
14. Vaintsvaig, M.N.: On the power of networks of functional elements (Russian). Dokl. Akad. Nauk SSSR **139**(2), 320–323 (1961)

Approximation Algorithms for Robust Clustering Problems Using Local Search Techniques

Chenchen Wu[1], Rolf H. Möhring[2,3], Yishui Wang[4], Dachuan Xu[5(✉)],
and Dongmei Zhang[6]

[1] Institute of Operations Research and Systems Engineering, College of Science,
Tianjin University of Technology, Tianjin 300384, People's Republic of China
[2] Institute for Applied Optimization,Department of Computer Science
and Technology, Hefei University,Hefei, People's Republic of China
[3] The Combinatorial Optimization and Graph Algorithms (COGA) group, Institute
for Mathematics, Technical University of Berlin, Berlin, Germany
[4] School of Mathematics and Physics, University of Science and Technology Beijing,
Beijing, People's Republic of China
[5] Department of Operations Research and Information Engineering, Beijing
University of Technology, Beijing 100124, People's Republic of China
xudc@bjut.edu.cn
[6] School of Computer Science and Technology, Shandong Jianzhu University,
Jinan 250101, People's Republic of China

Abstract. In this paper, we explore two robust models for the k-median and k-means problems: the outlier-version (k-MedO/k-MeaO) and the penalty-version (k-MedP/k-MeaP), enabling the marking and elimination of certain points as outliers. In k-MedO/k-MeaO, the count of outliers is restricted by a specified integer, while in k-MedP/k-MeaP, there's no explicit limit on outlier quantity, yet each outlier incurs a penalty cost.

We introduce a novel approach to evaluate the approximation ratio of local search algorithms for these problems. This involves an adapted clustering method that captures pertinent information about outliers within both local and global optimal solutions. For k-MeaP, we enhance the best-known approximation ratio derived from local search, elevating it from $25 + \varepsilon$ to $9 + \varepsilon$. The best-known approximation ratio for k-MedP is also obtained.

Regarding k-MedO/k-MeaO, only two bi-criteria approximation algorithms based on local search exist. One violates the outlier constraint (limiting outlier count), while the other breaches the cardinality constraint (restricting the number of clusters). We focus on the former algorithm, enhancing its approximation ratios from $17+\varepsilon$ to $3+\varepsilon$ for k-MedO and from $274 + \varepsilon$ to $9 + \varepsilon$ for k-MeaO.

Keywords: Clustering Problems · Approximation algorithms · Robust

X. Chen and B. Li (Eds.): TAMC 2024, LNCS 14637, pp. 197–208, 2024.
https://doi.org/10.1007/978-981-97-2340-9_17

1   Introduction

The utilization of extensive datasets for informed decision-making is increasingly vital and commonplace across various disciplines such as Operations Research, Management Science, Biology, Computer Science, and Machine Learning. In data analytics, the clustering of large datasets stands as a foundational challenge. Among numerous clustering methodologies, center-based clustering stands out as one of the most prevalent and widely employed approaches. Center-based clustering encompasses fundamental problems like the k-median and k-means problems, which represent some of the most fundamental and time-honored techniques. The objective of k-median/means clustering is to identify k cluster centers in such a way that minimizes the collective (squared) distance of each input data point to its nearest cluster center. Typically, k-median problems are considered within arbitrary metrics, while k-means problems operate within the Euclidean space $\mathbb{R}^d$.

Both the k-median and k-means problems are NP-hard to approximate beyond specific lower bounds, namely $1 + 2/e \approx 1.736$ [17] for k-median and 1.07 [10] for k-means. Numerous studies have concentrated on developing efficient approximation algorithms. The most notable known approximations stand at $2.675 + \varepsilon$ [4] for k-median and $6.357 + \varepsilon$ [1] for k-means, where ε represents a given parameter. When confined to a fixed-dimensional Euclidean space, both the k-median and k-means problems possess a PTAS (Polynomial-Time Approximation Scheme) as outlined in [2,11,14].

In the real world, datasets frequently include 'noises' that can significantly disrupt k-median/means clustering outcomes. Robust clustering techniques have emerged to mitigate the impact of these noises. Broadly, there are two types of robust formulations: k-median/means with penalties (k-MedP/k-MeaP) and k-median/means with outliers (k-MedO/k-MeaO). Let's formally define these problems.

In the *k-median problem with penalties* (k-MedP), the setup involves a client set $\mathcal{X}$ consisting of n points, a facility set $\mathcal{F}$ comprising m points, a metric space $(\mathcal{X} \cup \mathcal{F}, d)$, a positive integers $k < m$, and a penalty cost p_x for each client x. The problem aims to find at most k facilities to open and choose some clients to be penalty such that the total cost including the connection cost and penalty cost is minimized, that is,

$$\min_{S \subseteq \mathcal{F}, P \subseteq \mathcal{X}: |F| \leq k} \sum_{x \in \mathcal{X} \setminus P} \min_{s \in S} d(x, s) + \sum_{x \in P} p_x.$$

In the *k-means with penalties* (k-MeaP), the setup involves a data set $\mathcal{X} \subseteq \mathbb{R}^d$ consisting of n points, a positive integers $k < m$, and a penalty cost p_x for each client x. The distance between $u, v \in \mathcal{X}$ is the Euclidean distance, that is, $d(u, v) := \|u - v\|_2$. The problem aims to find at most k centers and choose at most z data as outliers such that the connection cost is minimized, that is,

$$\min_{S \subseteq \mathbb{R}^d, P \subseteq \mathcal{X}: |S| \leq k} \sum_{x \in \mathcal{X} \setminus P} \min_{s \in S} d(x, s)^2 + \sum_{x \in P} p_x.$$

In the *k-median problem with outliers* (*k*-MedO), the setup involves a client set $\mathcal{X}$ consisting of n points, a facility set $\mathcal{F}$ comprising m points, a metric space $(\mathcal{X} \cup \mathcal{F}, d)$, and two positive integers $k < m$ and $z < n$. The problem aims to find at most k facilities to open and choose at most z clients to be outliers such that the total connection cost is minimized, that is,

$$\min_{S \subseteq \mathcal{F}, P \subseteq \mathcal{X}: |F| \leq k, |P| \leq z} \sum_{x \in \mathcal{X} \setminus P} \min_{s \in S} d(x, s).$$

In the *k-means with outliers* (*k*-MeaO), the setup involves a data set $\mathcal{X} \subseteq \mathbb{R}^d$ consisting of n points, and a positive integers $k < m$ and $z < n$. The distance between $u, v \in \mathcal{X}$ is the Euclidean distance, that is, $d(u, v) := \|u - v\|_2$. The problem aims to find at most k centers and choose at most z data as outliers such that the connection cost is minimized, that is,

$$\min_{S \subseteq \mathbb{R}^d, P \subseteq \mathcal{X}: |S| \leq k, |P| \leq z} \sum_{x \in \mathcal{X} \setminus P} \min_{s \in S} d(x, s)^2.$$

A summary of the up-to-date approximation results for *k*-MedO/*k*-MeaO and *k*-MedP/*k*-MeaP along with their ordinary versions is given in Table 1.

Table 1. Comparion of (robust) clustering problems

Techniques and reference	*k*-median	*k*-MedO	*k*-MedP	*k*-means	*k*-MeaO	*k*-MeaP
LP rounding [5]	6.67					
Lagrangian relaxation [18]	6			108		
Lagrangian relaxation [6]			4			
Lagrangian relaxation [17]	4					
Local search [3]	$3 + \varepsilon$					
Local search + potential function [9]	$2.865 + \varepsilon$					
Local search [19]				$9 + \varepsilon$		
Successive local search [8]		constant				
Dependent LP rounding [7]	3.25					
Local search [16]			$3 + \varepsilon$			
Pseudo-approximation [21]	$2.732 + \varepsilon$					
Pseudo-approximation [4]	$2.675 + \varepsilon$					
Iterative LP rounding [20]		$7.081 + \varepsilon$			$53.002 + \varepsilon$	
Primal-dual [1]				$6.357 + \varepsilon$		
Local search [24]						$25 + \varepsilon$
Bipoint rounding [12]						$19.849 + \varepsilon$

Concerning time complexity, many local search algorithms have received attention and this technique has been well applied to clustering problems. However, the standard local search algorithm for *k*-MedO/*k*-MeaO cannot produce a feasible solution with a bounded approximation ratio [13]. So some research directions focus on bi-criteria approximation algorithms based on local search for these two problems. These algorithms have a bounded approximation ratio but violate either the k-constraint or the outlier constraint by a bounded factor. Gupta et al. [15] have developed a method for addressing outliers in a local search algorithm, yielding a bi-criteria $(274 + \varepsilon, O(\frac{k}{\varepsilon} \log n\delta))$-approximation algorithm (δ denotes the maximal distance between two points in the data set) that violates the outlier constraint. Friggstad et al. [13] have provided $(3 + \varepsilon, 1 + \varepsilon)$- and $(25 + \varepsilon, 1 + \varepsilon)$-local search bi-criteria approximation algorithms for k-MedO and k-MeaO respectively.

We will consider the standard local search algorithm for k-MedP/k-MeaP, and the outlier-based local search algorithm by Gupta et al. [15] for k-MedO/k-MeaO. Using our new technique, we will improve the approximation ratios for k-MeaP, k-MeaO and k-MedO. For k-MedP, we obtain the same approximation ratio which is the best one possible.

We list the related results about local search algorithms for k-MedO/k-MeaO and k-MedP/k-MeaP in Table 2.

Table 2. Local search algorithms for (robust) clustering problems. The # centers blowup means the factor by which the cardinality constraint is violated. The # outliers blowup means the factor by which the outlier constraint is violated.

Reference	Problem	Ratio	# centers blowup	# outliers blowup
[9]	k-median	$2.865 + \varepsilon$	none	none
[19]	k-means	$9 + \varepsilon$	none	none
[8]	k-MedO	constant	none	none
[16]	k-MedP	$3 + \varepsilon$	none	none
[24]	k-MeaP	$25 + \varepsilon$	none	none
[11]	k-median/k-means in minor-free metrics with fixed dimension	PTAS	none	none
[14]	k-median/k-means with fixed doubling dimension	PTAS	none	none
[13]	k-MedO	$3 + \varepsilon$	$1 + \varepsilon$	none
	k-MeaO	$25 + \varepsilon$	$1 + \varepsilon$	none
[15]	k-MedO	$17 + \varepsilon$	none	$O(k \log(n\delta)/\varepsilon)$
	k-MeaO	$274 + \varepsilon$	none	$O(k \log(n\delta)/\varepsilon)$
Our results	k-MedP	$3 + \varepsilon$	none	none
	k-MeaP	$9 + \varepsilon$	none	none
	k-MedO	$5 + \varepsilon$	none	$O(k \log(n\delta)/\varepsilon)$
		$3 + \varepsilon$	none	$O(k^2 \log(n\delta)/\varepsilon)$
	k-MeaO	$25 + \varepsilon$	none	$O(k \log(n\delta)/\varepsilon)$
		$9 + \varepsilon$	none	$O(k^2 \log(n\delta)/\varepsilon)$

1.1 Our Techniques

We focus on illustrating our techniques through k-MedP and k-MeaP. Extending these methods to their associated outlier versions is straightforward.

In the standard local search algorithm, the process begins with an arbitrary feasible solution. Operations like adding, deleting, or swapping centers define the neighborhood of the current feasible solution. A search then ensues within this neighborhood to find a locally optimal solution, which becomes the new current solution. The analysis is to establish valid inequalities by constructing swap operations. These operations establish "connections" between local and global optimal solutions. Traditionally, these connections are given individually for each point (that is, each point yields an inequality that gives a bound of its cost after the constructed swap operation). We call this type of analysis an "individual form". Another type is the "cluster form", in which the connections between the local and global optimal solutions are revealed for some clusters containing several points. Note that the cluster form is a generalization of the individual form (if the inequality (or equality) for a cluster is the sum of inequalities for each point in this cluster, then the cluster form is reduced to the individual form).

Our cluster form analysis establishes a bridge between local and global solutions for both robust and ordinary clusterings, and we obtain a clear and unified understanding of them. Furthermore, we believe that our technique can be generalized to other robust clustering problems such as the robust facility location and k-center problems.

In Sect. 2 presents the unified models and notations for k-MedP/k-MeaP and k-MedO/k-MeaO, and some useful technical lemmas. Section 3 then presents our standard local search algorithms for clustering problems and our corresponding theoretical results. The conclusions are given in Sect. 4. All proofs are deferred to the journal version.

2   Preliminaries

We use the following notation for the problems studied in this paper (in addition to the notation introduced in the introduction). Let $\mathcal{C}$ denote the candidate center set, and $\Delta(a, b)$ denotes the connection cost between two points a and b. For k-MedP and k-MedO, we have $\mathcal{C} = \mathcal{F}$ and $\Delta(a, b) = d(a, b)$; for k-MeaP and k-MeaO, we have $\mathcal{C} = \mathcal{X}$ and $\Delta(a, b) = d^2(a, b)$.

Considering k-MeaP and k-MedP, we assume that S is a set of k centers. It is obvious that the optimal penalized point set concerning S is $P = \{x \in \mathcal{X} | p_x \leq \min_{s \in S} d(s, x)\}$ for k-MedP and $P = \{x \in \mathcal{X} | p_x \leq \min_{s \in S} d^2(s, x)\}$ for k-MeaP, implying that S determines the corresponding k clusters $N(s) := \{x \in \mathcal{X} \backslash P | s_x = s\}$ for all $s \in S$, where s_x denotes the closest center in S to $x \in \mathcal{X} \backslash P$, i.e., $s_x := \arg \min_{s \in S} d(s, x)$. Thus, we also call S a feasible solution for k-MedP and k-MeaP.

Given a data subset $D \subseteq \mathcal{X}$ and a point $c \in \mathcal{C}$, we define $\Delta(c, D) := \sum_{x \in D} \Delta(c, x)$. Let $\text{cent}_\mathcal{C}(D)$ be a center point in $\mathcal{C}$ that optimizes the objective of the k-means/k-median problem, i.e., $\text{cent}_\mathcal{C}(D) := \arg \min_{c \in \mathcal{C}} \Delta(c, D)$. We remark that the notation $\arg \min$ ($\arg \max$) denotes an arbitrary element that minimizes (maximizes) the objective. From the well-known centroid lemma [19], we get $\text{cent}_C(D) = \text{cent}(D)$ for k-means, where $\text{cent}(D)$ is the centroid of D, that is defined as follows.

Definition 1 (Centroid). *Given a set $D \subseteq \mathbb{R}^d$, we call the point $\sum_{x \in D} x/|D|$ denoted by $\text{cent}(D)$ the centroid of D.*

Lemma 1 (Centroid Lemma [19]). *For any data subset $D \subseteq \mathcal{X}$ and a point $c \in \mathbb{R}^d$, we have $d^2(c, D) = d^2(\text{cent}(D), D) + |D|d^2(\text{cent}(D), c)$.*

So, the candidate center points of a k-means problem are the centroid points for all subsets of $\mathcal{X}$. Note that the total number of these candidate center points is $2^{|\mathcal{X}|} - 1$. To cut down this exponential magnitude, [23] introduces the concept of approximate centroid set shown in the following definition.

Definition 2. *A set $C' \subseteq \mathbb{R}^d$ is an ε-approximate centroid set for $\mathcal{X} \subseteq \mathbb{R}^d$ if for any set $D \subseteq \mathcal{X}$, we have $\min_{c \in C'} d^2(c, D) \leq (1 + \varepsilon) \min_{c \in \mathbb{R}^d} d^2(c, D)$.*

The following lemma shows the important observation that a polynomial size ε-approximate centroid set for $\mathcal{X}$ can be found in polynomial time. In the remainder of this paper, we restrict that the candidate center set of k-MeaP/k-MeaO is the $\hat{\varepsilon}$-approximate centroid set C', by utilizing this observation.

Lemma 2. *([23]) Given an n-point set $\mathcal{X} \subseteq \mathbb{R}^d$ and a real number $\varepsilon > 0$, an ε-approximate centroid set for $\mathcal{X}$, of size $O\left(n\varepsilon^{-d} \log(1/\varepsilon)\right)$, can be computed in time $O\left(n \log n + n\varepsilon^{-d} \log(1/\varepsilon)\right)$.*

3   Local Search Approximation Algorithm for k-MedP/k-MeaP

Let ρ be a fixed integer. For any feasible solution S, $A \subseteq S$ and $B \subseteq C \backslash S$ with $|A| = |B| \leq \rho$, we define the so-called multi-swap operation swap(A, B) such that all the centers in A are dropped from S and all centers in B are added to S.

We further denote the connection cost of the point $x \in \mathcal{X}$ by $\text{cost}_c(x)$, i.e., $\text{cost}_c(x) := \Delta(s_x, x)$, and denote by cost_c, cost_p, and $\text{cost}(S)$ the following expressions $\text{cost}_c := \sum_{x \in \mathcal{X} \backslash P} \text{cost}_c(x)$; $\text{cost}_p := \sum_{x \in P} p_x$; $\text{cost}(S) := \text{cost}_c + \text{cost}_p$, where P is the optimal penalized point set with respect to S.

Now we are ready to present our multi-swap local search algorithm **LS-Multi-Swap**($\mathcal{X}, C, k, \{p_j\}_{j \in \mathcal{X}}, \rho$).

Input: data set $\mathcal{X}$, candidate center set C, penalty cost p_j for all $j \in \mathcal{X}$, positive
   integers k and $\rho \leq k$.
Output: center set $S \subseteq C$.
 1: Arbitrarily choose a k-center subset S from C.
 2: Compute $(A, B) := \arg \min_{A \subseteq S, B \subseteq C \backslash S, |A| = |B| \leq \rho} \text{cost}(S \backslash A \cup B)$.
 3: **while** $\text{cost}(S \backslash A \cup B) < \text{cost}(S)$ **do**
 4:    Set $S := S \backslash A \cup B$.

5:    Compute $(A, B) := \arg\min_{A \subseteq S, B \subseteq C \setminus S, |A| = |B| \leq \rho} \text{cost}(S \setminus A \cup B)$.
6: **end while**
7: **return** S

For k-MedP, we run LS-Multi-Swap$(\mathcal{X}, \mathcal{F}, k, \{p_j\}_{j \in \mathcal{X}}, \rho)$; for k-MeaP, we first call the algorithm of [22] to construct an $\hat{\varepsilon}$-approximate centroid set $\mathcal{C}' \subseteq \mathcal{X}$, then run LS-Multi-Swap$(\mathcal{X}, \mathcal{C}', k, \{p_j\}_{j \in \mathcal{X}}, \rho)$. The values of ρ and $\hat{\varepsilon}$ will be determined in our analysis of the algorithm.

3.1   The Analysis

Let S^* be a global optimal solution with the penalized set $P^* = \{x \in \mathcal{X} | p_x \leq \min_{s^* \in S} \Delta(x, s^*)\}$. Similar to the feasible solution S, we introduce the corresponding notations s_x^*, $N^*(s^*)$, $\text{cost}_c^*(x)$, cost_c^*, cost_p^* and $\text{cost}(S^*)$.

We use the standard analysis for a local search algorithm, in which some swap operations between S and S^* are constructed, and then each point is reassigned to a center in the new solution. In the cluster form analysis, we try to bound the new cost for a set of points, rather than bounding the cost of each point individually and independently. To this end, we introduce the *adapted cluster* as follows.

$$N_q^*(s^*) := N^*(s^*) \setminus P, \qquad \forall s^* \in S^*.$$

With the adapted cluster, we set $\tilde{S}^* := \{\text{cent}_C(N_q^*(s^*)) | s^* \in S^*\}$. We introduce a mapping $\phi : \tilde{S}^* \to S$ and map each point $c \in \tilde{S}^*$ to $\phi(c) := \arg\min_{s \in S} d(c, s)$. We say that the center $\phi(\text{cent}_C(N_q^*(s^*))$ *captures* s^*. Considering one of all constructed swap operations, we will reassign some points to a center determined by the mapping ϕ (for instance, reassign the point x to $\phi(\text{cent}(N_q^*(s_x^*)))$. The details will be stated later.

Combining all swap operations, the sum of the costs of these points appears on the right-hand side of the inequality which is derived from the local optimality of S. For k-MeaP, we can bound this sum by the connection costs of S and S^*, see Lemma 3. Note that all these points are not outliers in both S and S^*. This is the reason why we need to use the adapted cluster rather than the cluster $N^*(s^*)$ which was used in the analysis for k-means [15].

In the proof of Lemma 3, we divide the set $\mathcal{X} \setminus (P \cup P^*)$ into some adapted clusters with respect to all $s^* \in S^*$, and apply the Centroid Lemma to each adapted cluster. Afterwards we bound the square of distances between a centroid c of the adapted cluster and its mapped point $\phi(c)$. This explains why the domain of the mapping ϕ is the set of centroids of adapted clusters.

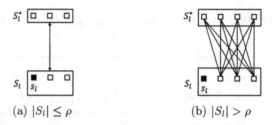

Fig. 1. Two cases for constructing the swap operations between S_l and S_l^* for $\rho = 3$. The solid squares belong to $\phi(\tilde{S}^*)$.

Lemma 3. *Let S and S^* be a local optimal solution and a global optimal solution of k-MeaP, respectively. Then,*

$$\sum_{x \in \mathcal{X} \setminus (P \cup P^*)} d^2(\phi(\text{cent}(N_q^*(s_x^*))), x) \leq \sum_{x \in \mathcal{X} \setminus (P \cup P^*)} (2\text{cost}_c^*(x) + \text{cost}_c(x)) +$$

$$2\sqrt{\sum_{x \in \mathcal{X} \setminus (P \cup P^*)} \text{cost}_c^*(x)} \cdot \sqrt{\sum_{x \in \mathcal{X} \setminus (P \cup P^*)} \text{cost}_c(x)}. \qquad (1)$$

Consider now $\phi(\tilde{S}^*)$, i.e., the image set of $\tilde{S}^*$ under ϕ. We list all elements of $\phi(\tilde{S}^*)$ as $\phi(\tilde{S}^*) = \{s_1, ..., s_m\}$ where $m := |\phi(\tilde{S}^*)|$. For each $l \in \{1, ..., m\}$, let $S_l := \{s_l\}$ and $S_l^* := \{s^* \in S^* | \phi(\text{cent}_C(N_q^*(s^*))) = s_l\}$. Thus, S^* is partitioned into $S_1^*, S_2^*, ..., S_m^*$. Noting that $|S| = |S^*| = k$, we can enlarge each S_l such that $S_1, S_2, ..., S_m$ is a partition of S with $|S_l| = |S_l^*|$ for each $l \in \{1, 2, ..., m\}$.

We will construct a swap operation between the points in S_l and S_l^* for each pair (S_l, S_l^*). Before doing this, we note that a center $s^* \in S^*$ need not belong to the candidate center set C' for k-MeaP. Thus, we introduce a center $\hat{s}^* \in C'$ associated with each $s^* \in S^*$ to ensure that the swap operation involved in s^* can be implemented in Algorithm 1. For each $s^* \in S^*$, let $\hat{s}^* := \arg\min_{c \in C'} d(c, N^*(s^*))$. Combined with Definition 2, we have (see [24])

$$\sum_{x \in N^*(s^*)} d^2(\hat{s}^*, x) = d^2(\hat{s}^*, N^*(s^*)) = \min_{c \in C} d^2(c, N^*(s^*))$$

$$\leq (1 + \hat{\varepsilon}) \min_{c \in \mathbb{R}^d} d^2(c, N^*(s^*)) = (1 + \hat{\varepsilon}) d^2(s^*, N^*(s^*))$$

$$= (1 + \hat{\varepsilon}) \sum_{x \in N^*(s^*)} d^2(s^*, x). \qquad (2)$$

The algorithm allows at most ρ points to be swapped. To satisfy this condition, we consider the following two cases to construct swap operations (cf. Fig. 1 for $\rho = 3$).

Case 1 (cf. Figure 1(a)). For each l with $|S_l| = |S_l^*| \leq \rho$, we consider the pair (S_l, S_l^*). Let $\hat{S}_l^* := \{\hat{s}^* | s^* \in S_l^*\}$. W.l.o.g., we assume that $\hat{S}_l^* \subseteq \mathcal{X} \setminus S$. For k-MedP, we consider the swap(S_l, S_l^*); for k-MeaP, we consider the swap$(S_l, \hat{S}_l^*)$. Utilizing these swap operations, we obtain the following result.

Lemma 4. *If $|S_l| = |S_l^*| \leq \rho$, then, for k-MedP, we have*

$$0 \leq \sum_{s \in S_l} \sum_{x \in N(s) \cap P^*} (p_x - \mathrm{cost}_c(x)) +$$

$$\sum_{s \in S_l} \sum_{x \in N(s) \backslash P^*} \big(d(\phi(\mathrm{cent}_{\mathcal{C}}(N_q^*(s_x^*))), x) - \mathrm{cost}_c(x)\big) +$$

$$\sum_{s^* \in S_l^*} \sum_{x \in N^*(s^*) \backslash P} (\mathrm{cost}_c^*(x) - \mathrm{cost}_c(x)) +$$

$$\sum_{s^* \in S_l^*} \sum_{x \in N^*(s^*) \cap P} (\mathrm{cost}_c^*(x) - p_x);$$

and for k-MeaP, we have

$$0 \leq \sum_{s \in S_l} \sum_{x \in N(s) \cap P^*} (p_x - \mathrm{cost}_c(x)) +$$

$$\sum_{s \in S_l} \sum_{x \in N(s) \backslash P^*} \big(d^2(\phi(\mathrm{cent}_{\mathcal{C}}(N_q^*(s_x^*))), x) - \mathrm{cost}_c(x)\big) +$$

$$\sum_{s^* \in S_l^*} \sum_{x \in N^*(s^*) \backslash P} ((1 + \hat{\varepsilon})\mathrm{cost}_c^*(x) - \mathrm{cost}_c(x)) +$$

$$\sum_{s^* \in S_l^*} \sum_{x \in N^*(s^*) \cap P} ((1 + \hat{\varepsilon})\mathrm{cost}_c^*(x) - p_x).$$

Case 2 (cf. Figure 1(b)). For each l with $|S_l| = |S_l^*| = m_l > \rho$, we consider $(m_l - 1)m_l$ pairs (s, s^*) with $s \in S_l \backslash \{s_l\}$ and $s^* \in S_l^*$. For k-MedP, we consider the swap(s, s^*); for k-MeaP, we consider the swap$(s, \hat{s}^*)$. Utilizing these swap operations, we obtain the following result.

Lemma 5. *Given any $s \in S_l \backslash \{s_l\}$ and $s^* \in S_l^*$, for k-MedP, we have*

$$0 \leq \sum_{x \in N(s) \cap P^*} (p_x - \mathrm{cost}_c(x)) +$$

$$\sum_{x \in N(s) \backslash P^*} \big(d(\phi(\mathrm{cent}_{\mathcal{C}}(N_q^*(s_x^*))), x) - \mathrm{cost}_c(x)\big) +$$

$$\sum_{x \in N^*(s^*) \backslash P} (\mathrm{cost}_c^*(x) - \mathrm{cost}_c(x)) + \sum_{x \in N^*(s^*) \cap P} (\mathrm{cost}_c^*(x) - p_x),$$

and for k-MenP, we have

$$0 \leq \sum_{x \in N(s) \cap P^*} (p_x - \mathrm{cost}_c(x)) +$$

$$\sum_{x \in N(s) \setminus P^*} \left(d^2(\phi(\mathrm{cent}_C(N_q^*(s_x^*))), x) - \mathrm{cost}_c(x) \right) +$$

$$\sum_{x \in N^*(s^*) \setminus P} ((1 + \hat{\varepsilon})\mathrm{cost}_c^*(x) - \mathrm{cost}_c(x)) +$$

$$\sum_{x \in N^*(s^*) \cap P} ((1 + \hat{\varepsilon})\mathrm{cost}_c^*(x) - p_x).$$

Lemma 4 shows a relationship between the sets S_l and S_l^*, while Lemma 5 shows a relationship between two points in S_l and S_l^* respectively. We remark that Lemma 5 holds for all pairs (S_l, S_l^*) (no matter whether $|S_l| > \rho$).

Combining Lemmas 4 and 5, we estimate the cost of S for k-MedP and k-MeaP in the following two theorems respectively.

Theorem 1. LS-Multi-Swap$(\mathcal{X}, \mathcal{F}, k, \{p_j\}_{j \in \mathcal{X}}, \rho)$ *for k-MedP produces a local optimal solution S satisfying* $\mathrm{cost}_c + \mathrm{cost}_p \leq (3 + 2/\rho)\mathrm{cost}_c^* + (1 + 1/\rho)\mathrm{cost}_p^*$.

Theorem 2. *Let $\mathcal{C}'$ be an $\hat{\varepsilon}$-approximate centroid set for $\mathcal{X}$. LS-Multi-Swap$(\mathcal{X}, \mathcal{C}', k, \{p_j\}_{j \in \mathcal{X}}, \rho)$ for k-MeaP produces a local optimal solution S satisfying* $\mathrm{cost}_c + \mathrm{cost}_p \leq (3 + 2/\rho + \hat{\varepsilon})^2 \mathrm{cost}_c^* + (3 + 2/\rho + \hat{\varepsilon})(1 + 1/\rho)\mathrm{cost}_p^*$.

We remark that Algorithm 1 can be adapted to a polynomial-time algorithm that only sacrifices ε in the approximation factor (see [3]). Combining this adaptation and Theorems 1 and 2, we obtain a $(3 + \varepsilon)$-approximation algorithm for k-MedP, and a $(9 + \varepsilon)$-approximation algorithm for k-MeaP, if ρ is sufficiently large and $\hat{\varepsilon}$ is sufficiently small.

Followed the above, we can obtain an adopted analysis of the local search algorithm for the k-MedO and k-MeaO. Via a subtle cluster form analysis, we obtain the following result.

Theorem 3. *The outlier-based local search algorithm yields bicriteria $(5 + \varepsilon, O(\frac{k}{\varepsilon} \log(n\delta)))$- and $(3 + \varepsilon, O(\frac{k^2}{\varepsilon} \log(n\delta)))$-approximations for k-MedO, and bicriteria $(25 + \varepsilon, O(\frac{k}{\varepsilon} \log(n\delta)))$- and $(9 + \varepsilon, O(\frac{k^2}{\varepsilon} \log(n\delta)))$-approximations for k-MeaO, where $O(\frac{k}{\varepsilon} \log(n\delta))$ and $O(\frac{k^2}{\varepsilon} \log(n\delta))$ are the factors by which the outlier constraint is violated, δ denotes the maximal distance between two points in the data set.*

4   Conclusions

The previous analyses of local search algorithms for the robust k-median/k-means, use only the individual form, in which the constructed connections between the local and global optimal solutions are individual for each point.

This has the disadvantage that the joint information about outliers remains hidden. In this paper, we develop a cluster form analysis and define the adapted cluster that captures the outlier information. We find that this new technique works better than the previous analysis methods of local search algorithms since it improves the approximation ratios of local search algorithms for k-MeaP, k-MeaO, and k-MedO, and obtains the same ratio which is the best for k-MedP.

We believe that our new technique will also work for the robust version of the facility location problem (FLP), since the structure of FLP is similar to k-median/k-means. Also, our technique seems to be promising for the robust k-center problem, even for any algorithm for robust clustering problems that is based on local search.

References

1. Ahmadian, S., Norouzi-Fard, A., Svensson, O., Ward, J.: Better guarantees for k-means and euclidean k-median by primal-dual algorithms. In: Proceedings of the 58th Annual Symposium on Foundations of Computer Science, pp. 61–72 (2017)
2. Arora, S., Raghavan, P., Rao, S.: Approximation schemes for Euclidean k-medians and related problems. In: Proceedings of the 30th Annual ACM Symposium on Theory of Computing, pp. 106–113 (1998)
3. Arya, V., Garg, N., Khandekar, R., Meyerson, A., Munagala, K., Pandit, V.: Local search heuristics for k-median and facility location problems. SIAM J. Comput. **33**(3), 544–562 (2004)
4. Byrka, J., Pensyl, T., Rybicki, B., Srinivasan, A., Trinh, K.: An improved approximation for k-median, and positive correlation in budgeted optimization, pp. 737–756 (2014)
5. Charikar, M., Guha, S., Tardos, E., Shmoys, D.B.: A constant-factor approximation algorithm for the k-median problem. J. Comput. Syst. Sci. **65**(1), 129–149 (2002)
6. Charikar, M., Khuller, S., Mount, D.M., Narasimhan, G.: Algorithms for facility location problems with outliers. In: Proceedings of the 12th Annual ACM-SIAM Symposium on Discrete Algorithms, pp. 642–651. Society for Industrial and Applied Mathematics (2001)
7. Charikar, M., Li, S.: A dependent LP-rounding approach for the k-median problem. In: Czumaj, A., Mehlhorn, K., Pitts, A., Wattenhofer, R. (eds.) Automata, Languages, and Programming. ICALP 2012, vol. 7391, pp. 194C205. Springer, Heidelberg (2012). https://doi.org/10.1007/978-3-642-31594-7_17
8. Chen, K.: A constant factor approximation algorithm for k-median clustering with outliers. In: Proceedings of the 19th Annual ACM-SIAM Symposium on Discrete Algorithms, pp. 826–835 (2008)
9. Cohen-Addad, V., Gupta, A., Hu, L., et al.: An improved local search algorithm for k-median. In: Proceedings of the 2022 Annual ACM-SIAM Symposium on Discrete Algorithms (SODA). Society for Industrial and Applied Mathematics, pp. 1556–1612 (2022)
10. Cohen-Addad, V., Karthik, C.: Inapproximability of clustering in l_p metrics. In: Proceedings of the 60th Annual Symposium on Foundations of Computer Science, pp. 519–539 (2019)
11. Cohen-Addad, V., Klein, P.N., Mathieu, C.: Local search yields approximation schemes for k-means and k-median in Euclidean and minor-free metrics. SIAM J. Comput. **48**(2), 644–667 (2019)

12. Feng, Q., Zhang, Z., Shi, F., Wang, J.: An improved approximation algorithm for the k-means problem with penalties. In: Chen, Y., Deng, X., Lu, M. (eds.) Frontiers in Algorithmics. FAW 2019. FAW 2019, vol. 11458, pp. 170–181. Springer, Cham (2019). https://doi.org/10.1007/978-3-030-18126-0_15

13. Friggstad, Z., Khodamoradi, K., Rezapour, M., Salavatipour, M.R.: Approximation schemes for clustering with outliers. ACM Trans. Algorithms **15**(2), 1–26 (2019)

14. Friggstad, Z., Rezapour, M., Salavatipour, M.R.: Local search yields a PTAS for k-means in doubling metrics. SIAM J. Comput. **48**(2), 452–480 (2019)

15. Gupta, S., Kumar, R., Lu, K., Moseley, B., Vassilvitskii, S.: Local search methods for k-means with outliers. In: Proceedings of the 43rd International Conference on Very Large Data Bases, vol. 10, no. (7), p. 757–768 (2017)

16. Hajiaghayi, M., Khandekar, R., Kortsarz, G.: Local search algorithms for the red-blue median problem. Algorithmica **63**(4), 795–814 (2012)

17. Jain, K., Mahdian, M., Markakis, E., Saberi, A., Vazirani, V.V.: Greedy facility location algorithms analyzed using dual fitting with factor-revealing LP. J. ACM **50**(6), 795–824 (2003)

18. Jain, K., Vazirani, V.V.: Approximation algorithms for metric facility location and k-median problems using the primal-dual schema and Lagrangian relaxation. J. ACM **48**(2), 274–296 (2001)

19. Kanungo, T., Mount, D.M., Netanyahu, N.S., Piatko, C.D., Silverman, R., Wu, A.Y.: A local search approximation algorithm for k-means clustering. Comput. Geom. **28**(2–3), 89–112 (2004)

20. Krishnaswamy, R., Li, S., Sandeep, S.: Constant approximation for k-median and k-means with outliers via iterative rounding. In: Proceedings of the 50th Annual ACM SIGACT Symposium on Theory of Computing, pp. 646–659 (2018)

21. Li, S., Svensson, O.: Approximating k-median via pseudo-approximation. SIAM J. Comput. **45**(2), 530–547 (2016)

22. Makarychev, K., Makarychev, Y., Sviridenko, M., Ward, J.: A bi-criteria approximation algorithm for k-means. In: Proceedings of the 19th International Workshop on Approximation Algorithms for Combinatorial Optimization Problems (APPROX), and the 20th International Workshop on Randomization and Computation (RANDOM), pp. 14:1–14:20 (2016)

23. Matousek, J.: On approximate geometric k-clustering. Discrete Comput. Geom. **24**(1), 61–84 (2000)

24. Zhang, D., Hao, C., Wu, C., Xu, D., Zhang, Z.: Local search approximation algorithms for the k-means problem with penalties. J. Comb. Optim. **37**(2), 439–453 (2019)

On the Power of Counting the Total Number of Computation Paths of NPTMs

Eleni Bakali[1], Aggeliki Chalki[2]([✉]) [iD], Sotiris Kanellopoulos[1,3],
Aris Pagourtzis[1,3] [iD], and Stathis Zachos[1]

[1] School of Electrical and Computer Engineering,
National Technical University of Athens, Athens, Greece
mpakali@corelab.ntua.gr, sotkanellopoulos@mail.ntua.gr,
{pagour,zachos}@cs.ntua.gr
[2] Department of Computer Science, Reykjavik University, Reykjavik, Iceland
angelikic@ru.is
[3] Archimedes Research Unit, Athena RC, 15125 Marousi, Greece

Abstract. In this paper, we define and study variants of several complexity classes of decision problems that are defined via some criteria on the number of accepting paths of an NPTM. In these variants, we modify the acceptance criteria so that they concern the total number of computation paths, instead of the number of accepting ones. This direction reflects the relationship between the counting classes #P and TotP, which are the classes of functions that count the number of accepting paths and the total number of paths of NPTMs, respectively. The former is the well-studied class of counting versions of NP problems, introduced by Valiant (1979). The latter contains all self-reducible counting problems in #P whose decision version is in P, among them prominent #P-complete problems such as NON-NEGATIVE PERMANENT, #PERF-MATCH and #DNF-SAT.

We show that almost all classes introduced in this work coincide with their '# accepting paths'-definable counterparts, thus providing an alternative model of computation for the classes ⊕P, Mod_kP, SPP, WPP, $\text{C}_=\text{P}$, and PP. Moreover, for each of these classes, we present a novel family of complete problems which are defined via problems that are TotP-complete under parsimonious reductions. This way, we show that all the aforementioned classes have complete problems that are defined via counting problems whose existence version is in P, in contrast to the standard way of obtaining completeness results via counting versions of NP-complete problems. To the best of our knowledge, prior to this work, such results were known only for ⊕P and $\text{C}_=\text{P}$.

We also build upon a result by Curticapean, to exhibit yet another way to obtain complete problems for WPP and PP, namely via the difference of values of the TotP function #PERFMATCH on pairs of graphs. Finally, for the so defined WPP-complete problem, we provide an exponential lower bound under the randomized Exponential Time Hypothesis, showcasing the hardness of the class.

Keywords: counting complexity · #P · number of perfect matchings · gap-definable classes

X. Chen and B. Li (Eds.): TAMC 2024, LNCS 14637, pp. 209–220, 2024.
https://doi.org/10.1007/978-981-97-2340-9_18

1   Introduction

Valiant introduced the complexity class #P in his seminal paper [28] to characterize the complexity of the permanent function. #P contains the counting versions of NP problems and equivalently, functions that count the accepting paths
of non-deterministic polynomial-time Turing machines (NPTMs). For example,
#SAT, i.e. the problem of counting the number of satisfying assignments of
a propositional formula, lies in #P. The class of functions that count the total
number of paths of NPTMs, namely TotP, was introduced and studied in [15,19].
Interestingly, TotP is the class of self-reducible problems in #P that have a decision version in P [19]; note that prominent #P-complete problems belong to
TotP, such as #PERFMATCH and #DNF-SAT. Complete problems under parsimonious reductions for TotP were provided in [3], e.g. SIZE-OF-SUBTREE [16].
The significance of TotP and its relationship with the class of approximable
counting problems have been investigated in [1,3,4,7,19].

The two classes #P and TotP imply two paradigms of counting computation
models that exhibit significant similarities and differences: on one hand, the
computational hardness of computing the exact function value is similar for
both models, as shown by the fact that they share complete problems under
polynomial-time Turing reductions [19]; on the other hand, checking whether a
solution exists is in P for all problems in TotP, while it is even NP-complete for
some problems in #P, a fact that shows that TotP is strictly included in #P,
unless $P = NP$.

In this paper, we build upon the comparison between these two paradigms by
studying well-known classes of decision problems, currently defined by means of
'accepting-path counting', under the perspective of the 'total counting' model. In
particular, we consider the complexity classes shown in Table 1, which are
defined using conditions on the number of accepting paths, or the difference—
which is called the *gap*—between accepting and rejecting paths of an NPTM [12];
for all these classes we introduce their 'TotP' counterparts, i.e. classes defined
by an analogous condition on functions that count the *total number of computation paths* of NPTMs. We compare each 'traditionally defined' class with its
counterpart, showing that many of them remain the same under both models.
We thus obtain alternative characterizations for these classes that lead to novel
insights and results on their computational complexity. Notably, we provide new
complete problems for $\oplus$P, Mod_kP, $C_=P$, PP, SPP and WPP, by using the 'total
counting' paradigm.

Related Work. Interestingly, several of the classes demonstrated in Table 1
have attracted attention, as either they have been essential for proving important theorems, or they contain significant problems. Specifically, the classes $\oplus$P
and PP have received much attention due to their relation to Toda's theorem.
Problems in $\oplus$P and PP can be decided with the information of the rightmost
and leftmost bit of a #P function, respectively. Toda's theorem consists of two
important results. First, PH can be reduced to $\oplus$P under probabilistic reductions, i.e. $PH \subseteq BPP^{\oplus P}$. Second, $BPP^{\oplus P}$ is contained in P^{PP}, which in turn
implies that is also contained in $P^{\#P}$, where one oracle call suffices. In fact,

Table 1. Classes UP [27], FewP [2], $\oplus$P [20], Mod_kP [8,9,14], SPP [12,18], WPP [12], $\text{C}_=\text{P}$ [25], and PP [13,25].

Class	Function f in:	If $x \in L$:	If $x \notin L$:
UP	#P	$f(x) = 1$	$f(x) = 0$
FewP	#P	$f(x) \leq p(\|x\|)$ for some polynomial p and $f(x) > 0$	$f(x) = 0$
$\oplus$P	#P	$f(x)$ is odd	$f(x)$ is even
Mod_kP	#P	$f(x) \not\equiv 0 \pmod{k}$	$f(x) \equiv 0 \pmod{k}$
SPP	GapP	$f(x) = 1$	$f(x) = 0$
WPP	GapP	$f(x) = g(x)$ for some $g \in$ FP with $0 \notin \text{range}(g)$	$f(x) = 0$
$\text{C}_=\text{P}$	GapP	$f(x) = 0$	$f(x) \neq 0$ [alt-def: $f(x) > 0$]
PP	GapP	$f(x) > 0$	$f(x) \leq 0$ [alt-def: $f(x) < 0$]

the oracle needs to compute the value of a #P function modulo 2^m, for some m. Another prominent result, preceding Toda's theorem, is the Valiant–Vazirani theorem [30], stating that $\text{NP} \subseteq \text{RP}^{\text{UP}}$, which implies that SAT remains hard even if the input instances are promised to have at most one satisfying assignment.

The complexity class UP has also been of great significance in cryptography, where the following statement holds: $\text{P} = \text{UP}$ if and only if there are no one-way functions. The class SPP attracted attention when GRAPH ISOMORPHISM was shown to lie in it [5]. The GRAPH ISOMORPHISM problem is believed to be one of the few NP-intermediate problems; there is no known polynomial-time algorithm for it, and there is strong evidence that it is not NP-complete [24]. SPP can be seen as the gap-analog of UP. In [12] it was shown that SPP is the smallest reasonable gap-definable class.

Valiant in [29] and Curticapean in [11] have provided complete problems for $\oplus$P and $\text{C}_=\text{P}$, respectively, which are defined by counting problems that are not #P-complete under parsimonious reductions (unless $\text{P} = \text{NP}$). In particular, Curticapean proved in [11] that the problem of determining whether two given graphs have the same number of perfect matchings is complete for $\text{C}_=\text{P}$. He also proved that this problem has no subexponential algorithm under the Exponential Time Hypothesis (ETH). The problem of counting perfect matchings in a graph, namely #PERFMATCH, is #P-complete and TotP-complete under poly-time Turing reductions [19,28]. However, it is not known to be complete for either of these classes under parsimonious reductions.

Our Contribution. In Sect. 3, we introduce the classes that are demonstrated in Table 2 which are defined via TotP functions. As TotP is a *proper* subclass of #P (unless P=NP), the first interesting question we answer is whether these classes are proper subclasses of the corresponding ones shown in Table 1. Our results exhibit a dichotomy; these classes are either equal to P, namely $\text{U}_{\text{tot}}\text{P} = \text{Few}_{\text{tot}}\text{P} = \text{P}$, or equal to their analogs definable by #P functions (Propositions 4, 5, 6, 8, and Corollary 2).

The results of Sect. 3 provide an alternative model of computation for $\oplus$P, Mod_kP, SPP, WPP, $C_=$P, and PP. These results also mean that the 'TotP' model captures the essence of the aforementioned classes, while the '#P' model turns out to be somewhat harder than necessary for defining them. As a consequence, in Sect. 4, for each of these classes, we obtain a new family of complete problems that are defined by TotP-complete problems under parsimonious reductions, which are not #P-complete under the same kind of reductions unless P = NP. Thus, we generalize the completeness results by Valiant and Curticapean for $\oplus$P and $C_=$P, respectively. In fact, an analogous model of computation and analogous complete problems are obtained for every gap-definable class, and not only the ones mentioned in this work.

We also present and study problems defined via the difference of the value of total counting problems. Building upon a relevant result by Curticapean [11], we show that such difference problems defined via the TotP function #PERFMATCH are complete for the classes PP and WPP, respectively (Propositions 11 and 12); we also show a hardness result for SPP (Proposition 13).

Finally, in Subsect. 4.2, we prove that under the randomized Exponential Time Hypothesis (rETH), there is no subexponential algorithm for the promise problem DIFFPERFMATCH$_{=g}$, which is the problem of determining whether the difference between the number of perfect matchings in two graphs is zero or equal to a specific value.

2　Preliminaries

Definition 1 ([12,19,28]). *(a)* #P $= \{acc_M : \Sigma^* \to \mathbb{N} \mid M$ *is an NPTM}*,
(b) FP $= \{f : \Sigma^* \to \mathbb{N} \mid f$ *is computable in polynomial time}*,
(c) TotP $= \{tot_M : \Sigma^* \to \mathbb{N} \mid M$ *is an NPTM}*,
(d) GapP $= \{\Delta M : \Sigma^* \to \mathbb{N} \mid M$ *is an NPTM}*,

where $acc_M(x) = $ #*(accepting paths of* M *on input* x*),* $tot_M(x) = $ #*(all computation paths of* M *on input* x*)* $- 1$*,* $\Delta M(x) = acc_M(x) - rej_M(x)$*, and* $rej_M(x) = $ #*(rejecting paths of* M *on input* x*), for every* $x \in \Sigma^*$*.*

Remark 1. Since every NPTM has at least one computation path, one is subtracted by the total number of paths in the definition of TotP, so that functions in the class can take the zero value. As a result, many natural counting problems lie in TotP.

Various kinds of reductions are used between counting problems. In particular, parsimonious reductions preserve the exact value of the two involved functions.

Definition 2. f *reduces to* g *under parsimonious reductions, denoted* $f \leq^p_{\text{par}} g$*, if and only if there is* $h \in$ FP*, such that for all* $x \in \Sigma^*$*,* $f(x) = g(h(x))$*.*

Proposition 1 ([12,19]). *(a)*FP $\subseteq$ TotP $\subseteq$ #P. *The inclusions are proper unless* P = NP.

(b) TotP *is the closure under parsimonious reductions of the class of self-reducible #P functions, whose decision version is in* P, *where the decision version of a function* $f : \Sigma^* \to \mathbb{N}$ *is* $L_f = \{x \in \Sigma^* \mid f(x) > 0\}$.
(c) GapP $= \#P - \#P = \#P - FP = FP - \#P$, *where the subtraction of a function class from another, denotes the class of functions that can be described as the difference of two functions, one from each class.*
(d) $P^{TotP} = P^{\#P} = P^{PP} = P^{GapP}$.

The rationale for introducing the class GapP is to capture variants of #P problems that take negative values as well. For example, the permanent of a matrix with non-negative integer entries is a #P problem [28], while the permanent of a matrix with arbitrary, possibly negative entries, lies in GapP [12].

Next, we provide known relationships among the classes of Table 1 in Proposition 2 and the Valiant–Vazirani and Toda's theorems in Theorem 1.

Proposition 2 ([23]). *(a)* UP $\subseteq$ FewP $\subseteq$ NP $\subseteq$ coC$_=$P $\subseteq$ PP.
(b) FewP $\subseteq$ SPP $\subseteq$ WPP $\subseteq$ C$_=$P $\subseteq$ PP.
(c) SPP $\subseteq$ $\oplus$P $\subseteq$ Mod$_k$P.

Theorem 1 ([26,30]). *(a) Valiant–Vazirani theorem:* NP $\subseteq$ RP^{UP}.
(b) Toda's theorem: PH $\subseteq$ BPP$^{\oplus P}$ $\subseteq$ P$^{\#P[1]}$.

In Subsect. 4.2 we use the Exponential Time Hypothesis and its randomized variant, namely rETH, which are given below.

- ETH: There is no deterministic algorithm that can decide 3-SAT in time $\exp(o(n))$.
- rETH: There is no randomized algorithm that can decide 3-SAT in time $\exp(o(n))$, with error probability at most $1/3$.

3   Classes Defined by Total Counting

3.1   The Class Gap$_{tot}$P

Definition 3. *A function f belongs to the class* Gap$_{tot}$P *iff it is the difference of two* TotP *functions.*

Proposition 3 demonstrates that Gap$_{tot}$P coincides with the class GapP. Corollary 1 provides alternative characterizations of GapP and Gap$_{tot}$P.

Proposition 3. Gap$_{tot}$P = GapP.

Proof. Gap$_{tot}$P $\subseteq$ GapP is straightforward, since TotP $\subseteq$ #P. For GapP $\subseteq$ Gap$_{tot}$P, note that for any #P function f, there exist NPTMs M, M' such that $f(x) = acc_M(x) = tot_{M'}(x) - tot_M(x)$, where we obtain M' by doubling the accepting paths of M. So for any $g \in$ GapP there exist NPTMs N, N', M, M' such that $g(x) = acc_N(x) - acc_M(x) = (tot_{N'}(x) - tot_N(x)) - (tot_{M'}(x) - tot_M(x)) = (tot_{N'}(x) + tot_M(x)) - (tot_N(x) + tot_{M'}(x))$. Since TotP is closed under addition [19], there are NPTMs M_1 and M_2 such that this last subtraction is equal to $tot_{M_1}(x) - tot_{M_2}(x)$. $\square$

Corollary 1. $\mathsf{Gap_{tot}P} = \mathsf{GapP} = \#\mathsf{P} - \#\mathsf{P} = \mathsf{TotP} - \mathsf{TotP} = \#\mathsf{P} - \mathsf{FP} = \mathsf{FP} - \#\mathsf{P} = \mathsf{FP} - \mathsf{TotP} = \mathsf{TotP} - \mathsf{FP}$.

The above corollary demonstrates, among other implications, that $\mathsf{Gap_{tot}P}$ contains problems in $\mathsf{FP} - \mathsf{TotP}$, such as counting the unsatisfying assignments of a formula in DNF, which is not in TotP unless $\mathsf{P} = \mathsf{NP}$.

3.2 The Classes $\mathsf{U_{tot}P}$, $\mathsf{Few_{tot}P}$, $\oplus_{tot}\mathsf{P}$, $\mathsf{Mod_{ktot}P}$, $\mathsf{SP_{tot}P}$, $\mathsf{WP_{tot}P}$, $\mathsf{C_{=tot}P}$, and $\mathsf{P_{tot}P}$

Table 2. Classes $\mathsf{U_{tot}P}$, $\mathsf{Few_{tot}P}$, $\oplus_{tot}\mathsf{P}$, $\mathsf{Mod_{ktot}P}$, $\mathsf{SP_{tot}P}$, $\mathsf{WP_{tot}P}$, $\mathsf{C_{=tot}P}$, $\mathsf{P_{tot}P}$.

Class	Function f in:	If $x \in L$:	If $x \notin L$:
$\mathsf{U_{tot}P}$	TotP	$f(x) = 1$	$f(x) = 0$
$\mathsf{Few_{tot}P}$	TotP	$f(x) \leq p(|x|)$ for some polynomial p and $f(x) > 0$	$f(x) = 0$
$\oplus_{tot}\mathsf{P}$	TotP	$f(x)$ is odd	$f(x)$ is even
$\mathsf{Mod_{ktot}P}$	TotP	$f(x) \not\equiv 0 \pmod{k}$	$f(x) \equiv 0 \pmod{k}$
$\mathsf{SP_{tot}P}$	$\mathsf{Gap_{tot}P}$	$f(x) = 1$	$f(x) = 0$
$\mathsf{WP_{tot}P}$	$\mathsf{Gap_{tot}P}$	$f(x) = g(x)$ for some $g \in \mathsf{FP}$ with $0 \notin \mathrm{range}(g)$	$f(x) = 0$
$\mathsf{C_{=tot}P}$	$\mathsf{Gap_{tot}P}$	$f(x) = 0$	$f(x) \neq 0$ [alt-def: $f(x) > 0$]
$\mathsf{P_{tot}P}$	$\mathsf{Gap_{tot}P}$	$f(x) > 0$	$f(x) \leq 0$ [alt-def: $f(x) < 0$]

The classes defined in Table 2 are the TotP-analogs of those contained in Table 1. First, we show that $\mathsf{P} = \mathsf{U_{tot}P} = \mathsf{Few_{tot}P}$ and next, that every other class of Table 2 coincides with its counterpart from Table 1.

Proposition 4. *(a)* $\mathsf{P} = \mathsf{U_{tot}P} \subseteq \mathsf{UP}$.

(b) If $\mathsf{UP} \subseteq \mathsf{U_{tot}P}$, then $\mathsf{P} = \mathsf{UP}$ *(and thus* $\mathsf{RP} = \mathsf{NP}$*)*.

Proof. (a) $\mathsf{U_{tot}P} \subseteq \mathsf{P}$: Let $L \in \mathsf{U_{tot}P}$. Then there exists an NPTM M such that $x \in L$ iff M has 2 paths, whereas $x \notin L$ iff M has 1 path. Define the polynomial time Turing machine M' that on any input, simulates either the unique path or the two paths of M deterministically, and it either rejects or accepts, respectively. To prove the inverse inclusion $\mathsf{P} \subseteq \mathsf{U_{tot}P}$, consider a language $L \in \mathsf{P}$ and define the NPTM N, which, on any input x, simulates the deterministic polynomial-time computation for deciding L and generates one or two paths if the answer is negative or positive, respectively. The inclusion $\mathsf{U_{tot}P} \subseteq \mathsf{UP}$ is immediate from $\mathsf{TotP} \subseteq \#\mathsf{P}$. (b) From (a), if $\mathsf{UP} \subseteq \mathsf{U_{tot}P}$, then $\mathsf{UP} \subseteq \mathsf{P}$ and $\mathsf{RP} = \mathsf{NP}$ by the Valiant–Vazirani theorem (Theorem 1(a)). $\qquad\square$

Proposition 5. *(a)* $P = U_{tot}P = Few_{tot}P$.
(b) If $FewP \subseteq Few_{tot}P$, *then* $P = FewP$.

The class $\oplus_{tot}P$ (odd-P or parity-P) is the class of decision problems, for which the acceptance condition is that the number of all computation paths of an NPTM is even (or the number of all computation paths minus 1 is odd).

Proposition 6. $\oplus_{tot}P = \oplus P$.

Valiant provided in [29] an $\oplus P$-complete problem definable by a TotP function. Let $\oplus PL$-RTW-MON-3CNF be the problem that on input a planar 3CNF formula where each variable appears positively and in exactly two clauses, accepts iff the formula has an odd number of satisfying assignments. The counting version of this problem, namely $\#PL$-RTW-MON-3CNF, is in TotP; it is self-reducible like every satisfiability problem and has a decision version in P, since every monotone formula has at least one satisfying assignment.

Proposition 7 ([29]). $\oplus PL$-RTW-MON-3CNF *is* $\oplus P$-*complete*.

Proposition 8. $Mod_{k tot}P = Mod_k P$.

Remark 2. So, we can say that if we have information about the rightmost bit of a TotP function is as powerful as having information about the rightmost bit of a $\#P$ function. Toda's theorem would be true if we used $\oplus_{tot}P$ instead of $\oplus P$. Moreover, it holds that $BPP^{\oplus P} \subseteq P^{TotP[1]}$, where it suffices to make an oracle call to a TotP function mod 2^m, for some m. However, $U_{tot}P$ is defined by a constraint on a TotP function that yields only NPTMs with polynomially many paths. This means that $U_{tot}P$ gives no more information than the class P and as a result, it cannot replace the class UP in the Valiant–Vazirani theorem.

By Proposition 3 and the definitions of the classes $SP_{tot}P$, $WP_{tot}P$, $C_{=tot}P$, and $P_{tot}P$, we obtain the following corollary.

Corollary 2. *(a)* $SP_{tot}P = SPP$, *(b)* $WP_{tot}P = WPP$, *(c)* $C_{=tot}P = C_=P$, *and* *(d)* $P_{tot}P = PP$.

A more general corollary of Proposition 3 is that every gap-definable class coincides with its TotP-analog.

4   Hardness Results for Problems Definable via TotP Functions

In this section, we introduce a new family of complete problems for $\oplus P$, $Mod_k P$, and gap-definable classes.

Definition 4. *Given a $\#P$ function $\#A : \Sigma^* \to \mathbb{N}$, we define the following decision problems associated with $\#A$:*

- $\oplus A$ *which on input* $x \in \Sigma^*$, *decides whether* $\#A(x)$ *is odd.*
- $\text{MOD}_k A$ *which on input* $x \in \Sigma^*$, *decides whether* $\#A(x) \not\equiv 0 \pmod{k}$.
- $\text{DIFFA}_{=0}$ *which on input* $(x,y) \in \Sigma^* \times \Sigma^*$, *decides whether* $\#A(x) = \#A(y)$.
- $\text{DIFFA}_{>0}$ *which on input* $(x,y) \in \Sigma^* \times \Sigma^*$, *decides whether* $\#A(x) > \#A(y)$.
- *the promise problem* $\text{DIFFA}_{=1}$ *which on input* $(x,y) \in I_{YES} \cup I_{NO}$, *decides whether* $(x,y) \in I_{YES}$, *where* $I_{YES} = \{(x,y) \mid \#A(x) = \#A(y) + 1\}$ *and* $I_{NO} = \{(x,y) \mid \#A(x) = \#A(y)\}$, $x, y \in \Sigma^*$.
- *the promise problem* $\text{DIFFA}_{=g}$ *which on input* $(x,y,k) \in I_{YES} \cup I_{NO}$, *decides whether* $(x,y,k) \in I_{YES}$, *where* $I_{YES} = \{(x,y,k) \mid \#A(x) = \#A(y) + k\}$ *and* $I_{NO} = \{(x,y,k) \mid \#A(x) = \#A(y)\}$, $x, y \in \Sigma^*$, $k \in \mathbb{N}_{>0}$.

Proposition 9. *For any function* $\#A \in \#P$, *it holds that:*

(a) $\oplus A \in \oplus P$, $\text{MOD}_k A \in \text{Mod}_k P$, $\text{DIFFA}_{=0} \in C_= P$, $\text{DIFFA}_{>0} \in PP$, $\text{DIFFA}_{=1} \in SPP$, *and* $\text{DIFFA}_{=g} \in WPP$.

(b) *If* $\#A$ *is* $\#P$-*complete or* TotP-*complete under parsimonious reductions, then* $\oplus A$, $\text{MOD}_k A$, $\text{DIFFA}_{=0}$, $\text{DIFFA}_{>0}$, $\text{DIFFA}_{=1}$, *and* $\text{DIFFA}_{=g}$ *are complete for* $\oplus P$, $\text{Mod}_k P$, $C_= P$, PP, SPP *and* WPP, *respectively.*

Example 1. For example, the problem $\text{DIFFSAT}_{=0}$ is complete for the class $C_= P$. Note that this problem was defined in [11], where it is denoted by $\text{SAT}_=$. We use a slightly different notation here, which we believe is more suitable for defining problems that lie in other gap-definable classes. The problem $\text{DIFFSAT}_{=g}$ takes as input two CNF formulas ϕ, ϕ' and a non-zero natural number k, such that either they have the same number of satisfying assignments or the first one has k more satisfying assignments than the second one. The problem is to decide which is the case. This is a generalization of the problem $\text{PROMISE-EXACT-NUMBER-SAT}$ defined in [21].

By Proposition 1(a) and the closure of TotP under parsimonious reductions ($\leq^p_{\text{par}}$), if $P \neq NP$, TotP-complete and $\#P$-complete problems under $\leq^p_{\text{par}}$ form disjoint classes. By combining that fact with Proposition 9(b), we obtain a family of complete problems for the classes $\oplus P$, $\text{Mod}_k P$, $C_= P$, PP, SPP, and WPP defined by functions that are TotP-complete under $\leq^p_{\text{par}}$. As a concrete example, consider the particularly interesting problem SIZE-OF-SUBTREE, first introduced by Knuth [16] as the problem of estimating the size of a backtracking tree, which is the tree produced by a backtracking procedure. This problem has been extensively studied from an algorithmic point of view (see e.g. [10,22]) and was recently shown to be TotP-complete under $\leq^p_{\text{par}}$ [3]. Proposition 9 implies that the six problems defined via SIZE-OF-SUBTREE as specified in Definition 4, are complete for $\oplus P$, $\text{Mod}_k P$, $C_= P$, PP, SPP and WPP, respectively.

Note that, these results provide the first complete problems for $\text{Mod}_k P$, SPP, WPP, and PP that are not definable via $\#P$-complete (under $\leq^p_{\text{par}}$) functions. Moreover, as every gap-definable class coincides with its TotP-analog, any such class has complete problems defined by TotP-complete problems under $\leq^p_{\text{par}}$. Alternatively, one can say that these complete problems are defined by problems in P, and not NP-complete ones (unless $P = NP$).

4.1 Problems Definable via the Difference of Counting Perfect Matchings

Curticapean proved in [11] that DIFFPERFMATCH$_{=0}$ is C$_=$P-complete. We provide analogous results for the classes PP and WPP. Note that #PERFMATCH is in TotP, and it is not known to be either #P-complete or TotP-complete under parsimonious reductions. This is yet another approach to obtain complete problems for PP and WPP, the counting versions of which are not even known to be TotP-complete.

Proposition 10 ([11]). DIFFPERFMATCH$_{=0}$ *is complete for* C$_=$P.

The proofs of Propositions 11 and 12 are established by adapting the proof of Proposition 10 given in [11].

Proposition 11. DIFFPERFMATCH$_{>0}$ *is complete for* PP.

Proof. The reduction from DIFFSAT$_{=0}$ to DIFFPERFMATCH$_{=0}$ provided in [11], is also a reduction from DIFFSAT$_{>0}$ to DIFFPERFMATCH$_{>0}$.   □

Proposition 12. DIFFPERFMATCH$_{=g}$ *is complete for* WPP.

Proof. By Proposition 9, DIFFSAT$_{=g}$ is WPP-complete. We show that DIFFSAT$_{=g}$ reduces to DIFFPERFMATCH$_{=g}$. Let (ϕ, ϕ', k) be an input to DIFFSAT$_{=g}$, such that (ϕ, ϕ', k) is a *yes* instance if #SAT(ϕ) − #SAT$(\phi') = k$, where $k \in \mathbb{N}_{>0}$. DIFFSAT$_{=g}$ reduces to DIFFPERFMATCH$_{=g}$ on input (G, G', l), where G, G' are the graphs described in the reduction that proves Proposition 10 and $l = 2^T \cdot k$, where T is computable in polynomial time. It holds that $(\phi, \phi', k) \in$ DIFFSAT$_{=g}$ if #SAT(ϕ) − #SAT$(\phi') = k$ iff #PERFMATCH(G) − #PERFMATCH$(G') = 2^T \cdot ($#SAT(ϕ) − #SAT$(\phi')) = 2^T \cdot k = l$. Also, $(\phi, \phi', k) \notin$ DIFFSAT$_{=g}$ if #SAT(ϕ) − #SAT$(\phi') = 0$ iff #PERFMATCH(G) − #PERFMATCH$(G') = 2^T \cdot ($#SAT(ϕ) − #SAT$(\phi')) = 0$.   □

In contrast, we cannot prove SPP-completeness for DIFFPERFMATCH$_{=1}$. However, we can prove hardness of DIFFPERFMATCH$_{=g}$ for SPP.

Proposition 13. *The problem* DIFFSAT$_{=1}$ *is reducible to* DIFFPERFMATCH$_{=g}$.

Proof. Consider two 3CNF formulas (ϕ, ϕ'), with n variables and $m = \mathcal{O}(n)$ clauses, such that either #SAT(ϕ) − #SAT$(\phi') = 1$ or #SAT(ϕ) − #SAT$(\phi') = 0$.

Then, using the polynomial-time reduction of [11] two graphs G, G' can be constructed such that

$$\#\text{PERFMATCH}(G) - \#\text{PERFMATCH}(G') = c^{|V|} \cdot (\#\text{SAT}(\phi) - \#\text{SAT}(\phi'))$$

where $|V| = \max\{|V(G)|, |V(G')|\}\} = \mathcal{O}(n + m)$ and $c \in (1, 2)$ is a constant depending on ϕ, ϕ' that can be computed in polynomial time. Moreover, the graphs G and G' have $\mathcal{O}(|V|)$ edges.

So, DIFFSAT$_{=1}$ on input (ϕ, ϕ') can be reduced to DIFFPERFMATCH$_{=g}$ on input $(G, G', c^{|V|})$, where $c \in (1, 2)$.   □

According to the proof of Proposition 13, the smallest possible non-zero difference between the number of satisfying assignments of two given 3CNF formulas can be translated to an exponentially large difference between the number of perfect matchings of two graphs. In addition, this exponentially large number depends on the input and it can be efficiently computed.

4.2   An Exponential Lower Bound for the Problem DIFFPERFMATCH$_{=g}$

Curticapean showed that under ETH, the problem DIFFPERFMATCH$_{=0}$ has no $2^{o(m)}$ time algorithm on simple graphs with m edges [11, Theorem 7.6]. The proof is based on the fact that the satisfiability of a 3CNF formula ϕ is reducible to the difference of #PERFMATCH on two different graphs, such that the number of perfect matchings of the graphs is equal iff ϕ is unsatisfiable. The reduction follows the steps of the reductions that are used in the proofs of Propositions 11 and 12. Using the reduction of Proposition 13, we prove the following corollary.

Corollary 3. *Under rETH there is no randomized* $\exp(o(m))$ *time algorithm for* DIFFPERFMATCH$_{=g}$ *on simple graphs with m edges.*

Remark 3. A different way to read Proposition 13 is the following: a positive result for #PERFMATCH would imply a corresponding positive result for DIFFPERFMATCH$_{=g}$ and therefore, for DIFFSAT$_{=1}$. Of course, this positive result would be an exponential-time algorithm for these problems! For example, a fully polynomial approximation scheme for #PERFMATCH would yield an algorithm that distinguishes between #PERFMATCH(G)−#PERFMATCH(G') = c^n and #PERFMATCH(G) − #PERFMATCH(G') = 0 with high probability in time $\mathcal{O}(\frac{2^m}{c^n})$, where $n = \max\{|V(G)|, |V(G')|\}$ and $m = \max\{|E(G)|, |E(G')|\} = \mathcal{O}(n)$. So, in time $\mathcal{O}(d_1^n)$, where $d_1 \in (1, 2)$. Note that d_1 depends on the input, so this is not a robust result, and that is why we state it just as a remark. The same kind of algorithm would then exist for all the problems in SPP. Among them is the well-studied GRAPH ISOMORPHISM [5], which is one of the NP problems that has been proven neither NP-complete, nor polynomial-time solvable so far [6,17], and all the problems in UP, since UP $\subseteq$ SPP (see Proposition 2).

5   Conclusion

Our work aims to gain more insights and a better understanding of aspects related to the power of the TotP model of computation. The contribution of this paper is primarily conceptual, but also illustrates how the introduction of appropriate definitions can lead to nontrivial results in a fairly straightforward way, circumventing complex and hard-to-read proofs. The introduction of TotP-analogs of the classes shown in Table 1 led to new characterizations and complete problems for ⊕P, Mod$_k$P, and all gap-definable classes. The TotP computational model was proven to be sufficient, and it is arguably more appropriate to define these classes. Moreover, to the best of our knowledge, for Mod$_k$P, SPP, WPP, and

PP, we present the first complete problems that are not defined via #P-complete (under $\leq^p_{par}$) problems. Finally, two significant results of our approach is (a) that if the randomized Exponential Time Hypothesis holds, the WPP-complete problem DIFFPERFMATCH$_{=g}$ has no subexponential algorithm and (b) every SPP problem (including GRAPH ISOMORPHISM) is decidable by the difference of #PERFMATCH on two graphs, which is promised to be either exponentially large or zero. We expect that our results may inspire further research on total counting functions, as well as on the complexity classes that can be defined via them, thus providing new tools for analyzing the computational complexity of interesting problems that lie in such classes.

Aknowledgements. Aggeliki Chalki has been funded by the project "Mode(l)s of Verification and Monitorability" (MoVeMnt) (grant no 217987). Sotiris Kanellopoulos and Aris Pagourtzis have been partially supported for this work by project MIS 5154714 of the National Recovery and Resilience Plan Greece 2.0 funded by the European Union under the NextGeneration EU Program.

References

1. Achilleos, A., Chalki, A.: Counting computations with formulae: logical characterisations of counting complexity classes. In: Proceedings of MFCS 2023. LIPICs, vol. 272, pp. 7:1–7:15 (2023). https://doi.org/10.4230/LIPICS.MFCS.2023.7
2. Allender, E., Rubinstein, R.S.: P-printable sets. SIAM J. Comput. **17**(6), 1193–1202 (1988). https://doi.org/10.1137/0217075
3. Antonopoulos, A., Bakali, E., Chalki, A., Pagourtzis, A., Pantavos, P., Zachos, S.: Completeness, approximability and exponential time results for counting problems with easy decision version. Theoret. Comput. Sci. **915**, 55–73 (2022). https://doi.org/10.1016/J.TCS.2022.02.030
4. Arenas, M., Muñoz, M., Riveros, C.: Descriptive complexity for counting complexity classes. Logical Methods Comput. Sci. **16**(1) (2020). https://doi.org/10.23638/LMCS-16(1:9)2020
5. Arvind, V., Kurur, P.P.: Graph isomorphism is in SPP. Inf. Comput. **204**(5), 835–852 (2006). https://doi.org/10.1016/j.ic.2006.02.002
6. Babai, L.: Graph isomorphism in quasipolynomial time [extended abstract]. In: Proceedings of STOC 2016, pp. 684–697 (2016). https://doi.org/10.1145/2897518.2897542
7. Bakali, E., Chalki, A., Pagourtzis, A.: Characterizations and approximability of hard counting classes below #P. In: Chen, J., Feng, Q., Xu, J. (eds.) TAMC 2020. LNCS, vol. 12337, pp. 251–262. Springer, Cham (2020). https://doi.org/10.1007/978-3-030-59267-7_22
8. Beigel, R., Gill, J.: Counting classes: thresholds, parity, mods, and fewness. Theoret. Comput. Sci. **103**(1), 3–23 (1992). https://doi.org/10.1016/0304-3975(92)90084-S
9. Cai, J.Y., Hemachandra, L.A.: On the power of parity polynomial time. In: Monien, B., Cori, R. (eds.) STACS 89. LNCS, vol. 349, pp. 229–239. Springer, Heidelberg (1989). https://doi.org/10.1007/BFb0028987
10. Chen, P.C.: Heuristic sampling: a method for predicting the performance of tree searching programs. SIAM J. Comput. **21**, 295–315 (1992). https://doi.org/10.1137/0221022

11. Curticapean, R.: The simple, little and slow things count : on parameterized counting complexity. PhD thesis (2015). https://doi.org/10.22028/D291-26612
12. Fenner, S.A., Fortnow, L., Kurtz, S.A.: Gap-definable counting classes. J. Comput. Syst. Sci. **48**(1), 116–148 (1994). https://doi.org/10.1016/S0022-0000(05)80024-8
13. Gill, J.: Computational complexity of probabilistic turing machines. SIAM J. Comput. **6**(4), 675–695 (1977). https://doi.org/10.1137/0206049
14. Hertrampf, U.: Relations among mod-classes. Theoret. Comput. Sci. **74**(3), 325–328 (1990). https://doi.org/10.1016/0304-3975(90)90081-R
15. Kiayias, A., Pagourtzis, A., Sharma, K., Zachos, S.: Acceptor-definable counting classes. In: Manolopoulos, Y., Evripidou, S., Kakas, A.C. (eds.) PCI 2001. LNCS, vol. 2563, pp. 453–463. Springer, Heidelberg (2001). https://doi.org/10.1007/3-540-38076-0_29
16. Knuth, D.: Estimating the efficiency of backtrack programs. Math. Comput. **29**, 122–136 (1974). https://doi.org/10.2307/2005469
17. Köbler, J., Schöning, U., Torán, J.: The Graph Isomorphism Problem: Its Structural Complexity. Progress in Theoretical Computer Science. Birkhäuser/Springer (1993). https://doi.org/10.1007/978-1-4612-0333-9
18. Ogiwara, M., Hemachandra, L.A.: A complexity theory for feasible closure properties. J. Comput. Syst. Sci. **46**(3), 295–325 (1993). https://doi.org/10.1016/0022-0000(93)90006-I
19. Pagourtzis, A., Zachos, S.: The complexity of counting functions with easy decision version. In: Proceedings of MFCS, vol. 2006, pp. 741–752 (2006). https://doi.org/10.1007/11821069_64
20. Papadimitriou, C.H., Zachos, S.K.: Two remarks on the power of counting. In: Cremers, A.B., Kriegel, H.P. (eds.) Theoretical Computer Science. LNCS, vol. 145, pp. 269–276. Springer, Heidelberg (1983). https://doi.org/10.1007/BFb0009651
21. Pay, T., Cox, J.L.: An overview of some semantic and syntactic complexity classes. In: Electronic Colloquium on Computational Complexity, TR18-166 (2018). https://eccc.weizmann.ac.il/report/2018/166
22. Purdom, P.W.: Tree size by partial backtracking. SIAM J. Comput. **7**(4), 481–491 (1978). https://doi.org/10.1137/0207038
23. Rao, R.P., Rothe, J., Watanabe, O.: Upward separation for FewP and related classes. Inf. Process. Lett. **52**(4), 175–180 (1994). https://doi.org/10.1016/0020-0190(94)90123-6
24. Schöning, U.: Graph isomorphism is in the low hierarchy. J. Comput. Syst. Sci. **37**(3), 312–323 (1988). https://doi.org/10.1016/0022-0000(88)90010-4
25. Simon, J.: On some central problems in computational complexity. PhD thesis (1975)
26. Toda, S.: PP is as hard as the polynomial-time hierarchy. SIAM J. Comput. **20**(5), 865–877 (1991). https://doi.org/10.1137/0220053
27. Valiant, L.G.: Relative complexity of checking and evaluating. Inf. Process. Lett. **5**(1), 20–23 (1976). https://doi.org/10.1016/0020-0190(76)90097-1
28. Valiant, L.G.: The complexity of computing the permanent. Theoret. Comput. Sci. **8**(2), 189–201 (1979). https://doi.org/10.1016/0304-3975(79)90044-6
29. Valiant, L.G.: Accidental algorithms. In: Proceedings of FOCS, vol. 2006, pp. 509–517 (2006). https://doi.org/10.1109/FOCS.2006.7
30. Valiant, L.G., Vazirani, V.V.: NP is as easy as detecting unique solutions. Theoret. Comput. Sci. **47**, 85–93 (1986). https://doi.org/10.1016/0304-3975(86)90135-0

The Parameterized Complexity of Maximum Betweenness Centrality

Šimon Schierreich[(✉)] and José Gaspar Smutný

Department of Theoretical Computer Science, Faculty of Information Technology,
Czech Technical University in Prague, Prague, Czechia
schiesim@fit.cvut.cz

Abstract. Arguably, one of the most central tasks in the area of social network analysis is to identify important members and communities of a given network. The importance of an agent is traditionally measured using some well-defined notion of *centrality*. In this work, we focus on *betweenness centrality*, which is based on the number of shortest paths that an agent intersects. This measure can be naturally generalized from a single agent to a group of agents.

Specifically, we study the computation complexity of the k-MAXIMUM BETWEENNESS CENTRALITY problem, which consists in finding a group of size k whose betweenness centrality exceeds a given threshold. Since this problem is NP-complete in general, we use the framework of parameterized complexity to reveal at least some tractable fragments. From this perspective, we show that the problem is W[1]-hard and in XP when parameterized by the group size k. As the threshold value is not a useful parameter in this context, we focus on the structural restrictions of the underlying social network. In this direction, we show that the problem admits FPT algorithms with respect to the vertex cover number, the distance to clique, or the twin-cover number and the group size combined.

Keywords: Maximum Betweenness Centrality · Centrality Measures · Fixed-Parameter Tractability · Structural Restrictions · Social Network Analysis · Vertex Cover · Distance to Clique · Twin-Cover

1 Introduction

In the area of social network analysis, the identification of the most important members and communities is one of the most intriguing and significant research directions. Traditionally, the relevance of a network member is measured using some well-defined notion of *centrality*. *Betweenness centrality*, introduced by Freeman [17], is probably one of the most studied measures of the centrality (or importance) of an agent in a network. It has found its application in many areas, including wireless [32], transportation [30], covert [33], and biological networks [34]. The generalization of the notion of betweenness centrality from a single agent to groups of agents is called *group-betweenness centrality* and is due to Everett and Borgatti [14]. Next, Puzis et al. [31] introduced a polynomial time algorithm that, given a group of agents C, computes the group-betweenness centrality for C. Moreover, they claimed that finding a group of a predefined size with group-betweenness centrality equal to at least a given threshold is NP-complete. The work of Puzis et al. [31] was then followed by many authors who focused mainly on approximation algorithms [1,16,26,28] and algorithms for restricted graph families such as trees [16].

Another important application of group-betweenness centrality comes from monitoring communication networks. This comes from an interpretation where communication between each pair of agents uses some shortest path selected uniformly at random from all shortest paths between these two agents. A group of vertices with maximum group-betweenness centrality then maximizes the probability of capturing such communication. A somewhat related problem is the TRACKING SHORTEST PATHS problem [3], which was also extensively studied from the viewpoint of approximation [4] and parameterized algorithms [2,5,10]. This problem differs in that we are given a graph, a pair of distinct vertices s and t, a number k, and our goal is to decide whether we can choose a k-sized set $T \subseteq V$ such that for each pair of shortest s,t-paths P_1 and P_2 we have $V(P_1) \cap T \neq V(P_2) \cap T$. That is, in TRACKING SHORTEST PATHS, we are interested in a subset of vertices that distinguish each shortest path from all other shortest s,t-paths and, moreover, we assume only the shortest paths between a fixed pair of vertices.

Our Contribution. In this paper, we study the computational complexity of the k-MAXIMUM BETWEENNESS CENTRALITY problem. We start by showing that the problem remains NP-complete even if the underlying network is a planar graph of maximum degree three. Then, in contrast to the hardness result, we show that the problem is solvable in polynomial time if the network is of maximum degree two.

As the first part of our paper clearly indicates, the problem is highly intractable in its full generality. Based on this, we initiate the study of the k-MAXIMUM BETWEENNESS CENTRALITY problem (k-MBC for short) from the viewpoint of parameterized algorithmics – a framework for a finer-grained complexity classification of "hard" computational problems [11]. Using this framework, we show that the k-MBC problem is W[1]-hard and in XP when parameterized by the group size k. The same classification holds also for $(n - q)$-MBC parameterized by q. Additionally, we show that our XP algorithms are optimal under the well-known Exponential-Time Hypothesis [24].

Finally, we turn our attention to the underlying social network and analyze how different restrictions of the networks affect the complexity of the problem. In this direction, we show that the problem is in FPT when parameterized by the vertex cover number, the distance to clique, or the twin-cover number and the group size k combined. Along the way to these algorithms, we also prove polynomial-time solvability for some restricted graph families.

2 Preliminaries

We use $\mathbb{N}$ to denote the set $\{1, 2, 3, \ldots\}$ of positive numbers and $\mathbb{Q}_{\geq 0}$ to denote the set of all nonnegative rational numbers. For a set S, we denote by 2^S the set of all subsets of S, and by $\binom{S}{\ell}$ we denote the set of all ℓ-sized subsets of S. For basic graph-theoretical notation, we follow the monograph of Diestel [12]. We assume the reader to be familiar with standard notions of the parameterized complexity theory, such as FPT, the W-hierarchy, or parameterized reductions. For a comprehensive introduction to parameterized complexity, we refer to the monograph by Cygan et al. [11].

Betweenness Centrality. Let $G = (V, E)$ be a graph, and let $u, v, w \in V$ be three distinct vertices. By $\sigma_{u,v}$ we denote the number of shortest u, v-paths and we use $\sigma_{u,v}(w)$ to denote the number of shortest paths that contain the vertex w. The *betweenness centrality* of a vertex w is defined as $\mathrm{BC}(w) = \sum_{u,v \in V, u \neq v} \frac{\sigma_{u,v}(w)}{\sigma_{u,v}}$. Let $C \subseteq V$ be a set of vertices. We extend the notion of $\sigma_{u,v}(\cdot)$ to sets such that $\sigma_{u,v}(C)$ is the number of shortest u, v-paths that contain at least one vertex of C. Then, the *group-betweenness centrality* $\mathrm{GBC}(C)$ of the set C is defined as $\mathrm{GBC}(C) = \sum_{u,v \in V, u \neq v} \frac{\sigma_{u,v}(C)}{\sigma_{u,v}}$.

3 The Problem

So far, the definition of the problem has been rather informal. We start with a formal definition of the computational problem, which we will study in the remainder of the paper.

k-MAXIMUM BETWEENNESS CENTRALITY (k-MBC for short)
Input: A simple, undirected, and connected graph $G = (V, E)$ called a *social network*, a *group size* $k \in \mathbb{N}$, and a *threshold* value $t \in \mathbb{Q}_{\geq 0}$.
Question: Is there a set $C \subseteq V$ of size k such that $\mathrm{GBC}(C) \geq t$?

In the first result of this work, we show that the k-MAXIMUM BETWEENNESS CENTRALITY problem is NP-complete, even if the underlying social network is significantly restricted.

Theorem 1. *The k-MAXIMUM BETWEENNESS CENTRALITY problem is NP-complete even if the input graph is planar and of maximum degree 3.*

Proof (sketch). We provide a reduction from the INDEPENDENT SET problem, which is known to be NP-complete even if the input graph H is planar and of maximum degree 3 [20]. Given an instance $\mathcal{I} = (H, \ell)$ of the INDEPENDENT SET problem where H is a planar graph of maximum degree 3, we construct an equivalent instance $\mathcal{J} = (G, k, t)$ of the k-MAXIMUM BETWEENNESS CENTRALITY problem as follows. We set $G = H$, $k = n - \ell$, and the threshold is set to $t = n^2 - n$. The threshold t is defined so that every solution S contains at least one vertex from every shortest path. Consequently, the vertices in $V \setminus S$ must form an independent set, as otherwise S would not exceed the threshold. $\square$

In contrast to the previous hardness, in the following result, we show that if the social network is of the maximum degree 2, the problem can be solved in polynomial time.

Theorem 2. *If the social network is a connected graph of maximum degree 2, then the k-MAXIMUM BETWEENNESS CENTRALITY problem can be solved in linear time.*

4   Natural Parameters

The primary purpose of our paper is to reveal at least some tractable fragments of the k-MAXIMUM BETWEENNESS CENTRALITY problem. We start our journey by studying natural parameters – these parameters directly restrict some parts of the input. The most widely studied parameter in this direction is the solution size, which, in our case, corresponds to the group size k. In other words, the question we study in the following lines is whether the problem becomes tractable if the required group size is small. In our first result, we give an algorithm that runs in polynomial time for every constant value of the group size k.

Theorem 3. *There is an algorithm solving the k-MAXIMUM BETWEENNESS CENTRALITY problem in $n^{\mathcal{O}(k)}$ time.*

Viewing this problem more from a computational complexity theorist perspective, Theorem 3 shows that the k-MBC problem parameterized by the group size k is in the complexity class XP. The natural question then is whether the algorithm can be turned into a fixed-parameter tractable one. We answer this question in a negative way in the following result. Moreover, and perhaps surprisingly, we show that the algorithm given in Theorem 3 is, under standard theoretical assumptions, asymptotically optimal.

Theorem 4. *The k-MAXIMUM BETWEENNESS CENTRALITY problem is W[1]-hard when parameterized by the group size k and, unless ETH fails, there is no algorithm running in $f(k) \cdot n^{o(k)}$ time for any computable function f.*

Proof (sketch). We prove the W[1]-hardness by giving a parameterized reduction from the PARTIAL VERTEX COVER problem. In this problem, we are given a connected graph H and two integers s and ℓ. The goal is to decide whether there exists a set $S \subseteq V(H)$ of size at most s such that at least ℓ edges of H are incident to at least one vertex of S. The PARTIAL VERTEX COVER problem is known to be W[1]-complete when parameterized by the cover size s [21]. Moreover, we assume that $s < V(H)$ as otherwise the instance is trivial.

Given an instance $\mathcal{I} = (H, s, \ell)$ of the PARTIAL VERTEX COVER problem, we construct an equivalent instance $\mathcal{J} = (G, k, t)$ of the k-MAXIMUM BETWEENNESS CENTRALITY problem as follows (see Fig. 1 for an illustration of the construction). The basic building block of the construction is a *selection gadget* X_v

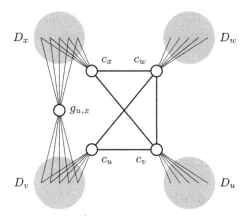

Fig. 1. An illustration of the construction used to prove Theorem 4. Gray circles schematically represent cliques of size d.

for every vertex $v \in V(H)$. This gadget consists of a clique D_v of size $d = n^5$ and one *core* vertex c_v, which is a neighbor of all the vertices of D_v. The specific value of d is not particularly important, as long as it is large compared to n. The purpose of the gadgets is to simplify the analysis of the reduction. If the cliques are large enough, we can focus only on the analysis of the shortest paths between them. Next, for each edge $\{u, v\} \in E(H)$, we add an edge $\{c_u, c_v\}$ connecting the core vertices of selection gadgets X_u and X_v. For every non-edge $\{u, v\} \notin E(H)$, we add a *guard* vertex $g_{u,v}$ and connect it to all vertices of D_u and D_v. To finalize the construction, we set $k = s$ and $t = 2\ell d^2$. The intuition behind the construction is to highlight the centrality of vertices adjacent to uncovered edges such that only these vertices can be selected for a solution.   □

Alternatively, one can ask whether the problem becomes easy to solve when we are allowed to assume a large number of agents. Clearly, if $k = n$, there is only one possible group, and we can decide the instance in constant time. How does the complexity of the problem change if the group size is only slightly below this guarantee?

Theorem 5. *There is an algorithm solving the $(n - q)$-*Maximum Between-ness Centrality* problem in $n^{\mathcal{O}(q)}$ time. Moreover, the $(n - q)$-*Maximum Betweenness Centrality* problem is W[1]-hard when parameterized by q, and, unless ETH fails, there is no algorithm solving the problem in $f(q) \cdot n^{o(q)}$ time for any computable function f.*

Another part of the input we can restrict is the value of the required threshold t. However, in this case, the situation is not as straightforward as when we parameterized by the budget. Specifically, if the value of the threshold is small compared to the instance size, the instance is trivial.

Lemma 1. *If* $t \leq \sum_{i=1}^{k} 2 \cdot (n-i)$, *then a solution of the* k-MAXIMUM BETWEEN-NESS CENTRALITY *problem is guaranteed to exist and can be found in linear time.*

5   Structural Parameters

As hinted at in the previous section, restricting the natural parameters does not lead to any fixed-parameter tractable algorithms. Therefore, in this section, we change our perspective and study the problem's complexity under several restrictions of the input social network.

5.1   Vertex Cover Number

We start this research direction with the well-known vertex cover number, which has been successfully used to give fixed-parameter algorithms for a wide variety of problems; e.g., [6–8,15,19,22].

First, we observe that if the group size k is at least the vertex cover number of G, then the given instance can be decided trivially.

Lemma 2. *Let* $\mathcal{I} = (G, k, t)$ *be an instance of* k-MBC. *If* $\mathrm{vc}(G) \leq k$ *and a vertex cover* M *of size* $\mathrm{vc}(G)$ *is given, then* $\mathcal{I}$ *can be decided in constant time.*

In the next auxiliary lemma, we show that twin vertices are interchangeable in the solution. Specifically, if we have a group C of vertices and replace a vertex with its twin (which initially is not in C), then the group-betweenness centrality of the new group is the same. The actual statement is, in fact, even stronger and is one of the crucial ingredients for most of our algorithmic results.

Lemma 3 (Twins Exchange Lemma). *Let* $X, Y \subseteq V(G)$ *be two sets of vertices such that* $|X| = |Y|$ *and* $\forall u, v \in X \cup Y \colon N(u) \setminus \{v\} = N(v) \setminus \{u\}$ *and* $C \subseteq V(G)$ *be a subset of vertices such that* $(X \cup Y) \cap C = \emptyset$. *Then it holds that* $\mathrm{GBC}(C \cup X) = \mathrm{GBC}(C \cup Y)$.

Proof. We begin by fixing an arbitrary ordering on the vertices of X and Y, that is, let $X = \{x_1, \ldots, x_\ell\}$ and $Y = \{y_1, \ldots, y_\ell\}$. Next, observe that the group-betweenness centrality of the set $C \cup X$ can be decomposed as follows

$$\mathrm{GBC}(C \cup X) = \mathrm{GBC}(C) + \sum_{u,v \in V, u \neq v} \frac{\sigma_{u,v}(C \cup X)}{\sigma_{u,v}} - \sum_{u,v \in V, u \neq v} \frac{\sigma_{u,v}(C)}{\sigma_{u,v}},$$

and identically for $C \cup Y$.

Clearly, the contribution of X and Y, respectively, is captured only in the middle term of the decomposition, since the first and last terms are independent of the particular choice of X and Y. Now, we show that $\sigma_{u,v}(C \cup Y)$ for every $u, v \in V$ is at least $\sigma_{u,v}(C \cup X)$. Let $u, v \in V$ be an arbitrary pair of distinct vertices. If $\sigma_{u,v}(C \cup X) = 0$, then the property is clearly valid. So, assume that $\sigma_{u,v}(C \cup X) \geq 1$ and let $\mathcal{Q}_x$ be the set of all shortest u, v-paths that contain at least one vertex of $C \cup X$. Then, every $Q \in \mathcal{Q}_x$ is one of the following three types.

1. The path Q contains at least one vertex of C. This path clearly also contains at least one vertex of $C \cup Y$ and therefore contributes by one to $\sigma_{u,v}(C \cup Y)$.
2. The path Q contains at least one vertex of Y. Again, it is immediate that this path increases $\sigma_{u,v}(C \cup Y)$ by one.
3. The path Q does not contain a vertex from $C \cup Y$. However, such a path necessarily contains at least one vertex from X. If we replace each $x_i \in X$, that appears in Q with a vertex $y_i \in Y$, we obtain a different path Q'. Since x_i and y_i are twins and $|X| = |Y|$, the path Q' is the same length as Q, all edges are preserved (up to the replacement of x_i with y_i), and the corresponding y_i always exists. Therefore, for each such path Q, there exists a shortest u,v-path Q' that contains at least one vertex of Y.

By the definition of the mapping, no shortest path Q' is assumed for $\sigma_{u,v}(C \cup Y)$ more than once. Consequently, $\sigma_{u,v}(C \cup Y)$ is at least $\sigma_{u,v}(C \cup X)$. Using the same arguments, we can show that $\sigma_{u,v}(C \cup X) \geq \sigma_{u,v}(C \cup Y)$. Hence, for every pair of distinct vertices $u, v \in V$, we have $\sigma_{u,v}(C \cup X) = \sigma_{u,v}(C \cup Y)$, and the lemma follows from the decompositions of $\mathrm{GBC}(C \cup X)$ and $\mathrm{GBC}(C \cup Y)$.□

Using Lemma 3, we can significantly restrict the space of potential solutions. This allows us to finally give a fixed-parameter tractable algorithm for k-MBC parameterized by the vertex cover number.

Theorem 6. *The k-Maximum Betweenness Centrality problem is fixed-parameter tractable when parameterized by the vertex cover number* $\mathrm{vc}(G)$.

Proof. Let $M \subseteq V$ be the minimum size vertex cover of G and $\tau = |M|$. Note that we can assume that an optimal vertex cover of a graph is given on input, as otherwise we can compute one in $\mathcal{O}(1.2738^\tau + \tau n)$ time [9]. If $k \geq \tau$, then we use Lemma 2 to resolve the instance directly in polynomial time. Therefore, we can suppose that $k < \tau$ for the rest of the proof.

As our first step, we partition the vertices in $V \setminus M$ based on their neighborhood in M. It is easy to see that this can be done in polynomial time and that there are at most 2^τ different parts. Then, for each part, we arbitrarily select k vertices and mark them blue. Additionally, we mark all vertices of M in blue. After these steps, there are at most $2^\tau \cdot k + \tau \leq 2^\tau \cdot \tau + \tau$ blue vertices. Finally, we exhaustively try all k-sized subsets C' of blue vertices, and for each C', we compute its group-betweenness centrality. If we find a group with $\mathrm{GBC}(C') \geq t$, we return *yes*. Otherwise, we return *no*. There are $\binom{2^\tau \cdot \tau + \tau}{k} = (2^\tau \cdot \tau + \tau)^{\mathcal{O}(\tau)}$ subsets C', and for each subset, we run the $\mathcal{O}(\tau^3)$ algorithm of Puzis et al. [31]. Therefore, the overall running time of the algorithm is $2^{\mathcal{O}(\tau^2)} + \mathcal{O}(nm)$.

For correctness, if the algorithm finds a solution, then the corresponding set C', which resulted in a *yes* response, is a certificate of correctness. Assume that our algorithm returned *no*, but there is a k-sized set $C^* \subseteq V$ such that $\mathrm{GBC}(C^*) \geq t$, that is, the response is incorrect. If C^* contains only blue vertices, then this situation is not possible since we examined all subsets of blue vertices of size k. Therefore, C^* clearly contains at least one unmarked vertex, and additionally, all unmarked vertices come from $V \setminus M$ as all vertex cover vertices are marked. However, by Lemma 3, we can replace the unmarked vertices with their twins without changing the value of the group-betweenness centrality of the set. Furthermore, there are always enough blue twins that can replace the unmarked vertices in C^* since we marked k of them. Hence, there exists an equivalent set C' consisting only of marked vertices, and our algorithm checked this set. This situation cannot, therefore, occur, and the algorithm is correct. □

5.2   Distance to Clique

The vertex cover number studied in the previous section can be seen as the minimum number of vertices we need to remove to obtain the sparsest possible graph – an edge-less graph. In some sense, the complete opposite of the vertex cover number is the distance to clique – the minimum number of vertices we need to remove to obtain a complete graph. Formally, we define the parameter of interest as follows.

Definition 1. *Let $G = (V, E)$ be a graph. A set of vertices $M \subseteq V$ is a modulator to clique if $G \setminus M$ is a complete graph. The distance to clique, denoted $\mathrm{dc}(G)$, is the minimum size of a modulator of G to clique.*

In the first result of this section, we show that this structural parameter is indeed worth studying, as the k-MBC problem becomes polynomial-time solvable if the underlying social network is a complete (bipartite) graph.

Theorem 7. *The k-MAXIMUM BETWEENNESS CENTRALITY problem can be solved in polynomial time if the social network G is a clique or a complete bipartite graph.*

Proof (sketch). In a clique, there is a unique shortest path of length one between each pair of vertices. Therefore, all sets of size k have the same group-betweenness centrality: $\sum_{i=1}^{k} 2 \cdot (n - i)$. Hence, we just compare the threshold value t with this bound and return the appropriate response.   □

The main result of this section is a fixed-parameter algorithm for k-MBC parameterized by the distance to clique. Unfortunately, one of the two main properties we exploited in the case of the vertex cover number no longer holds for this parameterization. Specifically, we did not find an easy way to bound the group-size k in terms of the parameter. It can even be shown that there are instances with $k > \mathrm{dc}(G)$ such that the only solution contains no modulator vertices.

Example 1. Let $G = (C \cup I, E)$ be a split graph where C is a clique, I is an independent set such that $\forall v \in I: \deg(v) = 1$, and $|C| \geq |I|^3$. It is clear that $\mathrm{dc}(G) = |I| = |N(I)|$. For an instance of the k-MBC problem where $k = |I|$ and with G as the social network, it holds that $\mathrm{GBC}(N(I)) > \mathrm{GBC}(I)$; vertices in I intersect all shortest paths between two vertices of I and between vertex of I

and a vertex of C, while $N(I)$ intersects the same shortest paths and additionally those between vertices in $N(I)$ and the rest of C. Therefore, there exists a threshold t such that $N(I)$ is a solution while I is not.

Therefore, to give a fixed-parameter algorithm, we use a different strategy. Again, we use Lemma 3 and combine it with an auxiliary lemma showing that the number of different kinds of shortest paths is small.

Lemma 4. *Let $u, v \in M$ be a pair of distinct modulator vertices and let P be a shortest u, v-path. Then P contains at most 2 vertices of $V \setminus M$.*

As the final ingredient, we need to find an optimal selection of vertices outside the modulator. To do so, we use Lemma 3 and a fixed-dimension ILP formulation of an appropriate subproblem.

Theorem 8. *The k-MAXIMUM BETWEENNESS CENTRALITY problem is fixed-parameter tractable when parameterized by the distance to clique $\mathrm{dc}(G)$.*

Proof (sketch). Let $M \subseteq V$ be the modulator to clique of size $\vartheta = \mathrm{dc}(G)$. As the first step of the algorithm, we guess (by guessing, we mean exhaustively trying all possible solutions) a set $M^* \subseteq M$ of modulator vertices for the solution. There are 2^ϑ such guesses. Let $|M^*| = k'$. Next, we partition the vertices outside the modulator according to their neighborhood in the modulator into $\mathfrak{T}$ vertex-types $T_1, \ldots, T_\mathfrak{T}$. For each vertex-type $i \in \{1, \ldots, \mathfrak{T}\}$, we determine the number n_i of vertices of this type. Then, for each pair of distinct modulator vertices $u, v \in M \setminus M^*$, we compute all their shortest paths. By Lemma 4, there are at most $2^\vartheta \cdot n^2$ such shortest paths for each pair u, v. Furthermore, the shortest paths can be partitioned into $2^{\mathcal{O}(\vartheta)}$ *kinds* based on the types of vertices in $V \setminus M$ they intersect, because all vertices of the same type are twins and, by Lemma 3, they are indistinguishable from the point of view of group-betweenness centrality. As the final step, we construct an ILP that finds an optimal number of vertices of each type to include in the solution. Specifically, we create a variable x_i for every vertex-type $i \in \{1, \ldots, \mathfrak{T}\}$. The constraints of the program ensure that no more than n_i vertices of type $i \in \{1, \ldots, \mathfrak{T}\}$, are selected and that exactly $k - k'$ vertices outside the modulator are selected. Most of the work is done in the optimization function, which computes the overall increase in the group-betweenness centrality of the selected vertices. Observe that there are $2^{\mathcal{O}(\vartheta)}$ variables (and $2^{\mathcal{O}(\vartheta)}$ constraints), and therefore we can use the FPT algorithm of Lenstra Jr. [27] to find an optimal selection of vertices outside M. The last step is to compare the group-betweenness centrality of the computed set with the threshold t and either return *yes* or continue with another M^*. □

5.3 Twin-Cover Number

The twin-cover number, introduced by Ganian [18], is one of the natural generalizations of the vertex cover number. This structural parameter is also based on modulator vertices; however, after removing modulator vertices, we obtain a disjoint union of cliques such that each clique consists of vertices that are twins in the original graph. Formally, the twin-cover number is defined as follows.

Definition 2 (*[18]*). *A set of vertices* $M \subseteq V(G)$ *is called a* twin-cover *of the graph* $G = (V, E)$ *if for every edge* $\{u, v\} \in E$ *it holds that either 1.* $u \in M$ *or* $v \in M$, *or 2.* $N(u) \setminus \{v\} = N(v) \setminus \{u\}$, *that is,* u *and* v *are twins. The* twin-cover number $\mathrm{tc}(G)$ *is the least size of a twin-cover of* G.

Observe that the twin-cover number is incomparable with the distance to clique studied in the previous section. On the other hand, there exists a common generalization of the twin-cover number and the distance to the clique called the *vertex cluster deletion number* [13], which is defined similarly to the twin-cover number but excludes the twins property. The twin-cover number was also used for the design of fixed-parameter algorithms in different scenarios [18,22,25,29].

Theorem 9. *The k-MAXIMUM BETWEENNESS CENTRALITY problem is fixed-parameter tractable when parameterized by the twin-cover number* $\mathrm{tc}(G)$ *and the group size k combined.*

Proof (sketch). The high-level idea of the algorithm is that we can again partition the vertices outside the modulator according to their neighborhood in the modulator. Then, we generalize Lemma 3 and show that vertices of the same type are indistinguishable from the point of view of other vertices. Next, we carefully mark $\mathcal{O}(k)$ vertices of each type that potentially increase the group-betweenness centrality the most. As a final step, we perform an exhaustive search through all subsets of marked vertices similar to the one in the proof of Theorem 6.   □

6   Conclusions

In this paper, we initiate the study of the k-MAXIMUM BETWEENNESS CENTRALITY problem from the perspective of parameterized complexity. In the first part, we study the complexity of the problem when the sought group size is restricted. This revealed that a simple brute-force algorithm trying all possible groups is, under standard theoretical assumptions, the best we can hope for. Then, we turn our attention to the structural restrictions of the underlying social network. There, the results are much more positive. Namely, we show that the problem is fixed-parameter tractable when parameterized by the vertex cover number, the distance to clique, or the twin-cover number and the group size combined.

We believe that our work opens avenues for further research. First, we are missing more lower bounds with respect to structural parameters. One of the most immediate questions in this direction is whether our algorithm for the vertex cover number parameterization is tight, e.g., assuming ETH, or can be improved. More generally, the only hardness result with respect to structural parameters is the following corollary of Theorem 1 and [23, Theorem 1].

Corollary 1. *The k-MAXIMUM BETWEENNESS CENTRALITY problem is para-NP-hard when parameterized by the twin-width and the maximum degree combined.*

Nevertheless, there is an overwhelming plethora of structural parameters between the twin-width and the vertex cover number. Consequently, many immediate questions arise. First of all, we would like to know whether k-MBC is in FPT or W[1]-hard when parameterized by the twin-cover number. Next, since it is known that k-MBC is solvable in polynomial time on trees, is the problem in FPT when parameterized by the treewidth?

Another promising direction is to study *kernelization* in the context of k-MBC. Specifically, in Sect. 5.1, we were able to significantly restrict the search space; on the other hand, the number of vertices we need to examine is still exponential in terms of the parameter. So, the natural question is whether we can preprocess the instance so that we end up with a polynomial number (w.r.t. the parameter) of marked vertices.

Acknowledgements. This work was supported by the Student Summer Research Program 2023 of FIT CTU in Prague. Šimon Schierreich acknowledges the support of the Czech Science Foundation Grant No. 22-19557S and of the Grant Agency of the CTU in Prague, grant No. SGS23/205/OHK3/3T/18.

References

1. Angriman, E., van der Grinten, A., Bojchevski, A., Zügner, D., Günnemann, S., Meyerhenke, H.: Group centrality maximization for large-scale graphs. In: ALENEX '20, pp. 56–69. SIAM (2020)
2. Banik, A., Choudhary, P., Raman, V., Saurabh, S.: Fixed-parameter tractable algorithms for tracking shortest paths. Theor. Comput. Sci. **846**, 1–13 (2020)
3. Banik, A., Katz, M.J., Packer, E., Simakov, M.: Tracking paths. Discrete Appl. Math. **282**, 22–34 (2020). https://doi.org/10.1016/j.dam.2019.11.013
4. Blažej, V., Choudhary, P., Knop, D., Křišťan, J.M., Suchý, O., Valla, T.: Constant factor approximation for tracking paths and fault tolerant feedback vertex set. Discrete Optim. **47**, 100756 (2023). https://doi.org/10.1016/j.disopt.2022.100756
5. Blažej, V., Choudhary, P., Knop, D., Křišťan, J.M., Suchý, O., Valla, T.: Polynomial kernels for tracking shortest paths. Inf. Process. Lett. **179**, 106315 (2023)
6. Blažej, V., Ganian, R., Knop, D., Pokorný, J., Schierreich, Š., Simonov, K.: The parameterized complexity of network microaggregation. In: AAAI '23, pp. 6262–6270 (2023). https://doi.org/10.1609/aaai.v37i5.25771
7. Bodlaender, H.L., Groenland, C., Pilipczuk, M.: Parameterized complexity of binary CSP: Vertex cover, treedepth, and related parameters. In: ICALP '23, pp. 27:1–27:20 (2023). https://doi.org/10.4230/LIPICS.ICALP.2023.27
8. Chapelle, M., Liedloff, M., Todinca, I., Villanger, Y.: Treewidth and pathwidth parameterized by the vertex cover number. Discrete Appl. Math. **216**, 114–129 (2017). https://doi.org/10.1016/j.dam.2014.12.012
9. Chen, J., Kanj, I.A., Xia, G.: Improved upper bounds for vertex cover. Theor. Comput. Sci. **411**(40), 3736–3756 (2010). https://doi.org/10.1016/j.tcs.2010.06.026
10. Choudhary, P., Raman, V.: Structural parameterizations of tracking paths problem. Theor. Comput. Sci. **934**, 91–102 (2022)
11. Cygan, M., et al.: Parameterized Algorithms. Springer, Cham (2015). https://doi.org/10.1007/978-3-319-21275-3

12. Diestel, R.: Graph Theory. GTM, vol. 173. Springer, Heidelberg (2017). https://doi.org/10.1007/978-3-662-53622-3

13. Doucha, M., Kratochvíl, J.: Cluster vertex deletion: a parameterization between vertex cover and clique-width. In: Rovan, B., Sassone, V., Widmayer, P. (eds.) MFCS 2012. LNCS, vol. 7464, pp. 348–359. Springer, Heidelberg (2012). https://doi.org/10.1007/978-3-642-32589-2_32

14. Everett, M., Borgatti, S.: The centrality of groups and classes. J. Math. Sociol. **23**, 181–201 (1999). https://doi.org/10.1080/0022250X.1999.9990219

15. Fiala, J., Golovach, P.A., Kratochvíl, J.: Parameterized complexity of coloring problems: treewidth versus vertex cover. Theor. Comput. Sci. **412**(23), 2513–2523 (2011). https://doi.org/10.1016/j.tcs.2010.10.043

16. Fink, M., Spoerhase, J.: Maximum Betweenness Centrality: approximability and tractable cases. In: Katoh, N., Kumar, A. (eds.) WALCOM 2011. LNCS, vol. 6552, pp. 9–20. Springer, Heidelberg (2011). https://doi.org/10.1007/978-3-642-19094-0_4

17. Freeman, L.C.: Centrality in social networks conceptual clarification. Soc. Netw. **1**(3), 215–239 (1978). https://doi.org/10.1016/0378-8733(78)90021-7

18. Ganian, R.: Improving vertex cover as a graph parameter. Discrete Math. Theor. Comput. Sci. **17**(2), 77–100 (2015). https://doi.org/10.46298/DMTCS.2136

19. Ganian, R., Hamm, T., Knop, D., Roy, S., Schierreich, Š., Suchý, O.: Maximizing social welfare in score-based social distance games. In: TARK '23, pp. 272–286 (2023). https://doi.org/10.4204/EPTCS.379.22

20. Garey, M.R., Johnson, D.S.: The rectilinear Steiner tree problem is NP-complete. SIAM J. Appl. Math. **32**(4), 826–834 (1977). https://doi.org/10.1137/0132071

21. Guo, J., Niedermeier, R., Wernicke, S.: Parameterized complexity of vertex cover variants. Theory Comput. Syst. **41**(3), 501–520 (2007)

22. Hanaka, T., Ono, H., Otachi, Y., Uda, S.: Grouped domination parameterized by vertex cover, twin cover, and beyond. In: Mavronicolas, M. (ed.) CIAC 2023. LNCS, vol. 13898, pp. 263–277. Springer, Cham (2023). https://doi.org/10.1007/978-3-031-30448-4_19

23. Hliněný, P., Jedelský, J.: Twin-width of planar graphs is at most 8, and at most 6 when bipartite planar. In: ICALP '23, pp. 75:1–75:18 (2023)

24. Impagliazzo, R., Paturi, R.: On the complexity of k-SAT. J. Comput. Syst. Sci. **62**(2), 367–375 (2001). https://doi.org/10.1006/jcss.2000.1727

25. Knop, D.: Local linear set on graphs with bounded twin cover number. Inf. Process. Lett. **170**, 106118 (2021). https://doi.org/10.1016/j.ipl.2021.106118

26. Lagos, T., Prokopyev, O.A., Veremyev, A.: Finding groups with maximum betweenness centrality via integer programming with random path sampling. J. Glob. Optim. (2023). https://doi.org/10.1007/s10898-022-01269-2

27. Lenstra, H.W., Jr.: Integer programming with a fixed number of variables. Math. Oper. Res. **8**(4), 538–548 (1983). https://doi.org/10.1287/MOOR.8.4.538

28. Mahmoody, A., Tsourakakis, C.E., Upfal, E.: Scalable betweenness centrality maximization via sampling. In: KDD '16, pp. 1765–1773. ACM (2016)

29. Misra, N., Mittal, H.: Imbalance parameterized by twin cover revisited. Theor. Comput. Sci. **895**, 1–15 (2021). https://doi.org/10.1016/j.tcs.2021.09.017

30. Puzis, R., Altshuler, Y., Elovici, Y., Bekhor, S., Shiftan, Y., Pentland, A.: Augmented betweenness centrality for environmentally aware traffic monitoring in transportation networks. J. Intell. Transp. Syst. **17**(1), 91–105 (2013)

31. Puzis, R., Elovici, Y., Dolev, S.: Fast algorithm for successive computation of group betweenness centrality. Phys. Rev. E **76**(5), 056709 (2007)

32. Tuzcu, A., Arslan, H.: Betweenness centrality in sparse real world and wireless multi-hop networks. In: Kahraman, C., Cebi, S., Cevik Onar, S., Oztaysi, B., Tolga, A.C., Sari, I.U. (eds.) INFUS 2021. LNNS, vol. 307, pp. 217–224. Springer, Cham (2022). https://doi.org/10.1007/978-3-030-85626-7_27
33. Waniek, M., Michalak, T.P., Wooldridge, M.J., Rahwan, T.: Hiding individuals and communities in a social network. Nat. Hum. Behav. **2**(2), 139–147 (2018)
34. Xiong, W., Xie, L., Zhou, S., Liu, H., Guan, J.: The centrality of cancer proteins in human protein-protein interaction network: a revisit. Int. J. Comput. Biol. Drug Des. **7**(2–3), 146–156 (2014). https://doi.org/10.1504/IJCBDD.2014.061643

Offensive Alliances in Signed Graphs

Zhidan Feng[1,2]([✉])[iD], Henning Fernau[2][iD], Kevin Mann[2][iD], and Xingqin Qi[1][iD]

[1] School of Mathematics and Statistic, Shandong University, 264209 Weihai, China
qixingqin@sdu.edu.cn
[2] Universität Trier, FB 4 – Informatikwissenschaften, 54286 Trier, Germany
{s4zhfeng,fernau,mann}@uni-trier.de

Abstract. Signed graphs have been introduced to enrich graph structures expressing relationships between persons or general social entities, introducing edge signs to reflect the nature of the relationship, e.g., friendship or enmity. Independently, offensive alliances have been defined and studied for undirected, unsigned graphs. We join both lines of research and define offensive alliances in signed graphs, hence considering the nature of relationships. Apart from some combinatorial results, mainly on k-balanced and k-anti-balanced signed graphs (a newly introduced family of signed graphs), we focus on the algorithmic complexity of finding smallest offensive alliances, looking at a number of parameterizations. While the parameter solution size leads to an FPT result for unsigned graphs, we obtain W[2]-completeness for the signed setting. We introduce new parameters for signed graphs, e.g., distance to weakly balanced signed graphs, that could be of independent interest. We show that these parameters yield FPT results. Here, we make use of the recently introduced parameter neighborhood diversity for signed graphs.

Keywords: Offensive Alliance · Parameterized Complexity · Fixed Parameter Tractable · Signed Graph

1 Introduction

An alliance is a grouping of entities that work together because of a mutual interest or a common purpose. The alliance might be formed to unite against potential attacks for self-defensive purposes or to establish active attacks against common enemies. An offensive alliance is a kind of alliance formed for the purpose of attacking, which has been widely studied in unsigned graphs [5,6,10,11,13,16,18,19]. More formally, if $G = (V, E)$ is an undirected, unsigned graph, then $S \subseteq V$ is an *offensive alliance* if for each $u \in V \setminus S$, if $\deg_S(u) > 0$, then $\deg_S(u) \geq \deg_{\overline{S}}(u) + 1$, where $\deg_X(v)$ counts the number of neighbors of vertex v in the vertex set X. Here, we intend to introduce the concept of offensive alliances into signed graphs, generalizing it from the unsigned setting.

We first list the necessary definitions. A signed graph is given by a triple $G = (V, E^+, E^-)$, where V is a finite vertex set and $E^+ \subseteq \binom{V}{2}$ is the positive

Supplementary Information The online version contains supplementary material available at https://doi.org/10.1007/978-981-97-2340-9_20.

edge set and $E^- \subseteq \binom{V}{2}$, with $E^+ \cap E^- = \emptyset$, is the negative edge set. Positive edges are interpreted as friendly connections, while negative edges are rather hostile relationships. For all $v \in V$, $N^+(v) = \{u \in V \mid vu \in E^+\}$ is the set of positive neighbors of v, $N^-(v) = \{u \in V \mid vu \in E^-\}$ is the set of negative neighbors of v. We call the set $N(v) = N^+(v) \cup N^-(v)$ the open neighborhood of v, and set $N[v] = N(v) \cup \{v\}$ the closed neighborhood of v. Accordingly, $\deg_G^+(v) = |N^+(v)|$, $\deg_G^-(v) = |N^-(v)|$ are the positive degree of v and the negative degree of v, respectively. Let $\deg_G(v) = \deg_G^+(v) + \deg_G^-(v)$ denote the (total) degree of v. Let $\delta^+(G)$ and $\Delta^+(G)$ denote the minimum and maximum positive degree of any vertex in G, respectively. Accordingly, $\delta^-(G)$ and $\Delta^-(G)$ are understood, as well as $\delta(G)$ and $\Delta(G)$. For any set $S \subseteq V$, $\deg_S^+(v) = |N^+(v) \cap S|$, $\deg_S^-(v) = |N^-(v) \cap S|$ are the positive degree and the negative degree of v with respect to S, respectively. Let $N(S) = \bigcup_{v \in S} N(v)$ denote the neighborhood of S, then the boundary of S is the set $\partial S = N(S) \setminus S$.

In this paper, we initiate the study of offensive alliances in signed graphs. Unlike unsigned setting, signed graphs distinguish the edges with the property of positive or negative signs, which models refined relationships in many real-world systems, like friendship or enmity. Consequently, this leads to different requirements on offensive alliance structures, and so we give a modified definition in the following, staying close to our proposal for a notion of defensive alliances in signed graphs, formulated in [1]. A subset $S \subseteq V$ of a signed graph $G = (V, E^+, E^-)$ is called an *offensive alliance* if, for each $v \in \partial S$,

1. $\deg_S^-(v) \geq \deg_S^+(v)$ and
2. $\deg_S^-(v) \geq \deg_{\bar{S}}^+(v) + 1$.

The first condition expresses that the offensive alliance is predominantly hostile (at least not friendly) to each vertex of the boundary. It makes sure that the alliance is ready for the purpose of attacking so that friends within the alliance are staying neutral and not coming to help their friends outside the alliance. The second condition is consistent with that of a traditional offensive alliance in unsigned graphs, which says that each attacked node should have at least as many attackers in the alliance than it has potential friends outside of the alliance, which guarantees a successful attack. It can be reformulated as $\deg_S(v) \geq \deg^+(v) + 1$. A vertex is said to be successfully attacked if the vertex satisfies these two offensive conditions. We illustrate these concepts in Fig. 1.

Note that a trivial case is that for each signed graph, the whole vertex set forms an offensive alliance according to the definition; more generally, a trivial alliance shall denote an alliance formed by a connected component. Unlike in the case of defensive alliance on signed graphs, the pure existence of an offensive alliance for a signed graph is therefore not a hard problem. In the following, we are going to study the following natural computational minimization problem:

OFFENSIVE ALLIANCE: Given a signed graph G and an integer $k \geq 1$, does there exist an offensive alliance S with $1 \leq |S| \leq k$?

The analogously defined computational problem in unsigned graph will be called OFFENSIVE ALLIANCE UG for distinction.

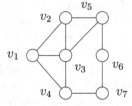

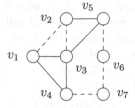

(a) Unsigned graph: $\{v_1, v_2, v_3\}$ is a minimum offensive alliance of size 3; it has various offensive alliances with four vertices, e.g., $\{v_1, v_3, v_5, v_7\}$, $\{v_1, v_3, v_4, v_6\}$, $\{v_1, v_4, v_5, v_7\}$, etc., but $\{v_1, v_3, v_4, v_5\}$ is not an offensive alliance.

(b) Signed graph: $\{v_1, v_3, v_4, v_5\}$ is a minimum offensive alliance of size 4, but it is not an offensive alliance in the unsigned setting; also, $\{v_1, v_2, v_3\}$ is not an offensive alliance in the signed setting, but $\{v_1, v_3, v_4, v_6\}$ still is.

Fig. 1. Offensive alliances on unsigned vs signed graph

Due to space limitations, proofs of statements marked with ($*$) are presented in the appendix.

2 Some Combinatorial Results

Here, we study the size of the smallest non-empty offensive alliance in a signed graph G, which we will also denote as $a_{so}(G)$.

Observations. Intuitively, the more negative edges there are, the more likely it is to form the non-trivial offensive alliance. An extreme example is that the signed graph is full of positive edges, in this case, the only offensive alliance is the trivial one, the whole vertex set. Our first observation shows a necessary relationship between negative degree and positive degree for the sake of existence of a non-trivial offensive alliance.

Observation 1. *(necessary condition for existence) If there exists a non-trivial offensive alliance in a signed connected graph G, then $\Delta^-(G) \geq \left\lceil \frac{\delta^+(G)+1}{2} \right\rceil$.*

Proof. Let S be a non-trivial offensive alliance of G. As G is connected, this means that $S \neq V$. Therefore, $\partial S \neq \emptyset$. Then for any vertex $v \in \partial S$, $\Delta^-(G) \geq \deg^-(v) \geq \deg_S^-(v) \geq \max\{\deg_S^+(v), \deg_{\overline{S}}^+(v)+1\} \geq \left\lceil \frac{\deg^+(v)+1}{2} \right\rceil \geq \left\lceil \frac{\delta^+(G)+1}{2} \right\rceil.\square$

From the previous proof, we can also deduce:

Corollary 1. *If a vertex v can be successfully attacked, (or in other words, v is in the boundary of an offensive alliance), then $\deg^-(v) \geq \left\lceil \frac{\deg^+(v)+1}{2} \right\rceil$.*

Proposition 1. ($*$) *For all connected signed graphs G, $a_{so}(G) \geq \delta^+(G) + 1$.*

Examples of equality in the previous upper bound are 2-balanced complete graphs K_{2n} (see below) with two partitions of the same size, i.e., $|V_1| = |V_2| = n$.

Proposition 2. *If S is a minimum-size offensive alliance in $G = (V, E^+, E^-)$, then $S \cup \partial S$ is connected in the underlying unsigned graph $(V, E^+ \cup E^-)$.*

Special Complete Signed Graphs. A signed graph $K = (V, E^+, E^-)$ is *complete* if $\binom{V}{2} = E^+ \cup E^-$. Hence, for each pair of objects from V, it is decided if they entertain a positive or a negative relationship. In the following, we determine a_{so} for two special complete signed graphs: (a) K is *(weakly) k-balanced ($k \geq 1$)* if the unsigned positive graph $K^+ = (V, E^+)$ has k connected components; there is no negative edge within any such component. This concept was introduced in [3,9] with a sociological motivation and is also a basic notion for *Correlation Clustering* [2]. Quite analogously, we introduce here the following (new) class of signed graphs as follows. (b) K is *(weakly) k-anti-balanced ($k \geq 1$)* if the unsigned negative graph $K^- = (V, E^-)$ has k connected components; there is no positive edge within any such component.

Theorem 1. *For any signed complete graph $K_n = (V, E^+, E^-)$, $n = |V|$, we can determine its offensive alliance number in the following cases.*

1) *If K_n is (weakly) k-balanced ($k \geq 1$) with partition $V = (V_1, \ldots, V_k), |V_1| \geq |V_2| \geq \cdots \geq |V_k|$, then $a_{so}(K_n) = |V_1|$; V_1 is a minimum offensive alliance. Moreover, if $S = S_1 \cup S_2 \cup \cdots \cup S_p, p \geq 2$, where $\emptyset \subsetneq S_1 \subseteq V_{i_1}, \emptyset \subsetneq S_2 \subseteq V_{i_2}, \ldots, \emptyset \subsetneq S_p \subseteq V_{i_p}, i_1 \leq i_2 \leq \cdots \leq i_p$, then $|S|$ is a minimum offensive alliance iff $|S| = |V_1| \geq 2|S_l|$, where $l = \underset{j \in \{1,\ldots,p\}}{argmax}\{|S_j| \mid V_{i_j} \setminus S_j \neq \emptyset\}$.*

2) *If K_n is (weakly) k-anti-balanced ($k \geq 1$) with partition $V = (V_1, \ldots, V_k)$, and $|V_1| \geq |V_2| \geq \cdots \geq |V_k|$, then we find:*

   i) *If $V = V_1 \cup V_2$, and $|V_1| \geq |V_2| \geq \lceil \frac{n+1}{3} \rceil$, then $a_{so}(K_n) = 2\lceil \frac{|V_1|+1}{2} \rceil$. Any subset S_1 of V_1 and any subset S_2 of V_2 with $|S_1| = |S_2| = \lceil \frac{|V_1|+1}{2} \rceil$, taken together, forms a minimum offensive alliance,*

   ii) *If $|V_1| \geq \frac{n}{2}$, then $a_{so}(K_n) = \max\{n - |V_1| + 1, 2(n - |V_1|)\}$. Any subset S_1 of V_1 with $|S| = n - |V_1|$ and $V \setminus V_1$ forms a minimum offensive alliance.*

   iii) *Otherwise, $a_{so}(K_n) = n$.*

Proof. 1) Let K_n be k-balanced ($k \geq 1$), and assume that the nonempty set $S = S_1 \cup S_2 \cup \cdots \cup S_p, p \geq 1$ is an offensive alliance of K_n, where $\emptyset \subsetneq S_1 \subseteq V_{i_1}, \emptyset \subsetneq S_2 \subseteq V_{i_2}, \ldots, \emptyset \subsetneq S_p \subseteq V_{i_p}, i_1 \leq i_2 \leq \cdots \leq i_p$. If $\partial S \cap V_1 = \emptyset$, clearly, $V_1 \subseteq S$, so $|S| \geq |V_1|$. Otherwise, consider some arbitrary $u \in \partial S \cap V_1$.

- $i_1 = 1$ implies $\deg_S^-(u) = |S| - |S_1| \geq \deg_{\overline{S}}^+(u) + 1 = |V_1| - |S_1|$, i.e., $|S| \geq |V_1|$.
- If $i_1 > 1$, then $\deg_S^-(u) = |S| \geq \deg_{\overline{S}}^+(u) + 1 = |V_1|$, i.e., $|S| \geq |V_1|$.

So in each case, $a_{so}(K_n) \geq |V_1|$.

<u>Case one:</u> Suppose $p = 1$, we can easily check that each boundary vertex can be successfully attacked, then $S = V_1$ is a minimum offensive alliance.

Case two: Suppose $p \geq 2$, if S is a minimum offensive alliance, then $|S| = |V_1|$ (as we already know that V_1 is a minimum offensive alliance). For each $u \in \partial S \cap V_{i_j}, j \in \{1, \ldots, p\}$, $\deg_S^-(u) = |S| - |S_j| \geq \deg_S^+(u) = |S_j|$, i.e., $|S| \geq 2|S_j|$, so that $|S| = |V_1| \geq 2|S_l|$. Conversely, assume that $|S| = |V_1| \geq 2|S_l|$, where $l = \underset{j \in \{1,2,\ldots,p\}}{argmax} \{|S_j| \mid V_{i_j} \setminus S_j \neq \emptyset\}$. We can verify that S is an offensive alliance. For each $u \in \partial S \cap V_{i_j}, j \in \{1, 2, \ldots, p\}$, $\deg_S^-(u) = |S| - |S_j| \geq |S| - |S_l| \geq |S_l| \geq |S_j| = \deg_S^+(u)$, $\deg_S^-(u) = |V_1| - |S_j| \geq |V_{i_j}| - |S_j| = \deg_{\overline{S}}^+(u) + 1$. For each $u \in \partial S \cap V_h, h \notin \{i_1, \ldots, i_p\}$, $\deg_S^-(u) = |V_1| \geq \max\{0, |V_h|\} = \max\{\deg_S^+(u), \deg_{\overline{S}}^+(u) + 1\}$.

2) For the analysis of the anti-balanced setting, we refer to the appendix. □

3   Classical and Fine-Grained Complexity Results

A natural transformation is to apply known results for unsigned graphs to the case of signed graphs. Here, we first show the hardness of OFFENSIVE ALLIANCE on signed graphs by a reduction from OFFENSIVE ALLIANCE UG, which has been known to be NP-complete [13]. This reduction allows us to further deduce parameterized hardness results with respect to a number of structural parameters.

Theorem 2. OFFENSIVE ALLIANCE *is NP-complete.*

Proof. For membership in NP, a simple guess-and-check approach will work. We will prove NP-hardness by a reduction from OFFENSIVE ALLIANCE UG. Let $G = (V, E)$ be an unsigned graph, and $k \in \mathbb{N}$, forming an instance of OFFENSIVE ALLIANCE UG. For each $v \in V$, define $d'(v) := \left\lceil \frac{\deg(v)+1}{2} \right\rceil$, and $M_v := \{v_{ij} \mid i \in \{1, \ldots, d'(v) - 1\}, j \in \{1, \ldots, 3k + 1\}\}$. Then, in polynomial time, one can construct from $\langle G, k \rangle$ an equivalent instance $\langle G', k' \rangle$ of OFFENSIVE ALLIANCE, with $G' = (V', E'^+, E'^-)$, as follows. This is also illustrated in Fig. 2.

(a) Vertex gadget: the diamond represents vertex v; the square represents the not-in-the-solution gadget.

(b) The not-in-the-solution gadget.

Fig. 2. Reduction construction for Theorem 2.

$$V' := V \cup \bigcup_{v \in V} M_v \cup \{v_1, v_2, v'_i \mid v \in V, i \in \{1, \ldots, d'(v) - 1\}\},$$

$$E'^+ := \{vv'_i, v'_i v_{i1}, v_{i1}v_{il}, v_{i2}v_{il} \mid v \in V, i \in \{1, \ldots, d'(v) - 1\}, l \in \{3, \ldots, 3k + 1\}\},$$

$$E'^- := E \cup \{v_1 v'_i, v_2 v'_i \mid v \in V, i \in \{1, \ldots, d'(v) - 1\}\};$$

$$k' := 3k.$$

In the following, for clarity we add a subscript G or G' when referring to degrees, for instance. If $S \subseteq V$ with $1 \le |S| \le k$ is an offensive alliance of G, then we prove $S' = S \cup \{v_1, v_2 | v \in S\}$ is an offensive alliance of G'. By definition, for each $u \in V \setminus S$ of G with $\deg_{G,S}(u) > 0$, we have $\deg_{G,S}(u) \ge \deg_{G,\overline{S}}(u) + 1$, that is, $\deg_{G,S}(u) \ge \left\lceil \frac{\deg_G(u)+1}{2} \right\rceil = d'(u)$. Therefore, for each $u \in \partial S$ of G', if $u \in V$, then $\deg^-_{G',S}(u) = \deg_{G,S}(u) \ge d'(u) = \deg^+_{G',\overline{S}}(u) + 1$. Moreover, $\deg^+_{G',S}(u) = 0 \le \deg^-_{G',S}(u) = \deg_{G,S}(u)$. Otherwise, if $u \in \partial S$ of G', then $u = v'_i$ for some $v \in S$. Now, $\deg^-_{G',S}(v'_i) = 2 = \deg^+_{G',\overline{S}}(v'_i) + 1$ and $\deg^-_{G',S}(v'_i) = 2 \ge \deg^+_{G',S}(v'_i) = 1$. So, S' is an offensive alliance on G' with $1 \le |S'| \le 3k$.

For the converse direction of this proof, we refer to the appendix. □

Next, we would like to display another simpler reduction from Vertex Cover, which shows the hardness of Offensive alliance on degree-bounded and planar signed graphs. Moreover, by this reduction, we can also get a lower-bound result on Offensive alliance based on the Exponential-Time Hypothesis ETH; see [20, 22]. A different reduction was used in [15] for unsigned graphs.

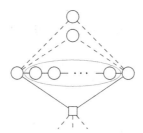

Fig. 3. Reduction construction for Theorem 3; the square represents vertex v; the set encircled in blue is serving as a not-in-the-solution gadget as in Fig. 2b. (Color figure online)

Theorem 3. Offensive Alliance *is NP-complete, even on planar graphs with maximum degree of* 5.

Proof. NP-membership is inherited from Theorem 2. We now give a simple reduction from Vertex Cover on planar cubic graphs (see [23]) to show the NP-hardness claim. Let $\langle G = (V, E), k \rangle$ be an instance of Vertex Cover on planar cubic graphs. We can get an instance $\langle G' = (V', E'^+, E'^-), k' \rangle$, $k' = 3k$, of Offensive alliance in polynomial time as follows, illustrated in Fig. 3:

$$V' := V \cup \{v^1, v^2, v_1, v_2, v_i' \mid v \in V, i \in \{1, \ldots, 3k+1\}\},$$
$$E'^+ := \{vv_1, vv_2, v_1v_1', v_i'v_{i+1}', v_2v_{4k}' \mid v \in V, i \in \{1, \ldots, 3k\}\},$$
$$E'^- := E \cup \{v_1v^1, v_1v^2, v_2v^1, v_2v^2 \mid v \in V\};$$

Notice that this construction is a simplified version that of in Theorem 2, as the input graph G with maximum degree of 3. Obviously, following the analogous analysis in Theorem 2, we can obtain $\langle G = (V, E), k \rangle$ is a yes-instance of VERTEX COVER on planar cubic graphs if and only if $\langle G' = (V', E'^+, E'^-), k' \rangle$ of is a yes-instance of OFFENSIVE ALLIANCE. In particular, if $S \subseteq V$ is a vertex cover in G, then $V \cup \{v^1, v^2 \mid v \in S\}$ is an offensive alliance of G'. Moreover, $\Delta(G') = 5$, and planarity is also shown in Fig. 3. $\square$

Theorem 4. *If ETH holds,* OFFENSIVE ALLIANCE *is not solvable in time* $\mathcal{O}(2^{o(n)})$.

Proof. We sightly modified the construction of Theorem 3 by sharing a common not-in-the-solution gadget:

$$V'_{new} := V \cup \{v^1, v^2, v_1, v_2 \mid v \in V\} \cup \{c_i' \mid i \in \{1, \ldots, 3k+1\}\},$$
$$E'^+_{new} := \{vv_1, vv_2, v_1c_1', c_i'c_{i+1}', v_2c_{4k}' \mid v \in V, i \in \{1, \ldots, 3k\}\},$$
$$E'^-_{new} := E'^- .$$

In this way, we get an equivalent construction which has at most $5n + 3k + 1 \leq 8n + 1$ many vertices, and then we have the ETH result by [21]. $\square$

4   Parameterized Complexity and Algorithms

It is known that OFFENSIVE ALLIANCE UG is in FPT when parameterized by solution size [12]. The analogy to signed graphs (as indicated by the previous results) now breaks down, as we show a W[2]-completeness result in Theorem 5. Here, we make use of HITTING SET, a vertex cover problem on hypergraphs.

Problem name: HITTING SET
Given: A hypergraph $\mathcal{H} = (V, E)$ consisting of a vertex set $V = \{v_1, \ldots, v_n\}$ and a hyperedge set $E = \{e_j \subseteq V \mid j \in \{1, \ldots, m\}\}$, and $k \in \mathbb{N}$
Parameter: k
Question: Is there a subset $S \subseteq V$ with $|S| \leq k$ such that for each $e_j \in E, e_j \cap S \neq \emptyset$?

For membership, we use a reduction using the problem SHORT BLIND NON-DETERMINISTIC MULTI-TAPE TURING MACHINE COMPUTATION which was introduced by Cattanéo and Perdrix in [4]. In that paper, they have also shown W[2]-completeness of this Turing machine problem.

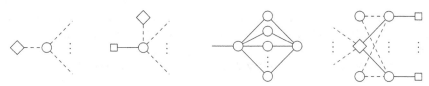

(a) Vertex gadget. (b) Edge gadget. (c) Not-in-the-solution (d) Universal gadget
                                        gadget.   around w.

Fig. 4. Reduction construction for Theorem 5: the diamond represents vertex w; the square represents the not-in-the-solution gadget.

Theorem 5. OFFENSIVE ALLIANCE *is* $W[2]$-*complete when parameterized by solution size.*

Proof. It is well-known that HITTING SET, parameterized by solution size, is $W[2]$-complete. We now describe a reduction that shows: HITTING SET $\leq_{FPT}$ OFFENSIVE ALLIANCE. Let $\langle \mathcal{H} = (V, E), k \rangle$ be any instance of HITTING SET. Then, in polynomial time, one can construct an equivalent instance $\langle G', k' \rangle$ of OFFENSIVE ALLIANCE, with $G' = (V', E'^+, E'^-)$, as follows.

$$M_v := \{v^j \mid j \in \{1, \ldots, 5k\}\},$$
$$V := V \cup \{v_1 \mid v \in V\} \cup \{w, p, q\} \cup \{v_e \mid e \in E\} \cup \bigcup_{e \in E} M_e \cup \bigcup_{v \in V} M_v,$$
$$E'^+ := \{wv_1, v_1v^1, v^1v^l, v^2v^l \mid v \in V, l \in \{3, \ldots, 5k\}\} \cup$$
$$\{v_e e^1, e^1 e^l, e^2 e^l \mid e \in E, l \in \{3, \ldots, 5k\}\},$$
$$E'^- := \{wv, wv_e \mid v \in V, e \in E\} \cup \{vv_e \mid v \in V, e \in E, v \in e\} \cup$$
$$\{pv_1, qv_1 \mid v \in V\},$$
$$k' := k + 3.$$

We first show that if $\langle \mathcal{H}, k \rangle$ is a yes-instance of HITTING SET, then $\langle G', k' \rangle$ is a yes-instance of OFFENSIVE ALLIANCE. If $S \subseteq V$ with $|S| \leq k$ is a hitting set of $\mathcal{H}$, then we will prove that $S' = \{v \mid v \in S\} \cup \{w, p, q\}$ is an offensive alliance. Notice that, because S is a hitting set,

$$\partial S' = \{v_e \mid e \in E\} \cup \{v \mid v \in V \setminus S\} \cup \{v_1 \mid v \in V\}.$$

Moreover, for each v_e, with $e \in E$, $w \in N^-(v_e)$ and $N^+(v_e) = \{e^1\}$, so that $\deg^-_{S'}(v_e) \geq 2 = \deg^+_{\overline{S'}}(v_e) + 1$ and $\deg^+_{S'}(v_e) = 0$. For each v, with $v \in V \setminus S$, we have $\deg^+(v) = 0$, so that $\deg^-_{S'}(v) = 1 = \deg^+_{\overline{S'}}(v) + 1$, and $\deg^+_{S'}(v) = 0$. Finally, for each v_1, with $v \in V$, $N^+(v_1) = \{w, v^1\}$ and $N^-(v_1) = \{p, q\}$, so that $\deg^-_{S'}(v_1) = 2 > 1 = \deg^+_{S'}(v_1)$, and $\deg^-_{S'}(v_1) = 2 = \deg^+_{\overline{S'}}(v_1) + 1$. Hence, S' is an offensive alliance of G with size at most $k' = k + 3$.

Secondly, we show that $\langle G', k' \rangle$ being a yes-instance of OFFENSIVE ALLIANCE implies that $\langle \mathcal{H}, k \rangle$ is a yes-instance of HITTING SET. Details can be found in the appendix.

The proof of membership in W[2] can be also found in the appendix. $\square$

Based on structural hardness results for OFFENSIVE ALLIANCE UG from [14], the reduction presented in Theorem 2 (or a slight variation thereof) shows:

Corollary 2. (*) OFFENSIVE ALLIANCE *on signed graphs is* W[1]-*hard when parameterized by any of the following parameters, defined in the appendix:*

- *feedback vertex set number of the underlying unsigned graph,*
- *treewidth and pathwidth of the underlying unsigned graph,*
- *treedepth of the underlying unsigned graph.*

As the preceding result implies that OFFENSIVE ALLIANCE problem parameterized by treewidth of the input graph is a hard problem, we look at a bigger parameter called domino treewidth. We will show that when parameterized by domino treewidth d, with a suitably given decomposition, the problem of finding smallest offensive alliance on signed graphs is fixed parameter tractable.

A tree decomposition $(T, \{X_t\}_{t \in V_T})$ is a *domino tree decomposition* if, for $i, j \in V_T$, where $i \neq j$ and $\{i, j\} \notin E_T$, we find $X_i \cap X_j = \emptyset$.

In other words, in a domino tree decomposition, every vertex of G appears in at most two bags in T. The *domino treewidth* $\mathsf{dtw}(G)$ is defined as the minimum *width* $\max_{t \in V(T)} |X_t| - 1$ over all domino tree decompositions.

Theorem 6. OFFENSIVE ALLIANCE *on signed graphs is* FPT, *parameterized by the width of a given domino tree decomposition of the underlying unsigned graph.*

Proof. Let $G = (V, E^+, E^-)$ be an instance of OFFENSIVE ALLIANCE and $(T, \{X_t\}_{t \in V_T})$ be a domino tree decomposition of $G_0 = (V, E^+ \cup E^-)$ of width d. We can assume that G is connected, as otherwise we can compute $a_{so}(G)$ by considering the connected components separately. Suppose $T = (V_T, E_T)$ is rooted at node r. For a node $t \in V_T$, let T_t be the subtree of T rooted at t, with vertex set $V(T_t)$, and $V_t = \bigcup_{t' \in V(T_t)} X_{t'}$ be the union of all bags of the subtree. We denote by $p(t)$ the parent node of t.

Let X_t be a non-leaf bag, then a vertex $v \in X_t$ can be of three types.
Type 1: v is also in the bag of one of the children of t;
Type 2: v is also in the bag of the parent of t;
Type 3: v is only in X_t.

The FPT algorithm will be a dynamic program. In this algorithm, we will inductively compute the values $c[t, A_t]$ for $t \in V_T$ and $A_t \subseteq X_t$. For the values of $c[t, A_t]$, we differentiate if $A_t = \emptyset$ or not in the following semantic definition.

If A_t is empty, $c[t, \emptyset]$ denotes the minimum cardinality of an offensive alliance S_t with $S_t \subseteq V_t \setminus X_{p(t)}$.

If $\emptyset \neq A_t \subseteq X_t$, we need more notation: we record a potential offensive alliance (pOA) on the subtree T_t, which is a smallest non-empty set $S_t \subseteq V_t$ with $S_t \cap X_t = A_t$, and for each $u \in \partial S_t \setminus X_{p(t)}$: 1) $\deg_{S_t}^-(u) \geq \deg_{S_t}^+(u)$; 2) $\deg_{S_t}^-(u) \geq \deg_{\overline{S_t}}^+(u) + 1$. That is, the pOA can successfully attack the boundary vertices of S_t except for those falling in the class of Type 2 with regard to X_t. Let $c[t, A_t] = |S_t|$ denote the size of S_t; if no such set exists, we set $c[t, A_t] = \infty$.

We can use bottom-up dynamic programming to obtain these table values:

- If t is a <u>leaf node</u>, then for each $A_t \subseteq X_t$, $A_t \neq \emptyset$, we define $c[t, A_t]$ as follows:

$$c[t, A_t] := \begin{cases} |A_t|, & \text{if } A_t \text{ can successfully attack all vertices in } \partial A_t \setminus X_{p(t)} \\ \infty, & \text{otherwise} \end{cases}$$

- If t is an <u>inner node</u> with children $t_1, t_2, \ldots, t_m$, for each $A_t \subseteq X_t$, $A_t \neq \emptyset$, let $A_{t_j} \subseteq X_{t_j}, j \in \{1, \ldots, m\}$, then A_t is said to be compatible with $A_{t_1}, A_{t_2}, \ldots, A_{t_m}$ if and only if:
  1) For $1 \leq j \leq m$, $A_t \cap X_t \cap X_{t_j} = A_{t_j} \cap X_t \cap X_{t_j}$;
  2) Type 1 and Type 3 vertices of $\partial S_t \cap X_t$ are successfully attacked by $A_t \cup A_{t_1} \cup \cdots \cup A_{t_m}$.
  For $A_t \subseteq X_t$, $A_t \neq \emptyset$, if there does not exist any $A_{t_j} \subseteq X_{t_j}$ that is compatible with A_t, then $c[t, A_t] = \infty$. Otherwise,

$$c[t, A_t] := |A_t| + \min\{ \sum_{j=1, A_{t_j} \neq \emptyset}^{m} (c[t_j, A_{t_j}] - |A_t \cap A_{t_j}|) \mid \tag{1}$$

$$\forall 1 \leq j \leq m, A_{t_j} \subseteq X_{t_j} \implies A_t \text{ is compatible with } A_{t_1}, \ldots, A_{t_m}\}.$$

Similarly the values of $c[t, \emptyset]$ can be computed. For details, see the appendix.

Claim. (∗) For every node t in T and every $A_t \subseteq X_t$, with $A_t \neq \emptyset$, $c[t, A_t]$ is the size of the smallest potential non-empty offensive alliance S_t where $S_t \subseteq V_t$ and $S_t \cap X_t = A_t$, while $c[t, \emptyset]$ is the size of smallest offensive alliance S_t where $S_t \subseteq V_t \setminus X_{p(t)}$ if it exists. Further, the size of the minimum non-empty offensive alliance in G is $c[r, \emptyset]$, where r is the root node of T.

Note that at an inner node, we compute (at most) 2^{d+1} many $c[\cdot, \cdot]$ values and the time needed to compute each of these values is $\mathcal{O}(2^{m(d+1)} \cdot d^2)$. The number of nodes in a domino tree decomposition is $\mathcal{O}(n)$. Also, $m \leq d$, because each child node has disjoint bags but must have common vertices with the bag of its parent since we assume that the signed graph G is connected. Hence, the total running time of the algorithm is $\mathcal{O}(d^2 2^{\mathcal{O}(d^2)} n)$. □

In [1], when studying the defensive alliance theory on signed graphs, a new structural parameter for signed graphs was introduced: The *signed neighborhood diversity* of a graph $G = (V, E^+, E^-)$, denoted by $\mathsf{snd}(G)$, is the least integer k for which we can partition the set V of vertices into k classes, such that for each class C_i, any pair of vertices $\{u, v\} \subseteq C_i$ satisfies $N^+(v) \setminus \{u\} = N^+(u) \setminus \{v\}$ and $N^-(v) \setminus \{u\} = N^-(u) \setminus \{v\}$. If the signed neighborhood diversity of a signed graph is k, there is a partition $\{C_1, C_2, ..., C_k\}$ of V into k equivalence classes. Notice that each class could be a positive clique, a negative clique or an independent set by definition. Moreover, it is not difficult to see that, between any two different classes C_i, C_j, the potential edges $\{u, v\}$ with $u \in C_i$ and $v \in C_j$ either all belong to E^+, or all belong to E^-, or all belong to $\binom{V}{2} \setminus (E^+ \cup E^-)$. Here, we will show that OFFENSIVE ALLIANCE on signed graphs is also FPT when parameterized by signed neighborhood diversity. The idea is to formulate an equivalent Integer Linear Program (ILP), with a linear number of variables in this parameter, and to then resort to [8, Theorem 6.5].

Theorem 7. $(*)$ OFFENSIVE ALLIANCE *is FPT when parameterized by the signed neighborhood diversity.*

For unsigned graphs, the distance to clique is a well-studied parameter. As we cannot expect to have cliques as a simple case for signed graphs, we rather propose the parameters *distance to k-balanced complete graph* and *distance to k-anti-balanced complete graph*. In both cases, technically speaking, we could consider the number k as a fixed constant or as another parameter.

Theorem 8. *Let $\mathcal{P}$ be a problem on signed graphs that can be solved in FPT-time when parameterized by neighborhood diversity. Then, $\mathcal{P}$ can also be solved in FPT-time when parameterized by any of the following parameters: 1) vertex cover number of the underlying graph; 2) distance to k-balanced complete graph and k; 3) distance to k-anti-balanced complete graph and k.*

Proof. Let $G = (V, E^+, E^-)$ be a signed graph. For 1), we refer to [1]. 2) Let $D \subseteq V$ be such that $|D|$ is the distance to k-balanced complete graph and $V - D = \bigcup_{i=1}^{k} V_i$, so that $(V_1, V_2, \dots, V_k)$ is the balanced partition of $G - D$. For each $v \in V_j$, $1 \le j \le k$, its neighbors in $V - D$ are the same, and v has at most $3^{|D|}$ different neighborhoods in D, so that $\mathsf{snd}(G) \le k \cdot 3^{|D|} + |D|$. 3) Let $D' \subseteq V$ be such that $|D'|$ is the distance to k-anti-balanced complete graph. Analogously to 2), we have $\mathsf{snd}(G) \le k \cdot 3^{|D'|} + |D'|$. $\square$

Consequently, we have the FPT results on a bunch of parameters.

Corollary 3. OFFENSIVE ALLIANCE *on signed graphs could be solved in FPT-time when parameterized by any of the parameters mentioned in Theorem 8.*

A natural research question is to find FPT algorithms with better running times for these structural parameters. Given $G = (V, E^+, E^-)$ and some $\ell \ge 0$, it can be decided in FPT-time if $\mathsf{vc}((V, E^+ \cup E^-)) \le \ell$. As the distance to 1-balanced complete graph equals $\mathsf{vc}((V, \binom{V}{2} \setminus E^+))$ and the distance to 1-anti-balanced complete graph equals $\mathsf{vc}((V, \binom{V}{2} \setminus E^-))$, these parameters can be computed in FPT-time, as well. If G is complete, then (by [17]) the distance to 2-balanced complete graph equals the odd cycle transversal number of (V, E^-), which can be computed in FPT-time by [7]. Further cases are still under investigation.

More generally speaking, very little research has been devoted to the parameterized complexity of problems on signed graphs. In particular, good structural parameters, genuinely designed for signed graphs, are lacking.

References

1. Arrighi, E., Feng, Z., Fernau, H., Mann, K., Qi, X., Wolf, P.: Defensive alliances in signed networks. Technical report, Cornell University (2023)
2. Bansal, N., Blum, A., Chawla, S.: Correlation clustering. Mach. Learn. **56**(1–3), 89–113 (2004)
3. Cartwright, D., Harary, F.: Structural balance: a generalization of Heider's theory. Psychol. Rev. **63**(5), 277–293 (1956)
4. Cattanéo, D., Perdrix, S.: The parameterized complexity of domination-type problems and application to linear codes. In: Gopal, T.V., Agrawal, M., Li, A., Cooper, S.B. (eds.) TAMC 2014. LNCS, vol. 8402, pp. 86–103. Springer, Cham (2014). https://doi.org/10.1007/978-3-319-06089-7_7
5. Chellali, M.: Trees with equal global offensive k-alliance and k-domination numbers. Opuscula Mathematica **30**, 249–254 (2010)
6. Chellali, M., Haynes, T.W., Randerath, B., Volkmann, L.: Bounds on the global offensive k-alliance number in graphs. Discussiones Mathematicae Graph Theory **29**(3), 597–613 (2009)
7. Cygan, M., et al.: On problems as hard as CNF-SAT. ACM Trans. Algorithms **12**(3), 41:1–41:24 (2016)
8. Cygan, M., et al.: Parameterized Algorithms. Springer, Heidelberg (2015). https://doi.org/10.1007/978-3-319-21275-3
9. Davis, J.A.: Clustering and structural balance in graphs. Human Relat. **20**, 181–187 (1967)
10. Dourado, M.C., Faria, L., Pizaña, M.A., Rautenbach, D., Szwarcfiter, J.L.: On defensive alliances and strong global offensive alliances. Disc. Appl. Math. **163**(Part 2), 136–141 (2014). https://doi.org/10.1016/j.dam.2013.06.029
11. Favaron, O.: Offensive alliances in graphs. Discussiones Mathematicae - Graph Theory **24**, 263–275 (2002)
12. Fernau, H., Raible, D.: Alliances in graphs: a complexity-theoretic study. In: Leeuwen, J., et al. (eds.) SOFSEM 2007, Proceedings, vol. II, pp. 61–70. Institute of Computer Science ASCR, Prague (2007)
13. Fernau, H., Rodríguez-Velázquez, J.A., Sigarreta, S.M.: Offensive r-alliances in graphs. Disc. Appl. Math. **157**, 177–182 (2009)
14. Gaikwad, A., Maity, S.: On structural parameterizations of the offensive alliance problem. In: Du, D.-Z., Du, D., Wu, C., Xu, D. (eds.) COCOA 2021. LNCS, vol. 13135, pp. 579–586. Springer, Cham (2021). https://doi.org/10.1007/978-3-030-92681-6_45
15. Gaikwad, A., Maity, S.: Offensive alliances in graphs. Technical Report. 2208.02992, Cornell University, ArXiv/CoRR (2022)
16. Gaikwad, A., Maity, S., Tripathi, S.K.: Parameterized complexity of defensive and offensive alliances in graphs. In: Goswami, D., Hoang, T.A. (eds.) ICDCIT 2021. LNCS, vol. 12582, pp. 175–187. Springer, Cham (2021). https://doi.org/10.1007/978-3-030-65621-8_11
17. Harary, F.: On the notion of balance of a signed graph. Michigan Math. J. **2**, 143–146 (1953–54)
18. Harutyunyan, A.: Global offensive alliances in graphs and random graphs. Disc. Appl. Math. **164**, 522–526 (2014)
19. Harutyunyan, A., Legay, S.: Linear time algorithms for weighted offensive and powerful alliances in trees. Theore. Comput. Sci. **582**, 17–26 (2015)

20. Impagliazzo, R., Paturi, R., Zane, F.: Which problems have strongly exponential complexity? J. Comput. Syst. Sci. **63**(4), 512–530 (2001)
21. Johnson, D.S., Szegedy, M.: What are the least tractable instances of Max Independent Set? In: Tarjan, R.E., Warnow, T.J. (eds.) Proceedings of the Tenth Annual ACM-SIAM Symposium on Discrete Algorithms, SODA, pp. 927–928. ACM/SIAM (1999)
22. Lokshtanov, D., Marx, D., Saurabh, S.: Lower bounds based on the exponential time hypothesis. EATCS Bull. **105**, 41–72 (2011)
23. Mohar, B.: Face cover and the genus problem for apex graphs. J. Comb. Theory Ser. B **82**, 102–117 (2001)

Quantum Path Parallelism:
A Circuit-Based Approach to
Text Searching

Simone Faro[1]([⊠])(iD), Arianna Pavone[2](iD), and Caterina Viola[1](iD)

[1] Department of Mathematics and Computer Science, University of Catania,
Catania, Italy
{simone.faro,caterina.viola}@unict.it
[2] Department of Mathematics and Computer Science, University of Palermo,
Palermo, Italy
ariannamaria.pavone@unipa.it

Abstract. Text searching problems are fundamental problems in computer science whose applications are found in a variety of fields. The *string matching problem* is the common framework for the class of text searching problems where the objective is to find all, possibly approximate, occurrences of a given pattern of length m within a larger text of length n. In this paper, we present the *quantum path parallelism* approach, a general strategy based on quantum computation that can be easily adapted to a variety of nonstandard text searching problems. Our method translates a text searching problem to an automata-based string recognition problem, associating each possible approximation of the pattern with a different path accepted by the automaton. Under favourable conditions, our proposed method solves the approximate search problem in $\mathcal{O}(\sqrt{nm}\log^2(n))$ time, offering a quadratic speed-up over classical solutions. In other cases such speed-up is achieved only for short patterns, i.e. assuming $m = \mathcal{O}(\log(n))$. As a demonstration of the flexibility of our method, we show how it can be adapted for solving some approximate string matching problems, obtaining a significant quantum advantage.

1 Introduction

Text processing, and string problems in particular, are among the areas that have recently got attention from quantum computer scientists. Over the years, innovations in tackling these problems have resulted in the development of foundational techniques like dynamic programming, hashing algorithms, automata algorithms and suffix trees. And it is worth noting that in recent times, quantum solutions have emerged as a promising avenue for addressing these problems

S. Faro and C. Viola are supported by the National Centre for HPC, Big Data and Quantum Computing, Project CN00000013, affiliated to Spoke 10., co-founded by the European Union - NextGenerationEU. A. Pavone is supported by PNRR project ITSERR - Italian Strengthening of the ESFRI RI RESILIENCE.

with the potential for significant speed-up. For example, efficient solutions have recently been given for exact string matching [4,14,15,18], approximate string matching [4], string comparison [6], edit distance [3,9], longest common substring [5,10], and longest palindrome substring [10]. These problems find widespread applications across diverse fields, including DNA sequencing, social media analysis, compiler design, antivirus software development, among others.

The object of this paper is the *string matching* problem and its non-standard variants. Formally, given a pattern x of length m and a text y of length n, the string matching problem may consist in (1) verifying if x occurs in y or in (2) finding all occurrences of x in y and their corresponding positions. In the case of approximate matching, we are also interested in (3) identifying the pattern configurations for each of its occurrences in the text.

In a classical model of computation, the exact matching problem has an $\Omega(n)$-time complexity, while the non-standard case can have different solutions depending on the allowed approximations. The use of finite-state automata has often been at the basis of original and efficient solutions [8]. In this direction, the simulation of nondeterministic finite-state automata realized by means of *bit-parallelism* [2] has been one of the fundamental breakthroughs in text searching, where the use of bitwise operations in a word RAM model allows the time complexity to be cut down by a factor equal to the word size.

Two decades later, quantum computation has given a new shake to text searching. The first quantum solution to exact string matching, due to Ramesh and Vinay [18], combines Grover search [12] with a parallel string matching technique. Their algorithm works in the quantum query model [7] and is designed to solve problem (1) making $\tilde{\mathcal{O}}(\sqrt{n})$ queries. Therefore, problem (2) is solved in $\tilde{\mathcal{O}}(\sqrt{\rho n})$ time,[1] where ρ is the number of occurrences of x in y, providing real quantum advantage when $\rho = o(n)$. For the average case, Montanaro [14] proved a super-polynomial separation between quantum and classical complexity. The first solution in the quantum circuit model [19] with a $\tilde{\mathcal{O}}(\sqrt{n})$-time complexity for (1) was presented by Niroula and Nam in 2021 [15], and recently extended [4] to swap matching with a $\tilde{\mathcal{O}}(\sqrt{\rho n 2^m})$ complexity for (3).

We generalize the basic model presented in [15] and introduce the *quantum path parallelism* (QPP) approach, a general strategy based on quantum computation that can be easily adapted to a variety of nonstandard text searching problems. Our method translates the text searching problem to an automata-based string recognition problem, associating each possible approximation of the pattern with a different accepted path in the automaton. Defining an *association rule* that identifies each path of the automaton with a binary string allows for using quantum computation to test, in parallel, the presence of all possible approximations of the pattern for each text position. This results in a quantum algorithm that solves the text searching problem (3) in $\mathcal{O}(\sqrt{\rho n 2^\omega} \log^2(n))$ time, where ω is a value strictly dependent on m and at least equal to the number of strings accepted by the automaton. Under favorable conditions, namely when $\omega = \mathcal{O}(\log(m))$, the time complexity reduces to $\mathcal{O}(\sqrt{nm} \log^2(n))$ for (1) and to

---

[1] Whenever the number of solutions is $\rho > 1$, and we would like to find all of them, $\Theta(\sqrt{\rho n})$ iterations of Grover's search are sufficient and necessary.

$\mathcal{O}(\sqrt{\rho nm}\log^2(n))$ for (3). In the case of short patterns, where one can assume $m = \mathcal{O}(\log(n))$, the complexity reduces to the quasi-linear $\mathcal{O}(n\log^2(n))$ for (1).

Regarding the quantum model that frames our work, it is fair to observe that the efficiency of many quantum algorithms is measured in the *query complexity* model [14,18]. In such a scenario, the input is presented as a *black box* that can be accessed by an *oracle* that, given a function f, returns the image of the input (or other variables depending on it) via f. The query complexity of an algorithm is defined as the number of *queries* that it makes to the oracle(s).

While the query model is intriguing and abstract enough to be valuable for purely theoretical approaches, it holds limited significance when it comes to designing algorithms intended for practical implementation on real hardware, since it is often unclear how one could implement a specific phase oracle efficiently.

On the other hand, there are different quantum models that easily and almost directly translate to concrete implementations on quantum computers. Perhaps, one of the most widespread among such models is the *quantum circuit model* [19], considering that there are several programming languages featuring the circuit formalism, such as IBM's `Qiskit`, Microsoft's `Q#`, and Google's `Cirq`, just to cite a few. In this context, a quantum circuit can be thought of as a Boolean circuit in which the basic gates represent elementary quantum operations, and the computational complexity of a quantum circuit is measured by its *depth*.

For the reasons discussed above we designed our algorithms in the circuit model of computation, rather than on the query complexity model, providing actual implementations of quantum circuits for solving text searching problems.

Regarding the quantum access to any input string, in many solutions based on the quantum query model [10,18] it is assumed a QRAM (Quantum Random Access Memory) model [17], where any input string s can be directly accessed by a random oracle at unit cost. However, the most efficient QRAM designs [11] exhibit a polylogarithmic time complexity for accessing the memory with respect to its size. In our scenario, the memory size is $\mathcal{O}(n)$, which implies that QRAM queries require $\mathcal{O}(\log^2(n))$ time. Howbeit, our algorithm does not rely on a random access oracle, and we assume that the input strings are available in quantum registers accessible by the circuit and which do not require any initialization.

As a demonstration of the flexibility of our method, we show how it can be adapted for solving some approximate text searching problems obtaining a significant quantum advantage. Specifically, we solved the string matching problems allowing for swaps and cyclic rotations[2]. Table 1 summarizes the time complexities obtained in this paper through the QPP approach in comparison with their respective classic solutions, which are generally designed to solve only problem (1). Problem (3) can be solved at the additional cost of $\mathcal{O}(\rho m)$.

Although in the worst case (i.e. when $\rho = n$) the quantum advantage is obtained only under suitable conditions, in the average case, when ρ tends to be a constant, the quantum advantage is obtained in many cases. Furthermore, many non-standard text searching techniques, that are highly efficient in practi-

---

[2] Our method can also be applied to solve the text searching problem with gaps.

Table 1. Some non-standard text searching problems, and their classical and QPP time complexity, where n and m are the length of the input strings, α is the approximation bound and ρ is the number of occurrences of x in y.

PROBLEM	CLASSICAL SOLUTION (1)	QPP SOLUTION (1)	QPP SOLUTION (3)
exact	$\mathcal{O}(n)$	$\mathcal{O}(\sqrt{n}\log^2(n))$	$\mathcal{O}(\sqrt{\rho n}\log^2(n))$
circular rotations	$\mathcal{O}(n)$	$\mathcal{O}(\sqrt{n}\log^2(n))$	$\mathcal{O}(\sqrt{\rho n}\log^2(n))$
swaps (unbounded)	$\mathcal{O}(n\sqrt{m}\log(m))$	$\mathcal{O}(\sqrt{n2^m}\log^2(n))$	$\mathcal{O}(\sqrt{\rho n2^m}\log^2(n))$
swaps (α-bounded)	$\mathcal{O}(n\sqrt{m}\log(m))$	$\mathcal{O}(\sqrt{nm^\alpha}\log^2(n))$	$\mathcal{O}(\sqrt{\rho nm^\alpha}\log^2(n))$
gaps (unbounded)	$\mathcal{O}(nm)$	$\mathcal{O}(\sqrt{n2^m}\log^2(n))$	$\mathcal{O}(\sqrt{\rho n2^m}\log^2(n))$
gaps (α-bounded)	$\mathcal{O}(nm)$	$\mathcal{O}(\sqrt{nm^\alpha}\log^2(n))$	$\mathcal{O}(\sqrt{\rho nm^\alpha}\log^2(n))$

cal implementation, and especially those implementing bit-parallelism, are based on filtering approaches, searching for a constant-length substring of the pattern(usually $m \leqslant 32$ is used). Under such conditions, ω is a constant and our approach requires $\tilde{\mathcal{O}}(\sqrt{n})$-time on average, yielding a real quantum advantage in all cases. However, an analysis of average cases goes beyond the purpose of this paper.

2   Quantum Path Parallelism

Nonstandard text searching concerns string matching problems where a substring of the text matches the pattern, based on general relationships between the corresponding symbols. Such a relationship can be manifold. To fit all the variants of the text searching problem in a single general framework, we assume the existence of a metric function for assessing how similar two strings are.

A *string metric* (or *string distance*) is a function $d : \Sigma^* \times \Sigma^* \to \mathbb{N}$ that measures the distance (or the *inverse similarity*) between two strings for approximate string matching. Given two strings $x, y \in \Sigma^*$, we generally assume that $d(x, y) = 0$ if they are equal, while $d(x, y) > 0$ in all other cases. Many formulations of the approximate string matching problem provide an upper bound, $k \geqslant 0$, so that x matches y if $d(x, y) \leqslant k$.

Given a string $x \in \Sigma^m$, a distance function d and a bound k, the *string matching automaton* associated with x, d and k, is a deterministic acyclic finite state automaton (DFA), denoted by the symbol $\mathcal{A}_{(x,d,k)}$, that recognizes the language $\mathcal{L} = \{y \in \Sigma^* : d(x, y) \leqslant k\}$. Formally the string matching DFA $\mathcal{A}_{(x,d,k)}$ is defined by the tuple $\mathcal{A} \equiv \{Q, \Sigma, I, F, \delta\}$, where Q is the set of states of the automaton, Σ is the set of input symbols, $I \in Q$ is the initial state, $F \subseteq Q$ is the set of final states and $\delta : Q \times \Sigma \to Q$ is the transition function. In this context we also make use of the generalized transition function $\delta^* : \Sigma^* \to Q$, which is recursively defined as $\delta^*(x.c) = \delta(\delta^*(x), c)$, for any $x \in \Sigma^*$ and $c \in \Sigma$, and $\delta^*(\varepsilon) = I$, for the base case. Along this paper, when the values of x, d, and k are clearly defined by context, we use the symbol $\mathcal{A}$ in place of $\mathcal{A}_{(x,d,k)}$. By definition, if the pattern occurrence is detected at position j of the text then we have that $d(y[j..j + r - 1], x) \leqslant k$, where the value r, which is the length of the *window*, i.e. the portion of the text recognized as an occurrences of the pattern, can vary within a limited range of values dependent on the function d and the bound k. For example, in exact string matching, as well as in swap string matching, we have $r = m$ for all possible occurrences. Differently, in the approximate string matching problem allowing for the presence of at most α gaps we have $m - \alpha \leqslant r \leqslant m$. We use the symbol $\mathcal{R}$ to indicate the set of all allowable lengths for any occurrence of the pattern in the text.

Since a quantum register of dimension $\log(n)$, like any binary sequence of the same dimension, can take on all values between 0 and $n - 1$, for simplicity we will assume that the text y has length $n = 2^p$, for some $p > 0$. We also assume that the text ends with a special character \$ not belonging to the alphabet Σ. These two assumptions can be made without loss of generality since it would suffice to take the smallest value p for which we have $n < 2^p$ and concatenate the text with $2^p - n$ copies of the symbol \$.

Thus, the resulting string has at most twice the length of the original string.

We also observe that any occurrence of length r can begin at any position j of the text, for $0 \leqslant j \leqslant n - r$. However, in this paper we also relax this condition by admitting values of j between 0 and $n - 1$ and assuming that a substring of the text can be obtained circularly. This assumption can also be made without loss of generality since the last character of the text is always the special character \$ that does not occur within the pattern.

Based on the assumptions above, the text searching problem, defined as an approach based on the use of the string matching DFA, can be defined as follows.

Definition 1 (Text-Searching Based on a String Matching DFA).
Let y be a text of length n and let x be a pattern of length m. Moreover, let $d : \Sigma^* \times \Sigma^* \rightarrow N$ be a distance function, let $k \geqslant 0$ be an approximation bound and let $\mathcal{R}$ be the set of all possible lengths of any occurrence of x in y. Finally, let $\mathcal{A} \equiv \{Q, \Sigma, I, F, \delta\}$ be the string matching DFA constructed on x. The text searching problem consists in finding all positions at which a substring of y with length $r \in \mathcal{R}$ accepted by $\mathcal{A}$ begins. More formally, we want to find the elements of the set $\Gamma(x, y)^{(\mathcal{R})} = \bigcup_{r \in \mathcal{R}} \Gamma(x, y)^{(r)}$, where

$$\Gamma(x, y)^{(r)} = \{0 \leqslant j < n \; : \; \delta^*(y[j..j + r - 1]) \in F\}. \tag{1}$$

2.1   The Ideas Behind the General Algorithm

A DFA can be seen, to some extent, as a generalization to acyclic graphs. In this context, graph nodes correspond to automaton states, and an edge labelled by c between two nodes, q_i and q_j, exists if $\delta(q_i, c) = q_j$. It is therefore possible to extend, in a natural way, the concept of a path to a DFA. The following definition underlies the idea on which the path parallelism approach is based.

Definition 2 (Complete Automaton Path). *Given a string matching DFA $\mathcal{A} \equiv \{Q, \Sigma, I, F, \delta\}$, defined for a string x, a distance function d and a bound k,* *we say that a path in the automaton is complete if it starts at the initial state I and ends at a final state. More formally, a complete automaton path of length r is a sequence of states* $\langle q_0, q_1, \ldots, q_{r-1} \rangle$, *where* $q_j \in Q$ *for* $0 \leqslant j < r$, $q_0 = I$, $q_{r-1} \in F$ *and* $\delta(q_j, c) = q_{j+1}$ *for a symbol* $c \in \Sigma$ *and* $0 \leqslant j < r - 1$.

We will denote by the symbol PATH($\mathcal{A}$) the set of all possible complete paths in $\mathcal{A}$. Any complete automaton path $p \in$ PATH($\mathcal{A}$) naturally induces a sequence of symbols associated with it and which we call *path labeling* of p. It roughly consists of the sequence of symbols that label the transitions in p. More formally, we get the following definition

Definition 3 (Path Labeling). *Let* $\mathcal{A} \equiv \{Q, \Sigma, I, F, \delta\}$ *be a string matching DFA, the path labeling of a path* $p = \langle q_0, q_1, \ldots, q_{r-1} \rangle$ *of length* r, *with* $p \in$ PATH($\mathcal{A}$), *is the sequence* $x_p = \langle c_0, c_1, \ldots, c_{r-2} \rangle$, *where* $c_i \in \Sigma$ *and* $\delta(q_i, c_i) = q_{i+1}$, *for* $0 \leqslant j < r - 1$.

Although the QPP approach can be adapted to handle most variants of nonstandard text searching, for lack of space in this paper we focus on the cases where a path labeling is a permutation of the sequence of characters in x, or at any rate in a subset of such sequence. We denote the path labeling of a path p by the symbol x_p to emphasize the fact that x_p is a sequence of symbols obtained from x by applying the modifications induced by p.

In light of Definition 2 and Definition 3, and because we are interested in the configuration of the occurrences, Eq. (1) can be rewritten as

$$\Gamma(x, y)^{(r)} = \{(j, p) \ : \ 0 \leqslant j < n \ \text{and} \ p \in \text{PATH}(\mathcal{A}) \ \text{and} \ y[j..j+r-1] = x_p\}. \quad (2)$$

Each complete path p of $\mathcal{A}$ is associated with a *path code*, i.e. a binary sequence π that uniquely identifies it within the set PATH($\mathcal{A}$).

Definition 4 (Path Codes Association Rule). *Let* $\mathcal{A}$ *be a string matching DFA, and* ω *a constant value. The path codes association rule defines a two-way association between the elements of the set* PATH($\mathcal{A}$) *and a subset* $\Pi_{\mathcal{A}} \subseteq \{0, 1\}^\omega$ *such that each complete path of* $\mathcal{A}$ *is uniquely identified by a code of length* ω.

We say that a binary sequence π of length ω is a *valid path code* if it is associated with a complete path in $\mathcal{A}$, i.e. if $\pi \in \Pi_{\mathcal{A}}$, where we use $\Pi_{\mathcal{A}}$ (or simply Π, if $\mathcal{A}$ is clear by context) to denote the set of all valid path codes of the automaton. If we denote by π_p the path code associated with p, we have $\Pi_{\mathcal{A}} = \{\pi_p \ : \ p \in \text{PATH}(\mathcal{A})\}$. Observe that $\Pi_{\mathcal{A}} \subseteq \{0, 1\}^\omega$, and since all path codes share the same length ω, we need $\omega \geqslant \log_2(|\text{PATH}(\mathcal{A})|)$.

Since there is a one-to-one correspondence between the sets $\Pi_{\mathcal{A}}$ and PATH(A), in what follows we will interchangeably use the notations x_p and x_π to denote the same path labeling, which is induced by p, whenever π is the path code of p.

The implementation of the QPP approach requires the following definitions of a transformation function and a test function.

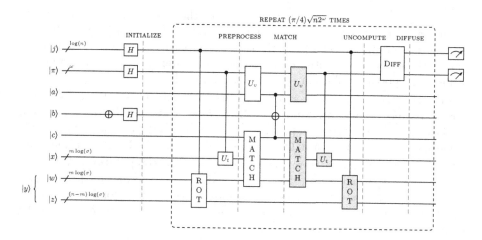

Fig. 1. A simplified diagram describing the circuit of the general algorithm based on the QPP approach for non-standard text searching.

Definition 5 (Transformation Function). *Let $\mathcal{A}$ be a string matching DFA defined for a string x of length m, a distance function d and a bound k, and let Π the set of all valid path codes defined by a given association rule. The transformation function, for any occurrence of length $r \in \mathcal{R}$, is a function $t : \Sigma^m \times \Pi \to \Sigma^r$, such that $t(x, \pi) = x_\pi$ for every $\pi \in \Pi$.*

Roughly speaking such a function maps the sequence of characters in x into the sequence of characters in x_π, applying the transformations induced by the path p associated with π.

Definition 6 (Test Function). *Let $\mathcal{A}$ be a string matching DFA defined for a string x, a distance function d and a bound k, and let Π be the set of all valid path codes defined by a given association rule. The test function is a map $v : \{0,1\}^\omega \to \{0,1\}$, such that, for any $\pi \in \{0,1\}^\omega$, $v(\pi) = 1$ iff $\pi \in \Pi_\mathcal{A}$.*

In other words the test function checks whether a path code $\pi \in \{0,1\}^\omega$ is valid. The definitions introduced allows us to rewrite Eq. (2) as follows.

$$\Gamma(x,y)^{(r)} = \{(j,\pi) \ : \ 0 \leqslant j < n, \ \pi \in \{0,1\}^\omega, \ y[j..j+r-1] = x_\pi \text{ and } v(\pi) = 1\}. \tag{3}$$

2.2   The General Quantum Algorithm

In this section, we describe how Eq. (3) translates into a quantum algorithm. Figure 1 provides a simplified diagram of the circuit that implements the QPP algorithm.

The circuit consists of the initialization, the preprocessing and the matching phases. The preprocessing and the match phases, with the subsequent uncomputing phase, implement the phase oracle of the function γ.

The register $|j\rangle$, of size $\log(n)$, is used to hold the value of any position j where a substring of the text may begin. The register starts with the value $|0\rangle^{\log(n)}$ and is then initialized to $|+\rangle^{\log(n)}$ through the application of $\log(n)$ Hadamard's gates. At the end of the initialization phase the register $|j\rangle$ contains the superposition of all possible values in $\{0, ..., n-1\}$. Formally $|j\rangle = \frac{1}{\sqrt{n}} \sum_{i \in \{0,1\}^{\log(n)}} |i\rangle$.

The register $|\pi\rangle$ has size ω and retains the path codes of the DFA constructed for the pattern x. Starting from $|0\rangle^{\omega}$, the register $|\pi\rangle$ is initialized to $|+\rangle^{\omega}$ and, at the end of the initialization phase, it contains the superposition of all possible values in $\{0, ..., 2^{\omega}-1\}$. Formally, $|\pi\rangle = \frac{1}{\sqrt{2^{\omega}}} \sum_{i \in \{0,1\}^{\omega}} |i\rangle$. Registers $|a\rangle$, $|b\rangle$, and $|c\rangle$ are single qubits and store the output of specific computations. While $|a\rangle$ and $|c\rangle$ are initialized to $|0\rangle$, $|b\rangle$ is initialized to $|-\rangle$. The register $|x\rangle$ has size $m \log(\sigma)$ and can be viewed as a sequence of m registers x_i (for $0 \leqslant i < m$) of size $\log(\sigma)$, where $\sigma = |\Sigma|$. Each $|x_i\rangle$ is initialized with the $\log(\sigma)$ bits of the binary representation of the i-th symbol of the pattern.

Similarly, the register $|y\rangle$ has size $n \log(\sigma)$ and can be viewed as a sequence of n registers y_i of size $\log(\sigma)$, such that $|y_i\rangle$ is initialized with the $\log(\sigma)$ bits of the binary representation of the i-th character of the text, for $0 \leqslant i < n$. For visualization purposes, in Fig. 1, the $|y\rangle$ register is divided into two registers, $|w\rangle$ and $|z\rangle$, of size $m \log(\sigma)$ and $(n-m) \log(\sigma)$, respectively. Each of the operators applied during this phase takes constant time, and since they can be applied in parallel, the initialization phase takes $O(1)$ time.

The preprocessing phase prepares the registers $|x\rangle$ and $|y\rangle$ for the upcoming match phase. The first step is to apply a *circular rotation operator* [16] (identified as ROT in Fig. 1) controlled by the register $|j\rangle$ to the register $|y\rangle$. Formally, the ROT gate applies a leftward circular rotation to $|y\rangle$ by $j \log(\sigma)$ positions, realizing the mapping $\text{ROT}|y_0, y_1, ..., y_{n-1}\rangle = |y_j, y_{j+1}, ..., y_{n-1}, y_0, y_2, ..., y_{j-1}\rangle$. In this context, we refer to j as the value identified by the $|j\rangle$ register of size $\log(n)$. Since $|j\rangle$ is initialized to the superposition of all possible values between 0 and $n-1$, at the end of the application of the ROT gate, the register $|y\rangle$ contains the superposition of all possible circular rotations of the text. Since the register $|w\rangle$ contains the first r symbols of $|y\rangle$, at the end of preprocessing it will contain the superposition of all circular substrings of the text with length r.

Although it is always possible to construct a quantum operator able to cyclically rotate a quantum register by a fixed number of positions with logarithmic depth [16], our operator must be able to rotate a register by a number of positions equal to the value stored in the register $|j\rangle$. We can prove that such operator can be implemented in depth $\mathcal{O}(\log^2(n))$ using only $\log(n)$ ancillæ qubits [15].

During the preprocessing phase we also apply the operator U_t, which is the boolean oracle implementing the transformation function $t: \Sigma^m \times \Pi \rightarrow \Sigma^r$. Note that U_t is controlled by the register $|\pi\rangle$, representing the path code associated with a specific path of the automaton. For all intents and purposes, the path code $|\pi\rangle$ governs the transformations to be applied to the sequence of characters in $|x\rangle$, depending on the type of approximation problem being considered. Formally, the boolean oracle U_t applies the transformation $U_t|\pi, x\rangle = |\pi, x_{\pi}\rangle$. Thus the register $|x\rangle$ will contain the superposition of the strings x_{π}, for $\pi \in \{0,1\}^{\omega}$.

After the preprocessing phase, the registers $|y\rangle$, $|w\rangle$ and $|x\rangle$ are as follows

$$|y\rangle = \frac{1}{\sqrt{n}} \sum_{j=0}^{n-1} \left[\bigotimes_{i=0}^{n-1} |y_{(j+i)}\rangle \right], \quad |w\rangle = \frac{1}{\sqrt{n}} \sum_{j=0}^{n-1} \left[\bigotimes_{i=0}^{r-1} |y_{(j+i)}\rangle \right], \quad |x\rangle = \frac{1}{\sqrt{2^\omega}} \sum_{\pi \in \{0,1\}^\omega} |x_\pi\rangle,$$

where the sums $(j+i)$ are computed cyclically, that is, modulo n.

If $D_t(\omega, m)$ is the depth of the circuit U_t, then the time complexity of the preprocessing phase is $\mathcal{O}\left(\max(\log^2(n), D_t(\omega, m))\right)$.

The match phase consists of the joint application of two operators allowing to check if the string w_j (for $0 \leqslant j < n$) matches the string x_π, for some $\pi \in \Pi$. The first operator applied in this phase is the boolean oracle U_v that implements the test function v. Such boolean oracle checks whether a given string $\pi \in \{0,1\}^\omega$ is a valid path code. Formally, for any $\pi \in \{0,1\}^\omega$, we have $U_v|\pi, 0\rangle = |\pi, 1\rangle$ if $\pi \in \Pi$, and $U_v|\pi, 0\rangle = |\pi, 0\rangle$ otherwise. We denote by $D_v(\omega)$ the depth of the circuit implementing the boolean oracle U_v.

The second operator (denoted by MATCH in Fig. 1) compares the two strings w_j and x_π in parallel, for all values $0 \leqslant j < n$ and $0 \leqslant \pi < 2^\omega$. Although the two strings are compared character by character, the comparisons between the two strings can be performed in parallel. More precisely, for $0 \leqslant i < r$, the characters $(w_j)_i$ and $(x_\pi)_i$ are compared in $O(\log(\sigma))$ time by means of $\log(\sigma)$ parallel CNOT gates and a multiple CNOT gate with $\log(\sigma)$ controls. Specifically, once the r characters of the two strings have been compared, the final comparison of the two strings can be performed in $O(\log(r))$ time by means of a multiple CNOT gate with r controls. Thus, since $r \in \mathcal{O}(m)$, the overall time complexity of the MATCH operator is $\mathcal{O}(\log(\sigma) + \log(m))$. After the match phase we have MATCH$|0, x_\pi, w_j\rangle = |1, x_\pi, w_j\rangle$ if $x_\pi = w_j$, and MATCH$|0, x_\pi, w_j\rangle = |0, x_\pi, w_j\rangle$ otherwise. The overall time complexity of the match phase is $\mathcal{O}\left(\max(\log(\sigma) + \log(m), D_v(\omega))\right)$.

The outputs of U_v and MATCH are stored into the qubits $|a\rangle$ and $|c\rangle$, respectively, and are finally combined by a CCNOT gate whose output is stored into the register $|b\rangle$. Since the preprocessing and matching phases must operate jointly within the phase oracle of the γ function, the output register $|b\rangle$ is initialized to $|-\rangle$, so that at the end of the phase we have

$$|b\rangle = \frac{1}{\sqrt{n2^\omega}} \sum_{\substack{0 \leqslant \pi < 2^\omega \\ 0 \leqslant j < n}} (-1)^{\gamma(x_\pi, w_j)} \left(|j\rangle \otimes |\pi\rangle \right).$$

The operators applied in the preprocessing and matching phases implement the boolean oracle of the function γ. The fact that the register $|b\rangle$, which holds the result of the computation at the end of the process, is initialized to $|-\rangle$ causes the computation to result in a phase oracle of the function γ. The computation ends with an *uncompute* process obtained by applying the inverses of the operators used in the previous two phases.

The search for a solution of the function γ is implemented through an application of Grover's algorithm. Multiple iterations of the entire computation process, followed by the application of the Grover's Diffuser to registers $|k\rangle$ and $|\pi\rangle$, may then be required. The following result on the complexity of the QPP algorithm holds.

Theorem 1. *The QPP algorithm, for searching a pattern of length m on a text of length n, both strings over an alphabet Σ of size σ, and with a path code size equal to ω, requires $\mathcal{O}(\sqrt{n2^\omega}(\max(\log^2(n), D_t(m)) + \max(\log(m) + \log(\sigma), D_v(m)) + \log(\log(m) + \omega))$-time.*

With reference to Theorem 1, the assumption that can be made most reliably is that the dominant term within the second multiplicative factor is $\log^2(n)$. This reduces, in most cases, the complexity to $\mathcal{O}(\sqrt{n2^\omega}\log^2(n))$. In some favourable cases, we are allowed to use a set of path codes of size $\omega = \mathcal{O}(\log(m))$, reducing the complexity to the sub-linear $\mathcal{O}(\sqrt{nm}\log^2(n))$ time. Alternatively, under the assumption that $\omega = \mathcal{O}(\log(n))$ (i.e. for short patterns) the running time of the algorithm reduces to the quasi-linear $O(n\log^2(n))$ time.

3   Solving Some Nonstandard Text Searching Problems

In this section we adapt the general algorithm presented in Sect. 2 to solve two nonstandard string matching problems, and specifically the string matching with cyclic rotations and the string matching with bounded and unbounded swaps. For space constraints, we present only the basic aspects of the proposed solutions without providing full details or proving some of our considerations.

In the *circular string matching* problem [13] (or string matching allowing for rotations) the pattern is a cyclic string, meaning that it can occur within the text in its original form or after undergoing a cyclic rotation. In the standard model of computation, the problem can be solved in $\mathcal{O}(n)$ time [13].

A cyclic string of length m has exactly m different configurations, thus the automaton for detecting all cyclic rotations of a pattern x can be obtained by triggering, at the initial state of the automaton, m distinct parallel paths such that the i-th path is labeled with the sequence of characters obtained from the leftward cyclic rotation of the pattern by s positions, for $0 \leqslant s < m$.

In this context, a path code can be defined as a binary sequence of dimension $\omega = \log(m)$ containing the binary representation of the amplitude s of the cyclic rotation applied to the pattern, with $0 \leqslant s < m$.

The corresponding transformation function t, takes a binary sequence $\pi \in \{0,1\}^{\log(m)}$ and maps a string x of length m to its leftward cyclic rotation by s positions, where $s = \sum_{i=0}^{\log(m)} \pi_i 2^i$. The boolean oracle U_t of the transformation function t can be implemented by the cyclic rotation operator [16], while the test function v can be dropped since any path code π consisting of $\log(m)$ bits represents a valid rotation amount, i.e. $v(\pi) = 1$ for any $\pi \in \{0,1\}^{\log(m)}$.

Since $2^\omega = m$, the resulting algorithm for solving problem (1) of cyclic string matching achieves $\mathcal{O}(\sqrt{nm}(\log^2(n) + \log(\sigma) + \log^2(m)))$ complexity, which reduces to $\mathcal{O}(\sqrt{nm}\log^2(n))$. In solving problem (3) we achieve $\mathcal{O}(\sqrt{\rho nm}\log^2(n))$-time complexity, yielding a quantum advantage for $\rho = o(n/m)$.

In the *string matching problem allowing for swaps* [1] (or *swap string matching*) we are interested in finding all occurrences of a pattern within a text by allowing two adjacent characters of the pattern to be swapped, assuming that each character can be involved in a single swap.

In the classical model of computation, we can find all ρ matching positions in $\mathcal{O}(n\sqrt[3]{m}\log(m)\log(\sigma))$ time [1]. But we require $\mathcal{O}(\rho m)$ additional time to retrieve all matching configurations.

For the case in which the number of allowed swaps is unbounded, we can define an association rule that associates each swap configuration of the pattern with a binary code of length $w = m - 1$, on the same line as what has been done in [4]. It would therefore be easy to get an algorithm with $\mathcal{O}(\sqrt{n2^m}\log^2(m))$-time complexity for problem (1), obtaining a quantum advantage for $m = O(\log(n))$. In the case of problem (3) we achieve a $\mathcal{O}(\sqrt{\rho n2^m}\log^2(m))$-time complexity, with a quantum advantage for $\rho = o(n)$ and $m = O(\log(n))$.

We now consider the α-bounded variant of the problem, where an upper bound $\alpha > 0$ is given to the number of swaps involved in a match. We first analyze in more detail the simplest case in which $\alpha = 1$, and discuss later how to generalize it to the cases with $\alpha > 1$.

Let $\mathcal{A}$ be the DFA for the 1-bounded swap matching problem constructed over a pattern of length m. It is easy to observe that the number of complete paths of $\mathcal{A}$ is equal to m. Thus, the set Π of all path codes is defined so that any $p \in \text{PATH}(\mathcal{A})$ is associated to a binary sequence $\pi \in \{0,1\}^{\log(m)}$, which identifies (in binary notation) the position s of the unique swap (if any) in the pattern. Formally, given the value $s = \sum_{i=0}^{\log(m)}\pi_i 2^i$, the characters $x[s]$ and $x[s+1]$ need to be swapped as to match the corresponding substring of the text. Since the value $m - 1$ does not identify a valid location for a swap, we agree that this value is used for the case where no swap is present in the text substring.

Accordingly, the corresponding transformation function t maps a pair (π, x), where π is a binary string of length $\log(m)$ and x is a string of length m, to the string x_π of length $r = m$ such that $x_\pi[i] = x[i]$ if $i \notin \{s, s+1\}$, $x_\pi[s] = x[s+1]$ and $x_\pi[s+1] = x[s]$.

The boolean oracle U_t implementing the transformation function t acts on the pattern register $|x\rangle$ by swapping the qubits $|x_s\rangle$ and $|x_{s+1}\rangle$ whenever the path code $|\pi\rangle$ contains the value s. Such a transformation can be obtained through $\log(m)$ applications of the rotation operator [16], controlled by the $|\pi\rangle$ register, which acts on an additional register of m ancillæ qubits. The latter is then used to control the swap operations, executable in constant time. It is easy to check that the overall depth of U_t is $\mathcal{O}(\log^2(m))$.

Since any path code π consisting of $\log(m)$ bits represents a valid swap position between 0 and $m-1$, the test function v can be dropped again, that is, it can be implemented as the constant function which returns 1 for any $\pi \in \{0,1\}^{\log(m)}$.

In the general case where $\alpha > 0$, the problem can be solved by considering π as the concatenation of α binary sequences, $\langle \pi_0, \pi_1, ..., \pi_{\alpha-1}\rangle$, all of size $\log(m)$, and where π_i identifies the position of the i-th swap, with $0 \leqslant i < \alpha$. A path codes is then a binary sequence $\pi \in \{0,1\}^{\alpha\log(m)}$. It is agreed that two sequences

π_i and π_j, with $i \neq j$, must represent different positions in the pattern, with the exception of position $m - 1$, which identifies the case where the correspondent swap operation does not apply. This allows for a number of swaps between 0 and α. Specifically, the case where the occurrence of the text involves no swap operation is obtained by posing $\pi_i = m - 1$, for $0 \leqslant i < \alpha$. Therefore, the boolean oracle U_t implementing the transformation function t has a depth equal to $\mathcal{O}(\alpha \log^2(m))$. The test function υ must check for the presence of a pair of values i and j, for which $i \neq j$ and $\pi_i = \pi_j$. This verification can be done for each pair in time $\log(\log(m))$. Since it must be done for each pair of values, the full check will take $\mathcal{O}(m^2 \log(\log(m)))$ time.

The resulting algorithm has time-complexity $\mathcal{O}(\sqrt{nm^\alpha}(\log^2(n) + \log^2(m) + m^2 \log\log(m)))$ for problem (1), which reduces to $O(\sqrt{nm^\alpha} \log^2(n))$ time, yielding a quantum advantage over the classical $\mathcal{O}(n\sqrt[3]{m} \log(m) \log(\sigma))$ solution. In the case of problem (3) the algorithm achieves an $O(\sqrt{\sqrt{\rho n m^\alpha}} \log^2(n))$ time, obtaining a quantum advantage for $\rho = o(n)$ or for $\alpha \log(m) = o(m)$.

References

1. Amir, A., Aumann, Y., Landau, G.M., Lewenstein, M., Lewenstein, N.: Pattern matching with swaps. J. Algorithms **37**(2), 247–266 (2000). https://doi.org/10.1006/jagm.2000.1120

2. Baeza-Yates, R.A., Gonnet, G.H.: A new approach to text searching. Commun. ACM **35**(10), 74–82 (1992). https://doi.org/10.1145/135239.135243

3. Boroujeni, M., Ehsani, S., Ghodsi, M., Hajiaghayi, M., Seddighin, S.: Approximating edit distance in truly subquadratic time: quantum and mapreduce. J. ACM **68**(3) (2021). https://doi.org/10.1145/3456807

4. Cantone, D., Faro, S., Pavone, A.: Quantum string matching unfolded and extended. In: Kutrib, M., Meyer, U. (eds.) RC 2023. LNCS, vol. 13960, pp. 117–133. Springer, Cham (2023). https://doi.org/10.1007/978-3-031-38100-3_9

5. Cantone, D., Faro, S., Pavone, A., Viola, C.: Longest common substring and longest palindromic substring in $\tilde{O}(\sqrt{n})$ time. CoRR, abs/2309.01250 (2023). arXiv:2309.01250

6. Cantone, D., Faro, S., Pavone, A., Viola, C.: Quantum circuits for fixed substring matching problems. CoRR, abs/2308.11758 (2023). arXiv:2308.11758

7. Cleve, R.: An introduction to quantum complexity theory. In: Quantum Computation and Quantum Information Theory, pp. 103–127. World Scientific (2001). https://doi.org/10.1142/9789810248185_0004

8. Crochemore, M., Rytter, W.: Text Algorithms. Oxford University (1994)

9. Equi, M., van de Griend, A.M., Mäkinen, V.: Quantum linear algorithm for edit distance using the word qram model (2023). arXiv:2112.13005

10. Le Gall, F., Seddighin, S.: Quantum meets fine-grained complexity: sublinear time quantum algorithms for string problems. Algorithmica **85**(5), 1251–1286 (2023). https://doi.org/10.1007/s00453-022-01066-z

11. Giovannetti, V., Lloyd, S., Maccone, L.: Quantum random access memory. Phys. Rev. Lett. **100**(16) (2008). https://doi.org/10.1103/physrevlett.100.160501

12. Grover, L.K.: A fast quantum mechanical algorithm for database search. In: Proceedings of the Twenty-Eighth Annual ACM Symposium on Theory of Computing, STOC 1996, pp. 212–219. ACM (1996). https://doi.org/10.1145/237814.237866

13. Lothaire, M.: Applied Combinatorics on Words. Cambridge University Press, Cambridge (2005)
14. Montanaro, A.: Quantum pattern matching fast on average. Algorithmica **77**(1), 16–39 (2017). https://doi.org/10.1007/s00453-015-0060-4
15. Niroula, P., Nam, Y.: A quantum algorithm for string matching. NPJ Quantum Inf. **7**, 37 (2021). https://doi.org/10.1038/s41534-021-00369-3
16. Pavone, A., Viola, C.: The quantum cyclic rotation gate. In: Proceedings of the 24th Italian Conference on Theoretical Computer Science (ITCTS 2023), Italy, 13–15 September 2023, vol. 3587, pp. 206–218. University of Palermo (2023)
17. Phalak, K., Chatterjee, A., Ghosh, S.: Quantum random access memory for dummies (2023). arXiv:2305.01178
18. Ramesh, H., Vinay, V.: String matching in $\mathcal{O}(n + m)$ quantum time. J. Discret. Algorithms **1**(1), 103–110 (2003). https://doi.org/10.1016/S1570-8667(03)00010-8
19. Yao, A.C.-C.: Quantum circuit complexity. In: 34th Annual Symposium on Foundations of Computer Science, pp. 352–361. IEEE Computer Society (1993). https://doi.org/10.1109/SFCS.1993.366852

Space-Efficient Graph Kernelizations

Frank Kammer[iD] and Andrej Sajenko[(✉)][iD]

THM, University of Applied Sciences Mittelhessen, Giessen, Germany
{Frank.Kammer,Andrej.Sajenko}@mni.thm.de

Abstract. Let n be the size of a parameterized problem and k the parameter. We present kernels for FEEDBACK VERTEX SET and PATH CONTRACTION whose sizes are all polynomial in k and that are computable in polynomial time and with $O(\text{poly}(k) \log n)$ bits (of working memory). By using kernel cascades, we obtain the best known kernels in polynomial time with $O(\text{poly}(k) \log n)$ bits.

Keywords: path contraction · feedback vertex set · space-efficient algorithm · full kernel

1 Introduction

With the rise of big data the focus on algorithms that treat space as a valuable resource becomes increasingly important. Large inputs may cause "standard" solutions to fail their execution due to out-of-memory errors, or cause them to spend a significant amount of time for memory swapping due to cache faults.

Within the last ten years, there is a new research direction called *space-efficient algorithms* where one tries to solve a problem with as little space as possible while "almost" maintaining the same running time of a standard solution for the problem under consideration. Space-efficient algorithms are mostly designed for problems that already run in polynomial time, e.g., we have algorithms for several graph problems [5,10].

Our goal is to combine the research on space-efficient algorithms with parameterized algorithms. In the classical literature a *parameterized problem* $P \subseteq \Sigma^* \times \mathbb{N}$ is a language where Σ is a finite alphabet and the second part $k \in \mathbb{N}$ is called *parameter*. In addition, P is called *fixed-parameter tractable* (FPT) if there exists an algorithm $\mathcal{A}$ (called *FPT algorithm*) and a computable function f such that, given an *instance* $(I, k) \in \Sigma^* \times \mathbb{N}$, the algorithm $\mathcal{A}$ correctly decides whether $(I, k) \in P$ in a time bounded by $f(k) \text{poly}(|I|)$. A popular way to find an FPT algorithm is to find a so-called kernelization algorithm. Given a parameterized problem P, a *kernelization algorithm* (or simply called *kernelization*) is a polynomial-time algorithm $\mathcal{A} : (\Sigma^* \times \mathbb{N}) \to (\Sigma^* \times \mathbb{N})$ such that, given an instance $(I, k) \in \Sigma^* \times \mathbb{N}$, then $(I', k') = \mathcal{A}(I, k)$ is another instance (called *kernel*) of the problem with the property that $(I', k') \in P$ if and only

Funded by the DFG – 379157101; A full version can be found in [17].

if $(I, k) \in P$ and $|I'|, k' \leq g(k)$ (where usually $k' \leq k$) for some function g. Then the kernel can usually be solved by a brute-force algorithm in $O(f(g(k))$ time, for some function f. Furthermore, we call (I, k) a *yes*-instance exactly if $(I, k) \in P$. Otherwise, we call it a *no*-instance. A yes- or no-instance of constant size is called *trivial*. The modification steps of the kernelization are called *reduction rules* and such a rule is *safe* if, given input (I, k), it produces an output (I', k') such that (I', k') is a yes-instance if and only if (I, k) is a yes-instance.

In this paper, we describe *space-efficient kernelizations*, which we define as a kernelization $\mathcal{A}$ (as above) with the additional property of using $O(h(k) \log |I|)$ bits for some computable function h. Following this definition, a *space-efficient FPT algorithm* is an FPT algorithm that runs within $O(h(k) \log |I|)$ bits of space. By using our kernel of $g(k)$ vertices and edges as an intermediate kernel, which needs $O(g(k) \log |I|)$ bits to be stored, we then can apply the best known solution to further reduce the kernel size via *kernel cascades* (i.e., the consecutive application of a kernelization on the kernel). Alternatively, we can use the kernel to easily build a space-efficient FPT algorithm $\mathcal{A}$ that produces an optimal solution S for the instance (I, k) in time $f(k) \operatorname{poly}(|I|)$ using $O((g(k) + h(k)) \log |I|)$ bits for some functions f and h. We focus on graph problems, i.e., an instance I is a graph $G = (V, E)$ as well as $|I| = |V| + |E|$, and our space-efficient kernelizations are *space-efficient graph kernelizations*. For the remainder of the paper, let n and m be the vertices and edges, respectively, of the graph under consideration.

We also recognize the need for *full kernels* [9], i.e., a kernel that contains the vertices/edges of all minimal solutions in a yes-instance (G, k). Such a full kernel allows us to enumerate all minimal solutions of size at most k. Those kernels are, e.g., necessary for the application of frameworks such as shown in [14] and for parameterized enumeration [9]. Our computation model is based on a read-only word-RAM with a word size of $w = \Omega(\log n)$ bits, enabling constant-time arithmetic operations and bit-shift operations on w-bit sequences. The input is divided into three types: read-only *input memory*, write-only *output memory*, and read-write *working memory*. Space-efficient algorithm space bounds are typically in bits and refer to the working memory. When expressed in words, these space bounds include an extra factor of $\Theta(\log w)$ or $\Theta(\log n)$ depending on the specific implementation, making them less precise or more complex to describe. In contrast, we express kernel sizes in terms of vertices/edges (i.e., in words), following the conventional approach for describing kernel size.

Parameterized Space Complexity. In classical research, the focus is often on achieving minimal space bounds, but this comes at the cost of significantly increased running times, rendering them impractical. *Parameterized space complexity* is a research area where one mainly classifies problems based on the amount of memory required to solve them. Two important classes in this field are para–L (aka. logspace + advice) and XL (aka. slicewise logspace). para–L contains problems that can be solved with $f(k) + O(\log n)$ bits, while XL contains problems that can be solved with $O(f(k) \log n)$ bits [4].

We next give an overview over parameterized space complexity restricted to graph problems. An early work on parameterized space complexity is due to Cai et al. [4] who showed that vertex cover is in para–L. Elberfeld et al. [12] showed that FEEDBACK VERTEX SET (FVS) is in XL. Bannach et al. [1] studied packing, covering and clustering problems and show (among other results) that TRIANGLE PACKING, (EXACT PARTIAL) VERTEX COVER, and MANY CLUSTER EDITING are in para–L (more precisely in a class called para–AC^0 $\subseteq$ para–L) and CLUSTER EDITING is in para–L (more precisely in para–TC^0 $\subseteq$ para–L). Fafianie and Kratsch [13] showed that several graph deletion problems where the target classes have finite forbidden sets are also in para–L. This result was recently generalized to infinite forbidden sets by Biswas et al. [3], who showed that deletion problems like deletion to linear forest and deletion to pathwidth 1 are also in para–L.

To our knowledge neither the membership of para–L nor a lower bound for FVS was discovered yet. While FVS can be formulated as a deletion problem, Biswas et al. [3] mentioned that the techniques required are not "easily" applicable to FVS (and other deletion problems). They attacked the problem by using different parameterizations and showed that (among other problems) FVS is in para–L if parameterized by vertex cover. Moreover, they presented a space-efficient FVS algorithm, parameterized by solution size k, that runs in $5^k n^{O(1)}$ time and with $O(k \log n)$ bits, an improvement over the previously known best space-efficient FVS algorithm of Elberfeld et al. [12], which runs in $O(k^k n^5)$ time and uses $O(k \log n)$ bits.

Approach and Contribution. Our approach departs from the conventional method of computing a kernel by applying reduction rules globally to the entire graph, a process that can be resource-intensive in terms of space. Aiming for a memory usage of $O(\text{poly}(k) \log n)$ bits, we show a process that involves systematically modifying and condensing disjoint subgraphs of the input graph $G = (V, E)$ while preserving poly(k) vertices and edges in the resulting kernel. The main idea here is to make use of a separator set U of size poly(k), which partitions graph G into disjoint subgraphs. Each subgraph undergoes a reduction process under consideration of the separator U to efficiently shrink it to poly(k) vertices and edges. Subsequently the reduced subgraph is then carefully integrated into an initial kernel $G' = G[U]$ under construction. G' is then repeatedly reduced to ensure we stay within our space bound. To make our algorithm work we need to show that both, computing the separator and applying the reduction rules, must be implemented with $O(\text{poly}(k) \log n)$ bits. This often means that we are not able to run all known reduction rules or to run the rules in a restricted setting.

In Sect. 2, we present a simple PATH-CONTRACTION kernelization that runs in $O(n \log k + \text{poly}(k))$ time using $O(\text{poly}(k) \log n)$ bits. To find a kernel for PATH CONTRACTION, one usually searches for *bridges* (i.e., edges whose removal disconnect the graph) and merges the endpoints of such a bridge to a single vertex [18]. Bridges are usually found by running a DFS—to the authors knowledge, all polynomial-time DFS need $\Omega(\sqrt{n})$ bits and a polynomial-time depth-first search (DFS) with $O(\sqrt{n})$ bits is due to Izumi and Otachi [16]. Instead of this reduction rule, we use a separator U (which is the queue of a breadth-first search (BFS)) and iteratively expand our kernel while shrinking *induced degree-2-chains* (paths whose vertices all have degree 2) as long as they consist of more than $k + 1$ edges. To achieve our space bound we show that a yes-instance of PATH CONTRACTION cannot have a tree as an induced subgraph with more than $k + 2$ leaves. This bounds the size of U by $O(k)$ and makes it possible to construct a BFS algorithm that stores at most $O(k)$ vertices at a time, allowing us to construct a kernel with $O(k^2)$ vertices and edges in $O((n + k^2) \log k)$ time using $O(k^2 \log n)$ bits. To get the current best kernel size we subsequently apply Li et al.'s [18] polynomial-time kernelization and so get a kernel of $3k + 4$ vertices in $O(n \log k + \text{poly}(k))$ time and using $O(\text{poly}(k) \log n)$ bits. Li et al.'s kernelization for PATH CONTRACTION uses $\Theta((n + m) \log n)$ bits due to searches for bridges and to store the modification in adherence to their reduction rules.

Our main result is a new kernelization for FEEDBACK VERTEX SET (FVS) in Sect. 3. Our idea is to compute an approximate minimum feedback vertex set U as a separator whose removal partitions the graph into several trees. We use a so-called Loop Rule and a restricted Flower Rule as well as a so-called Leaf Rule and Chain Rule. If the kernel is still too large, we follow ideas from Thomassé [19]. For details on these rules, see Sect. 3. We want to remark that a solution for PATH CONTRACTION of size k implies one for FVS of size $2k$ by simply taking the endpoints of the contracted edges. Our kernelization for FVS runs in $O(n^5 \text{poly}(k))$ time, uses $O(k^4 \log n)$ bits and outputs a kernel of $n' = 2k^2 + k$ vertices. After computing our kernel we can use the deterministic algorithm of Iwata and Kobayashi to solve it in $O(3.46^k n') = O(3.5^k)$ time. In total, we can solve FVS in $O(n^5 \text{poly}(k) + 3.5^k)$ time and with $O(k^4 \log n)$ bits.

Compared to Elberfeld et al.'s algorithm [12], which solves FVS in time $O(k^k n^5)$, we are faster, but we use $O(k^4 \log n)$ instead of $O(k \log n)$ bits. Concurrently and independently to our result, Biswas et al. [3] presented an iterative compression algorithm (based on the Chen et al. [6] algorithm) that maintains the space bound of $O(k \log n)$ bits and has a runtime of $O(5^k n^{O(1)})$. The degree of the polynomial is not mentioned explicitly, presumably due to the fact that some of the used log-space auxiliary results do not mention their "exact" running time either. However, we and they use the $O(\log n)$ bit realizations of the so-called Leaf and Chain Rules (aka. Degree-2 Rule) from Elberfeld et al. [12]. The non space-efficient version of the rules removes vertices from the input graph that are not relevant for solving FVS, but due to the space restriction, the information of the graph resulting from the removal is computed on demand. By our analysis (Lemma 2) this results in a running time factor of $\Theta(n^5 k \log k)$ for each vertex/edge access. Moreover, Biswas et al. have a nested loop where $\Theta(n^2)$ connectivity tests have to be performed, which increases the running time to $\Theta(n^7 \text{poly}(k) + 5^k n^{\Theta(1)})$. Based on that our running time is faster in k and in n, but they use only $O(k \log n)$ bits.

The proofs for the lemmas and theorems can be found in a long version of the paper [17].

2   Path Contraction

Let G be an n-node m-edge graph and let C be a subset of edges of G. We write G/C for the graph obtained from G by contracting each edge in C. *Contracting an edge* is done by merging its endpoints and removing any loops or parallel edges afterwards. In the parameterized PATH CONTRACTION problem, a connected graph $G = (V, E)$ is given together with a parameter k and the task is to find a set $C \subseteq E$ with $|C| \leq k$ such that G/C is a path. In particular, G/C is a connected graph with $n' \in \mathbb{N}$ vertices and $n' - 1$ edges. One reduction rule used in Li et al.'s [18] kernelization is an iterative contraction of a bridge for which no polynomial-time $O(\mathrm{poly}(k) \log n)$-bits algorithm is known. (Bridges are found by running a DFS and the best-known polynomial-time DFS with a minimum of space uses $\Omega(\sqrt{n})$ bits [16].) Instead of computing bridges, we introduce two new reduction rules below. In the following, a *subtree* T of G is a subgraph of G that is a tree. Moreover, let a *degree-2 chain* be a maximal simple path $P = v_1, \ldots, v_\ell$ ($\ell \in \mathbb{N}$) whose vertices $v_1, \ldots, v_\ell$ are all of degree 2. Observe, if P is not a cycle, then v_1 and v_ℓ must each have a neighbor that is not of degree 2.

Rule 1. If there exists a degree-2 chain P with more than $k+1$ edges, contract all except $k+1$ arbitrary edges of P.

Rule 2. After k applications of Rule 1, if G contains more than $k^2 + 4k + 1$ vertices, more than $(3k^2 + 13k)/2$ edges, or a subtree with more than $k + 2$ leaves, then output "no-instance".

The bound on the number of leaves in Rule 2 helps us to guarantee that our kernelization works in our space bound. We want to remark that we do not apply Rule 2 exhaustively; more precisely, we do not explicitly search for subtrees with more than $k + 2$ leaves, which is NP-hard.

Lemma 1. *Rule 1 and 2 are safe and produce a full kernel.*

For the kernel construction we use a BFS. We shortly sketch a usual BFS and the construction of a so-called *BFS tree*. The BFS visits the vertices of an input graph round-wise. As a preparation of the first round it puts some vertex v into a queue Q, marks it as visited, and starts a round. In a round it dequeues every vertex u of Q, and marks u as visited. Moreover, it puts every unvisited neighbor $w \in N(u)$ of u into a queue Q' and marks it as visited. We then say that w was first discovered from u and add the edge $\{u, w\}$ to an initial empty BFS tree. If Q' is empty at the end of the round, the BFS finishes. Otherwise, it proceeds with the next round with $Q := Q'$ and $Q' := \emptyset$.

As the BFS progresses, its exploration adheres to the subtree structure of G. The BFS queue size is restricted to a maximum of $k + 2$ vertices; otherwise, we can determine a no-instance by Rule 2 and halt the BFS process. With regards to marking vertices as visited, since Q effectively delineates between previously visited and unvisited vertices after each BFS round, only the last BFS round's queue is necessary to verify the visited vertices. Notably, in this context, separator U consists of vertices in Q and is thus dynamic, adapting

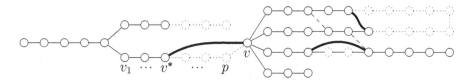

Fig. 1. Our adapted BFS starts from the leftmost vertex, removing dotted vertices on a degree-2 chain with over $k+1$ predecessors and connecting the neighbors of removed vertices with bold edges. Dashed edges are skipped by the BFS.

with each BFS iteration. So far this approach ensures that every yes-instances is traversed and any no-instance is identified in $O(n \log k)$ time utilizing only $O(k \log n)$ bits.

To realize Rule 1 we need additional information. Instead of storing just vertices v on the BFS queue we store quadruples that we use to identify degree-2 chains and apply Rule 1—see also Fig. 1. Each quadruple (v, p, i, v^*) consists of a vertex v and its predecessor p if v is not the root, the counter $i \in \{0, \ldots, n\}$ with $i > 0$ being v's position on a degree-2 chain, and the vertex v^* with $v^* \neq$ null being the $(k+1)$th vertex on a degree-2 chain that contains v. So we can easily check Rule 1 as shown in the proof of Theorem 1. By Rule 2, we can guarantee our space bound by maintaining $O(k^2)$ vertices and edges.

Theorem 1. *Given an n-vertex instance (G, k) of* PATH CONTRACTION, *there is an $O((n + k^2) \log k)$-time $O(k^2 \log n)$-bits kernelization that outputs a full kernel of $O(k^2)$ vertices and edges, or outputs that (G, k) is a no-instance. The result can be used to find a (possibly not full) kernel of at most $3k + 4$ vertices in $O(n \log k + \mathrm{poly}(k))$ time using $O(\mathrm{poly}(k) \log n)$ bits.*

3 Feedback Vertex Set

Given an n-vertex m-edge graph $G = (V, E)$ a set $F \subseteq V$ is called *feedback vertex set* if the removal of the vertices of F from G turns G into an acyclic graph (also called *forest*). In the parameterized FEEDBACK VERTEX SET problem, a tuple (G, k) is given where G is a graph, and k is a parameter. We are searching for a feedback vertex set F of size at most k in G. In kernelization, it is common to identify vertices that must be in every minimal feedback vertex set of size at most k, remove them from the instance, and restart the kernelization. To avoid modifying the given instance, we simulate this by starting with an empty set F and adding vertices to F when we determine they must be in every solution of size at most k. Subsequently, we realize the graph $G[V \setminus F]$ by considering G and disregarding vertices in F.

Iwata showed a kernelization for FEEDBACK VERTEX SET that produces a kernel consisting of at most $2k^2 + k$ vertices and $4k^2$ edges and runs in $O(k^4 m)$ time [15]. He mentions that all other kernelizations for FEEDBACK VERTEX SET exploit an exhaustive application of the three basic rules below and the so-called v-flower rule. A *v-flower of order d* is a set of d cycles pairwise intersecting exactly on vertex v.

Loop Rule. Remove a vertex v with a loop and reduce to $(G - v, k - 1)$ and $F := F \cup \{v\}$.

Leaf Rule. Remove a vertex v with $\deg(v) \leq 1$.

Chain Rule. Remove a vertex v that has only two incident edges $\{v, u\}$ and $\{v, w\}$ (possibly $u = w$), and add the edge $\{u, w\}$.

Flower Rule. Remove a vertex v if a v-flower of order $k + 1$ exists and reduce to $(G - v, k - 1)$ and $F := F \cup \{v\}$.

By allowing $O(k \log n)$ bits for the algorithm, Elberfeld et al. also showed how to find a cycle of $2k$ vertices. To realize the flower rule at a vertex v we need to run along up to $k + 1$ cycles and check if they intersect at vertices other than v. If the given graph is reduced with respect to the Leaf and Chain Rule and does not contain vertices with self-loops, then it can be guaranteed that the smallest cycle is of length at most $2k$ (maximum girth of a graph with the mentioned restrictions, minimum degree 3 and a feedback vertex set of size at most k [11]). However, the length of the remaining cycles can be bound only by a function depending on n, not on k. So it seems to be hard to find and verify a flower with $O(\text{poly}(k) \log n)$ bits. As shown by Iwata, one does not need the Flower Rule to find a kernel for FVS. Instead he uses a so-called *s-cycle cover reduction* [15, Section 3] where he has to know which edges incident to a vertex s are bridges in the graph. (For space bounds to find bridges, recall Sect. 2.) Since we have no solution to find a v-flower or an s-cycle with $O(\text{poly}(k) \log n)$ bits, we show how to construct a kernel without using both rules exhaustively.

Thomassé [19] introduced the rule below to compute a kernel consisting of $4k^2$ vertices. As input we assume a simple graph. Since his rule introduces double edges, our kernel G' is a multi graph where every multi edge is a double edge.

Thomassé's Rule [19]. Let X be a set of vertices, let $x \in V \setminus X$ and let $\mathcal{C}$ be a set of connected components of $G \setminus (X \cup \{x\})$ (not necessarily all the connected components) such that
  - G is loopless, with degree ≥ 3 and all multi-edges are double-edges,
  - there is exactly one edge between x and every $C \in \mathcal{C}$,
  - every $C \in \mathcal{C}$ induces a tree, and
  - for every subset $Z \subseteq X$, the number of trees of $\mathcal{C}$ having some neighbor in Z is at least $2|Z|$.

Then reduce to (G', k) where G' is the graph obtained by joining x to every vertex of X by double edges and by removing the edges between x and the components of $\mathcal{C}$.

Thomassé applies the rule to the whole graph G. This means, he has to store graph changes over the whole graph G. This is too expensive for us. As discussed in the introduction, we utilize a separator U to break down the graph into manageable components. This allows us to construct a kernel by processing and gradually incorporating these components. In the next subsection, we outline the construction of an approximate minimum feedback vertex set as separator U, so that the graph divides into trees. In the subsequent subsection, we show how to iterate over the trees in $\mathcal{T} = G[V \setminus U]$. In a third subsection, we iteratively add these trees into a graph G' (initially $G' = G[U]$) while upholding our desired space bound of G' having $O(k^4)$ vertices and edges, i.e., $O(k^4 \log n)$ bits. However, the size of the trees in $\mathcal{T}$ is unbounded in k and we need to perform an on-the-fly tree size reduction of the tree. For this, we first make sure that every tree T has not too many edges to U (or we either find a vertex for the solution F and restart, or conclude a no-instance). Afterwards, we have to traverse T and put exactly those vertices of T into G' that are not removed by an exhaustive application of the Leaf and Chain Rule. To keep the size of G' within our space bound we show in a fourth subsection how to shrink G' again. To shrink the size of G' to $O(k^2)$ vertices, we apply Thomassé's Rule, which has a precondition requiring that G' has a minimum degree of 3. However, we cannot satisfy this precondition for vertices in U within G'. Nevertheless, we can demonstrate that violating the precondition only for the vertices in U still allows the rule to function if we adjust the bounds accordingly (see Lemma 6). Finally, we show that our construction of a kernel of $O(k^2)$ vertices and $O(k^3)$ edges runs in $O(n^5 \operatorname{poly}(k))$-time and with $O(k^4 \log n)$ bits with this approach.

Separator U of Size $3k^2$. Becker and Geiger [2] presented a 2-approximation algorithm for feedback vertex set in which they extend an $(2 \log d)$-approximation algorithm (where d is the maximum degree of the graph) by a phase that iteratively removes a vertex v from the computed feedback vertex set S if all cycles that intersect v in G also intersect $S \setminus \{v\}$. It is unlikely that this can be done with $O(\operatorname{poly}(k) \log n)$ bits or even with $O(f(k) \log n)$ bits for some function f since a cycle in G can consists of $\Theta(n)$ vertices and there can be $\Theta(n)$ cycles.

We instead present only an $O(k)$-approximation algorithm, but it runs with $O(\operatorname{poly}(k) \log n)$ bits. For this we use the following well-known rule. Given a loopless graph G of minimum degree 3 and without self-loops, every FVS in G of size at most k contains at least one vertex of the $3k$ vertices of largest degree [8, Lemma 3.3]. A graph with such properties can be computed by an exhaustive application of the Loop, Leaf and Chain Rule. Elberfeld et al. [12, Theorem 4.13] showed how to implement the rules with $O(\log n)$ bits. The graph obtained does not actually have to be stored. Instead we compute the required information on demand with Lemma 2, which is similar to parts of the proof of [12, Theorem 4.13].

Lemma 2. *Assume that an n-vertex m-edge graph $G = (V, E)$ and a set $U \subseteq V$ consisting of $k^{O(1)}$ vertices is given. Let G' be the graph obtained by an exhaustive application of the Loop, Leaf and Chain Rule on $G[V \setminus U]$. We can provide a structure that allows the iteration over the edges of every vertex of G' in $O(n^5 k \log k)$ total time by using $O(\log n)$ bits. In particular, we do not store G'.*

Since by Lemma 2 we have access to a loopless graph of minimum degree three, we iterate k-times over a graph $G[V \setminus U]$ (where initially $U = \emptyset$) and in each iteration select the $3k$ vertices of largest degree into U. We so get an $O(k)$ approximate minimum feedback vertex set.

Theorem 2. *Given an n-vertex m-edge instance (G, k) of FEEDBACK VER-TEX SET, there is an $O(n^5 k^2 \log k)$-time, $O(k^2 \log n)$-bits algorithm that either returns a feedback vertex set U consisting of at most $3k^2$ vertices or answers that (G, k) is a no-instance.*

Iterations over Trees. We want to output every tree T of $G[V \setminus U]$ once. For this, we iterate over all $u \in U$ and, intuitively speaking, output those trees T adjacent to u, i.e., every T having a vertex v such that u and v are adjacent in G. Moreover, note that several vertices of U can have edges to the same tree. We show in the proof of the next lemma how to avoid outputting a tree multiple times. To distinguish the trees, we partition the trees T as follows (also see Fig. 2). T_0 is the set of trees in T that have at most one edge to a single vertex of U. T_1 is the set of trees in T where each tree has at most one edge to at least two vertices of U. T_2 is the set of the remaining trees in T with least two edges to some vertex of U.

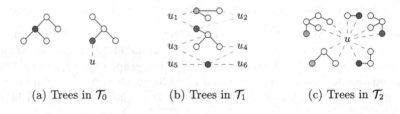

(a) Trees in T_0 (b) Trees in T_1 (c) Trees in T_2

Fig. 2. Our partition of trees with edges to $U = \{u_1, u_2, u_3, \ldots\}$. The non-white vertices are the vertices with the smallest ID in the trees.

Lemma 3. *Given an n-vertex m-edge graph $G = (V, E)$ and a set U of $O(k^2)$ vertices, there is an algorithm that outputs a single vertex w of T as a representative for each tree T in $T_1 \cup T_2$ and some trees of T_0 once. The algorithm runs in $O(n^3 k^3 \log^2 k)$ time and with $O(\log n)$ bits.*

Observe that we do not need to add trees of T_0 to G' since they can be removed by the Leaf Rule anyway. If we identify such a tree, we skip over it.

Tree Size Reduction. We now want to shrink each tree T of $G[V \setminus U]$ so that we can add them to G' without exceeding our space bound of $O(k^4 \log n)$ bits. Due to Cook and McKenzie [7], $O(\log n)$ bits DFS exists which suffice to find out for each tree T to which set T_0, T_1 or T_2 it belongs.

By definition a tree $T \in T_1$ can have at most one edge to every vertex of U. Thus T has at most $|U| = O(k^2)$ edges into U. By the lemma below, we can add T into G'.

Lemma 4. *Given U and an $\bar{n}$-vertex tree $T = (V_T, E_T)$ in $G[V \setminus U]$ such that T has ℓ edges to U. After applying the Leaf and Chain Rule to $G[V_T \cup U]$ while forbidding the removal of vertices of U, T has at most $O(\ell)$ vertices. This can be done in $O(\bar{n}^3)$ time using $O(\log n)$ bits.*

For a tree $T = (V_T, E_T) \in \mathcal{T}_2$, a vertex of U can be connected to multiple vertices of T. Hence we need to run the following two steps to bound the number of edges, where Step 1 is a precondition for Step 2. By Step 2 we get subtrees T' of T such that T' and U have at most $(k + 1)^2$ edges in between. Thus, we can add T' with $O(k^2)$ vertices to G' by Lemma 4.

Step 1: Consider a bipartite graph $Y = (U \cup V_T, E')$ where $\{u, w\} \in E'$ exactly if there are at least $k + 2$ internally vertex disjoint paths between u and w. (For the ongoing algorithm observe: To find a solution for feedback vertex set of size k in Y, we need a vertex cover of size at most k in Y. Furthermore, any vertex in any vertex cover of size k in Y must also be in any feedback vertex set of size k in G.) If a vertex $u \in U$ has degree at least $k + 1$ in Y, take u into our solution set F and restart. If Y has more than $k^2 + k$ vertices, output no-instance. Otherwise, define the common vertices of Y and V_T as set U'. Note that $|U'| \leq k^2$. Temporarily take $U \cup U'$ as separator. This splits T in several *small trees* T', which we will process iteratively in Step 2.

Step 2: If a small tree T' has at least $(k + 1)^2$ edges to a vertex $u \in U \cup U'$, take u into our solution F and restart. Otherwise add T' to our graph G'.

Lemma 5. *Steps 1 and 2 are safe and both steps run in $O(n^3 \log k)$ time and with $O(k^3 \log n)$ bits.*

Finally note that, by Step 2 above, we temporarily have a separator $U \cup U'$ of size $4k^2$. After adding all subtrees T' of a tree T in $\mathcal{T}_2$ to G', we can also add U' to G' and we are back to a separator of size $3k^2$.

Shrink the Kernel Again. After adding several trees to G' we have to ensure that the size of G' does not exceed our space bound. We run into two issues.

1. A naive application of a reduction rule may "unsafely" remove vertices since only subgraphs are considered, e.g., a vertex of U could be considered as being a leaf in G' because several vertices that are connected to U were not added to G' yet.
2. Thomassé's Rule requires G' to be a loopless graph of minimum degree 3, and we cannot ensure that the vertices of U in G' are of minimum degree 3.

We address issue (1) with respect to all reduction rules. The application of the Loop and the Flower is still safe to use because whenever they apply, vertices are selected into a solution F for feedback vertex set and we restart. To deal with the Leaf and the Chain Rule we forbid that vertices of U are removed in G'. Thomassé's Rule does not remove vertices. Instead, it only removes existing edges and adds new ones within our subgraph G'. Moreover as stated in the rule, it does not have to consider all connected components, which makes the rule safe for usage in G'.

Concerning issue (2) we show in the next lemma that vertices of U can be exempted from fulfilling the property of being of minimum degree 3.

Lemma 6. *Let $G = (V, E)$ be is a subgraph of G' where G' is a loopless graph and multi-edges are double edges, such that G has $n > 16k^2$ vertices and only a* subset $U \subseteq V$ with $|U| \leq 4k^2$ have degree 0, 1 or 2 in G. If G has a feedback vertex set of size k, then we can apply the Flower Rule or Thomassé's Rule in polynomial time and with $O(k^2 \log n)$ bits.

Construct a Kernel: By Theorem 2, we have access to a approximate minimum feedback vertex set U of size $3k^2$. U divides G into several trees, enabling us to construct our initial kernel $G' = G[U]$ under construction. We iterate over each tree T of $\mathcal{T}_1 \cup \mathcal{T}_2$ using Lemma 3 and determine its type. If $T \in \mathcal{T}_1$, we add it to G' using Lemma 4. Otherwise, if $T \in \mathcal{T}_2$, it must be integrated into G' by first breaking it down into smaller trees. To achieve this, we execute Step 1 and 2 and get another separator U'. Combining U' with U divides T into several subtrees $T_1, T_2, \ldots$. We then utilize Lemma 3 with $G[V \setminus (U \cup U')]$, but use U' to iterate over the trees $T_i = T_1, T_2, \ldots$ (since they are all adjacent to U') and add one tree at a time with Lemma 4. While adding a tree apply the Loop, Leaf, Chain Rule and shrink G' by the Flower and Thomassé's Rule. If all trees of T_i were added, add also U' into G'. After all trees of G have been added to G', we apply the best kernelization [15] to get a kernel of $2k^2 + k$ vertices.

Theorem 3. *Given an n-vertex instance (G, k) of* FEEDBACK VERTEX SET, *there is an $O(n^5 \operatorname{poly}(k))$-time, $O(k^4 \log n)$-bits kernelization that either outputs a kernel consisting of $2k^2 + k$ vertices or returns that (G, k) is a no-instance.*

References

1. Bannach, M., Stockhusen, C., Tantau, T.: Fast parallel fixed-parameter algorithms via color coding. In: Proceedings of the 10th International Symposium on Parameterized and Exact Computation (IPEC 2015). LIPIcs, vol. 43, pp. 224–235. Schloss Dagstuhl - Leibniz-Zentrum für Informatik (2015). https://doi.org/10.4230/LIPIcs.IPEC.2015.224

2. Becker, A., Geiger, D.: Approximation algorithms for the loop cutset problem. In: Proceedings of the 10th Annual Conference on Uncertainty in Artificial Intelligence (UAI 1994), pp. 60–68. Morgan Kaufmann (1994)

3. Biswas, A., Raman, V., Satti, S.R., Saurabh, S.: Space-efficient FPT algorithms. CoRR, abs/2112.15233 (2021). arXiv:2112.15233

4. Cai, L., Chen, J., Downey, R.G., Fellows, M.R.: Advice classes of parameterized tractability. Ann. Pure Appl. Log. **84**(1), 119–138 (1997). https://doi.org/10.1016/S0168-0072(95)00020-8

5. Chakraborty, S., Sadakane, K., Satti, S.R.: Optimal in-place algorithms for basic graph problems. In: Gąsieniec, L., Klasing, R., Radzik, T. (eds.) IWOCA 2020. LNCS, vol. 12126, pp. 126–139. Springer, Cham (2020). https://doi.org/10.1007/978-3-030-48966-3_10

6. Chen, J., Fomin, F.V., Liu, Y., Songjian, L., Villanger, Y.: Improved algorithms for feedback vertex set problems. J. Comput. Syst. Sci. **74**(7), 1188–1198 (2008). https://doi.org/10.1016/j.jcss.2008.05.002

7. Cook, S.A., McKenzie, P.: Problems complete for deterministic logarithmic space. J. Algorithms **8**(3), 385–394 (1987). https://doi.org/10.1016/0196-6774(87)90018-6

8. Cygan, M., et al.: Parameterized Algorithms. Springer, Cham (2015). https://doi.org/10.1007/978-3-319-21275-3

9. Damaschke, P.: Parameterized enumeration, transversals, and imperfect phylogeny reconstruction. Theor. Comput. Sci. **351**(3), 337–350 (2006). https://doi.org/10.1016/j.tcs.2005.10.004

10. Datta, S., Kulkarni, R., Mukherjee, A.: Space-efficient approximation scheme for maximum matching in sparse graphs. In: Proceedings of the 41st International Symposium on Mathematical Foundations of Computer Science (MFCS 2016). LIPIcs, vol. 58, pp. 28:1–28:12. Schloss Dagstuhl – Leibniz-Zentrum für Informatik (2016). https://doi.org/10.4230/LIPIcs.MFCS.2016.28

11. Downey, R.G., Fellows, M.R.: Parameterized Complexity. Monographs in Computer Science, Springer, Cham (1999). https://doi.org/10.1007/978-1-4612-0515-9

12. Elberfeld, M., Stockhusen, C., Tantau, T.: On the space and circuit complexity of parameterized problems: classes and completeness. Algorithmica **71**(3), 661–701 (2015). https://doi.org/10.1007/s00453-014-9944-y

13. Fafianie, S., Kratsch, S.: A shortcut to (sun)flowers: kernels in logarithmic space or linear time. In: Italiano, G.F., Pighizzini, G., Sannella, D.T. (eds.) MFCS 2015. LNCS, vol. 9235, pp. 299–310. Springer, Heidelberg (2015). https://doi.org/10.1007/978-3-662-48054-0_25

14. Heeger, K., Himmel, A.-S., Kammer, F., Niedermeier, R., Renken, M., Sajenko, A.: Multistage graph problems on a global budget. Theor. Comput. Sci. **868**, 46–64 (2021). https://doi.org/10.1016/j.tcs.2021.04.002

15. Iwata, Y.: Linear-time kernelization for feedback vertex set. In: Proceedings of the 44th International Colloquium on Automata, Languages, and Programming, (ICALP 2017). LIPIcs, vol. 80, pp. 68:1–68:14. Schloss Dagstuhl - Leibniz-Zentrum für Informatik (2017). https://doi.org/10.4230/LIPIcs.ICALP.2017.68

16. Izumi, T., Otachi, Y.: Sublinear-space lexicographic depth-first search for bounded treewidth graphs and planar graphs. In: Proceedings of the 47th International Colloquium on Automata, Languages, and Programming, (ICALP 2020). LIPIcs, vol. 168, pp. 67:1–67:17. Schloss Dagstuhl - Leibniz-Zentrum für Informatik (2020). https://doi.org/10.4230/LIPICS.ICALP.2020.67

17. Kammer, F., Sajenko, A.: Space-efficient graph kernelizations. CoRR, abs/2007.11643 (2020). arXiv:2007.11643

18. Li, W., Feng, Q., Chen, J., Shuai, H.: Improved kernel results for some FPT problems based on simple observations. Theor. Comput. Sci. **657**, 20–27 (2017). https://doi.org/10.1016/j.tcs.2016.06.012

19. Thomassé, S.: A $4k^2$ kernel for feedback vertex set. ACM Trans. Algorithms, **6**(2), 32:1–32:8 (2010). https://doi.org/10.1145/1721837.1721848

Counting on Rainbow k-Connections

Robert D. Barish$^{(\boxtimes)}$ (iD) and Tetsuo Shibuya (iD)

Division of Medical Data Informatics, Human Genome Center,
Institute of Medical Science, University of Tokyo,
4-6-1 Shirokanedai, Minato-ku, Tokyo 108-8639, Japan
`rbarish@ims.u-tokyo.ac.jp`, `tshibuya@hgc.jp`

Abstract. For an undirected graph imbued with an edge coloring, a
rainbow path (resp. *proper path*) between a pair of vertices corresponds
to a simple path in which no two edges (resp. no two adjacent edges) are
of the same color. In this context, we refer to such an edge coloring as
a *rainbow k-connected w-coloring* (resp. *k-proper connected w-coloring*)
if at most w colors are used to ensure the existence of at least k internally vertex disjoint rainbow paths (resp. k internally vertex disjoint
proper paths) between all pairs of vertices. At present, while there have
been extensive efforts to characterize the complexity of finding rainbow
1-connected colorings, we remark that very little appears to known for
cases where $k \in \mathbb{N}_{>1}$.

In this work, in part answering a question of (Ducoffe et al.; *Discrete Appl. Math.* **281**; 2020), we first show that the problems of counting rainbow k-connected w-colorings and counting k-proper connected
w-colorings are both linear time treewidth Fixed Parameter Tractable
(FPT) for every $(k, w) \in \mathbb{N}^2_{\geq 0}$. Subsequently, and in the other direction,
we extend prior NP-completeness results for deciding the existence of a
rainbow 1-connected w-coloring for every $w \in \mathbb{N}_{>1}$, in particular, showing
that the problem remains NP-complete for every $(k, w) \in \mathbb{N}_{>0} \times \mathbb{N}_{>1}$.
This yields as a corollary that no Fully Polynomial-time Randomized
Approximation Scheme (FPRAS) can exist for approximately counting
such colorings in any of these cases (unless $NP = RP$). Next, concerning counting hardness, we give the first #P-completeness result we are
aware of for rainbow connected colorings, proving that counting rainbow
k-connected 2-colorings is #P-complete for every $k \in \mathbb{N}_{>0}$.

Keywords: edge colorings · rainbow connected colorings · proper
connected colorings · fixed-parameter tractability · FPT · treewidth
FPT · approximation hardness · NP · #P

1 Introduction

In circa 2008, Chartrand, Johns, McKeon, and Zhang [8,9] introduced the notion
of *rainbow k-connected w-colorings*, generalizing proper coloring problems to

This work was supported by a Grant-in-Aid for JSPS Research Fellow (18F18117 to R.
D. Barish), and by JSPS Kakenhi grants {20K21827, 20H05967, 21H04871, 21H05052
23H03345, 23K18501} to T. Shibuya.

X. Chen and B. Li (Eds.): TAMC 2024, LNCS 14637, pp. 272–283, 2024.
https://doi.org/10.1007/978-981-97-2340-9_23

treat conflicts over communication channels in networks. Here, letting G be a simple undirected graph with an edge coloring $\mathcal{C}$ using at most w distinct colors, we call $\mathcal{C}$ a rainbow k-connected w-coloring of G if and only if there exist at least k internally vertex disjoint *rainbow paths* between each pair of vertices in G, where rainbow paths are simple paths having no two edges of the same color. Chartrand et al. [8,9] originally conceptualized G as an *ad hoc* network with layered-encryption akin to TOR or I2P [21], wherein one has agents (vertices) communicating packets of information over various firewalled channels (edges), where each channel has a specified key (color), and where no key is used twice by the same data packet. In this context, the existence of a rainbow k-connected w-coloring would imply the existence of a feasible layered encryption scheme using at most w keys and allowing for k internally vertex disjoint redundant paths between all pairs of agents.

Borozan et al. [6] and Andrews et al. [2] would later moderate and generalize *rainbow connectivity* with the notion of k-*proper connected w-colorings* of graphs. Here, once again letting G be a simple undirected graph imbued with an edge coloring $\mathcal{C}$ having at most w distinct colors, we say that $\mathcal{C}$ is a k-*proper connected w-coloring* if and only if there exists at least k internally vertex disjoint *proper paths* between each pair of vertices in G, where proper paths are simple paths having no two adjacent edges of the same color. Akin to rainbow connectivity, proper connectivity would also prove to have direct application to managing conflicts over channels of communication. As noted by Li and Magnant [20], the aforementioned graph G can be conceptualized as a wireless communication network with the natural constraint that incoming and outgoing signals from a given receiver and emitter node must have distinct frequencies. In this context, a k-proper connected w-coloring for G would imply the existence of a message passing scheme using at most w frequencies, and allowing for a k internally vertex disjoint redundant paths between all pairs of nodes.

However, despite the clear importance of having redundant pathways for information transmission in the aforementioned applications, little is known concerning the complexity of deciding the existence of and counting rainbow k-connected w-colorings or k-proper connected w-colorings. For rainbow connected colorings in particular, while it is known that deciding if a graph admits a rainbow $(k = 1)$-connected w-coloring is NP-complete for $w = 2$ [7], and more generally, for every $w \in 2\mathbb{N}_{>0}$ [7] and for every $w \in 2\mathbb{N}_{>0} + 1$ [1], we are aware of no similar findings for cases where $k \in \mathbb{N}_{>1}$.

To address this gap, in this work we begin by showing that the problems of counting rainbow k-connected w-colorings and counting k-proper connected w-colorings are both linear time treewidth FPT for every $(k, w) \in \mathbb{N}_{>0}^2$ (Proposition 1 and Proposition 2). This answers in the affirmative a question posed by Ducoffe, Marinescu-Ghemeci, and Popa [16] as to whether a linear time treewidth FPT algorithm exists for the problem of deciding if a given graph admits a $(k = 1)$-proper connected w-coloring for any $w \in \mathbb{N}_{>1}$. Pertaining to hardness results, we first prove that deciding the existence of and counting rainbow k-connected $(w = 2)$-colorings of a graph is NP-complete and $\#P$-complete (under many-one counting reductions), respectively, for every $k \in \mathbb{N}_{>1}$ (Theorem 1). Next, we extend the aforementioned NP-completeness result to deciding the existence of a rainbow k-connected w-coloring for any $(k, w) \in \mathbb{N}_{>0} \times \mathbb{N}_{>1}$ (Theorem 2), and furthermore show that, unless $NP = RP$, no FPRAS can exist for counting such colorings for any $(k, w) \in \mathbb{N}_{>0} \times \mathbb{N}_{>1}$ (Corollary 1).

Finally, we remark that we were able to extend our Theorem 1 proof argument to establish the NP-completeness and $\#P$-completeness (under many-one counting reductions) of deciding the existence of and counting, respectively, orientations of graphs in which all pairs of vertices are unilaterally connected by some fixed number $k \in \mathbb{N}_{>0}$ of directed paths ≤ 2 edges in length. However, due to space constraints, an elaboration will be deferred to a future version of the current work.

2   Preliminaries

2.1   Graph Theoretic Terminology

We will generally follow the terminology of Diestel [14], or where appropriate, Bondy and Murty [5]. Unless stated otherwise, all graphs in this work should be assumed to be simple (i.e., loop and multi-edge-free), all digraphs should be assumed to be *orientations* of simple undirected graphs (i.e., the product of assigning directions to the edges of an undirected graph), and all path lengths should be understood in terms of their edge-wise lengths (e.g., a path from v_a to v_c given by the edges $\{v_a \leftrightarrow v_b, v_b \leftrightarrow v_c\}$ has length 2).

Concerning vertex identification operations, let G be a graph with vertex set V_G. Recall that we may *identify* a pair of vertices $v_a, v_b \in V_G$ by constructing a new graph in which we delete v_a and v_b, then create a vertex $u \notin V_G$ with an adjacency to a vertex $v_c \in V_G \setminus \{v_a, v_b\}$ if and only if v_c was formerly adjacent to v_a or v_b. Letting H be a graph with vertex set V_H, it is also possible to identify a vertex $u \in V_G$ with H. In this context, we construct a new graph, with vertex set $(V_G \cup V_H) \setminus \{u\}$, by deleting $u \in V_G$ and then adding edges to make every vertex $v_i \in V_H$ adjacent to every vertex $v_j \in V_G \setminus \{u\}$ formerly adjacent to $u \in V_G$.

2.2   Fixed-Parameter Tractability

Recall that a language $\mathcal{L}$ is Fixed-Parameter Tractable (FPT) if, letting $x \in \mathcal{L}$ be an arbitrary problem instance and k be some parameter, we have that a witness for x can be found in $f(k) \cdot |x|^{\mathcal{O}(1)}$ time for some computable function $f(k)$.

2.3   Fully Polynomial-Time Randomized Approximation Scheme (FPRAS)

Letting $x \in \mathcal{L}$ be a word in some language $\mathcal{L}$, and letting $f : \mathcal{L} \to \mathbb{N}_0$ be the problem of counting the number of witnesses $f(x)$ for x, a Randomized Approximation Scheme (RAS) is a procedure that outputs a value $\hat{f}_{(\epsilon, \delta)}(x)$ under the constraint that $Prob\left[\left(\frac{|\hat{f}_{(\epsilon, \delta)}(x) - f(x)|}{f(x)}\right) > \epsilon\right] < \delta$ [19]. If the running time for the RAS is a polynomial function of $|x|$, ϵ^{-1}, and $\ln\left(\delta^{-1}\right)$, then we may refer to it as a Fully Polynomial-time Randomized Approximation Scheme (FPRAS).

3 Treewidth Fixed-Parameter Tractability Results

Proposition 1. *For any fixed* $(k, w) \in \mathbb{N}_{>0}^2$, *counting rainbow k-connected w-colorings of graphs is linear time treewidth FPT.*

Proof. Our method of proof will be to show that counting rainbow k-connected w-colorings is expressible in Monadic Second-order (MS$_2$) logic.

Following "pg. 265–266" of Downey and Fellows [15], in MS$_2$ logic we are restricted to using: (1) the logical operators $\wedge$, $\vee$, $\neg$; (2) variables for vertices, edges, sets of vertices, and sets of edges; (3) the quantifiers $\forall$ and $\exists$; (4) the equality binary relation $(=)$ for both variables and sets; (5) the binary relation $x \in X$, where x is a variable corresponding to either a vertex or an edge; (6) the binary relation $inc(v_i, e_j)$ to express that a vertex v_i is incident to an edge e_j; and (7) the binary relation $adj(v_i, v_j)$ to express that a pair of vertices v_i and v_j are adjacent.

Here, by the standard proof argument for Courcelle's theorem [10–12], any MS$_2$ expressible problem can be translated into a finite Bottom-Up Tree Automaton (BUTA) that recognizes tree decompositions of an input graph, yielding a linear time treewidth FPT algorithm. Extensions of Courcelle's theorem can moreover guarantee the existence of a linear time treewidth FPT algorithms for MS$_2$ definable counting and optimization problems [3, 13]. In particular, we can observe the following lemma:

Lemma 1 *("Theorem 32" of Courcelle, Makowsky, and Rotics [13]). Any MS$_2$-definable counting problem ϕ for a graph G with vertex set V, edge set E, and treewidth $tw(G)$, can be solved in time $f\left(tw(G)\right) \cdot \mathcal{O}\left(|V| + |E|\right)$ where f is a function that depends only on $tw(G)$.*

To begin, letting G be an input graph with vertex set V_G and edge set E_G, we will first proceed along the lines of Eiben et al.'s [18] MS$_2$ formulation of the rainbow connectivity problem in writing the following MS$_2$ sentence, ψ_1, to express the decomposition of E_G into w distinct color classes, $\{C_1, \ldots C_w\}$:

$$\psi_1 := \exists C_1, \ldots, C_w \subseteq E_G \left(\forall e_i \in E_G \left(e_i \in C_1 \vee \ldots \vee e_i \in C_w\right)\right) \wedge$$
$$\forall i, j \in \{1, \ldots, w\} \left(i = j \vee C_i \cap C_j = \emptyset\right)$$

To express the property of there being k internally vertex disjoint proper paths between all pairs of vertices in V_G, we will use a modified version of Courcelle's notion of a *quasipath*. Following Courcelle [11], we have that an edge set X_i is a quasipath connecting a vertex v_x to a vertex v_y if and only if the following three constraints are met: (1) $v_x \neq v_y$; (2) both v_x and v_y are incident to a single unique edge in X_i; and (3) any vertex v_z not in the set $\{v_x, v_y\}$ is incident to exactly two edges in X_i. In the current context, to eliminate the possibility of disjoint cycles, we can define a *connected quasipath* as a quasipath satisfying an additional constraint, (4), that every vertex incident to an edge in X_i is reachable from v_x (or from v_y). As constraints (1) through (4) are expressible in MS_2 logic [11], we can accordingly use an auxiliary predicate $ConnectedQuasipath(X_i, v_x, v_y)$ to check whether an edge set X_i is a connected quasipath between v_x and v_y. We also write an auxiliary predicate, $Rainbow(X_i)$, to check if a given set of edges X_i defines a rainbow path between a pair of vertices v_x and v_y:

$$Rainbow(X_i) := \forall e_a, e_b \in X_i(e_a \neq e_b \implies \exists r \in \{1, \dots, w\}(e_a \in C_r \wedge e_b \notin C_r))$$

Finally, we can write an MS_2 sentence, ψ_2, to express the property of G having k internally vertex disjoint proper paths between all pairs of vertices $v_x, v_y \in V_G$, as follows (note that the constraints in quotes are defined in [11]):

$$\psi_2 := \forall v_x, v_y \in V_G(\exists X_1, \dots, X_k \subseteq E_G(v_x = v_y \vee$$
$$(ConnectedQuasipath\,(X_1, v_x, v_y) \wedge Rainbow(X_1) \wedge \dots \wedge$$
$$ConnectedQuasipath\,(X_k, v_x, v_y) \wedge Rainbow(X_k) \wedge \text{``}X_1, \dots, X_k$$
$$are\ pairwise\ disjoint\text{''} \wedge \text{``}no\ vertex\ except\ v_x\ and\ v_y\ belongs\ to\ an\ edge$$
$$of\ X_i\ and\ one\ of\ X_j\ for\ i \neq j\text{''})))$$

Putting everything together, we have that the sentence $\psi_1 \wedge \psi_2$ yields an MS_2 formula expressing the property of a graph possessing a rainbow k-connected w-coloring, and by Lemma 1, that there therefore exists a treewidth FPT algorithm for counting all such edge colorings in linear time.   □

Proposition 2. *For any fixed* $(k, w) \in \mathbb{N}^2_{>0}$, *counting* k-*proper connected* w-*colorings of graphs is linear time treewidth FPT.*

Proof. Observe that we can write an auxiliary predicate, $PP(X_i)$, to check if a given set of edges X_i defines a proper path between a pair of vertices v_x and v_y:

$$PP(X_i) := \forall e_a, e_b \in X_i, \forall v_z \in V_G(e_a \neq e_b \wedge inc(e_a, z) \wedge inc(e_b, z)$$
$$\implies \exists r \in \{1, \dots, w\}(e_a \in C_r \wedge e_b \notin C_r))$$

Now, otherwise following exactly the Proposition 1 proof argument, we can substitute the auxiliary predicate $Rainbow(X_i, v_x, v_y)$ with $PP(X_i)$ for each $1 \leq i \leq k$. This will check that each connected quasipath given by the edge set X_i, for each $1 \leq i \leq k$, is a proper path, yielding an MS_2 formulation for the property of a graph having a k-proper connected w-coloring. As in the Proposition 1 proof argument, it now suffices to appeal to Lemma 1.   □

4 Hardness Results

Proposition 3. *Unless* $NP = RP$, *no FPRAS can exist for counting rainbow* $(k = 1)$-*connected* w-*colorings of graphs for any fixed* $w \in \mathbb{N}_{>1}$.

Proof. Recall that the problem of deciding the existence of a rainbow $(k = 1)$-connected w-coloring is NP-complete for every $w \geq 2$ [1,7]. Here, by "Theorem 1" of Dyer et al. [17], for any NP-complete problem Q, we have that the corresponding counting version of this problem, $\#Q$, is complete for $\#P$ with respect to AP-reducibility. We therefore have that $\#Q$ is AP-interreducible with $\#SAT$, and thus cannot admit an FPRAS unless $NP = RP$ [17,22]. Accordingly, no FPRAS can exist for counting rainbow $(k = 1)$-connected w-coloring of graphs for any $w \geq 2$ unless $NP = RP$. $\qquad\square$

Theorem 1. *Deciding the existence of and counting rainbow k-connected $(w = 2)$-colorings of graphs is NP-complete and $\#P$-complete under many-one counting reductions, respectively, for every fixed $k \in \mathbb{N}_{>0}$.*

Proof. Recall that Monotone-3-SAT is a variant of 3-SAT where clauses have either all positive literals or all negative literals, and that the problem of deciding the existence of and counting solutions for such formula is NP-complete and $\#P$-complete under many-one counting reductions, respectively, due to the existence of a trivial parsimonious reduction from $\#SAT$ (see, e.g., the proof argument for "Theorem 2.2" of [4]). Letting $k \in \mathbb{N}_{>0}$ be a fixed parameter, we will establish the current theorem by giving a polynomial time many-one counting reduction from $\#$Monotone-3-SAT to counting rainbow k-connected $(w = 2)$-colorings of a graph for any fixed $k \in \mathbb{N}_{>0}$, where we will ensure that this graph will possess a rainbow k-connected $(w = 2)$-coloring if and only if the encoded instance of Monotone-3-SAT is satisfiable.

To begin, let ϕ be an arbitrary instance of Monotone-3-SAT with a set of variables $w_1, w_2, \ldots, w_n \in W$ and a set of clauses $c_1, c_2, \ldots, c_m \in C$. To slightly abuse notation, we will write $w_i \in c_j$ to indicate that the variable w_i exists in a clause c_j. Additionally, let $C_{pos} \subseteq C$ and $C_{neg} = C \setminus C_{pos}$ be the sets of clauses for ϕ having all positive and all negative literals, respectively.

We now assemble what we refer to as an "α construct", and remark that this step of the reduction is similar in nature to the reduction from 3-SAT (as opposed to Monotone-3-SAT) used by Chakraborty et al. [7] to establish the NP-completeness of deciding if a rainbow $(k = 1)$-connected $(w = 2)$-coloring exists for a given graph. The vertex set for the "α construct" is specified as $V_\alpha = \{v_a, v_b, v_c, v_q\} \cup W \cup B \cup C$, where $b_1, b_2, \ldots, b_n \in B$ are a set of auxiliary vertices for the clause vertices $c_1, c_2, \ldots, c_n \in C$, such that each $b_i \in B$ is uniquely associated with a $c_i \in C$, and B can accordingly be partitioned into sets $B_{pos} \subseteq B$ and $B_{neg} = B \setminus B_{pos}$. The edge set for the "$\alpha$ construct" can be specified as $E_\alpha = \{v_a \leftrightarrow v_c, v_b \leftrightarrow v_c\} \cup \{v_a \leftrightarrow b_i \mid \forall b_i \in B_{pos}\} \cup \{v_b \leftrightarrow b_i \mid \forall b_i \in B_{neg}\} \cup \{b_i \leftrightarrow c_i \mid \forall i \text{ s.t. } 1 \leq i \leq m\} \cup \{w_i \leftrightarrow c_j \mid \forall w_i, c_j \text{ s.t. } w_i \in c_j\} \cup \{w_i \leftrightarrow v_q \mid \forall i \text{ s.t. } 1 \leq i \leq n\}$.

Next, we proceed by finding an edge 2-coloring for the smallest possible number of edges in the "α construct" which is sufficient to ensure that: (constraint 1) there is a rainbow path connecting the vertices v_a and v_b; (constraint 2) there is a rainbow path connecting the vertex v_a to each of the vertices $c_i \in C_{pos}$; (constraint 3) there is a rainbow path connecting the vertex v_b to each of the vertices $c_i \in C_{neg}$; (constraint 4) there is a rainbow path connecting each $b_i \in B$ to each variable vertex w_r whenever we have that w_r occurs in c_i. Let P be the set of all vertex pairs in the "α construct", let $S \subset P$ be the set of all vertex pairs that must be connected by a rainbow path under (constraint 1) through (constraint

4), and observe that the elements in the sets P and S can trivially be listed out in polynomial time. Here, letting E_1 and E_2 be the two sets into which E_α can be partitioned by an edge 2-coloring, under these constraints we have that $v_a \leftrightarrow v_c \in E_2$ implies $\{v_b \leftrightarrow v_c\} \cup \{v_a \leftrightarrow b_i \mid \forall b_i \in B_{pos}\} \cup \{b_i \leftrightarrow c_i \mid \forall i \; s.t. \; 1 \leq i \leq m \; and \; b_i \in B_{neg}\} \cup \{w_i \leftrightarrow c_j \mid \forall w_i, c_j \; s.t. \; w_i \in c_j \; and \; c_j \in C_{pos}\} \subset E_1$ and $\{v_a \leftrightarrow v_c\} \cup \{v_b \leftrightarrow b_i \mid \forall b_i \in B_{neg}\} \cup \{b_i \leftrightarrow c_i \mid \forall i \; s.t. \; 1 \leq i \leq m \; and \; b_i \in B_{pos}\} \cup \{w_i \leftrightarrow c_j \mid \forall w_i, c_j \; s.t. \; w_i \in c_j \; and \; c_j \in C_{neg}\} \subset E_2$.

At this point, observe that deciding if all pairs in the set $R = S \cup \{(c_i, v_q) \mid \forall i \; s.t. \; 1 \leq i \leq m\}$ can be connected via a rainbow path is polynomial time equivalent to deciding if ϕ is satisfiable. In particular, this will require partitioning the edges in the set $\{w_i \leftrightarrow v_q \mid \forall i \; s.t. \; 1 \leq i \leq n\}$ into E_1 and E_2, where partitioning an edge $w_i \leftrightarrow v_q$, for some $1 \leq i \leq n$, into E_1 (resp. E_2) is equivalent to assigning the value "True" (resp. "False") to the variable corresponding to w_i in ϕ.

The subsequent step of our reduction will be to create a "β construct" from the "α construct" which will ensure the existence of k instances of 2-edge rainbow paths connecting each pair of vertices in P if and only if ϕ is satisfiable. Here, add $(k-1) \cdot |R|$ new vertices and $2(k-1) \cdot |R|$ new edges to allow for exactly $k-1$ instances of 2-edge rainbow path "bridges" connecting all vertex pairs in the set $R \subset P$, and add $\lceil (k^3 \cdot (|P|+3)^6) \rceil$ new vertices (and twice this number of new edges) to allow for the existence of $\geq k$ length ≤ 2 edge rainbow path "bridges" connecting all vertex pairs in the set $P \setminus R$. Let $V_{(\beta,\Delta)}$ be the set of $(k-1) \cdot |R| + \lceil (k^3 \cdot (|P|+3)^6) \rceil$ new vertices added to create the aforementioned rainbow path "bridges" between the vertex pairs in P. Finally, complete the "β construct" by adding $\frac{1}{2} (|V_{(\beta,\Delta)}| - 1) \cdot |V_{(\beta,\Delta)}|$ edges to form a clique out of the vertices in $V_{(\beta,\Delta)}$, and let $E_{(\beta,\Delta)}$ be the set of edges for the "β construct" disjoint from the set of edges for the "α construct". Observe that the aforementioned rainbow path "bridges" will allow for the existence of k rainbow paths connecting all pairs of vertices in the set P if and only if ϕ is satisfiable.

To complete the current proof, it remains to show that, provided an oracle that counts rainbow k-connected 2-colorings of arbitrary graphs, we can determine the number of satisfying assignments for ϕ in time polynomial in the size of the "β construct". To do this, letting Ψ_β be the total number of rainbow k-connected 2-colorings of the "β construct", and letting Υ be a lower bound for the number of rainbow k-connected 2-colorings of the "β construct" per witness for ϕ, we will show that it is possible to estimate Υ well enough that the total number of witnesses for ϕ will exactly correspond to $\lfloor (\frac{\Psi_\beta}{\Upsilon}) \rfloor$.

Here, let $\omega_1, \omega_2, \ldots \omega_{(|P \setminus R|)} \in \Omega$ be a set of subgraphs of the "β construct", in which the ith subgraph is induced by the ith pair of vertices in $P \setminus R$, together with the exactly $\lceil (k^3 \cdot (|P|+3)^6) \rceil$ vertices in $V_{(\beta,\Delta)}$ adjacent to both vertices in this pair. In addition, let Φ be the clique subgraph of the "β construct" induced by the vertices in the set $V_{(\beta,\Delta)}$, and let Λ be the set of $2(k-1) \cdot |R|$ "bridges" connecting all vertex pairs in the set $R \subset P$. In this context, Υ can be expressed as a product of the following terms:

- ($\varsigma_1 = 2$) the number of edge 2-colorings of the "α construct" per witness for ϕ which are consistent with (constraint 1) through (constraint 4);
- ($\varsigma_2 = 2^{((k-1)\cdot|R|)}$) the number of edge 2-colorings of the edges in Λ which are consistent with the existence of a rainbow k-connected 2-coloring of the "β construct";
- ($\varsigma_3 = 2^{(|E_{(\beta,\Delta)}|-((k-1)\cdot|R|))}$) all possible 2-colorings of edges adjacent to at least one vertex in $V_{(\beta,\Delta)}$ and disjoint from Λ;
- (ς_4) a lower bound for the fraction of edge 2-colorings counted by ($\varsigma_1 \cdot \varsigma_2 \cdot \varsigma_3$) (i.e., a superset of all possible edge 2-colorings potentially consistent with the existence of a rainbow k-connected for the "β construct") which ensure that the subgraphs in $M \cup \{\Phi\}$ are at least rainbow k-connected.

To find an appropriate value for ς_4, we will make use of the following lemma:

Lemma 2. *For $k \leq n^{1/3}$, $w \geq 2$, $n \geq 7677106$, and letting G be an order n graph with $\geq \lceil(\sqrt{n})\rceil$ internally vertex disjoint paths of length ≤ 2 edges between all pairs of vertices, an edge w-coloring of G, chosen uniformly from the set of all possible edge w-colorings, is rainbow $(\geq k)$-connected with probability $\geq \left(1 - 2^{(-n^{1/3})}\right)$.*

Proof Sketch. Assume $w = 2$, let G be an arbitrary order n graph with $\geq \lceil(\sqrt{n})\rceil$ internally vertex disjoint paths of length ≤ 2 edges between all pairs of vertices, and imbue G with an edge $(w = 2)$-coloring chosen uniformly from the set of all possible such edge colorings. Here, observe that the probability a given pair of vertices in G are connected by at most $k - 1$ internally vertex disjoint rainbow paths is $\geq \left(\frac{1}{2}\right)^{(\lceil(\sqrt{n})\rceil)} \cdot \sum_{j=0}^{k-1}\binom{\lceil(\sqrt{n})\rceil}{j} \leq \left(\frac{1}{2}\right)^{(\sqrt{n})} \cdot \sum_{j=0}^{k-1}(e \cdot \sqrt{n})^j$. Accordingly, assuming $n \geq 3$ and $k \geq 2$, by Boole's inequality (i.e., the "union bound") we can express an upper bound for the probability, $\mathcal{P}_{(\alpha,n\geq3,k\geq2)}$, that some pair of vertices in G is connected by at most $k - 1$ internally vertex disjoint rainbow paths as $\mathcal{P}_{(\alpha,n\geq3,k\geq2)} = \binom{n}{2} \cdot \left(\frac{1}{2}\right)^{(\sqrt{n})} \cdot \sum_{j=0}^{k-1}(e \cdot \sqrt{n})^j$. While we omit the details due to space constraints, here, it can be shown that $2^{(-n^{1/3})} \cdot \left(\mathcal{P}_{(\alpha,n\geq3,k\geq2)}\right)^{-1} \geq 1$ for $k \leq n^{1/3}$ and $n \geq 7677106$, yielding the current lemma. $\qquad\square$

For $|P|$ sufficiently large (e.g., $|P| \geq 7677106$), invoking Boole's inequality (i.e., the "union bound") together with Lemma 2 allows us to specify that $\varsigma_4 = 1 - \left(1 + |P|^2\right) \cdot \left(2^{\left(-\left(\sqrt{(|P|+3)^6}\right)^{1/3}\right)}\right) = 1 - \left(1 + |P|^2\right) \cdot 2^{(-|P|-3)}$. In particular,

the term $2^{(-|P|-3)}$ from the expression for ς_4 corresponds to an upper bound for the fraction of edge 2-colorings counted by ($\varsigma_1 \cdot \varsigma_2 \cdot \varsigma_3$) – again, corresponding to the cardinality of a superset of all possible edge 2-colorings allowing the "β construct" to be rainbow k-connected – in which a specific subgraph $m_i \in M$ fails to be rainbow k-connected. We are then multiplying this term by $\left(1 + |P|^2\right)$ to compute a further upper bound (via Boole's inequality) for the fraction of colorings in which any instance of a subgraph in the set $M \cup \{\Phi\}$ fails to be rainbow k-connected. Accordingly, ς_4 will give the desired lower bound for the fraction of colorings in which every subgraph in the set $M \cup \{\Phi\}$ is rainbow k-connected, implying that the larger "β construct" will have a proper k-connected coloring. To now show that $\left\lfloor \left(\frac{\Psi_\beta}{\gamma}\right) \right\rfloor$ will recover the correct number of witnesses for ϕ, and letting $2\sqrt{|P|}$ correspond to a trivial upper bound for the number of witnesses for ϕ, it suffices to observe that:

$$2^{\sqrt{|P|}} \cdot (1 - \varsigma_4) < \varsigma_4$$
$$\Longleftarrow 2^{\sqrt{|P|}} \cdot (1 + |P|^2) \cdot 2^{(-|P|-3)} < 1 - \left((1 + |P|^2) \cdot 2^{(-|P|-3)}\right)$$
$$\Longleftarrow \left(2^{\sqrt{|P|}} + 1\right)(|P|^2 + 1) < 2^{(|P|+3)} \Longleftarrow |P| \geq 1$$

Putting everything together, for any $k \in \mathbb{N}_{>0}$ we have given a many-one counting reduction from counting witnesses for Monotone-3-SAT to counting rainbow k-connected 2-colorings of a graph. We have moreover done so in such a manner that a rainbow k-connected 2-coloring can exist if and only if the encoded Monotone-3-SAT formula has at least one witness. Accordingly, deciding the existence of and counting rainbow k-connected 2-colorings is NP-hard and $\#P$-hard under many-one counting reductions, respectively. As the decision problem is trivially in NP, and as the counting problem is trivially in $\#P$, this yields the theorem.   □

Theorem 2. *For any fixed* $(k, w) \in \mathbb{N}_{>0} \times \mathbb{N}_{>1}$, *the problem of deciding the existence of a rainbow* k-*connected* w-*coloring of a graph is* NP-*complete.*

Proof. As the proof argument for Theorem 1 establishes the current theorem in all cases where $k \in \mathbb{N}_{>0}$ and $w = 2$, we need only concern ourselves with the cases where $w \geq 3$. Here, for each $w \geq 3$, we proceed via reduction from the problem of deciding the existence of a rainbow 1-connected w-coloring of an arbitrary graph, which is once again known to be NP-complete for every $w \in 2\mathbb{N}_{>0}$ [7] and for every $w \in 2\mathbb{N}_{>0} + 1$ [1].

To begin, let G be a graph with vertex set $v_1, v_2, \ldots v_n \in V_G$ and edge set E_G, where $|V_G| \geq k$. If $k \in 2\mathbb{N}_{>0} + 1$, retaining the original labels for vertices in V_G (i.e., such that we may later refer to a vertex $v_i \in V_G$ in the modification of G), construct a graph H with the following vertex and edge sets:

$$V_{H_{odd}} = V_G \cup \{h_{(1,1)}\} \cup \left(\bigcup_{i=2}^{(k+1)/2} \left(\bigcup_{j=1}^{n} h_{(i,j)}\right)\right) \cup \left(\bigcup_{i=1}^{k-1} b_i\right)$$
$$E_{H_{odd}} = E_G \cup \left(\bigcup_{i=1}^{n} h_{(1,1)} \leftrightarrow h_{(2,i)}\right) \cup \left(\bigcup_{i=2}^{((k-1)/2} \left(\bigcup_{j=1}^{n} h_{(i,j)} \leftrightarrow h_{(i+1,j)}\right)\right)$$
$$\cup \left(\bigcup_{i=1}^{n} v_i \leftrightarrow h_{((k+1)/2,i)}\right) \cup \left(\bigcup_{i=1}^{k-1} \left(\bigcup_{j=1}^{n} b_i \leftrightarrow v_j\right)\right)$$
$$\cup \left(\bigcup_{i=1}^{k-1} \left(\bigcup_{j=1}^{n} b_i \leftrightarrow h_{((k+1)/2,j)}\right)\right)$$

Alternatively, if $k \in 2\mathbb{N}_{>1}$, likewise retaining the original labels for vertices in V_G, construct a graph H with the following vertex and edge sets:

$$V_{H_{even}} = V_G \cup \left(\bigcup_{i=1}^{k/2} \left(\bigcup_{j=1}^{n} h_{(i,j)} \right) \right) \cup \left(\bigcup_{i=1}^{k-1} b_i \right)$$

$$E_{H_{even}} = E_G \cup \left(\bigcup_{i=1}^{n-1} \left(\bigcup_{j=(i+1)}^{n} h_{(1,i)} \leftrightarrow h_{(1,j)} \right) \right)$$

$$\cup \left(\bigcup_{i=1}^{((k-2)/2)} \left(\bigcup_{j=1}^{n} h_{(i,j)} \leftrightarrow h_{(i+1,j)} \right) \right) \cup \left(\bigcup_{i=1}^{n} v_i \leftrightarrow h_{(k/2,i)} \right)$$

$$\cup \left(\bigcup_{i=1}^{k-1} \left(\bigcup_{j=1}^{n} b_i \leftrightarrow v_j \right) \right) \cup \left(\bigcup_{i=1}^{k-1} \left(\bigcup_{j=1}^{n} b_i \leftrightarrow h_{(k/2,j)} \right) \right)$$

Finally, letting $\zeta_{(k,w)}$ be a graph gadget generated by identifying every vertex of a clique K_{2k} with an independent set (i.e., empty graph) of order w, create a graph H' from H, with vertex set $V_{H'}$ and edge set $E_{H'}$, by identifying every vertex $h_{(i,j)}$, $\forall i \in [1, (k+1)/2]$ if $k \in 2\mathbb{N}_{>0} + 1$ or $\forall i \in [1, k/2]$ if $k \in 2\mathbb{N}_{>1}$ and $\forall j \in [1, n]$, with an instance of the $\zeta_{(k,w)}$ gadget. In this context, for some vertex $v \in V_{H'}$, we will write $v \in V_G$ to express that v is one of the original vertices in the graph G, we will write $v \in \mathcal{B}$ to express that v is a vertex with a label of the form b_i for some $i \in [1, k-1]$, and we will write $v \in \mathcal{Z}$ to express that v belongs to one of the $\zeta_{(k,w)}$ graph gadgets used to construct H' from H.

We can now observe the following lemma:

Lemma 3. *Following the notation of the Theorem 2 proof argument, H' will admit a rainbow k-connected w-coloring if and only if G admits a rainbow 1-connected w-coloring.*

Proof Sketch. To first treat the "only if" direction, observe that only $k - 1$ internally vertex disjoint rainbow paths between a pair of vertices $v_a, v_b \in V_G$ can include vertices internal to the one of the $\zeta_{(k,w)}$ gadgets. Observe further that each such path must also include a distinct vertex in $\mathcal{B}$, where $|\mathcal{B}| = k - 1$. Here, this is a consequence of the fact that any (not necessarily rainbow) path between $v_a, v_b \in V_G$, which includes no vertex in $\mathcal{B}$, and at least one vertex in $\mathcal{Z}$, must have a (edge-wise) length of at least $w + 1$. Accordingly, there must exist at least one rainbow path between v_a and v_b including only vertices in V_G.

To next treat the "if" direction, it suffices to detail a partial edge coloring of the graph H', using a set of colors given by integers 1 through w, which satisfies the following requirements for paths with edges strictly belonging to the partial coloring: (requirement 1) there exist $k - 1$ internally vertex disjoint rainbow paths connecting each pair of vertices in V_G; (requirement 2) there exist at least k internally vertex disjoint rainbow paths connecting each vertex in V_G to each vertex in $\mathcal{B}$; (requirement 3) there exist at least k internally vertex disjoint rainbow paths connecting each pair of vertices in $\mathcal{B}$; (requirement 4) there exist at least k internally vertex disjoint rainbow paths connecting each pair of vertices in $\mathcal{Z}$; (requirement 5) there exist at least k internally vertex disjoint rainbow paths connecting each vertex in V_G to each vertex in $\mathcal{Z}$; (requirement 6) there exist at least k internally vertex disjoint rainbow paths connecting each vertex in $\mathcal{B}$ to each vertex in $\mathcal{Z}$.

To address (requirement 1) through (requirement 3), we assign the colors 1, 2, and 3 to edges connecting vertices in V_G to vertices in $\mathcal{Z}$, connecting vertices in $\mathcal{B}$ to vertices in $\mathcal{Z}$, and connecting vertices in V_G to vertices in $\mathcal{B}$, respectively.

On account of the fact that any vertex in V_G or $\mathcal{B}$ will have $(2k \cdot w)$ edges to distinct vertices in $\mathcal{Z}$, it is straightforward to check that these colored edges will satisfy (requirement 1) by ensuring the existence of exactly $k-1$ internally vertex disjoint rainbow paths between all pairs of vertices in V_G. More specifically, these paths will be ordered sets of edges of the form $\{v_a \leftrightarrow v_b, v_b \leftrightarrow v_c, v_c \leftrightarrow v_d\}$, where $v_a, v_d \in V_G$, $v_b \in \mathcal{Z}$, and $v_c \in \mathcal{B}$. It is also straightforward to check that this scheme will satisfy (requirement 2) by ensuring the existence of at least k internally vertex disjoint rainbow paths of length 2 connecting any vertex $v_a \in V_G$ to any vertex in $\mathcal{B}$, and satisfy (requirement 3) by ensuring at least k internally vertex disjoint rainbow paths of length 3 connecting any pair of vertices in $\mathcal{B}$.

To address (requirement 4) through (requirement 6), for each of the $2k$ independent sets of order w in each instance of a $\zeta_{(k,w)}$ gadget, we independently and arbitrarily assign each vertex a distinct integer "ordering" label from the interval $[1, w]$ (i.e., such that a vertex in $V_{H'}$ is assigned an "ordering" label if and only if it belongs to the set $\mathcal{Z}$). In this context, for each vertex $v \in \mathcal{Z}$, we let $order(v)$ be a function which returns the "ordering" label of v. We now strictly extend the earlier partial coloring of H' (i.e., used to treat (requirement 1) through (requirement 3)) by assigning each edge $v_a \leftrightarrow v_b \in E_{H'}$, where $v_a, v_b \in \mathcal{Z}$, the color $((order(v_a) + order(v_b)) \mod w) + 1$. Letting $\mathcal{I}_1$ and $\mathcal{I}_2$ be the sets of vertices in any two order w independent sets (i.e., empty graphs) belonging to the same or adjacent instances of the $\zeta_{(k,w)}$ gadget in H', observe that this will ensure the complete bipartite graph $K_{w,w}$ induced by $\mathcal{I}_1 \cup \mathcal{I}_2$ is proper edge colored.

At this point, due to space constraints, we leave it to the reader to verify that this coloring scheme satisfies (requirement 4) through (requirement 6), yielding the current lemma.   □

Provided Lemma 3, it remains to observe for every $(k, w) \in \mathbb{N}_{>0} \times \mathbb{N}_{>1}$ that the problem of deciding the existence of a rainbow k-connected w-coloring is in NP. Here, this is a consequence of the fact that, for any fixed $w \in \mathbb{N}_{>0}$, a rainbow path can be of length at most w. Accordingly, for a graph on n vertices, we have that there will be at most $\mathcal{O}\left(n^{w+1}\right)$ rainbow paths of any length in the graph (connecting any pair of vertices), allowing us to likewise verify that a given edge coloring is a rainbow k-connected w-coloring in time polynomial in the order of the graph. Putting everything together, as we have that the problem of deciding the existence of a rainbow k-connected w-coloring is both NP-hard and in NP for every $(k, w) \in \mathbb{N}_{>0} \times \mathbb{N}_{>1}$, we have that all such decision problems are NP-complete, yielding the current theorem.   □

Corollary 1. *For any fixed* $(k, w) \in \mathbb{N}_{>0} \times \mathbb{N}_{>1}$, *unless* $NP = RP$, *no FPRAS can exist for counting rainbow k-connected w-colorings of graphs.*

Proof. By Theorem 1 and Theorem 2, the problem of deciding if a graph G admits a rainbow k-connected w-coloring is NP-complete for each $(k, w) \in \mathbb{N}_{>0} \times \mathbb{N}_{>1}$. Accordingly, the current proposition follows directly from "Theorem 1" of Dyer et al. [17] and the proof argument for Proposition 3.   □

References

1. Ananth, P., Nasre, M., Sarpatwar, K.K.: Rainbow connectivity: hardness and tractability. In: Proceedings of the 31st FSTTCS, pp. 241–251 (2011)
2. Andrews, E., Lumduanhom, C., Laforge, E., Zhang, P.: On proper-path colorings in graphs. J. Comb. Math. Comb. Comput. **97**, 189–207 (2016)
3. Arnborg, S., Lagergren, J., Seese, D.: Easy problems for tree-decomposable graphs. J. Algorithms **12**(2), 308–340 (1991)
4. Barrett, C., et al.: Predecessor existence problems for finite discrete dynamical systems. Theoret. Comput. Sci. **386**(1–2), 3–37 (2007)
5. Bondy, J.A., Murty, U.S.R.: Graph Theory with Applications, 1st edn. Macmillan Press, New York (1976)
6. Borozan, V., et al.: Proper connection of graphs. Discrete Math. **312**(17), 2550–2560 (2012)
7. Chakraborty, S., Fischer, E., Matsliah, A., Yuster, R.: Hardness and algorithms for rainbow connection. J. Comb. Optim. **21**(3), 330–347 (2011)
8. Chartrand, G., Johns, G.L., McKeon, K.A., Zhang, P.: Rainbow connection in graphs. Math. Bohem. **133**(1), 85–98 (2008)
9. Chartrand, G., Johns, G.L., McKeon, K.A., Zhang, P.: The rainbow connectivity of a graph. Networks **54**(2), 75–81 (2009)
10. Courcelle, B.: The monadic second-order logic of graphs. I. Recognizable sets of finite graphs. Inform. Comput. **85**(1), 12–75 (1990)
11. Courcelle, B.: The monadic second-order logic of graphs XII: planar graphs and planar maps. Theoret. Comput. Sci. **237**(1–2), 1–32 (2000)
12. Courcelle, B.: Graph structure and monadic second-order logic: language theoretical aspects. In: Aceto, L., Damgård, I., Goldberg, L.A., Halldórsson, M.M., Ingólfsdóttir, A., Walukiewicz, I. (eds.) ICALP 2008. LNCS, vol. 5125, pp. 1–13. Springer, Heidelberg (2008). https://doi.org/10.1007/978-3-540-70575-8_1
13. Courcelle, B., Makowsky, J.A., Rotics, U.: On the fixed parameter complexity of graph enumeration problems definable in monadic second-order logic. Discrete Appl. Math. **108**(1–2), 23–52 (2001)
14. Diestel, R.: Graph Theory, 5th edn. Springer, Heidelberg (2016). https://doi.org/10.1007/978-3-662-53622-3
15. Downey, R.G., Fellows, M.R.: Fundamentals of Parameterized Complexity, 1st edn. Springer, New York (2013). https://doi.org/10.1007/978-1-4471-5559-1
16. Ducoffe, G., Marinescu-Ghemeci, R., Popa, A.: On the (di)graphs with (directed) proper connection number two. Discrete Appl. Math. **281**, 203–215 (2020)
17. Dyer, M., Goldberg, L.A., Greenhill, C., Jerrum, M.: The relative complexity of approximate counting problems. Algorithmica **38**(3), 471–500 (2004)
18. Eiben, E., Ganian, R., Lauri, J.: On the complexity of rainbow coloring problems. Discrete Appl. Math. **246**, 38–48 (2018)
19. Karp, R.M., Luby, M.: Monte-Carlo algorithms for enumeration and reliability problems. In: Proceedings of the 24th FOCS, pp. 56–64 (1983)
20. Li, X., Magnant, C.: Properly colored notions of connectivity – a dynamic survey. Theory Appl. Graphs (1), 1–16 (2015)
21. Reed, M.G., Syverson, P.F., Goldschlag, D.M.: Anonymous connections and onion routing. IEEE J. Sel. Area. Commun. **16**(4), 482–494 (1998)
22. Zuckerman, D.: On unapproximable versions of NP-complete problems. SIAM J. Comput. **25**(6), 1293–1304 (1996)

Some Combinatorial Algorithms on the Edge Cover Number of k-Regular Connected Hypergraphs

Zhongzheng Tang[1,2], Yaxuan Li[3], and Zhuo Diao[3(✉)]

[1] Key Laboratory of Mathematics and Information Networks (Beijing University of Posts and Telecommunications), Ministry of Education, Beijing, China
tangzhongzheng@amss.ac.cn
[2] School of Science, Beijing University of Posts and Telecommunications, Beijing 100876, China
[3] School of Statistics and Mathematics, Central University of Finance and Economics, Beijing 100081, China
liyaxuanlee@email.cufe.edu.cn, diaozhuo@amss.ac.cn

Abstract. For $k \geq 3$, let H be a k-regular connected hypergraph on n vertices and m edges. The edge cover number $\rho(H)$ is the minimum number of edges that intersect every vertex. We prove the following inequality: $\rho(H) \leq \frac{(k-1)n+1}{k}$. Furthermore, the extremal hypergraphs with equality holds are exactly k-star hypertrees. Based on the proofs, some combinatorial algorithms on the edge cover number are designed.

Keywords: edge cover · k-regular · extremal hypergraph · k-star hypertree

1 Introduction

A hypergraph is a generalization of a graph in which an edge can join any number of vertices. The loop is an edge which contains only one vertex. The multiple edges are edges which contain same vertices. A multiple hypergraph is a hypergraph allowing loops and multiple edges. As for a graph, the order of H, denoted by n, is the number of vertices. The number of edges is denoted by m. The studies of this paper are applicable to multiple hypergraphs.

For each vertex $v \in V$, the degree $d(v)$ is the number of edges in E that contains v. We say v is an isolated vertex of H if $d(v) = 0$. Hypergraph H is *k-regular* if each vertex's degree is k. Hypergraph H is *k-uniform* if each edge contains exactly k vertices. The rank $r(H)$ is the maximal cardinality of an edge in H.

For any vertex set $S \subseteq V$, we write $H \backslash S$ for the subhypergraph of H obtained from H by deleting all vertices in S and all edges incident with some vertices in

Supported by National Natural Science Foundation of China under Grant No. 11901605, No. 12101069, Program for Innovation Research in Central University of Finance and Economics, the disciplinary funding of Central University of Finance and Economics.

X. Chen and B. Li (Eds.): TAMC 2024, LNCS 14637, pp. 284–295, 2024.
https://doi.org/10.1007/978-981-97-2340-9_24

S. For any edge set $A \subseteq V$, we write $H \setminus A$ for the subhypergraph of H obtained from H by deleting all edges in A and keeping vertices. If S is a singleton set s, we write $H \setminus s$ instead of $H \setminus \{s\}$. For any vertex set $S \subseteq V$, the hypergraph $H_S = (S, \{e \cap S \mid e \in E\})$ is called the subhypergraph induced by the vertex set S.

Let $k \geq 2$ be an integer. A cycle of length k, denoted as k-cycle, is a vertex-edge sequence $C = v_1 e_1 v_2 e_2 \cdots v_k e_k v_1$ with: (1) $\{e_1, e_2, \ldots, e_k\}$ are distinct edges of H. (2) $\{v_1, v_2, \ldots, v_k\}$ are distinct vertices of H. (3) $\{v_i, v_{i+1}\} \subseteq e_i$ for each $i \in [k]$, here $v_{k+1} = v_1$. We consider the cycle C as a subhypergraph of H with vertex set $\{v_i, i \in [k]\}$ and edge set $\{e_j, j \in [k]\}$.

Similarily, a path of length $k \geq 1$, denoted as k-path, is a vertex-edge sequence $P = v_1 e_1 v_2 e_2 \cdots v_k e_k v_{k+1}$ with: (1) $\{e_1, e_2, \ldots, e_k\}$ are distinct edges of H. (2) $\{v_1, v_2, \ldots, v_{k+1}\}$ are distinct vertices of H. (3) $\{v_i, v_{i+1}\} \subseteq e_i$ for each $i \in [k]$. We consider the path P as a subhypergraph of H with vertex set $\{v_i, i \in [k+1]\}$ and edge set $\{e_j, j \in [k]\}$.

A hypergraph $H = (V, E)$ is called *connected* if any two of its vertices are linked by a path in H. For two distinct vertices u and v, the distance between u and v is the length of a shortest path connecting u and v, denoted by $d(x, y)$. A hypergraph $H = (V, E)$ is called a *hypertree* if H is connected and acyclic, denoted by $T(V, E)$.

Given a hypergraph $H(V, E)$, a set of vertices $S \subseteq V$ is an independent set if every two distinct vertices in S are not adjacent. The independent number is the maximum cardinality of an independent set, denoted by $\alpha(H)$. A set of edges $A \subseteq E$ is an *edge cover* if every vertex is incident to at least an edge in A. The edge cover number is the minimum cardinality of an edge cover, denoted by $\rho(H)$. In this paper, we consider the edge cover number in k-regular connected hypergraphs.

1.1 Known Results

Edge cover number is also called the edge covering number. The edge covering problem of hypergraphs is a generalization of the edge covering problem of graphs, which was first proposed by Norman and Rabin in 1959 [8]. Berger and Ziv [4] proved that for a hypergraph H with m edges and rank $r > 2$, its edge covering number $\rho(H)$ satisifies $\rho(H) \leq \frac{(r-2)m+\alpha}{r-1}$, where α is the independence number of H. Aharoni, Berger and Ziv [1] studied a conjecture on the edge covering number of the intersection of two matroids at the level of hypergraphs. Aharoni, Holzman and Zerbib [2] studied the d-interval hypergraph and bound the edge covering number in terms of a parameter expressing independence of systems of partitions of the d unit intervals. Khan and Kally [5] proved a sufficient condition to ensure that for fixed $k \geq 2$, H is a k-bounded hypergraph, and $t : H \to \mathbb{R}^+$ is a fraction cover, the edge covering number $\rho(H)$ of H is asymptotically at most $\sum_{A \in H} t(A)$. Olejnik [9] proved that for an r-uniform, n-order hypergraph H and its complement $\overline{H}$ where the edge covering numbers are denoted by $\rho(H)$ and $\rho(\overline{H})$, there existed $\lfloor \frac{n}{r} \rfloor \leq \rho(H) + \rho(\overline{H}) \leq 2\lfloor \frac{n}{r} \rfloor$.

The edge covering problem of hypergraphs is equivalent to the set covering problem, the algorithms used to solve the set covering problem can often be used to solve the edge covering problem of hypergraph. Nieminen [7] designed a combinatorial algorithm to find an upper bound of the edge covering number for a hypergraph. Peleg, Schechtman and Wool [10] designed several approximation algorithms to solve the minimum edge covering problem on r-uniform hypergraphs, and gave the approximate ratios. Krivelevich [6] proposed a new method to solve the minimum edge covering problem on a r-uniform hypergraph with a maximum degree Δ, which achieves the approximate ratio of $r(1-\frac{c}{\Delta^{\frac{1}{r-1}}})$.

1.2   Our Results

In this paper, for every k-regular connected hypergraph, we give a sharp upper bound for the edge cover number. The result is stated in Theorem 1.

Theorem 1. *For $k \geq 3$, let $H(V, E)$ be a k-regular connected hypergraph with n vertices. Then $\rho(H) \leq \frac{(k-1)n+1}{k}$. Furthermore, $\rho(H) = \frac{(k-1)n+1}{k}$ holds if and only if H is a k-star hypertree.*

For a non-regular hypergraph H with maximum degree Δ, we can construct a Δ-regular hypergraph H' by adding some loops to the vertices and these operations keep the edge cover number. According to Theorem 1, the next corollary is instant:

Corollary 1. *For $\Delta \geq 3$, let $H(V, E)$ be a connected hypergraph with maximum degree Δ on n vertices, then $\rho(H) \leq \frac{(\Delta-1)n+1}{\Delta}$. Furthermore, $\rho(H) = \frac{(\Delta-1)n+1}{\Delta}$ holds if and only if the hypergraph resulting from adding some loops on H can be a Δ-star regular hypertree.*

The main content of the article is organized as follows:

- In Sect. 2, we introduce a *breaking-cycle* operation and prove $\rho(H) \leq [(k-1)n + 1]/k$ holds for k-regular connected hypergraphs. Based on the proof, a polynomial time algorithm is designed to compute an edge cover set with cardinality at most $[(k-1)n + 1]/k$.
- In Sect. 3, we characterize the extremal k-regular connected hypergraphs with $\rho(H) = [(k-1)n+1]/k$. By structure analysis, we prove the extremal hypergraphs are exactly k-star hypertrees.

2   The Upper Bound for Edge Cover Number

In this section, we prove $\rho(H) \leq \frac{(k-1)n+1}{k}$ holds for k-regular connected hypergraphs, as stated in Theorem 2.

Theorem 2. *For $k \geq 3$, let $H(V, E)$ be a k-regular connected hypergraph with n vertices. Then $\rho(H) \leq \frac{(k-1)n+1}{k}$.*

The content is organized as follows:

- The conception and property of *breaking-cycle* operations are introduced by Definition 1 and Lemma 1.
- The upper bound for the edge cover number is proved in Lemma 2, Lemma 3, Lemma 4 and Theorem 2.
- Computing the edge cover with cardinality no more than the upper bound is shown in Algorithm 2.

Definition 1. *For a k-regular connected hypergraph $H(V,E)$, let $C = v_1 e_1 v_2 e_2 \cdots v_t e_t v_1, t \geq 2$ be a cycle in H. The breaking-cycle operation of C is eliminating an adjacent vertex v_i in e_i of the cycle C and adding a loop for v_i, as shown in Fig. 1.*

Lemma 1. *The breaking-cycle operation does not decrease the edge cover number.*

Proof. Denote $\widetilde{H}$ as the hypergraph obtained by the *breaking-cycle* operation of C. Obviously, $\widetilde{H}$ is k-regular and connected. v_i is the breaking adjacent vertex in H and $\widetilde{e}_i$ is the corresponding loop in $\widetilde{H}$; $e_i(v_i, v_{i+1})$ is the breaking edge in H, as shown in Fig. 1.

Take a minimum edge cover $\widetilde{A}$ of $\widetilde{H}$. If $\widetilde{e}_i \in \widetilde{A}$, then replace $\widetilde{e}_i$ by e_{i-1}. Thus there is a minimum edge cover $\widetilde{A}$ of $\widetilde{H}$ and $\widetilde{e}_i \notin \widetilde{A}$. Take $A = \widetilde{A}$ and A is an edge cover of H. Thus $\rho(H) \leq \rho(\widetilde{H})$. □

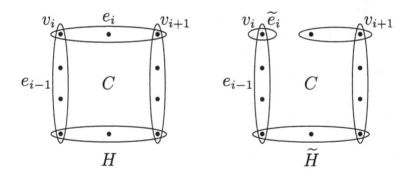

Fig. 1. The schematic diagram of *breaking-cycle* operation

Lemma 2. *For every k-regular connected hypergraph $H(V,E)$ with n vertices and m edges, we have $m \leq (k-1)n + 1$.*

Proof. We prove this lemma by induction on the vertex number n. When $n = 1$, H is a single vertex and $m \leq k$ holds. Assume the proposition holds for $n \leq t$. When $n = t + 1$, take arbitrarily a vertex v and consider the subhypergraph induced by the vertex set $V \setminus v$, denoted by H'. Let $c(H')$ denote the number of components of H' and $l(v)$ denote the number of edges containing exactly v. Then we have $c(H') + l(v) \leq k$. All of the components are still k-regular, and by induction we have $m_i \leq (k-1)n_i + 1$ where n_i and m_i are vertex number and edge number of the i-th component for $i = 1, \ldots, c(H')$. Thus, we have

$$
\begin{aligned}
m &= \sum_{i=1}^{c(H')} m_i + l(v) \leq \sum_{i=1}^{c(H')} [(k-1)n_i + 1] + l(v) \\
&= (k-1)(n-1) + c(H') + l(v) \leq (k-1)(n-1) + k = (k-1)n + 1. \quad \square
\end{aligned}
$$

Lemma 3. *For every k-regular connected hypergraph $H(V,E)$ on n vertices and m edges, $m = (k-1)n+1$ if and only if $H(V,E)$ is a hypertree.*

Proof. **Sufficiency:** If H is a hypertree, we prove $m = (k-1)n+1$ by induction on the vertex number n. When $n = 1$, H is a single vertex and $m = k$ holds. Assume the lemma holds for $n \leq t$. When $n = t+1$, take arbitrarily a vertex v and consider the subhypergraph induced by the vertex set $V \setminus v$, denoted by H'. Let $c(H')$ denote the number of components of H' and $l(v)$ denote the number of edges containing exactly v. Then we have $c(H') + l(v) = k$. All of the components are k-regular hypertrees, and by induction we have $m_i = (k-1)n_i + 1$ where n_i and m_i are vertex number and edge number of the i-th component for $i = 1, \ldots, c(H')$. Thus, we have

$$
\begin{aligned}
m &= \sum_{i=1}^{c(H')} m_i + l(v) = \sum_{i=1}^{c(H')} [(k-1)n_i + 1] + l(v) \\
&= (k-1)(n-1) + c(H') + l(v) = (k-1)(n-1) + k = (k-1)n + 1.
\end{aligned}
$$

Necessity: We prove the necessity part by contradiction. If H is not a hypertree, H contains a cycle C. Take arbitrary one vertex v in C and consider the subhypergraph induced by the vertex set $V \setminus v$, denoted by H'. Let $c(H')$ denote the number of components of H' and $l(v)$ denote the number of edges containing exactly v. Then we have $c(H') + l(v) \leq k-1$. According to the proof of Lemma 2, we derive $m \leq (k-1)n$, which contradicts with $m = (k-1)n+1$. Thus H is a hypertree. $\square$

Lemma 4. *For every k-regular hypertree $T(V,E)$ on m edges, $\rho(T) \leq \frac{m}{k}$.*

Proof. We prove this lemma by induction on the number of vertices.
When the number of vertices is 1, we have $m = k, \rho(T) = \frac{m}{k} = 1$. Assume this result holds when the vertex number is smaller than n. When T contains $n \geq 2$ vertices, take a longest path $P = v_1 e_1 v_2 \ldots v_t e_t v_{t+1}$ from T for $t \geq 1$. Considering the vertices in edge e_t without taking loops into account, v_t is the unique vertex with degree no less than 2. The subhypergraph induced by the vertex set $V \setminus e_t$ is denoted by T' and the components of T' are $\{T_i \mid 1 \leq i \leq c\}$.

Notice that if T' is empty, all vertices are in e_t, so $\rho(T) = 1 \leq \frac{m}{k}$. Assume that T_i contains m_i edges and n_i vertices. Because T_i is a k-regular hypertree, we have $\rho(T_i) \leq \frac{m_i}{k}$ by induction. Thus,

$$\rho(T') = \sum_{i=1}^{c} \rho(T_i) \leq \sum_{i=1}^{c} \frac{m_i}{k} \leq \frac{m-k}{k}$$

where the last inequality holds as the k edges incident to $v_{t+1} \in e_t$ are eliminated.

For every edge cover set A of T', $A \cup \{e_t\}$ is an edge cover set in T. Then, we have

$$\rho(T) = \rho(T') + 1 \leq \frac{m-k}{k} + 1 \leq \frac{m}{k}. \qquad \square$$

Based on the proofs, for a hypertree T, an optimal edge cover can be found in polynomial time as follows.

Algorithm 1. Optimal Edge Cover of a Hypertree

Input: A hypertree $T(V, E)$.

Output: ALG1(T), an optimal edge cover of H.

1: **if** $|V| = 1$ **then**

2:　　Let e be an arbitrary edge in E.

3:　　**return** $\{e\}$

4: Find a longest path $P = v_1 e_1 v_2 \ldots v_t e_t v_{t+1}$ in T.

5: Let $T_1, T_2, \ldots, T_l$ be all components of subhypergraph induced by $V \setminus e_t$.

6: **return** $\{e_t\} \cup \bigcup_{i=1}^{l} \text{ALG1}(T_i)$

Theorem 3. *For $k \geq 3$, let $H(V, E)$ be a k-regular connected hypergraph with n vertices. Then $\rho(H) \leq \frac{(k-1)n+1}{k}$.*

Proof. Take arbitrarily a sequences of *breaking-cycle* operations. There is a sequence of k-regular connected hypergraphs $\widetilde{H}_1, \ldots, \widetilde{H}_t$. All these hypergraphs have n vertices. $\widetilde{H}_t$ is a hypertree and the number of edges is m_t. According to Lemma 1, we have the following inequalities:

$$\rho(H) \leq \rho(\widetilde{H}_1) \leq \cdots \leq \rho(\widetilde{H}_t).$$

According to Lemma 4, we have the following inequalities:

$$\rho(\widetilde{H}_t) \leq \frac{m_t}{k} = \frac{(k-1)n+1}{k},$$

which means $\rho(H) \leq \frac{(k-1)n+1}{k}$. $\qquad \square$

Algorithm 2. Edge Cover of a k-Regular Connected Hypergraph

Input: A k-regular connected hypergraph H with n vertices.

Output: An edge cover of H with cardinality at most $\frac{(k-1)n+1}{k}$.

1: Set $\widetilde{H}_1 = H$ and $i = 1$.

2: **while** $\widetilde{H}_i$ contains some cycle C_i **do**

3:     Do breaking-cycle operation on vertex v_i in cycle C_i.

4:     Denote the resulting hypergraph as $\widetilde{H}_{i+1}$.

5:     Set $i = i + 1$.

6: Compute an optimal edge cover E_i of the hypertree $\widetilde{H}_i$ by Algorithm 1.

7: **for** $j = i$ down to 2 **do**

8:     **if** $\widetilde{e}_{j-1} \in E_j$ **then**

9:         Take $E_{j-1} = E_j \setminus \{\widetilde{e}_{j-1}\} \cup \{e_{j-1}\}$.

10:     **else**

11:         Take $E_{j-1} = E_j$.

12: **return** E_1

The proof of Theorem 2 implies a combinatorial algorithm for computing an edge cover of k-regular connected n-vertex hypergraph H with cardinality at most $\frac{(k-1)n+1}{k}$.

Remark 1. A slight variation on Algorithm 2 can be applied to a non-regular connected hypergraph H with maximum degree Δ on n vertices and get an edge cover with cardinality at most $\frac{(\Delta-1)n+1}{\Delta}$. Furthermore, an analysis demonstrates that the algorithm achieves the approximation ratio of $\frac{(\Delta-1)r+1}{\Delta}$, here r is the rank of H.

3   Extremal k-Regular Connected Hypergraphs

In this subsection, we characterize the extremal hypergraphs achieving the bound in Theorem 2. The content is organized as follows:

– A class of hypergraphs called k-star hypertrees are introduced in Definition 2.
– We prove k-star hypertrees are exactly the extremal hypergraphs with $\rho(H) = \frac{(k-1)n+1}{k}$ in Theorem 5.

Definition 2. *A hypertree $T(V, E)$ is a star hypertree if there is a vertex set $S \subseteq V$ satisfying:*

– *For each vertex $v \in S$, the degree of v is at least 3.*

- S is an independent set in T.
- The incident edges of S are all the edges in E.

Specially, for $k \geq 3$, if each vertex in S is k-degree, $T(V, E)$ is a k-star hypertree (Fig. 2).

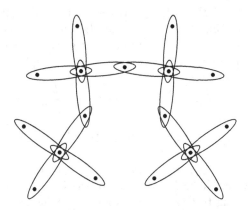

Fig. 2. 4-star hypertree

The independent sets and the edge covers are related in a prime-dual way. The property $\rho(H) = \alpha(H)$ is called the dual *König Property*. The next theorem is useful in our main proof:

Theorem 4 [3]. *$T(V, E)$ is a hypertree, $\rho(T) = \alpha(T)$.*

Theorem 5. *Let $H(V, E)$ be a connected k-regular hypergraph with n vertices. Then $\rho(H) = \frac{(k-1)n+1}{k}$ if and only if $H(V, E)$ is a k-star hypertree.*

Proof. Sufficiency: If H is a k-star hypertree, then according to Lemma 3 and Theorem 4, we have next equalities:

$$\rho(H) = \alpha(H) = \frac{m}{k} = \frac{(k-1)n+1}{k}.$$

Necessity: When $\rho(H) = \frac{(k-1)n+1}{k}$, we need to prove H is a k-star hypertree. It is enough to prove H is acyclic. Actually, if H is acyclic, according to Lemma 3 and Theorem 4, we have next inequalities:

$$\rho(H) = \alpha(H) \leq \frac{m}{k} = \frac{(k-1)n+1}{k}.$$

Combined with $\rho(H) = \frac{(k-1)n+1}{k}$, we have next equalities, which says H is a k-star hypertree.

$$\rho(H) = \alpha(H) = \frac{m}{k} = \frac{(k-1)n+1}{k}.$$

By contradiction, let us take out a counterexample H with minimum vertices. Then $\rho(H) = \frac{(k-1)n+1}{k}$ and H contains cycles. We have two key claims:

Claim 1. *Every two distinct cycles in H share common vertices.*

Actually, for every two distinct cycles $C_1 = u_1e_1u_2e_2...u_te_tu_1$ and $C_2 = w_1e_1w_2e_2...w_se_sw_1$, denote $\{u_1, u_2, ..., u_t\}$ as U and $\{w_1, w_2, ..., w_s\}$ as W. If $U \cap W = \varnothing$, then we can partition the set of vertices $V(H)$ into two parts V_1 and V_2 such that $U \subseteq V_1, W \subseteq V_2$ and the vertex-induced subhypergraphs H_1 with n_1 vertices and H_2 with n_2 vertices are both k-regular and connected. Furthermore, H_1 contains the cycle C_1 and H_2 contains the cycle C_2.

Since H is a counterexample with minimum vertices, we have next inequalities, which contradict with the assumption $\rho(H) = \frac{(k-1)n+1}{k}$.

$$\rho(H_1) \leq \frac{(k-1)n_1}{k}, \quad \rho(H_2) \leq \frac{(k-1)n_2}{k}$$

$$\Rightarrow \rho(H) \leq \rho(H_1) + \rho(H_2) \leq \frac{(k-1)n_1}{k} + \frac{(k-1)n_2}{k} = \frac{(k-1)n}{k}.$$

Let us take out a shortest cycle $C = v_1e_1v_2e_2...v_le_lv_1$ and denote $\{v_1, v_2, ..., v_l\}$ as S. We have $V \setminus S \neq \varnothing$, otherwise $\rho(C) \leq \frac{n+1}{2} < \frac{(k-1)n+1}{k}$. Furthermore, according to Claim 1, we know $V \setminus S$ induces a k-regular hyperforest. The next claim is essential.

Claim 2. *Every hypertree induced by $V \setminus S$ must be a k-regular vertex.*

We assume that there exists a hypertree T induced by $V \setminus S$ with $| V(T) | \geq 2$. Then let us take arbitrarily a vertex $v \in S$ and $e \in T$ such that $v \in e$ in H. T' is the hypertree by adding the vertex v into e in T and denote the farthest vertex from v in T' as v'. We have next two cases.

Case 1: distance $d(v, v') = 1$ in T', we have a partial structure in Fig. 3. Because e is the only edge incident to v, e contains all vertices of T. Thus $\rho(T) = 1$. Denote all the vertices of T as X and other vertices in H as Y. Denote the vertex-induced subhypergraph by X as H_1 and the vertex-induced subhypergraph by Y as H_2 with n_2 vertices. Then H_1 is T. Due to $S \subseteq Y$, H_2 contains the cycle C. Each hypertree induced by the vertex set $V \setminus S$ is connected to C by some vertices in S. Thus H_2 is also connected. H_1 and H_2 are both k-regular and connected.

Considering $n_2 = n - | V(T) | \leq n - 2$, we have next inequalities, which contradict with $\rho(H) = \frac{(k-1)n+1}{k}$.

$$\rho(H) \leq \rho(H_1) + \rho(H_2) \leq 1 + \frac{(k-1)n_2 + 1}{k}$$

$$\leq 1 + \frac{(k-1)(n-2)+1}{k} = \frac{(k-1)n - k + 3}{k} \leq \frac{(k-1)n}{k}.$$

Case 2: distance $d(v, v') \geq 2$ in T', we have a partial structure in Fig. 4. Take out the unique v-v' path in T' as P and the vertex incident to v' in P as u. Because T is k-regular, there are k edges incident to u. Ignoring some loops, there are t non-loop edges $\{e_1(u), e_2(u), ..., e_t(u)\}$ incident to u with $t \leq k$. For

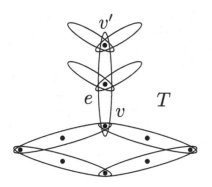

Fig. 3. distance $d(v, v') = 1$ in T

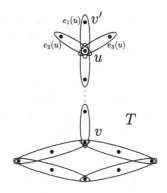

Fig. 4. distance $d(v, v') \geq 2$ in T

each $e_i(u), 1 \leq i \leq t$, $e_i(u)$ has at least one vertex other than u. Without loss of generality, we assume $e_1(u)$ and $e_t(u)$ are the edges incident to u in P and $e_1(u)$ contains v'. For each $1 \leq i \leq t - 1, v_i \in e_i(u), v_i \neq u$, the k edges incident to v_i are $k - 1$ loops and $e_i(u)$.

We can partition all the vertices V as follows: Denote the vertices in $e_1(u)$ as X_1; For each $2 \leq i \leq t - 1$, denote the vertices in $e_i(u) \setminus u$ as X_i; Denote the other vertices as X_t.

Denote the vertex-induced subhypergraph by X_i as H_i with n_i vertices for $1 \leq i \leq t$. H_1 is the edge $e_1(u)$ and some loops. For each $2 \leq i \leq t - 1$, H_i is the edge $e_i(u) \setminus u$ and some loops. Due to $S \subseteq X_t$, H_t contains the cycle C. Each hypertree induced by the vertex set $V \setminus S$ is connected to C by some vertices in S. Thus H_t is also connected. For each $1 \leq i \leq t$, H_i is k-regular and connected. Furthermore, we have $\rho(H_i) = 1, 1 \leq i \leq t - 1$ and $\rho(H_t) \leq \frac{(k-1)n_t}{k}$.

We have the next inequalities, which contradict with $\rho(H) = \frac{(k-1)n+1}{k}$.

$$n_1 \geq 2, n_i \geq 1, 2 \leq i \leq t - 1 \Rightarrow \sum_{1 \leq i \leq t-1} n_i \geq t, n_t = n - \sum_{1 \leq i \leq t-1} n_i \leq n - t$$

$$\rho(H) \leq \sum_{1 \leq i \leq t-1} \rho(H_i) + \rho(H_t) \leq t - 1 + \frac{(k-1)n_t}{k}$$

$$\leq \frac{(k-1)n_t + k(t-1)}{k} \leq \frac{(k-1)(n-t) + k(t-1)}{k} \leq \frac{(k-1)n + t - k}{k} \leq \frac{(k-1)n}{k}.$$

Finally, let us consider the set of k-regular vertices induced by $V \setminus S$.

Case 1: the shortest cycle $C = v_1 e_1 v_2 e_2 ... v_l e_l v_1$ has some non-join vertices besides the join vertices $\{v_1, v_2, ..., v_l\}$. Without loss of generality, we can assume there is a non-join vertex $u \in e_1$. A partial structure is shown in Fig. 5. Because v_1 is k-regular in H, there are at most $k - 2$ vertices in H incident to v_1 besides the cycle C. The similar analysis applies to v_2. Besides the cycle C, let us denote the distinct vertices in H incident to v_1 as $X = \{x_1, x_2, ..., x_{t_1}\}$ and the distinct vertices in H incident to v_2 as $Y = \{y_1, y_2, ..., y_{t_2}\}$. Here we can assume that the two vertex sets X and Y do not intersect because any common vertex is placed into X or Y. Thus we have $|X| = t_1 \leq k - 2, |Y| = t_2 \leq k - 2$.

We can partition all the vertices V as follows: Besides the vertex sets X and Y, denote the vertices $\{v_1, v_2, u\}$ as U and the other vertices as W. For each vertex $x_i \in X$, $1 \leq i \leq t_1$, the vertex-induced subhypergraph X_i is a k-regular vertex x_i. For each vertex $y_i \in Y, 1 \leq i \leq t_2$, the vertex-induced subhypergraph Y_i is a k-regular vertex y_i. The vertex-induced subhypergraph H_U by $U = \{v_1, v_2, u\}$ contains the edge e_1 which covers all the vertices in $U = \{v_1, v_2, u\}$. Thus we have

$$\rho(X_i) = 1,\ 1 \leq i \leq t_1,\ \rho(Y_i) = 1,\ 1 \leq i \leq t_2,\ \rho(H_U) = 1.$$

The vertex-induced subhypergraph H_W with n_w vertices by W contains all the join vertices of the cycle C other than v_1 and v_2, the k-regular vertices induced by $V \setminus S$ besides X and Y are connected to C by some vertices in $S \setminus \{v_1, v_2\}$. Thus H_W is also connected. Furthermore, we have $\rho(H_W) \leq \frac{(k-1)n_w+1}{k}$.

We have the next inequalities, which contradict with $\rho(H) = \frac{(k-1)n+1}{k}$.

$$|X| = t_1 \leq k - 2, |Y| = t_2 \leq k - 2, t = t_1 + t_2 \leq 2k - 4, n_w = n - t - 3$$

$$\rho(H) \leq \sum_{1 \leq i \leq t_1} \rho(X_i) + \sum_{1 \leq i \leq t_2} \rho(Y_i) + \rho(H_U) + \rho(H_W) \leq t_1 + t_2 + 1 + \frac{(k-1)n_w + 1}{k}$$

$$\leq t + 1 + \frac{(k-1)n_w + 1}{k} = \frac{(k-1)n + t - 2k + 4}{k} \leq \frac{(k-1)n}{k}.$$

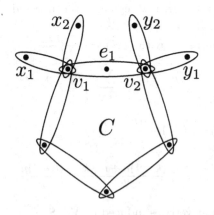

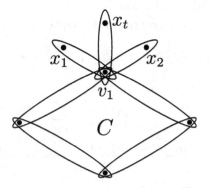

Fig. 5. partial structure in Case 1   Fig. 6. partial structure in Case 2

Case 2: the shortest cycle $C = v_1 e_1 v_2 e_2 ... v_l e_l v_1$ has no non-join vertices besides the join vertices $\{v_1, v_2, ..., v_l\}$. Without loss of generality, we can assume there are some k-regular vertices induced by $V \setminus S$, which are incident to the join vertex v_1 in H. A partial structure is shown in Fig. 6. Because v_1 is k-regular in H, there are at most $k - 2$ vertices in H incident to v_1 besides the cycle C. Denote the distinct vertices in H incident to v_1 as $X = \{x_1, x_2, ..., x_t\}$. Thus we have $|X| = t \leq k - 2$.

We can partition all the vertices V as follows: For each $1 \leq i \leq t - 1$, denote $\{x_i\}$ as V_i; Denote $\{x_t, v_1\}$ as V_t; Denote the other vertices as V_{t+1}. For each vertex $x_i \in X, 1 \leq i \leq t - 1$, the vertex-induced subhypergraph H_i is a k-regular vertex x_i. The vertex-induced subhypergraph H_t by $V_t = \{x_t, v_1\}$ contains an incident edge to v_1 which covers all the vertices in $V_t = \{x_t, v_1\}$. Thus we have $\rho(H_i) = 1, 1 \leq i \leq t$.

The vertex-induced subhypergraph H_{t+1} with n_{t+1} vertices by V_{t+1} contains all the join vertices of the cycle C other than v_1, the k-regular vertices induced by $V \setminus S$ besides X are connected to C by some vertices in $S \setminus \{v_1\}$. Thus H_{t+1} is also connected. Furthermore, we have $\rho(H_{t+1}) \leq \frac{(k-1)n_{t+1}+1}{k}$.

We have the next inequalities, which contradict with $\rho(H) = \frac{(k-1)n+1}{k}$.

$$|X| = t \leq k - 2, |\, V_i \,| = 1, 1 \leq i \leq t - 1, |\, V_t \,| = 2, |\, V_{t+1} \,| = n_{t+1} = n - t - 1$$

$$\rho(H) \leq \sum_{1 \leq i \leq t} \rho(H_i) + \rho(H_{t+1}) \leq t + \frac{(k-1)n_{t+1}+1}{k}$$

$$= t + \frac{(k-1)(n-t-1)+1}{k} = \frac{(k-1)n+t-k+2}{k} \leq \frac{(k-1)n}{k}.$$

Above all, in whatever case, there always exists a contradiction. Thus, $H(V, E)$ is acyclic and a k-star hypertree. $\qquad \square$

References

1. Aharoni, R., Berger, E., Ziv, R.: The edge covering number of the intersection of two matroids. Discret. Math. **312**(1), 81–85 (2012)
2. Aharoni, R., Holzman, R., Zerbib, S.: Edge-covers in d-interval hypergraphs. Discrete Comput. Geom. **58**(3), 650–662 (2017)
3. Berge, C.: Hypergraphs. North-Holland, Paris (1989)
4. Berger, E., Ziv, R.: A note on the edge cover number and independence number in hypergraphs. Discret. Math. **308**(12), 2649–2654 (2008)
5. Kahn, J., Kayll, P.: Fractional v. integral covers in hypergraphs of bounded edge size. J. Combin. Theory Ser. A **78**(2), 199–235 (1997)
6. Krivelevich, M.: Approximate set covering in uniform hypergraphs. J. Algorithms **25**(1), 118–143 (1997)
7. Nieminen, J.: Some applications of graph theory in the theory of hypergraphs. Appl. Math. **4**(14), 607–612 (1975)
8. Norman, R.Z., Rabin, M.O.: An algorithm for a minimum cover of a graph. Proc. Am. Math. Soc. **10**(2), 315–319 (1959)
9. Olejník, F.: On edge independence numbers and edge covering numbers of k-uniform hypergraph. Math. Slovaca **39**(1), 21–26 (1989)
10. Peleg, D., Schechtman, G., Wool, A.: Approximating bounded 0-1 integer linear programs. In: The 2nd Israel Symposium on Theory and Computing Systems, pp. 69–70. IEEE Computer Society (1993)

Time Efficient Implementation for Online K-Server Problem on Trees

Kamil Khadiev$^{(\boxtimes)}$ and Maxim Yagafarov

Kazan Federal University, Kazan, Russia
kamilhadi@gmail.com

Abstract. We consider online algorithms for the k-server problem on trees of size n. Chrobak and Larmore proposed a k-competitive algorithm for this problem that has the optimal competitive ratio. However, the existing implementations have $O\left(k^2 + k \cdot \log n\right)$ or $O\left(k(\log n)^2\right)$ time complexity for processing a query, where n is the number of nodes. We propose a new time-efficient implementation of this algorithm that has $O(n)$ time complexity for preprocessing and $O\left(k \log k\right)$ time for processing a query. The new algorithm is faster than both existing algorithms and the time complexity for query processing does not depend on the tree size.

Keywords: online algorithms · k-server problem on trees · time complexity · time efficient implementation

1 Introduction

Online optimization is a field of optimization theory that deals with optimization problems not know the future [32]. An online algorithm reads an input piece by piece and returns an answer piece by piece immediately, even if the optimal answer can depend on future pieces of the input. The goal is to return an answer that minimizes an objective function (the cost of the output). The most standard method to define the effectiveness of an online algorithm is the competitive ratio [20,34]. That is the worst-case ratio between the cost of the solution found by the algorithm and the cost of an optimal solution. If the ratio is c, then the online algorithm is called c-competitive. In the general setting, online algorithms have unlimited computational power. Nevertheless, many papers consider them with different restrictions. Some of them are restrictions on memory [1,5,8,12,17,21–27,29,31], others are restrictions on time complexity [16,19,30,33]. This paper focuses on efficient online algorithms in terms of time complexity. One of the well-known online minimization problems is the k-server problem on trees [13]. Other related well-known problems are the matching problem, r-gathering problem, and facility assignment problem [2,3,18]. The k-server problem on trees is the following. We have a tree with n nodes and k servers which are in some nodes. In online fashion, we receive queries that are nodes of the tree. Servers move by the graph and one of the servers should come to the query node. Our goal is a

This paper has been supported by the Kazan Federal University Strategic Academic Leadership Program ("PRIORITY-2030").

minimization of the total moving distance for all servers in all queries. There is a k-competitive deterministic algorithm for the k-server problem on trees [13], and the algorithm has the best competitive ratio. The expected competitive ratio for a best-known randomized algorithm [6,7] is $O(\log^3 n \log^2 k)$. In this paper, we are focused on the deterministic one. So, the competitive ratio of the deterministic algorithm is the best possible. At the same time, the naive implementation has $O(n)$ time complexity for each query and preprocessing. There is a time-efficient algorithm for general graphs [33] that uses a min-cost-max-flow algorithm, but it is too slow in the case of a tree. In the case of a tree, there are two efficient algorithms. The first one was introduced in [19]. It has time complexity $O(n \log n)$ for preprocessing and $O\left(k^2 + k \log n\right)$ for query processing. Another one was presented in [30]. It has time complexity $O(n)$ for preprocessing and $O\left(k(\log n)^2\right)$ for query processing.

We suggest a new time-efficient implementation of the algorithm from [13]. It has $O(n)$ time complexity for preprocessing and $O(k \log k)$ for processing a query. It is based on the compression of tree technique called virtual tree and fast algorithms for computing Lowest Common Ancestor (LCA) [9,11] and Level Ancestor (LA) Problem [4,10,15]. The presented algorithm has a better time complexity of query processing compared to all existing algorithms. The query processing is asymptotically better than both algorithms. At the same time, the query processing complexity of our algorithm does not depend on n. If k is constant, then the complexity is $O(1)$. The time complexity of preprocessing is the same as for the Algorithm from [30], and better than the preprocessing complexity of the Algorithm from [19].

2   Preliminaries

2.1   Online Algorithms

An online minimization problem consists of a set $\mathcal{I}$ of inputs and a cost function. Each input $I = (x_1, \ldots, x_M)$ is a sequence of requests, where $M = |I|$ is the length of the input. Furthermore, a set of feasible outputs (or solutions) $\widetilde{O}(I)$ is associated with each I; an output is a sequence of answers $O = (y_1, \ldots, y_M)$. The cost function assigns a positive real value $cost(I, O)$ to $I \in \mathcal{I}$ and $O \in \widetilde{O}(I)$. An optimal solution for $I \in \mathcal{I}$ is $O_{opt}(I) = \arg\min_{O \in \widetilde{O}(I)} cost(I, O)$.

Let us define an online algorithm for this problem. **A deterministic online algorithm** A computes the output sequence $A(I) = (y_1, \ldots, y_M)$ such that y_i is computed based on $x_1, \ldots, x_i$. We say that A is c-*competitive* if there exists a constant $\alpha \geq 0$ such that, for every n and for any input I of size n, we have: $cost(I, A(I)) \leq c \cdot cost(I, O_{Opt}(I)) + \alpha$. The minimal c that satisfies the previous condition is called the **competitive ratio** of A.

2.2   Rooted Trees

Consider a rooted tree $T = (V, E)$, where V is the set of nodes/vertices, and E is the set of undirected edges. Let $n = |V|$ be the number of nodes, or equivalently the size of the tree. We denote by 1 the root of the tree. A path P is a sequence of nodes $(v_1, \ldots, v_h)$ that are connected by edges, i.e. $(v_i, v_{i+1}) \in E$ for all $i \in \{1, \ldots, h-1\}$, such that there are no duplicates among $v_1, \ldots, v_h$. Here $h-1$ is the length of the path. Between any two nodes v and u on the tree, there is a unique path. The distance $\text{dist}(v, u)$ is the length of this path. For each node v we can define a parent node $\text{PARENT}(v)$ which is the first node on the unique path from v to root 1. We have $\text{dist}(1, \text{PARENT}(v)) + 1 = \text{dist}(1, v)$. Additionally, we can define the set of children $\text{CHILDREN}(v) = \{u : \text{PARENT}(u) = v\}$. Any node y on the unique path from root 1 to node v is an ancestor of node v.

Lowest Common Ancestor (LCA). Given two nodes u and v of a rooted tree, the Lowest Common Ancestor is the node w such that w is an ancestor of both u and v, and w is the closest one to u and v among all such ancestors. The following result is well-known.

Lemma 1 ([9]). *There is an algorithm for the LCA problem with the following properties: (i) The time complexity of the preprocessing step is $O(n)$; (ii) The time complexity of computing LCA for two nodes is $O(1)$.*

We call $\text{LCA\_PREPROCESSING}()$ the subroutine that does the preprocessing for the algorithm and $\text{LCA}(u, v)$ that computes the LCA of two nodes u and v.

Level Ancestor (LA) Problem. Given a node v of a rooted tree and a distance d, we should find a vertex u that is the ancestor of the node v and has distance d from the root. The following result is well-known. There are several algorithms with required complexity [4,10,15].

Lemma 2 ([4,10,15]). *There is an algorithm for the LA problem with the following properties: (i) The time complexity of the preprocessing step is $O(n)$; (ii) The time complexity of computing the required node is $O(1)$.*

We call $\text{LA\_PREPROCESSING}()$ the subroutine that does the preprocessing for the algorithm and $\text{LA}(v, d)$ that computes the ancestor of the node v and has distance d from the root.

2.3 k-Server Problem on Trees

Let $T = (V, E)$ be a rooted tree, and we are given k servers that can move among nodes of T. The tree and initial positions of the servers are given. At each time slot, a request $q \in V$ appears as an input. We have to "serve" this request, that is, to choose one of the k servers and move it to q. The other servers are also allowed to move. The cost function is the distance by which we move the servers. In other words, if before the request, the servers are at nodes $v_1, \ldots, v_k$ and after the request they are at $v_1', \ldots, v_k'$, then $q \in \{v_1', \ldots, v_k'\}$ and the cost of the move is $\sum_{i=1}^{k} \text{dist}(v_i, v_i')$. The cost of a sequence of requests is the sum of the costs of serving all requests. The problem is to design a strategy that minimizes the cost of servicing a sequence of requests given online.

3   Algorithm

Firstly, we describe Chrobak-Larmore's k-competitive algorithm for k-server problem on trees from [13]. Assume that we have a request on a node q, and the servers are on the nodes $v_1, \ldots, v_k$. We say that a server i is *active* if there are no other servers on the path from v_i to q. If there are several servers in a node, then we assume that the server with the minimal id in initial enumeration is *active* and others are not *active*. In each phase, we move every *active* server one step towards the node q. After each phase, the set of *active* servers can change. We repeat this phase (moving the active servers) until one of the servers reaches the queried node q.

In this section, we present an effective implementation of Chrobak-Larmore's algorithm with preprocessing. The preprocessing part is done once and has $O(n)$ time complexity (Theorem 1). The request processing part is done for each request and has $O\left(k \log k\right)$ time complexity (Theorem 2).

3.1   Preprocessing

We do the following steps for the preprocessing : (i) We invoke preprocessing for the LCA algorithm (Sect. 2.2). (ii) We invoke preprocessing for the LA problem (Sect. 2.2). (iii) For each node v we compute the distance from the root to v, i.e. $\text{dist}(1, v)$. This can be done using a depth-first search (DFS) algorithm [14]. During the DFS algorithm, we store in and out time for a node. We use a common counter variable as a timer and increase it in each event of entering or leaving a node. For a node v we store values of the timer on entering the node event in $t_{in}(v)$ and on leaving the node event in $t_{out}(v)$. We use two arrays for storing the values of t_{in} and t_{out} for every node. The implementation of step is COMPUTEDISTANCE() procedure which implementation is presented in arXiv version of the paper [28]. The implementation of the whole preprocessing procedure is presented in Algorithm 1, and the time complexity is discussed in Theorem 1.

Algorithm 1. PREPROCESSING. The preprocessing procedure.

LCA_PREPROCESSING()
LA_PREPROCESSING()
$\text{dist}(1,1) \leftarrow 0$
COMPUTEDISTANCE(1)

Theorem 1. *The preprocessing has time complexity $O(n)$.*

Proof. The time complexity of the preprocessing phase is $O(n)$ for LCA, $O(n)$ for the LA problem, and $O(n)$ for COMPUTEDISTANCE() as a complexity of depth-first search algorithm [14]. Therefore, the total time complexity is $O(n)$.   □

3.2　Query Processing

Assume that we have a query on a node q, and the servers are on the nodes $v_1, \ldots, v_k$. As a first step, we construct a Virtual Tree. We describe this process in the Virtual Tree subsection of the current Section. It is a compressed version of the original tree that contains the query node, server nodes, and their LCA nodes. We show that we can use only this tree for processing the query. The size of the compressed tree is $O(k)$. So, it allows us to obtain a small time complexity for processing the query. The whole algorithm of processing a query is presented in Sect. 3.2 (Algorithm for Query Processing).

Virtual Tree. Let $V^s = \{v_1, \ldots, v_k, q\}$ be a set of nodes that contains servers and the query. Let $V^{lca} = \{LCA(v, w) : v, w \in V^s\}$ be a set of all LCA nodes for pairs from V^s. Let us sort nodes from V^s according to the DFS traversal order. In other words, we have $v_{i_1}, \ldots, v_{i_{k+1}}$ such that $t_{in}(v_{i_j}) \leq t_{in}(v_{i_{j+1}})$. Let $V^{olca} = \{LCA(v_{i_j}, v_{i_{j+1}}) : 1 \leq j \leq k\}$ be set of all LCA nodes for sequential pairs of server nodes in the DFS traversal order. Let us show two properties: (i) $V^{lca} = V^s \cup V^{olca}$ (in Lemma 3); (ii) $|V^{lca}| \leq 2|V^s|$ (in Lemma 4). These two properties show that we can easily compute the set V^{lca}, and it is not much larger than the original set V^s.

Lemma 3. $V^{lca} = V^s \cup V^{olca}$.

Proof. By definition, $V^{olca} \subseteq V^{lca}$. At the same time, $LCA(v, v) = v$. Therefore, each node $v \in V^s$ belongs to V^{lca}. So, $V^s \subseteq V^{lca}$. Hence, $V^s \cup V^{olca} \subseteq V^{lca}$. To finish the proof, we should show that $V^{lca} \subseteq V^s \cup V^{olca}$. Assume that we have a node $par \in V^{lca}$ such that $par \notin V^s \cup V^{olca}$. In other words, we have $u, v \in V^s$ such that $LCA(u, v) \notin V^s \cup V^{olca}$. W.l.o.g. we can assume that u has a smaller index than the index of v in the order $i_1, \ldots, i_{k+1}$. Let T_u be a sub-tree of par and $u \in T_u$. Let T_v be a sub-tree of par and $v \in T_v$. Let T_m be a set of all other sub-trees of par that are between T_u and T_v in the DFS traversal order. See Fig. 1 for illustration.

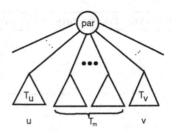

Fig. 1. Subtrees of the $par = LCA(u, v)$ node.

Let j be the minimal index such that $v_{i_j} \in T_v$. Therefore, $v_{i_{j-1}} \in T_u \cup T_m$ because we have at least u from these sub-trees that has smaller index in the

DFS traversal order. The nodes v_{i_j} and $v_{i_{j-1}}$ are in different sub-trees of par. Hence, $par = LCA(v_{i_{j-1}}, v_{i_j})$, and $par \in V^{olca}$. We obtain a contradiction with the assumption. Therefore, any $par \in V^{lca}$ belongs to $V^s \cup V^{olca}$. So, we have $V^{lca} \subseteq V^s \cup V^{olca}$. Finally, we obtain the claim of the lemma. $\square$

Lemma 4. $|V^s| \le |V^{lca}| \le 2|V^s|$.

Proof. Due to construction of V^{olca}, we have $|V^{olca}| \le |V^s| - 1$. Due to Lemma 3, we have $V^{lca} \subseteq V^s \cup V^{olca}$. Therefore, $|V^{lca}| \le |V^s| + |V^{olca}| \le |V^s| + |V^s| - 1 \le 2|V^s|$. At the same time, $V^s \subseteq V^{lca}$. So, $|V^s| \le |V^{lca}|$. $\square$

A virtual tree T^{virt} is a weighted tree constructed by the tree T and the set V^s that has the following structure. The set of the virtual tree's nodes is V^{lca}. There is an edge (u, v) in T^{virt} iff there is no $x \in V^{lca}$ such that $x \ne u, x \ne v$ and $dist(x, u) + dist(x, v) = dist(u, v)$. Note that $dist$ is the distance between nodes in the original tree T. In other words, there is no $x \in V^{lca}$ such that it is on the path between u and v in the original tree T. The weight of the edge (u, v) in T^{virt} is $w^{virt}(u, v) = dist(u, v)$. Let us present an example of the virtual tree in Fig. 2.

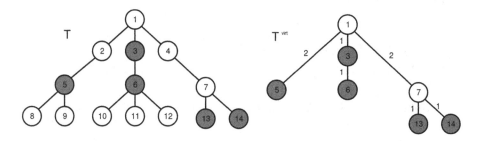

Fig. 2. The virtual tree T^{virt} for the tree T and set $V^s = \{3, 5, 6, 13, 14\}$.

Let us present the algorithm for constructing the virtual tree T^{virt}.

Step 1. Let us sort all nodes of V^{lca} in the DFS traversal order. In the algorithm, we assume that V^{lca} is sorted.

Step 2. Let us use a stack S for nodes from V^{lca}. Initially, we push the first node v from V^{lca}.

Step 3. For each node v from V^{lca} we do Steps 4, 5 and 6.

Step 4. We remove all nodes from the top of S that are not ancestors of v. We can check the property in constant time using arrays t_{in} and t_{out}. A node u is an ancestor of a node v iff $t_{in}(u) \le t_{in}(v)$ and $t_{out}(v) \le t_{out}(u)$. We assume that we have ISANCESTOR(u, v) procedure that returns *True* iff u is an ancestor of v.

Step 5. If after Step 4 the stack is not empty, then we take the top node (without removing it from the stack). Let it be u. Then, we add an edge (u, v) to the tree T^{virt}. If the stack is empty, then we do nothing.

Step 6. We push the node v to the stack.

Let the algorithm be CONSTRUCTVIRTUALTREE$(T, v_1, \ldots, v_k, q)$ procedure and its implementation is presented in Algorithm 2. We assume that we have the following procedures:

ADDEDGE(T^{virt}, u, v) procedure for adding the edge (u, v) to the tree T^{virt}.

INITSTACK$(\mathcal{S})$ procedure for initialization an empty stack.

PUSH$(\mathcal{S}, v)$ procedure for putting a node v to the top of the stack.

POP$(\mathcal{S})$ procedure for removing a node from the top of the stack.

PICK$(\mathcal{S})$ procedure that returns the top element of the stack but does not remove it.

ISEMPTY$(\mathcal{S})$ procedure that returns $True$ iff the stack is empty; and otherwise $False$.

Algorithm 2. CONSTRUCTVIRTUALTREE$(T, v_1, \ldots, v_k, q)$ procedure for constructing virtual tree T^{virt} for the original tree T and the set of server nodes V^s

---

$(v_{i_1}, \ldots, v_{i_{k+1}}) \leftarrow$ SORT$(v_1, \ldots, v_k, q)$ ▷ Sorting V^s in the DFS traversal order

$V^{olca} \leftarrow \{\}$ ▷ Initially V^{olca} is empty

for $j \in \{1, \ldots, k\}$ **do**

     $V^{olca} \leftarrow V^{olca} \cup \{LCA(v_{i_j}, v_{i_{j+1}})\}$

$V^{lca} \leftarrow V^s \cup V^{olca}$

$(b_1, \ldots b_{|V^{lca}|}) \leftarrow$ SORT(V^{lca})

INITSTACK$(\mathcal{S})$

PUSH$(\mathcal{S}, v_{b_1})$

for $j \in \{2, \ldots, |V^{lca}|\}$ **do**

     while ISEMPTY$(\mathcal{S}) = False$ **and** ISANCESTOR$($PICK$(\mathcal{S}), v_{b_j}) = False$ **do**

         POP$(\mathcal{S})$

     if ISEMPTY$(\mathcal{S}) = False$ **then**

         ADDEDGE$(T^{virt},$ PICK$(\mathcal{S}), v_{b_j})$

     PUSH$(\mathcal{S}, v_{b_j})$

The complexity of the algorithm is presented in the next lemma.

Lemma 5. *The time complexity of Algorithm 2 is $O(k \log k)$.*

Proof. Sorting k nodes in V^s has $O(k \log k)$ time complexity. The procedure LCA has $O(1)$ time complexity due to Lemma 1. We can implement sets V^{lca} and V^{olca} as Self-Balanced Search Tree [14]. Due to Lemma 4, $|V^s|, |V^{lca}|, |V^{olca}| = O(k)$. So, the complexity of adding an element to a set has $O(\log k)$ time complexity [14]. Therefore, constructing V^{lca} has $O(k \log k)$ time complexity. The complexity of Step 1 is $O(k \log k)$ because of the complexity of sorting. We add each node of V^{lca} to the stack once and remove it once. Procedures ISANCESTOR, ISEMPTY, POP, PICK, ADDEDGE, and PUSH have $O(1)$ time complexity. Therefore, the total time complexity of the for-loop is $O(k)$. So, the total time complexity of the algorithm is $O(k \log k) + O(k) = O(k \log k)$.   □

Algorithm for Query Processing. We have several events. Each event is moving an active server or deactivating an active server. Note that during such events, active servers can be only in nodes from V^{lca}. Only at the deactivating moment, a server can move to a node that does not belong to V^{lca}. So, we implement an algorithm that moves servers by the virtual tree T^{virt}. At the same moment, we have a procedure $\text{MOVE}(u, z)$ that moves a server from a node v to the distance z in the direction to the query q in the original tree T. We discuss the $\text{MOVE}(u, z)$ procedure in Sect. 3.3.

Let us discuss the algorithm. Let us fix the node q as a root of the virtual tree T^{virt}. The algorithm contains two main phases.

Phase 1. For each node $v \in V^{lca}$, we compute two values that are a node $\text{CLOSEST}(v)$ and an integer $\text{DISTTOCLOSEST}(v)$ that are

- $\text{CLOSEST}(v)$ is the closest server to the node v from the subtree with the root v. If there are several such servers, then $\text{CLOSEST}(v)$ is the server with the smallest id in initial enumeration among them. In other words, it is the server that comes to the node v if we have the only subtree with the root v.
- $\text{DISTTOCLOSEST}(v)$ is the distance from $\text{CLOSEST}(v)$ to v

We can compute these two values for each node using the depth-first search algorithm. Assume that we are processing v and already invoked the procedure for children and we know values of CLOSEST and DISTTOCLOSEST for them. In that case $\text{DISTTOCLOSEST}(v) = min\{\text{DISTTOCLOSEST}(u) + w^{virt}(u, v) : u \in \text{CHILDREN}(v)\}$, and $\text{CLOSEST}(v) = \text{CLOSEST}(u)$ for the corresponding child. Note that if the node contains a server ($v \in \{v_1, \ldots, v_k\}$), then $\text{CLOSEST}(v) = v$ and $\text{DISTTOCLOSEST}(v) = 0$. Note that we can implement the set of servers using a Self-balanced Binary Search Tree (for example, Red-Black tree [14]) and check the property with $O(\log k)$ time complexity. Let the implementation of the phase be a $\text{CLOSESTCOMPUTING}(v)$ procedure that is presented in Algorithm 3. We assume that we have $\text{ID}(v_i)$ function that returns id in initial enumeration for a server v_i.

Phase 2. In this phase, we compute new positions of servers. This phase is also based on the depth-first search algorithm. We start with the root node r. Note that we fix the node q as a root of the virtual tree. We know that the $\text{CLOSEST}(r)$ server comes to this node in $\text{DISTTOCLOSEST}(r)$, and it processes the query. It means that (i) the new position of server $\text{CLOSEST}(r)$ is q; (ii) any other servers become inactive at most after $\text{DISTTOCLOSEST}(r)$ steps.

Moreover, we can say that any server $\text{CLOSEST}(u)$, for $u \in \text{CHILDREN}(r)$ become inactive exactly after $\text{DISTTOCLOSEST}(r)$ steps. It is correct because no other server comes to u in their subtree, but $\text{CLOSEST}(r)$ comes faster to q than others. So, we can move all servers $v \in \{\text{CLOSEST}(u) : u \in \text{CHILDREN}(r)\}\backslash\{\text{CLOSEST}(r)\}$ to the distance $\text{DISTTOCLOSEST}(r)$, and it is the final position of these servers.

Let us look at the children of a node u, where $u \in \text{CHILDREN}(r)$. There are two options:

- all servers from children becomes inactive because of $\text{CLOSEST}(r)$ server;

Algorithm 3. CLOSESTCOMPUTING(v) procedure for the first phase. Here $v \in V^{lca}$ is the processed node.

---

for $u \in$ CHILDREN(v) **do**
    CLOSESTCOMPUTING(u)
if $v \in \{v_1, \ldots, v_k\}$ **then**
    CLOSEST(v) $\leftarrow v$
    DISTTOCLOSEST(v) $\leftarrow 0$
else
    $v_c \leftarrow NULL$
    $d_c \leftarrow NULL$
    for $u \in$ CHILDREN(v) **do**
        if $d_c = NULL$ or $d_c >$ DISTTOCLOSEST(u) $+ w^{virt}(u,v)$ or ($d_c =$ DISTTOCLOSEST(u) $+ w^{virt}(u,v)$ and ID(v_c) $>$ ID(CLOSEST(u))) **then**
            $d_c \leftarrow$ DISTTOCLOSEST(u) $+ w^{virt}(u,v)$
            $v_c \leftarrow$ CLOSEST(u)
    CLOSEST(v) $\leftarrow v_c$
    DISTTOCLOSEST(v) $\leftarrow d_c$

- all servers from children becomes inactive because of CLOSEST(u) server. In that case, the CLOSEST(u) server can come to u faster than in DISTTOCLOSEST(r) steps. At the same time, the way of the CLOSEST(u) server to r takes more than DISTTOCLOSEST(r) steps.

So, we take $b \leftarrow \min\{$CLOSEST(r), CLOSEST(u)$\}$, and be sure that any child of u becomes inactive in b steps. Let us describe the general case. Assume that we process a node v and b is the number of steps for deactivating CLOSEST(v).

Firstly, we update $b \leftarrow \min\{b,$ DISTTOCLOSEST(v)$\}$. Then, we consider all children of v that are $u \in$ CHILDREN(v). If CLOSEST(u) = CLOSEST(v), i.e. this server comes to the node, then we ignore it (we already moved them on processing the parent of v). For other servers, we move CLOSEST(u) to b steps.

Finally, we invoke the same function for all children of v and the new value of b. Let the implementation of the phase be a procedure UPDATEPOSITIONS(v, b) that is presented in Algorithm 4.

Algorithm 4. UPDATEPOSITIONS(v, b) procedure for second phase. Here v is the processed node, and in b steps CLOSEST(v) becomes inactive.

$b \leftarrow \min\{b,$ DISTTOCLOSEST(v)$\}$
for $u \in$ CHILDREN(v) **do**
    if CLOSEST(v) $\neq$ CLOSEST(u) **then**
        MOVE(CLOSEST(u), b)
for $u \in$ CHILDREN(v) **do**
    UPDATEPOSITIONS(u, b)

Let us present the final algorithm for processing a query q in Algorithm 5. Assume that $\text{ROOT}(T^{virt})$ is the root node of the virtual tree T^{virt}.

Algorithm 5. $\text{PROCESSINGAQUERY}(q, T, v_1, \ldots, v_k)$ procedure for processing a query q on the tree T and initial positions of servers $v_1, \ldots, v_k$.

$\text{CONSTRUCTVIRTUALTREE}(T, v_1, \ldots, v_k, q)$
$r \leftarrow \text{ROOT}(T^{virt})$
$\text{CLOSESTCOMPUTING}(r)$
$\text{MOVE}(\text{CLOSEST}(r), \text{DISTTOCLOSEST}(r))$
$\text{UPDATEPOSITIONS}(r, \text{DISTTOCLOSEST}(r))$

3.3 Moving a Server

We now consider the following problem: given a server v and a distance z, how to efficiently compute the new position of the server after moving it z steps towards q. As a subroutine, we use the solution of the Level Ancestor (LA) Problem from Sect. 2.2.

Let $l = LCA(v, q)$. If $\text{dist}(l, v) \geq z$, then the result node is on the path between v and l. Note that l is an ancestor of v, therefore, the target node is also an ancestor of v. We can say that the result node is on the distance $d = dist(1, v) - z$ from the root node. So, the target node is the result of $LA(v, dist(1, v) - z)$ procedure from Sect. 2.2. Otherwise, we should move the server first to l. Then, we move it $z - \text{dist}(l, v)$ steps down towards q from l. In this case, the target node is an ancestor of q and the distance from the root is $d = dist(1, l) + z - \text{dist}(l, v)$. So, we can find it using the result of $LA(v, dist(1, l) + z - \text{dist}(l, v))$ procedure. The algorithm is presented in Algorithm 6.

Algorithm 6. $\text{MOVE}(v, z)$. Moves of a server from v to distance z on a path from v to q.

$l = \text{LCA}(v, q)$
if $z \leq \text{dist}(l, v)$ **then**
    $Result \leftarrow \text{LA}(v, dist(1, v) - z)$
if $z > \text{dist}(l, v)$ **then**
    $Result \leftarrow \text{LA}(q, dist(1, l) + z - \text{dist}(l, v))$
return $Result$

Lemma 6. *The time complexity of* MOVE *is* $O(1)$.

Proof. The time complexity of the LA procedure is $O(1)$ due to Lemma 2. The time complexity of the LCA procedure is $O(1)$ due to Lemma 1. Furthermore, we can compute the distance between any two nodes in $O(1)$ thanks to the preprocessing. Therefore, the total complexity is $O(1)$. ☐

Theorem 2. *The time complexity of the query processing phase is $O(k \log k)$.*

Proof. The complexity of constructing a virtual tree is $O(k \log k)$. The complexity of the two depth-first search algorithms is $O(k \log k)$. The complexity of moving all servers is $O(k)$. So, the total complexity is $O(k \log k)$. ☐

4   Conclusion

In the current work, we present a new algorithm with $O(n)$ time complexity for preprocessing and $O(k \log k)$ for processing a query. Notably, other existing algorithms have worse complexities. Additionally, we proposed that the query processing complexity does not depend on n. The complexity of preprocessing seems the best possible. At the same time, the existence of an algorithm with $O(k)$ query processing time complexity is still an open question.

References

1. Ablayev, F., Ablayev, M., Khadiev, K., Vasiliev, A.: Classical and quantum computations with restricted memory. In: Böckenhauer, HJ., Komm, D., Unger, W. (eds.) Adventures Between Lower Bounds and Higher Altitudes. LNCS, vol. 11011, pp. 129–155. Springer, Cham (2018). https://doi.org/10.1007/978-3-319-98355-4_9
2. Ahmed, A.R., Rahman, M.S., Kobourov, S.: Online facility assignment. Theoret. Comput. Sci. **806**, 455–467 (2020)
3. Akagi, T., Nakano, S.i.: On r-gatherings on the line. In: Wang, J., Yap, C. (eds.) Frontiers in Algorithmics. FAW 2015. LNCS, vol. 9130, pp. 25–32. Springer, Cham (2015). https://doi.org/10.1007/978-3-319-19647-3_3
4. Alstrup, S., Holm, J.: Improved algorithms for finding level ancestors in dynamic trees. In: Automata, Languages and Programming: 27th International Colloquium, ICALP 2000, pp. 73–84 (2000)
5. Baliga, G.R., Shende, A.M.: On space bounded server algorithms. In: Proceedings of ICCI'93: 5th International Conference on Computing and Information, pp. 77–81. IEEE (1993)
6. Bansal, N., Buchbinder, N., Madry, A., Naor, J.: A polylogarithmic-competitive algorithm for the k-server problem. In: 2011 IEEE 52nd Annual Symposium on Foundations of Computer Science, pp. 267–276. IEEE (2011)
7. Bansal, N., Buchbinder, N., Madry, A., Naor, J.: A polylogarithmic-competitive algorithm for the k-server problem. J. ACM (JACM) **62**(5), 1–49 (2015)
8. Becchetti, L., Koutsoupias, E.: Competitive analysis of aggregate max in windowed streaming. In: Albers, S., Marchetti-Spaccamela, A., Matias, Y., Nikoletseas, S., Thomas, W. (eds.) Automata, Languages and Programming. ICALP 2009. LNCS, vol. 5555, pp. 156–170. Springer, Heidelberg (2009). https://doi.org/10.1007/978-3-642-02927-1_15

9. Bender, M.A., Farach-Colton, M.: The lca problem revisited. In: Gonnet, G.H., Viola, A. (eds.) LATIN 2000: Theoretical Informatics. LATIN 2000, LNCS, vol. 1776, pp. 88–94. Springer, Heidelberg (2000). https://doi.org/10.1007/10719839_9

10. Bender, M.A., Farach-Colton, M.: The level ancestor problem simplified. Theoret. Comput. Sci. **321**(1), 5–12 (2004)

11. Berkman, O., Vishkin, U.: Recursive star-tree parallel data structure. SIAM J. Comput. **22**(2), 221–242 (1993)

12. Boyar, J., Larsen, K.S., Maiti, A.: The frequent items problem in online streaming under various performance measures. Int. J. Found. Comput. Sci. **26**(4), 413–439 (2015)

13. Chrobak, M., Larmore, L.L.: An optimal on-line algorithm for k servers on trees. SIAM J. Comput. **20**(1), 144–148 (1991)

14. Cormen, T.H., Leiserson, C.E., Rivest, R.L., Stein, C.: Introduction to Algorithms. McGraw-Hill, New York (2001)

15. Dietz, P.F.: Finding level-ancestors in dynamic trees. In: Dehne, F., Sack, JR., Santoro, N. (eds.) Algorithms and Data Structures. WADS 1991. LNCS, vol. 519, pp. 32–40. Springer, Heidelberg (1991). https://doi.org/10.1007/BFb0028247

16. Flammini, M., Navarra, A., Nicosia, G.: Efficient offline algorithms for the bicriteria k-server problem and online applications. J. Discrete Algorithms **4**(3), 414–432 (2006)

17. Hughes, S.: A new bound for space bounded server algorithms. In: Proceedings of the 33rd Annual on Southeast Regional Conference, pp. 165–169 (1995)

18. Kalyanasundaram, B., Pruhs, K.: Online weighted matching. J. Algorithms **14**(3), 478–488 (1993)

19. Kapralov, R., Khadiev, K., Mokut, J., Shen, Y., Yagafarov, M.: Fast classical and quantum algorithms for online k-server problem on trees. In: CEUR Workshop Proceedings, vol. 3072, pp. 287–301 (2022)

20. Karlin, A.R., Manasse, M.S., Rudolph, L., Sleator, D.D.: Competitive snoopy caching. In: 27th Annual Symposium on FOCS 1986, pp. 244–254. IEEE (1986)

21. Khadiev, K., Khadieva, A., Kravchenko, D., Mannapov, I., Rivosh, A., Yamilov, R.: Quantum versus classical online streaming algorithms with logarithmic size of memory. Lobachevskii J. Math. **44**(2), 687–698 (2023)

22. Khadiev, K.: Quantum request-answer game with buffer model for online algorithms. application for the most frequent keyword problem. In: CEUR Workshop Proceedings, vol. 2850, pp. 16–27 (2021)

23. Khadiev, K., Khadieva, A.: Two-way quantum and classical machines with small memory for online minimization problems. In: International Conference on Micro- and Nano-Electronics 2018. Proc. SPIE, vol. 11022, p. 110222T (2019)

24. Khadiev, K., Khadieva, A.: Two-way quantum and classical automata with advice for online minimization problems. In: Sekerinski, E., et al. (eds.) Formal Methods. FM 2019 International Workshops. FM 2019. LNCS, vol. 12233, pp. 428–442. Springer, Cham (2020). https://doi.org/10.1007/978-3-030-54997-8_27

25. Khadiev, K., Khadieva, A.: Quantum online streaming algorithms with logarithmic memory. Int. J. Theor. Phys. **60**, 608–616 (2021)

26. Khadiev, K., Khadieva, A., Mannapov, I.: Quantum online algorithms with respect to space and advice complexity. Lobachevskii J. Math. **39**(9), 1210–1220 (2018)

27. Khadiev, K., et al.: Two-way and one-way quantum and classical automata with advice for online minimization problems. Theoret. Comput. Sci. **920**, 76–94 (2022)

28. Khadiev, K., Yagafarov, M.: Time efficient implementation for online k-server problem on trees (2024). arXiv preprint, arXiv:2402.14633

29. Khadiev, K., Khadieva, A.: Quantum and classical log-bounded automata for the online disjointness problem. Mathematics **10**(1), 143 (2022). https://doi.org/10.3390/math10010143
30. Khadiev, K., Yagafarov, M.: The fast algorithm for online k-server problem on trees. In: Kulikov, A.S., Raskhodnikova, S. (eds.) Computer Science – Theory and Applications. CSR 2022. LNCS, vol. 13296, pp. 190–208. Springer, Cham (2022). https://doi.org/10.1007/978-3-031-09574-0_12
31. Khadiev, K., Lin, D.: Quantum online algorithms for a model of the request-answer game with a buffer. Uchenye Zapiski Kazanskogo Universiteta. Seriya Fiziko-Matematicheskie Nauki **162**(3), 367–382 (2020)
32. Komm, D.: An Introduction to Online Computation. Determinism, Randomization, Advice. Springer, Cham (2016). https://doi.org/10.1007/978-3-319-42749-2
33. Rudec, T., Baumgartner, A., Manger, R.: A fast work function algorithm for solving the k-server problem. CEJOR **21**(1), 187–205 (2013)
34. Sleator, D.D., Tarjan, R.E.: Amortized efficiency of list update and paging rules. Commun. ACM **28**(2), 202–208 (1985)

Improved Approximation Algorithm for the Distributed Lower-Bounded k-Center Problem

Ting Liang[1,4], Qilong Feng[1,4(✉)], Xiaoliang Wu[1,4], Jinhui Xu[3], and Jianxin Wang[2,4]

[1] School of Computer Science and Engineering, Central South University, Changsha 410083, China
{tingliang,xiaoliangwu}@csu.edu.cn, csufeng@mail.csu.edu.cn
[2] Hunan Provincial Key Lab on Bioinformatics, Central South University, Changsha 410083, China
jxwang@mail.csu.edu.cn
[3] Department of Computer Science and Engineering, State University of New York at Buffalo, New York 14260-1660, USA
jinhui@cse.buffalo.edu
[4] Xiangjiang Laboratory, Changsha 410205, China

Abstract. Clustering large data is a fundamental task with widespread applications. The distributed computation methods have received greatly attention in recent years due to the increasing size of data. In this paper, we consider a variant of the widely used k-center problem, i.e., the lower-bounded k-center problem, and study the lower-bounded k-center problem in the Massively Parallel Computation (MPC) model. The lower-bounded k-center problem takes as input a set C of points in a metric space, the desired number k of centers, and a lower bound L. The goal is to partition the set C into at most k clusters such that the number of points in each cluster is at least L, and the k-center clustering objective is minimized. The current best result for the above problem in the MPC model is 16-approximation algorithm with 4 rounds. In this paper, we obtain a 2-round $(7+\epsilon)$-approximation algorithm for this problem in the MPC model.

Keywords: k-center · lower-bounded k-center · approximation algorithm

1 Introduction

Clustering is a fundamental unsupervised machine learning task that has been extensively studied in various fields of applications, such as customer grouping [15], image classification [20] and facility location [12], etc. The goal of clustering is to partition a given set of points into several disjoint clusters such that

This work was supported by National Natural Science Foundation of China (62172446), Open Project of Xiangjiang Laboratory (22XJ02002), and Central South University Research Programme of Advanced Interdisciplinary Studies (2023QYJC023).

© The Author(s), under exclusive license to Springer Nature Singapore Pte Ltd. 2024
X. Chen and B. Li (Eds.): TAMC 2024, LNCS 14637, pp. 309–319, 2024.
https://doi.org/10.1007/978-981-97-2340-9_26

similar points end up in the same cluster, and dissimilar points are separated into different clusters [19]. Among different types of clustering problems, the k-center problem is one of the most popular clustering problems, which aims to minimize the maximum within-cluster distance of the given dataset. For the k-center problem, it is known to be NP-hard [10], and admits a 2-approximation algorithm [11,14]. The variants of the k-center problem have received greatly attention with the increasing of data size and applications, including the lower-bounded k-center problem [1,3], the capacitated k-center problem [4,8], the k-center with outliers [6,7], the fair k-center problem [5,16,17], etc.

In this paper, we consider a widely used variant of the k-center problem, i.e., the lower-bounded k-center problem. Given a set C of n points in a metric space, the desired number k of centers, and a lower bound $L \in \mathbb{N}^+$, the goal of the lower-bounded k-center problem is to find a subset $S \subseteq C$ with at most k centers and a mapping $\phi : C \to S$ such that for any $s \in S$, $|\{v \in C \mid \phi(v) = s\}| \geq L$, and the cost $\max_{v \in C} d(v, \phi(v))$ is minimized. Aggarwal et $al.$ [1] studied the lower-bounded k-center problem using the threshold method and maximum flow technique, and proposed a polynomial time 2-approximation algorithm. Angelidakis et $al.$ [5] studied the fairness version of the lower-bounded k-center problem, and presented a 15-approximation algorithm with $O(nk^2 + k^5)$. Rösner and Schmidt [21] considered the lower-bounded k-center problem under outliers and capacity constraints, and obtained a 4-approximation algorithm and 11-approximation algorithm, respectively.

In this paper, we focus on the lower-bounded k-center problem in the massively parallel computational (MPC) distributed model. The computational model is used when dealing with massive-scale data. In the MPC setting, the points in the input set C are partitioned across l machines, and a subset $M_j \subseteq C$ of data points is assigned to machine j where $j \in \{1, \ldots, l\}$. The computation of the MPC model occurs in rounds. In each round, each machine is allowed to run a local computation without machine communication, and communications happen in rounds between the coordinator and the machines. The goal of the MPC model is to optimize the communication cost and the number of rounds. Malkomes et $al.$ [18] considered k-center problem and k-center problem with outlier in the MPC model, and achieved a 2-round 4-approximation algorithm and 2-round 13-approximation algorithm, respectively. For k-center problem in the MPC model, Haqi and Hamid [13] achieved an approximation factor of $2 + \epsilon$ for any constant $\epsilon > 0$, which improves the best previously-known approximation factor in [18]. Aghamolaei, Ghodsi and Miri [2] first considered the lower-bounded k-center problem in the MPC model, and gave a 4-round 16-approximation algorithm, and a $(3 + \epsilon)$-approximation algorithm with $O(1/\epsilon)$ rounds. In addition, Epasto et $al.$ [9] considered the lower-bounded k-center problem for euclidean space in the MPC model, and achieved an $O(1)$-approximation algorithm with $O(\log^\epsilon n)$ rounds.

There exist some obstacles to obtain better MPC approximation algorithm with only 2 rounds for the lower-bounded k-center problem. For the k-center problem in the MPC model, Malkomes et $al.$ [18] achieved a 4-approximation algorithm with 2 rounds. One natural idea of solving the lower-bounded problem in the MPC model is firstly to apply the method in [18], and then adjust the obtained result to satisfy the lower bound constraints, which is consistent with the idea of the 4-round 16-approximation algorithm in [2]. In [2], after obtaining a subset S using the 2-round algorithm in [18], a cluster assignment of S is run on each machine in the third round, and then the cluster assignments are picked to satisfy the lower bound constraints in the final round. However, the approximation result of this algorithm is rough, since it only selects the centers from the subset S obtained by the algorithm in [18]. To overcome this obstacle, in [2], they also gave a $(3 + \epsilon)$-approximation algorithm with $O(1/\epsilon)$ rounds. However, the number of communications is very large such that the time complexity is higher than the previous algorithm in [2].

1.1 Our Result

In this paper, we obtain the following result for the lower-bounded k-center problem in the MPC model.

Theorem 1. *Assuming that the data is initially partitioned equally across different machines, there exists an MPC algorithm that requires 2 rounds of communication among the machines, communicates $O(kl)$ amount of data, and achieves $(7 + \epsilon)$-approximation for the lower-bounded k-center problem, where $\epsilon > 0$ is a parameter.*

We now give the high-level idea of our algorithm. Given an instance $\mathcal{I} = (C, k, (\mathcal{X}, d), L)$ of the lower-bounded k-center problem, our algorithm consists of two rounds. In the first round, our algorithm starts with a greedy algorithm presented in [11] to obtain a set $S_j \subseteq M_j$ with k points on machine j where $j \in \{1, ..., j\}$. Then, based on S_j, a weighted set U_j of points is constructed, where each point in U_j is associate with an integer weight. For each machine, we send all messages obtained to the coordinator. In the second round, we first use the threshold method obtain a set of at most k centers. Finally, we try to assign the weighted points to the obtained centers satisfying the lower bound constraints based on the flow network.

2 Preliminaries

Throughout this work, for any $m \in \mathbb{N}^+$, let $[m]$ denote $\{1, \ldots, m\}$. Given a set C of points in a metric space $(\mathcal{X}, d)$, for a point $v \in C$ and a set $S \subseteq C$, let $d(v, S) = \min_{s \in S} d(v, s)$ be the smallest distance between v and S. The k-center problem is formally defined as follows.

Definition 1 (the k-center problem). *Given a set C of points in a metric space $(\mathcal{X}, d)$ and an integer k, the goal is to find a subset $S \subseteq C$ of at most k centers such that the cost $\max_{v \in C} d(v, S)$ is minimized.*

Given an instance $(C, k, (\mathcal{X}, d))$ of the k-center problem, we call S a feasible solution of this instance if S is a subset of C with at most k centers. Moreover, let $\text{cost}_k(S) = \max_{v \in C} d(v, S)$ denote the cost of the feasible solution S.

The lower-bounded k-center problem can be formally defined as follows.

Definition 2 (the lower-bounded k-center problem). *Given a set C of n points in a metric space $(\mathcal{X}, d)$ and two integers k, L, the goal is to find a subset $S \subseteq C$ of at most k centers and a mapping $\phi : C \to S$ such that for any $s \in S$*

$$|\{v \in C \mid \phi(v) = s\}| \geq L, \tag{1}$$

and the cost $\max_{v \in C} d(v, \phi(v))$ is minimized.

Given an instance $\mathcal{I} = (C, k, (\mathcal{X}, d), L)$ of the lower-bounded k-center problem, we call (S, ϕ) a feasible solution of this instance if S is a subset of C with at most k centers, and $\phi : C \to S$ is a mapping satisfying constraints (1). Further, let $\text{cost}_L(S, \phi) = \max_{v \in C} d(v, \phi(v))$ denote the cost of the feasible solution (S, ϕ).

3   Distributed Algorithm for the Lower-Bounded k-Center Problem

In this section, we study the lower-bounded k-center problem in the MPC model, and propose a distributed algorithm, called DISTRIBUTED-SOLVER, which achieves a 2-round $(7 + \epsilon)$-approximation with communication cost $O(kl)$. We now give the general idea of DISTRIBUTED-SOLVER. Given an instance $\mathcal{I} = (C, k, (\mathcal{X}, d), L)$ of the lower-bounded k-center problem, assume that the points in C are distributed among l machines, where M_j is the set of points stored on machine $j \in [l]$. DISTRIBUTED-SOLVER has two rounds. In the first round, on each machine $j \in [l]$, we run separately algorithm GREEDY (see Subsect. 3.1) on M_j to obtain a set $S_j = \{s_j^1, \ldots, s_j^k\}$ of k points. Based on the set S_j, we now construct a weighted set of points, denoted as U_j, where each point in U_j is associate with an integer weight. Specifically, for each point $s_j^i \in S_j$ ($i \in [k]$), a point u_j^i with the same position as s_j^i with weight w_j^i is added to U_j, where w_j^i is equal to the total number of points in M_j that have s_j^i as their nearest point in S_j. Meanwhile, on each machine $j \in [l]$, we also compute a lower bound r_j of the optimal cost of $\mathcal{I}$ (see Subsect. 3.2 with more details). Then, we send (S_j, U_j, r_j) to the coordinator for each machine $j \in [l]$. Let $U = \cup_{j=1}^l U_j$. In the second round, we first use the threshold method [14] with parameter R on $\cup_{j=1}^m S_j$ to obtain a set $S = \{s_1, \ldots, s_{k'}\}$ of k' ($k' \leq k$) centers in the coordinator, where R is a guess of the optimal cost of $\mathcal{I}$. Then, based on the center set S and the weighted set U, we try to assign the weighted points in U to the centers in S satisfying the lower bound constraints, which can be formulated as the weighted assignment problem (see Definition 3). Finally, we obtain the final solution by solving the weighted assignment problem (see Subsect. 3.2). The pseudocode is given in Algorithm 1.

Algorithm 1. DISTRIBUTED-SOLVER

Input: An instance $\mathcal{I} = (C, k, (\mathcal{X}, d), L)$ of the lower-bounded k-center problem, $\{M_1, \ldots, M_j\}$ distributed across l machines, and a parameter $\epsilon > 0$
Output: A feasible solution of $\mathcal{I}$
  1: /*** The first round: run in parallel on individual machines ***/;
  2: **for** $j = 1$ **to** l **do**
  3:     $S_j \leftarrow$ GREEDY$(M_j, k, (\mathcal{X}, d))$;
  4:     $r_j \leftarrow$ compute a lower bound of the optimal cost of $\mathcal{I}$;
  5:     $U_j \leftarrow \emptyset$;
  6:     **for** $i = 1$ **to** k **do**
  7:         Define a weighted point u_j^i with the same position as s_j^i with weight w_j^i;
  8:         $U_j \leftarrow U_j \cup \{u_j^i\}$;
  9:     **end for**
 10: **end for**
 11: Send (S_j, U_j, r_j) to the coordinator for each machine $j \in [l]$;
 12: /*** The second round ***/;
 13: $lb \leftarrow \max_{j \in [l]} r_j$;
 14: **for** $R = \{lb, lb(1 + \epsilon), lb(1 + \epsilon)^2, \ldots\}$ **do**
 15:     $S \leftarrow$ use the threshold algorithm with parameter R;
 16:     Construct network flow $G = (V, E)$ based on S and $U = \cup_{j=1}^l U_j$;
 17:     **if** there exists a flow of value $L \cdot |S|$ on G **then**
 18:         $\gamma \leftarrow$ find a weighted relation based on G;
 19:         **return** (S, γ).
 20:     **else**
 21:         Continue to next R;
 22:     **end if**
 23: **end for**

3.1 The First Round

In this section, we show how DISTRIBUTED-SOLVER works in the first round. Recall that in the first round, we first use algorithm GREEDY on M_j to obtain a set S_j of k centers on each machine $j \in [l]$. GREEDY is an algorithm presented in [11] used for solving the k-center problem. Here, we briefly introduce the process of GREEDY. Given an instance $(C, k, (\mathcal{X}, d))$ of the k-center problem, GREEDY first chooses an arbitrary point from C as center. Then, it iteratively chooses the next center that is the farthest point from all chosen centers until k centers are chosen. Algorithm 2 gives the specific process of GREEDY. Moreover, we have the following theorem.

Theorem 2 ([11]). GREEDY *is a 2-approximation algorithm for the k-center problem.*

Given an instance $\mathcal{I} = (C, k, (\mathcal{X}, d), L)$ of the lower-bounded k-center problem, let S_j be the set returned by the step 3 of Algorithm 1 for each machine $j \in [l]$, τ and τ_l denote the costs of optimal solutions of the k-center problem instance and the lower-bounded k-center problem instance, respectively. Then, we have the following property about the set $\cup_{j=1}^l S_j$.

Algorithm 2. GREEDY

Input: A set C of points, a positive integer k, and a metric space $(\mathcal{X}, d)$
Output: A set $S \subseteq C$ of k centers

1: $S \leftarrow \emptyset$;
2: Choose a point c_1 from C arbitrarily;
3: $S \leftarrow \{c_1\}$;
4: **for** $i = 2$ **to** k **do**
5: $\quad c_i \leftarrow \arg \max_{c \in C} d(c, S)$;
6: $\quad S \leftarrow S \cup \{c_i\}$;
7: **end for**
8: **return** S.

Lemma 1. *For any $v \in C$, $d(v, \cup_{j=1}^{l} S_j) \leq 2\tau_l$.*

Proof. We have $\tau \leq \tau_l$ since the feasible solution of the lower-bounded k-center problem instance is also a feasible solution of the k-center problem instance. Recall that for any $j \in [l]$, M_j is the set of points stored on machine j. By Theorem 2, for any $v \in M_j$, we have $d(v, S_j) \leq 2\tau$. Combining all l machines, for any $v \in \cup_{j=1}^{l} M_j = C$, we have $d(v, \cup_{j=1}^{l} S_j) \leq 2\tau \leq 2\tau_l$. $\qquad\square$

We now give the specific process of obtaining r_j on each machine $j \in [l]$, and show that r_j is a lower bound of τ_l. By Theorem 2, on each machine $j \in [l]$, we have that for any $v \in M_j$, $d(v, S_j) \leq 2\tau \leq 2\tau_l$. After obtaining the set $S_j = \{s_j^1, \ldots, s_j^k\}$ of k points, we continue to use the greedy strategy of GREEDY to find $(k+1)$-th point in M_j, denoted as s_j^{k+1}. Note that the point s_j^{k+1} is the farthest point in M_j from the set S_j. Let $r_j = d(s_j^{k+1}, S_j)/2$. Formally, we state the following result with the lower bound of τ_l.

Lemma 2. *For any machine $j \in [l]$, $r_j \leq \tau_l$.*

Proof. For any machine $j \in [l]$, let S_j be the set of k centers returned by GREEDY. Let s_j^{k+1} be the $(k+1)$-th point selected by GREEDY on M_j. Recall that τ is the cost of optimal solution of the k-center problem instance. We now show that for any machine $j \in [l]$, $r_j \leq \tau$. Assume that $r_j > \tau$. Under this assumption, we have $r_j = d(s_j^{k+1}, S_j)/2 > \tau$. Thus, $d(s_j^{k+1}, S_j) > 2\tau$. By Theorem 2, for any $v \in M_j$ we have $d(v, S_j) \leq 2\tau$, a contradiction. Thus, for any machine $j \in [l]$, we have $r_j \leq \tau \leq \tau_l$. $\qquad\square$

3.2 The Second Round

In this section, we show how DISTRIBUTED-SOLVER works in the second round. We first describe the role of parameter R in DISTRIBUTED-SOLVER (step 14 of the while-loop). In fact, the parameter R is used for guessing the cost of optimal solution of $\mathcal{I}$. More precisely, let lb and up be a lower bound and an upper bound of τ_l, respectively, where τ_l is the cost of optimal solution of $\mathcal{I}$. Then, for a given arbitrary small parameter $\epsilon > 0$, we can guess the optimal cost τ_l

---

Algorithm 3. The Threshold Method

Input: A set C of points, a parameter R, and a metric space $(\mathcal{X}, d)$
Output: A set $S \subseteq C$ of centers
1: $X \leftarrow C$, and $S \leftarrow \emptyset$;
2: **while** $X \neq \emptyset$ **do**
3:     Choose a point s from X arbitrarily;
4:     $S \leftarrow S \cup \{s\}$;
5:     $B(s, 2R) \leftarrow \{x \in X \mid d(x, s) \leq 2R\}$;
6:     $X \leftarrow X \setminus B(s, 2R)$;
7: **end while**
8: **return** S.

---

from $\{lb, lb(1 + \epsilon), lb(1 + \epsilon)^2, \ldots, up\}$, which has at most $\log_{1+\epsilon}(up/lb)$ choices. By Lemma 2, we have that for each machine $j \in [l]$, r_j is a lower bound of τ_l. Thus, we can run our algorithm (steps 14–22 of DISTRIBUTED-SOLVER) in the coordinator with a parameter R starting at $lb = \max_{j \in [l]} r_j$ until it successfully finds a feasible solution of $\mathcal{I}$. When the algorithm returns a feasible solution, the value of parameter R is at least τ_l and at most $(1 + \epsilon)\tau_l$. Therefore, ϵ does not influence the analysis of the approximation ratio. Moreover, the above guessing process is executed in the coordinator without additional communication cost. For simplicity, assume that we have guessed the R with $\tau_l \leq R \leq (1 + \epsilon)\tau_l$.

We now show how to use the threshold algorithm [14] with parameter R on $\cup_{j=1}^m S_j$ to obtain a set $S = \{s_1, \ldots, s_{k'}\}$ of k' ($k' \leq k$) centers in the coordinator. Initially, let all points in $\cup_{j=1}^m S_j$ be unmarked, and let $S = \emptyset$. We first arbitrarily select a point s from $\cup_{j=1}^m S_j$, and add it to S. Then, we mark all unmarked points within distance $2R$ of s (including s), which forms a cluster with center at s. Then, it iteratively chooses the next center from unmarked points until all points in $\cup_{j=1}^m S_j$ are marked. Obviously, there are at most k clusters generated by above steps. For completeness, we give the process in Algorithm 3.

Lemma 3. *For any $v \in C$, we have $d(v, S) \leq 2\tau_l + 2R$.*

Proof. By lemma 1, for any $v \in C$, we have $d(v, \cup_{j=1}^l S_j) \leq 2\tau_l$. By the process of obtaining S, it is easy to know that for any $s \in \cup_{j=1}^l S_j$, there must exist a center in S within distance $2R$ from s. Thus, by the triangle inequality, for any $v \in C$, $d(v, S) \leq 2\tau_l + 2R$. □

Let $S = \{s_1, \ldots, s_{k'}\}$ be the set of k' centers returned by Algorithm 3. The next goal is to assign the weighted points in U to the centers in S. Note that a point in U may be assigned to more than one center in S, since the points in U are weighted. For simplicity, we start with some notions. Let U and S denote the weighted points and centers, respectively. We use a weighted relation $\gamma : U \times S \to \mathbb{N}^+ \cup \{0\}$ to represent the assignment of the weighted points to the centers. Formally, for any $u \in U$ and $s \in S$, $\gamma(u, s)$ stands for the number of points assigned to s. In particular, $\gamma(u, s) = 0$ means that there is no points assigned to s. Moreover, we define the cost of γ as $\text{cost}_w(\gamma) =$

$\max_{u \in U} \max_{s \in S, \gamma(u,s) > 0} d(u,s)$, is the maximum distance between a point $u \in U$ and a center $s \in S$ with $\gamma(u,s) > 0$. Note that a point $u \in U$ is assigned to a center $s \in S$ if and only if $\gamma(u,s) > 0$. Therefore, we now consider the weighted assignment problem, which can be formally defined as follows.

Definition 3. (the weighted assignment problem). *Given a set U of points in a metric space $(\mathcal{X}, d)$, where each point $u \in U$ is associated with an integer weight w_u, two integers k, L, a set S of k centers, the goal is to find a weighted relation $\gamma : U \times S \to \mathbb{N}_0$ satisfying $\sum_{s \in S} \gamma(u,s) = w_u$ for any $u \in U$, and*

$$\sum_{u \in U} \gamma(u,s) \geq L, \forall s \in S \tag{2}$$

and the cost $\max_{u \in U} \max_{s \in S, \gamma(u,s) > 0} d(u,s)$ is minimized.

Lemma 4. *Given an instance $\mathcal{I} = (C, k, (\mathcal{X}, d), L)$ of the lower-bounded k-center problem, let U be the set of weighted points obtained in the coordinator, and let S be the set of centers obtained by Algorithm 3. Then, there must exist a weighted relation $\gamma : U \to S$ satisfying constraints (2) with cost at most $5\tau_l + 2R$.*

Proof. We first prove that based on S, there exists a mapping ϕ satisfying lower bound constraints that assigns each point in C to a center in S. Let (S^*, ϕ^*) be an optimal solution of $\mathcal{I}$. Let $S^* = \{s_1^*, \ldots, s_k^*\}$ be the set of k optimal centers, and let $P^* = \{P_1^*, \ldots, P_k^*\}$ be the corresponding k optimal clusters under mapping ϕ^*. By Lemma 3, we have that for any $s_i^* \in S^*$ ($i \in [k]$), there must exist a center in S within distance $2\tau_l + 2R$ to s_i^*. For $i \in [k]$, let $\pi(s_i^*) = \arg\min_{s \in S} d(s, s_i^*)$ be the nearest center in S to s_i^*. Thus, we have $d(s_i^*, \pi(s_i^*)) \leq 2\tau_l + 2R$ due to the definition of π. For any $v \in C$, let $\phi(v) = \pi(\phi^*(v))$. We now prove that (S, ϕ) is a feasible solution of $\mathcal{I}$ with cost at most $3\tau_l + 2R$. We first show that the cost of (S, ϕ) is at most $3\tau_l + 2R$. For any $v \in P_i^*$ ($i \in [k]$), by the triangle inequality, we get that

$$d(v, \phi(v)) \leq d(v, \pi(\phi^*(v))) \leq d(v, \phi^*(v)) + d(\phi^*(v), \pi(\phi^*(v)))$$
$$\leq \tau_l + 2\tau_l + 2R \leq 3\tau_l + 2R.$$

We now prove that (S, ϕ) is a feasible solution of $\mathcal{I}$, i.e., ϕ satisfies the lower bound constraints. Recall that since (S^*, ϕ^*) is a feasible solution of $\mathcal{I}$, for any $i \in [k]$, we have

$$|\{v \in C \mid \phi(v) = s_i^*\}| = |P_i^*| \geq L.$$

Note that some optimal centers may have the same nearest center in S. For any $s \in S$, let $\text{nrst}(s) = \{s^* \in S^* \mid \pi(s^*) = s\}$ be all centers in S^* for which s is the nearest center. Thus, for any $s \in S$, we have

$$|\{v \in C \mid \phi(v) = s\}| = |\{v \in C \mid \pi(\phi^*(v)) = s\}|$$
$$= \sum_{s^* \in \text{nrst}(s)} |\{v \in C \mid \phi(v) = s^*\}|$$
$$= \sum_{s^* \in \text{nrst}(s)} |P^*| \geq L.$$

Thus, (S, ϕ) is a feasible solution of $\mathcal{I}$.

Based on (S, ϕ), we now construct a weighted relation $\gamma : U \to S$ satisfying constraints (2). For any $v \in C$, suppose that point v is processed by machine $j \in [l]$, i.e., $v \in M_j$. Let $\sigma(v)$ be the nearest point in S_j to v. Assume that $\sigma(v) \in S_j$ is i-th ($i \in [k]$) point selected by GREEDY. By Theorem 2, we have $d(v, \sigma(v)) \leq 2\tau_l$. Recall that the weighted set U_j has a point u_j^i with the same position as $\sigma(v) \in S_j$. In the solution (S, ϕ), the point v is assigned to a center $\phi(v) \in S$. Then, we assign 1 unit of weight of the point u_j^i to $\phi(v)$, i.e., set $\gamma(u_j^i, \phi(v)) = 1$. By the above process, we have that the total weight of points in U assigned to a center $s \in S$ is exactly equal to the number of points assigned to s in the solution (S, ϕ), i.e., for any $s \in S$, $\sum_{u \in U} \gamma(u, s) = |\{v \in C \mid \phi(v) = s\}|$. Since (S, ϕ) is a feasible solution of $\mathcal{I}$, the weighted relation γ satisfies constraints (2). We now bound the cost of the weighted relation γ. By the triangle inequality, we have

$$d(\sigma(v), \phi(v)) \leq d(\sigma(v), v) + d(v, \phi(v)) \leq 2\tau_l + 3\tau_l + 2R \leq 5\tau_l + 2R,$$

Thus, the cost of the weighted relation γ is at most $5\tau_l + 2R$. □

Lemma 4 shows the existence of a weighted relation with cost at most $5\tau_l + 2R$. To obtain such a weighted relation, we now construct a flow network $G = (V, E)$ as follows.

- $V = V_1 \cup V_2 \cup V_3 \cup V_4$, where $V_1 = \{so\}, V_2 = \{v_s \mid s \in S\}, V_3 = \{w_u$ identical nodes $v_u \mid u \in U\}$, and $V_4 = \{si\}$.
- $E = E_1 \cup E_2 \cup E_3$, where $E_1 = \{(so, v_s) \mid s \in S\}$ with capacity L, $E_2 = \{(v_s, v_u) \mid s \in S, u \in U, d(s, u) \leq 5\tau_l + 2R\}$ with capacity 1, and $E_3 = \{(v_u, si) \mid u \in U\}$ with capacity 1.

Lemma 5. *If a flow of value $L \cdot |S|$ can be found on $G = (V, E)$, then we can obtain a weighted relation γ satisfying constraints (2) with cost at most $5\tau_l + 2R$ in polynomial time.*

Proof. By lemma 4, we have that there must exist a weighted relation $\gamma : U \to S$ satisfying constraints (2) with cost at most $5\tau_l + 2R$. By the construction process of flow network $G = (V, E)$, there exists a flow of value $L \cdot |S|$ on $G = (V, E)$. We now construct a weighted relation γ as follows. For any $s \in S$ and $u \in U$, if the flow value of edge $(v_s, v_u) \in E_2$ is non-zero, we set $\phi(s, u) = 1$. For any remaining node $v_u \in V_3$ that has not receive a flow, we set $\phi(s, u) = 1$, where $s \in S$ is the nearest centers to u within distance $5\tau_l + 2R$. By the above process, we have that the weighted relation obtained satisfies the lower bound constraints. Moreover, the cost of the weighted relation γ is at most $5\tau_l + 2R$. □

Recall that when the step 18 of DISTRIBUTED-SOLVER of is executed, we have $R \leq (1 + \epsilon)\tau_l$. By lemma 5, the cost of solution returned by the algorithm DISTRIBUTED-SOLVER is at most $(7 + \epsilon)\tau_l$. Moreover, since the size of (S_j, U_j, r_j) is at most $O(k)$ for each machine $j \in [l]$, the algorithm DISTRIBUTED-SOLVER communicates $O(kl)$ amount of data. By the above discussion, Theorem 1 can be proved.

4 Conclusions

In this paper, we consider the lower-bounded k-center problem in the MPC model and propose a $(7 + \epsilon)$-approximation algorithm. In the future, we plan to explore the MPC approximation algorithms for the k-center problem under other constraints.

References

1. Aggarwal, G., et al.: Achieving anonymity via clustering. ACM Trans. Algorithms **6**(3), 49:1–49:19 (2010)
2. Aghamolaei, S., Ghodsi, M., Miri, S.: A mapreduce algorithm for metric anonymity problems. In: Proceedings of the 31st Canadian Conference on Computational Geometry, pp. 117–123 (2019)
3. Ahmadian, S., Swamy, C.: Approximation algorithms for clustering problems with lower bounds and outliers. In: Proceedings of the 43rd International Colloquium on Automata, Languages, and Programming, pp. 69:1–69:15 (2016)
4. An, H., Bhaskara, A., Chekuri, C., Gupta, S., Madan, V., Svensson, O.: Centrality of trees for capacitated k-center. Math. Program. **154**(1–2), 29–53 (2015)
5. Angelidakis, H., Kurpisz, A., Sering, L., Zenklusen, R.: Fair and fast k-center clustering for data summarization. In: Proceedings of the 39th International Conference on Machine Learning, pp. 669–702 (2022)
6. Chakrabarty, D., Goyal, P., Krishnaswamy, R.: The non-uniform k-center problem. In: Proceedings of the 43rd International Colloquium on Automata, Languages, and Programming, pp. 67:1–67:15 (2016)
7. Charikar, M., Khuller, S., Mount, D.M., Narasimhan, G.: Algorithms for facility location problems with outliers. In: Proceedings of the 12th Annual ACM-SIAM Symposium on Discrete Algorithms, pp. 642–651 (2001)
8. Cygan, M., Hajiaghayi, M., Khuller, S.: LP rounding for k-centers with non-uniform hard capacities. In: Proceedings of the 53rd Annual Symposium on Foundations of Computer Science, pp. 273–282 (2012)
9. Epasto, A., Mahdian, M., Mirrokni, V., Zhong, P.: Massively parallel and dynamic algorithms for minimum size clustering. In: Proceedings of the 33rd Annual ACM-SIAM Symposium on Discrete Algorithms, pp. 1613–1660 (2022)
10. Garey, M.R., Johnson, D.S.: Computers and Intractability: A Guide to the Theory of NP-Completeness. WH Freeman, New York (1979)
11. Gonzalez, T.F.: Clustering to minimize the maximum intercluster distance. Theoret. Comput. Sci. **38**, 293–306 (1985)
12. Hansen, P., Brimberg, J., Urošević, D., Mladenović, N.: Solving large p-median clustering problems by primal-dual variable neighborhood search. Data Min. Knowl. Disc. **19**, 351–375 (2009)
13. Haqi, A., Zarrabi-Zadeh, H.: Almost optimal massively parallel algorithms for k-center clustering and diversity maximization. In: Proceedings of the 35th ACM Symposium on Parallelism in Algorithms and Architectures, pp. 239–247 (2023)
14. Hochbaum, D.S., Shmoys, D.B.: A best possible heuristic for the k-center problem. Math. Oper. Res. **10**(2), 180–184 (1985)
15. Ip, W., Mou, W.: Customer grouping for better resources allocation using GA based clustering technique. Expert Syst. Appl. **39**(2), 1979–1987 (2012)

16. Jones, M., Lê Nguyên, H., Nguyen, T.: Fair k-centers via maximum matching. In: Proceedings of the 37th International Conference on Machine Learning, pp. 4940–4949 (2020)
17. Kleindessner, M., Samadi, S., Awasthi, P., Morgenstern, J.: Guarantees for spectral clustering with fairness constraints. In: Proceedings of the 36th International Conference on Machine Learning, pp. 3458–3467 (2019)
18. Malkomes, G., Kusner, M.J., Chen, W., Weinberger, K.Q., Moseley, B.: Fast distributed k-center clustering with outliers on massive data. In: Advances in Neural Information Processing Systems, vol. 28, pp. 1063–1071 (2015)
19. Pan, W., Shen, X., Liu, B.: Cluster analysis: unsupervised learning via supervised learning with a non-convex penalty. J. Mach. Learn. Res. **14**(7), 1865–1889 (2013)
20. Rollet, R., Benie, G., Li, W., Wang, S., Boucher, J.: Image classification algorithm based on the RBF neural network and k-means. Int. J. Remote Sens. **19**(15), 3003–3009 (1998)
21. Rösner, C., Schmidt, M.: Privacy preserving clustering with constraints. In: Proceedings of the 45th International Colloquium on Automata, Languages, and Programming, pp. 96:1–96:14 (2018)

Parameterized Complexity of Weighted Target Set Selection

Takahiro Suzuki[1($\boxtimes$)], Kei Kimura[2], Akira Suzuki[1]📷, Yuma Tamura[1], and Xiao Zhou[1]

[1] Graduate School of Information Sciences, Tohoku University, Sendai, Japan
takahiro.suzuki.q4@dc.tohoku.ac.jp, {akira,tamura,zhou}@tohoku.ac.jp
[2] Faculty of Information Science and Electrical Engineering, Kyushu University,
Fukuoka, Japan
kkimura@inf.kyushu-u.ac.jp

Abstract. Consider a graph G where each vertex has a threshold. A vertex v in G is activated if the number of active vertices adjacent to v is at least as many as its threshold. A vertex subset A_0 of G is a target set if eventually all vertices in G are activated by initially activating vertices of A_0. The TARGET SET SELECTION problem (TSS) involves finding the smallest target set of G with vertex thresholds. This problem has already been extensively studied and is known to be NP-hard even for very restricted conditions. In this paper, we analyze TSS and its weighted variant, called the WEIGHTED TARGET SET SELECTION problem (WTSS) from the perspective of parameterized complexity. Let k be the solution size and ℓ be the maximum threshold. We first show that TSS is W[1]-hard for split graphs when parameterized by $k + \ell$, and W[2]-hard for cographs when parameterized by k. We also prove that WTSS is W[2]-hard for trivially perfect graphs when parameterized by k. On the other hand, we show that WTSS can be solved in $O(n \log n)$ time for complete graphs. Additionally, we design FPT algorithms for WTSS when parameterized by $nd + \ell$, $tw + \ell$, ce, and vc, where nd is the neighborhood diversity, tw is the treewidth, ce is the cluster editing number, and vc is the vertex cover number of the input graph.

Keywords: Parameterized complexity · Graph algorithms · Target set selection

1 Introduction

Let $G = (V, E)$ be an undirected graph with a threshold function $\mathrm{thr} : V \to \mathbb{Z}_+$ and A_0 be a subset of V. The vertices in A_0 are considered to be *active* at

A. Suzuki—Partially supported by JSPS KAKENHI Grant Number JP20K11666, Japan.

Y. Tamura—Partially supported by JSPS KAKENHI Grant Number JP21K21278, Japan.

X. Chen and B. Li (Eds.): TAMC 2024, LNCS 14637, pp. 320–331, 2024.
https://doi.org/10.1007/978-981-97-2340-9_27

step 0. A vertex $v \notin A_0$ of G is activated at step 1 if at least $\text{thr}(v)$ vertices in A_0 are adjacent to v. We denote by A_1 the set of all the active vertices at step 1. Then, depending on A_1 and thr, some vertices are additionally activated at step 2. Iterating this process produces a sequence $A_0, A_1, A_2, \ldots$ of sets of active vertices, called an *activation process starting at* A_0. Suppose that this process stops at step $i \in \mathbb{N}$, that is, $A_j = A_i$ for every non-negative integer $j \geq i$. (Note that $A_i \subseteq A_{i+1}$ for any $i \geq 0$ and any activation process stops within step $n = |V|$.) We say that A_0 is a *target set* of G if $A_i = V$, that is, all vertices in G are activated. Given an undirected graph $G = (V, E)$ with a threshold function $\text{thr} : V \to \mathbb{Z}_+$, the TARGET SET SELECTION problem (TSS for short) asks for finding a target set $A_0 \subseteq V$ of the minimum size.

A natural application of TSS is word-of-mouth marketing in a social network. Suppose that a company wants to launch a new product. One way to promote the product to consumers is to advertise it on television or streaming media services, but this would be very expensive. Therefore, the company promotes the product by word of mouth instead. The company selects some consumers and gives them free samples. Assuming they like the product, they will recommend it to their friends by word of mouth. If the friends receive more recommendations than the individual's threshold, they will also buy the product and recommend it to other friends. This diffusion of the product in the social network corresponds to an activation process in the graph. The company's goal is to select key consumers in order to get all the people in the social network to buy the product. The company also wants to select as few key consumers as possible in order to save advertising costs. This situation can be modeled by TSS.

It is known that TSS is intractable even for restricted situations. TSS is NP-hard for split graphs [16], planar graphs [6], graphs with clique cover number 2 [16], and graphs with diameter 2 [8]. Moreover, TSS is NP-hard for bipartite graphs even if the threshold of vertices is at most 2 [6]. On the other hand, TSS is solvable in polynomial time for graphs of bounded treewidth [3], block-cactus graphs [7], chordal graphs with threshold at most 2 [7]. TSS has also been studied from the viewpoint of parameterized complexity. TSS is W[1]-hard when parameterized by feedback vertex set number [16], pathwidth [8], neighborhood diversity [11], twin cover number [11], and W[P]-hard by the solution size [2], while TSS is fixed parameter tractable when parameterized by vertex cover number [16], feedback edge set number [16], cluster editing number [16], and bandwidth [8]. In addition, the fixed parameter tractability by a combination of two parameters is known: TSS is fixed parameter tractable when parameterized by $\text{cw}+\ell$ [13] and $\text{cw}+t_{\max}$ [9], where cw is the clique-width of the given graph, ℓ is the maximum threshold of vertices, and $t_{\max}$ is the maximum activation time.

In word-of-mouth marketing, it is natural that each consumer has a different advertising cost. In [17], Raghavan and Zhang proposed the WEIGHTED TARGET SET SELECTION problem (WTSS for short). In this variant, we are given an undirected graph $G = (V, E)$ with a threshold function $\text{thr} : V \to \mathbb{Z}_+$ and a weight function $w : V \to \mathbb{Z}_+$. Our task is to find a target set A_0 of G such that $\sum_{v \in A_0} w(v)$ is minimized. Clearly, WTSS is a generalization of TSS and hence all the hardness results of TSS are taken over to WTSS. Raghavan and Zhang showed that WTSS is solvable in linear time for trees and cycles [17].

1.1 Our Contributions

In this paper, we tackle with WTSS from the viewpoint of parameterized complexity. We first discuss the hardness of WTSS with respect to several parameters. Note that some of our results in fact show the hardness of TSS. Let k be the solution size of TSS and ℓ be the maximum threshold of vertices in a given graph. We show that TSS parameterized by $k + \ell$ is W[1]-hard for split graphs and TSS parameterized by k is W[2]-hard for cographs. It is worth noting that cographs have bounded modular-width mw. Thus, our result implies that TSS is para-NP-hard for mw, that is, NP-hard even for graphs of bounded modular-width. By modifying our proof, we also show that WTSS parameterized by k is W[2]-hard for trivially perfect graphs, which is a subclass of cographs.

We then investigate tractable cases of WTSS. Let n be the number of vertices of a given graph G. In contrast to the hardness for split graphs and trivially perfect graphs, we design an $O(n \log n)$-time algorithm for complete graphs. We then provide an FPT algorithm for WTSS when parameterized by neighborhood diversity nd $+ \ell$. The FPT algorithm is also an XP algorithm parameterized by nd and directly implies the polynomial-time solvability of WTSS for complete bipartite graphs and complete split graphs because they have neighborhood diversity 2. Note that, as complete graphs have neighborhood diversity 1, the XP algorithm solves WTSS for complete graphs in polynomial time. We emphasize that our $O(n \log n)$-time algorithm for complete graphs uses a different method to achieve the faster running time. Furthermore, we provide an FPT algorithm for WTSS parameterized by tw $+ \ell$ and an XP algorithm by tw, where tw is the treewidth of G. Finally, we give FPT algorithms for WTSS parameterized by cluster editing number ce and vertex cover number vc. We summarize the complexity of WTSS in Figs. 1 and 2.

Due to the space limitation, several proofs marked ♠ are omitted in this paper.

2 Preliminaries

Let $G = (V, E)$ be a graph. We also denote by $V(G)$ and $E(G)$ the vertex set and the edge set of G, respectively. All the graphs considered in this paper are finite, simple, and undirected. For a vertex v of G, the *open neighborhood* $N(G; v)$ and the *degree* $d_G(v)$ of v are defined as $N(G; v) = \{w \in V \mid vw \in E\}$ and $d_G(v) = |N(G; v)|$ respectively. We also define the *closed neighborhood* $N[G; v] = N(G; v) \cup \{v\}$. For a vertex subset $S \subseteq V(G)$, we denote by $G[S]$ the subgraph induced by S. For two positive integers i and j such that $i \leq j$, we denote $[i, j] = \{i, i+1, \ldots, j\}$. We use the shorthand $[j] = [1, j]$. An *independent set* I of G is a vertex subset such that there are no edges between any pair of

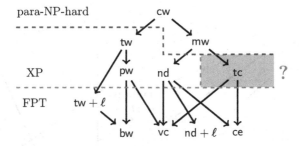

Fig. 1. The complexity of WTSS from the perspective of graph parameters. Each arrow represents the relationship between parameters: $A \to B$ means that the parameter A is bounded by some function of the parameter B.

vertices in I. A *clique* C of G is a vertex subset such that any pair of vertices in C are joined by an edge.

For a graph $G = (V, E)$ with a weight function $w : V \to \mathbb{Z}_+$ and a subset $V' \subseteq V$, we denote $w(V') = \sum_{v \in V'} w(v)$. The definition of WEIGHTED TARGET SET SELECTION is as follows.

WEIGHTED TARGET SET SELECTION (WTSS)
Input: A graph $G = (V, E)$ with a threshold function $\mathsf{thr} : V \to \mathbb{Z}_+$ and a weight function $w : V \to \mathbb{Z}_+$.
Task: Find a target set A_0 of G such that $w(A_0)$ is minimized.

If every vertex has weight 1 in WTSS, then we call the problem TARGET SET SELECTION (TSS). We recall here the standard reduction rule, which was used in the literature, for example [1,16].

Reduction rule 1. Let $G = (V, E)$ and $v \in V$. If $\mathsf{thr}(v) > d_G(v)$, then delete v from G, decrease the threshold of all its neighbors by one, and then add v into an optimal target set A_0 of G.

The following variant of WTSS sometimes helps us to design algorithms for WTSS.

WEIGHTED POSET TARGET SET SELECTION (WPTSS)
Input: A graph $G = (V, E)$ with a threshold function $\mathsf{thr} : V \to \mathbb{Z}_+$, a weight function $w : V \to \mathbb{Z}_+$, and a partially ordered set $(S, \prec)$ such that $S \subseteq V$.
Task: Find a target set A_0 of G such that $w(A_0)$ is minimized, there is a bijection $\pi : V \setminus A_0 \to [|V \setminus A_0|]$ with $|\{u \in N(G; v) \mid \pi(u) < \pi(v)\} \cup (A_0 \cap N(G; v))| \geq \mathsf{thr}(v)$ for every $v \in V \setminus A_0$, and for any pair of vertices $u, v \in S \setminus A_0$, $\pi(u) < \pi(v)$ if $u \prec v$.

For the basic concepts and definitions in parameterized complexity theory, see some textbooks, for example [10]. We also note that the definitions of several structural graph parameters are omitted. For the omitted definitions, we refer the reader to [8,11,14].

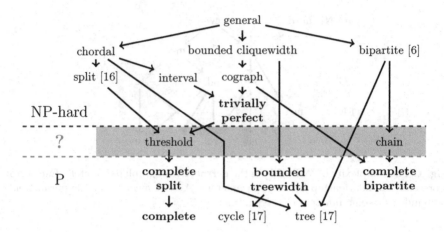

Fig. 2. The complexity of WTSS from the perspective of graphs classes. Each arrow represents the inclusion relationship between graph classes: $A \to B$ means that the graph class B is a subclass of the graph class A. The graph classes for which new results are presented in this paper are marked in bold.

3   Hardness

3.1   Split Graphs

A graph $G = (V, E)$ is a *split graph* if V can be partitioned into a clique C and an independent set I.

Theorem 1 (♠). *Let k be the solution size and ℓ be an upper bound of thresholds of vertices in a given graph. Then TSS parameterized by $k + \ell$ is W[1]-hard for split graphs.*

3.2   Cographs

We first define the class of cographs (also known as P_4-free graphs) [4]. For two graphs $G_1 = (V_1, E_1)$ and $G_2 = (V_2, E_2)$, we say that the graph $G_1 \oplus G_2 = (V_1 \cup V_2, E_1 \cup E_2)$ is the *disjoint union* of G_1 and G_2, and the graph $G_1 \otimes G_2 = (V_1 \cup V_2, E_1 \cup E_2 \cup \{v_1 v_2 \mid v_1 \in V_1, v_2 \in V_2\})$ is the *join* of G_1 and G_2. A graph $G = (V, E)$ is a *cograph* if G can be constructed using the following three rules: (i) a graph with exactly one vertex is a cograph; (ii) for two cographs G_1 and G_2, the disjoint union $G_1 \oplus G_2$ is also a cograph; and (iii) for two cographs G_1 and G_2, the join $G_1 \otimes G_2$ is also a cograph.

Theorem 2. *TSS parameterized by solution size k is W[2]-hard for cographs.*

Proof. We reduce the HITTING SET problem to the decision variant of TSS. In HITTING SET, we are given a universe $U = \{1, 2, \ldots, n\}$, a collection $\mathcal{T} = \{T_1, T_2, \ldots, T_m\}$ of m subsets of U, and an integer k. Then we are asked to determine whether there is a subset $S \subseteq U$ such that $S \cap T_i \neq \emptyset$ for any $i \in [m]$. HITTING SET is known to be W[2]-complete when parameterized by solution size k [10].

We construct an instance (G, thr, k) of the decision variant of TSS from an instance $(U, \mathcal{T}, k)$ of HITTING SET as follows. For each $j \in [m]$, we define a_j as the sum of the sizes of $T_1, T_2, \ldots, T_j$, that is, $a_j = |T_1| + |T_2| + \cdots + |T_j|$. For convenience, we let $a_0 = 0$. For each $i \in [n]$, we define b_i as the number of subsets in $\mathcal{T}$ that contain i, that is, $b_i = |\{T \in \mathcal{T} \mid i \in T\}|$.

The graph G consists of the join of two cographs G_1 and G_2. The graph G_1 is composed of n disjoint stars $S_1, S_2, \ldots, S_n$, where a star is a tree with exactly one internal vertex called the *center*. We denote by c_i the center vertex of S_i and let $C = \{c_i \mid i \in [n]\}$. For each $i \in [n]$, S_i consists of the center vertex c_i and $b_i(k+1)$ leaves. The leaves are separated into b_i vertex sets of size $k+1$, denoted by $V_{i,j}$ for each j such that $i \in T_j$. For convenience, we define $V_{i,j} = \emptyset$ if $i \notin T_j$, and $V_{i,0} = \emptyset$ for every $i \in [n]$. Next we define the threshold for each vertex in G_1. Let $\text{thr}(c_i) = (b_i + m + 1)(k+1)$ for each $i \in [n]$, and let $\text{thr}(v) = j(k+1)+1$ for each $v \in V_{i,j}$.

The graph G_2 consists of an independent set of $(m+2)(k+1)$ vertices. These vertices are separated into $m+2$ vertex sets of size $k+1$, denoted by $V_0, V_1, \ldots, V_{m+1}$. We define the threshold for each vertex in G_2 as follows: $\text{thr}(v) = k$ for each $v \in V_0$; $\text{thr}(v) = (a_{j-1}+2)(k+1)-1$ for each pair of $j \in [m]$ and $v \in V_j$; and $\text{thr}(v) = a_m(k+1)+k$ for each $v \in V_{m+1}$.

This completes the construction of G. The correctness of our reduction is omitted here due to the space limitation. $\qquad\qquad\qquad\qquad\qquad\qquad\qquad\quad\square$

Trivially perfect graphs are also known to be (P_4, C_4)-free graphs [4]. By slightly modifying the above graph, we can obtain the following theorem.

Theorem 3 (♠). *Let k be the solution size and $w_{\max}$ be maximum weight of a given graph. WTSS parameterized by $k + w_{\max}$ is W[2]-hard for trivially perfect graphs.*

4 Polynomial-Time Algorithm for Complete Graphs

Theorem 4. *Let G be an n-vertex complete graph. Then WTSS is solvable in $O(n \log n)$ time.*

We give the pseudocode of our algorithm for complete graphs in Algorithm 1. The correctness of Algorithm 1 is omitted here due to the space limitation.

5 Parameterized Algorithms

5.1 FPT Algorithm Parameterized by nd + ℓ

Two distinct vertices u and v in a graph $G = (V, E)$ are called *twins* if $N(G; u) \setminus \{v\} = N(G; v) \setminus \{u\}$. The *neighborhood diversity* nd of G is the minimum positive

---

Algorithm 1. An algorithm for complete graphs.

1: **Input**: An n-vertex complete graph $G = (V, E)$ with a threshold function thr :
$V \to \mathbb{Z}_+$ and a weight function $w : V \to \mathbb{Z}_+$
2: **Output**: An optimal target set A_0^* of G
3: Sort vertices of G in non-decreasing order by their threshold, and then label the
sorted vertices with $v_1, v_2, \ldots, v_n$
4: **for** $i = 1$ to n **do**
5:　　$s(v_i) \leftarrow \text{thr}(v_i) - i + 1$
6: **end for**
7: $k \leftarrow 1$ and $A_0^* \leftarrow \emptyset$
8: Construct a min-heap P with no element
9: **for** $i = n$ downto 1 **do**
10:　　Push v_i into P using its weight as a key
11:　　**if** $s(v_i) = k$ **then**
12:　　　　Pop a vertex v with minimum weight from P
13:　　　　$A_0^* \leftarrow A_0^* \cup \{v\}$
14:　　　　$k \leftarrow k + 1$
15:　　**end if**
16: **end for**
17: **return** A_0^*

---

integer such that V can be partitioned into $\mathcal{P} = \{P_1, P_2, \ldots, P_{\text{nd}}\}$, where all vertices in P_i are twins for each $i \in [\text{nd}]$. The partition can be obtained in $O(n + m)$ time via modular decomposition [15], where $n = |V|$, and $m = |E|$. It is not hard to see that P_i for each $i \in [\text{nd}]$ forms either an independent set or a clique of G. Moreover, for any pair of distinct sets P_i and P_j with $i, j \in [\text{nd}]$, one of the following conditions holds: there is an edge between any pair of vertices $u \in P_i$ and $v \in P_j$; and there is no edge between any pair of vertices $u \in P_i$ and $v \in P_j$.

Theorem 5. *Let G be a graph with n vertices and m edges. Suppose that G has neighborhood diversity* nd *and there is an integer ℓ such that* thr$(v) \leq \ell$ *for every vertex v in G. Then* WTSS *is solvable in* $O(m + n \log n + \text{nd}^2((2\ell + 1)(\ell + 1))^{\text{nd}})$ *time. Moreover,* WTSS *admits a polynomial kernel with* $O(\ell \cdot \text{nd})$ *vertices.*

We first bound the number of vertices in each $P \in \mathcal{P}$ by using the following reduction rule.

Reduction rule 2. Let v be a vertex in $P \in \mathcal{P}$. Suppose that there are vertex sets $X \subseteq P$ and $Y \subseteq P \setminus X$ such that $|X| = |Y| = \ell$, $v \notin X \cup Y$, thr$(x) \leq$ thr(v) for any $x \in X$, and $w(y) \leq w(v)$ for any $y \in Y$. Then remove v from G.

Lemma 1 (♠). *Reduction rule 2 is safe.*

We apply Reduction rule 2 until it is no longer applicable. It is not hard to see that this reduction can be completed in $O(n \log n)$ time by sorting vertices of G twice according to their threshold and weight. Following this reduction, each

$P \in \mathcal{P}$ has at most 2ℓ vertices. This immediately leads to a polynomial kernel with $2\ell \cdot$ nd vertices and thus an FPT algorithm parameterized by nd $+ \ell$ that solves WTSS in $2^{O(\ell \cdot \mathsf{nd})} n^{O(1)}$ time.

In the remainder of this section, we improve this running time. Assume that a graph G is the graph after applying Reduction rule 2 until it is no longer applicable. The *quotient graph* $Q_\mathcal{P}$ is a graph such that its vertex set is $\mathcal{P}$ and there is an edge between P_i and P_j for distinct integers $i, j \in [\mathsf{nd}]$ if and only if there are vertices $u \in P_i$ and $v \in P_j$ of G such that $uv \in V(G)$. The edge of $Q_\mathcal{P}$ between P_i and P_j means that there is an edge of G between any pair of vertices $u \in P_i$ and $v \in P_j$. For each $i \in [\mathsf{nd}]$, we denote $n_i = |P_i|$ and $P_i = \{v_i^1, v_i^2, \ldots, v_i^{n_i}\}$, assuming that $\mathsf{thr}(v_i^1) \leq \mathsf{thr}(v_i^2) \leq \cdots \leq \mathsf{thr}(v_i^{n_i}) \leq \ell$.

To make it easier to illustrate our algorithmic idea, we consider WPTSS instead of WTSS. Let $(V(G), \prec)$ be the partially ordered set such that, for each integer $i \in [\mathsf{nd}]$ and each pair of distinct integers $x, y \in [n_i]$, $v_i^x \prec v_i^y$ if and only if $x < y$. Note that v_i^x and v_j^y are not comparable when $i \neq j$. We denote by (G, thr, w) and $(G, \mathsf{thr}, w, (V(G), \prec))$ the instances of WTSS and WPTSS, respectively. Then, (G, thr, w) has a target set A_0 of G if and only if $(G, \mathsf{thr}, w, (V(G), \prec))$ has a target set A_0 of G. This is because, in both problems, for each integer $i \in [\mathsf{nd}]$ and each pair of distinct integers $x, y \in [n_i]$ such that $x < y$ and $x, y \notin A_0$, we may assume that v_i^x is activated before v_i^y from the fact that $\mathsf{thr}(x) \leq \mathsf{thr}(y)$ and x and y are twins.

Let $\boldsymbol{a} = (a_1, a_2, \ldots, a_{\mathsf{nd}})$ and $\boldsymbol{b} = (b_1, b_2, \ldots, b_{\mathsf{nd}})$ be vectors of length nd, where $a_i \in [0, 2\ell]$ and $b_i \in [0, \ell]$ for each $i \in [\mathsf{nd}]$. We denote $V_i = \{v_i^j \mid j \in [a_i]\}$ and $V_{\boldsymbol{a}} = \bigcup_{i \in [\mathsf{nd}]} V_i$. We also denote by $\boldsymbol{0}$ the vector of length nd whose entries are 0. Note that if $\boldsymbol{a} = \boldsymbol{0}$, then $V_{\boldsymbol{a}} = \emptyset$. Let A_0 be an optimal target set of G and $A_0, A_1, \ldots$ be an activation process starting at A_0 (under the activation rule of WPTSS). For each pair of possible vectors $\boldsymbol{a}$ and $\boldsymbol{b}$, we guess that the step t satisfies $A_t = A_0 \cup V_{\boldsymbol{a}}$ and $|A_0 \cap P_i \setminus V_i| = b_i$ for each $i \in [\mathsf{nd}]$. Recall that we assumed $b_i \leq \ell$ for each $i \in [\mathsf{nd}]$. Any target set A_0 of G with more than ℓ vertices in $A_0 \cap P_i$ activates all vertices in $N(Q_\mathcal{P}; P_i)$ because every vertex has threshold at most ℓ. As the same holds for the set $A_0 \setminus \{v\}$ for any $v \in A_0 \cap P_i$, one can observe that $A_0 \setminus \{v\}$ is also a target set.

Formally, we compute the function $\alpha(\boldsymbol{a}, \boldsymbol{b})$ that is defined as the minimum value of $w(A \cap V_{\boldsymbol{a}})$ among all subsets $A \subseteq V(G)$ that satisfy the following three conditions:

(1) $|A \cap P_i \setminus V_i| = b_i$ for each $i \in [\mathsf{nd}]$;
(2) there is a bijection $\pi : V_{\boldsymbol{a}} \setminus A \to [|V_{\boldsymbol{a}} \setminus A|]$ with $|\{u \in N(G; v) \mid \pi(u) < \pi(v)\} \cup (A \cap N(G; v))| \geq \mathsf{thr}(v)$ for every $v \in V_{\boldsymbol{a}} \setminus A$; and
(3) for any pair of vertices $u, v \in V_{\boldsymbol{a}} \setminus A$, $\pi(u) < \pi(v)$ if $u \prec v$.

If there is no subset A that satisfies the above three conditions, then let $\alpha(\boldsymbol{a}, \boldsymbol{b}) = +\infty$. Our algorithm computes $\alpha(\boldsymbol{a}, \boldsymbol{b})$ for all possible pairs $(\boldsymbol{a}, \boldsymbol{b})$ by means of dynamic programming. Computing $\alpha((n_1, n_2, \ldots, n_{\mathsf{nd}}), \boldsymbol{0})$, we obtain $w(A_0)$ for an optimal target set A_0 of G.

If $\boldsymbol{a} = \boldsymbol{0}$, we set $\alpha(\boldsymbol{a}, \boldsymbol{b}) = 0$ for every possible vector $\boldsymbol{b}$. Suppose that $\boldsymbol{a} \neq \boldsymbol{0}$. We define the subfunction $f_i(\boldsymbol{a}, \boldsymbol{b})$ that computes $\alpha(\boldsymbol{a}, \boldsymbol{b})$ with the additional constraint that $v_i^{a_i}$ is contained in A. If $a_i = 0$, $v_i^{a_i}$ does not exist and hence $f_i(\boldsymbol{a}, \boldsymbol{b}) = +\infty$. We also conclude that $f_i(\boldsymbol{a}, \boldsymbol{b}) = +\infty$ if $b_i = \ell$ because in this case, P_i contains $\ell+1$ vertices in A and thus A is not optimal. Otherwise, $f_i(\boldsymbol{a}, \boldsymbol{b})$ is computed from an optimal solution that activates all vertices in $V_a \setminus \{v_i^{a_i}\}$. Thus, we have

$$f_i(\boldsymbol{a}, \boldsymbol{b}) = \begin{cases} \alpha(\boldsymbol{a}', \boldsymbol{b}') + w(v_i^{a_i}), & \text{if } a_i > 0 \text{ and } b_i < \ell, \\ +\infty & \text{otherwise.} \end{cases}$$

where $\boldsymbol{a}' = (a_1', a_2', \ldots, a_{\mathsf{nd}}')$ and $\boldsymbol{b}' = (b_1', b_2', \ldots, b_{\mathsf{nd}}')$ such that $a_i' = a_i - 1$, $b_i' = b_i + 1$, and $a_j' = a_j$ and $b_j' = b_j$ for each $j \in [\mathsf{nd}] \setminus \{i\}$.

We also define the subfunction $g_i(\boldsymbol{a}, \boldsymbol{b})$ that computes $\alpha(\boldsymbol{a}, \boldsymbol{b})$ with the additional constraint that $\pi^{-1}(|V_a \setminus A|) = v_i^{a_i}$. It should be noted that $\pi^{-1}(|V_a \setminus A|) \in \{v_i^{a_i} \mid i \in [\mathsf{nd}] \text{ and } a_i > 0\}$, unless $v_i^{a_i}$ is in A for every $i \in [\mathsf{nd}]$, because of the condition (3). Moreover, any vertex in $P_i \setminus A$ is not yet active in the current step. This allows us to count active vertices adjacent to $v_i^{a_i}$ according to the vectors $\boldsymbol{a}$ and $\boldsymbol{b}$. To compute $g_i(\boldsymbol{a}, \boldsymbol{b})$, we consider the two cases: P_i is an independent set of G; and P_i is a clique of G. Let $\boldsymbol{a}' = (a_1', a_2', \ldots, a_{\mathsf{nd}}')$ be a vector such that $a_i' = a_i - 1$ and $a_j' = a_j$ for each $j \in [\mathsf{nd}] \setminus \{i\}$. If P_i is an independent set, then we have

$$g_i(\boldsymbol{a}, \boldsymbol{b}) = \begin{cases} \alpha(\boldsymbol{a}', \boldsymbol{b}) & \text{if } \sum_{P_j \in N(Q_{\mathcal{P}}; P_i)} (a_j + b_j) \geq \mathsf{thr}(v_i^{a_i}) \text{ and } a_i > 0, \\ +\infty & \text{otherwise.} \end{cases}$$

If P_i is a clique, then we have

$$g_i(\boldsymbol{a}, \boldsymbol{b}) = \begin{cases} \alpha(\boldsymbol{a}', \boldsymbol{b}) & \text{if } \sum_{P_j \in N[Q_{\mathcal{P}}; P_i]} (a_j + b_j) - 1 \geq \mathsf{thr}(v_i^{a_i}) \text{ and } a_i > 0, \\ +\infty & \text{otherwise.} \end{cases}$$

After obtaining $f_i(\boldsymbol{a}, \boldsymbol{b})$ and $g_i(\boldsymbol{a}, \boldsymbol{b})$ for each $i \in [\mathsf{nd}]$, we take their minimum value, that is, $\alpha(\boldsymbol{a}, \boldsymbol{b}) = \min_{i \in [\mathsf{nd}]} \{f_i(\boldsymbol{a}, \boldsymbol{b}), g_i(\boldsymbol{a}, \boldsymbol{b})\}$. Our algorithm runs in time $O(m + n \log n + \mathsf{nd}^2((2\ell + 1)(\ell + 1))^{\mathsf{nd}})$ because, after preprocessing, $\alpha(\boldsymbol{a}, \boldsymbol{b})$ is computed for each pair $(\boldsymbol{a}, \boldsymbol{b})$ and there are $(2\ell + 1)^{\mathsf{nd}} \cdot (\ell + 1)^{\mathsf{nd}}$ possible pairs of $(\boldsymbol{a}, \boldsymbol{b})$.

5.2   FPT Algorithm Parameterized by tw + ℓ

Theorem 6 (♠). *Let G be an n-vertex graph. Suppose that G has treewidth* tw *and there is an integer ℓ such that* $\mathsf{thr}(v) \leq \ell$ *for any vertex v in G. Then* WTSS *is solvable in* $\ell^{O(\mathsf{tw})} n$ *time.*

5.3   FPT Algorithm Parameterized by ce

Theorem 7 (♠). *Let $G = (V, E)$ be an n-vertex graph. Suppose that G has cluster editing number* ce. *Then there exists an FPT algorithm of* WTSS *parameterized by* ce.

5.4   FPT Algorithm Parameterized by vc

Theorem 8. *Let $G = (V, E)$ be an n-vertex graph. Then there exists an FPT algorithm of* WTSS *parameterized by* vc.

Let $C \in V$ be a vertex cover of G with size vc. It is known that a vertex cover with size vc can be found in $O(1.2738^{\text{vc}} + \text{vc} \cdot n)$ time [5].

Consider an instance (G, thr, w) of WTSS. Toward getting an FPT algorithm to solve the instance, we give a Turing reduction from WTSS to a variant of HITTING SET which is defined below.

WEIGHTED MULTI-HITTING SET

Input: A universe $U = \{1, 2, \ldots, n\}$, a collection $\mathcal{T} = \{T_1, T_2, \ldots, T_m\}$ of m subsets of U, a function $\lambda : [m] \to \mathbb{Z}_+$ such that $\lambda(i) \leq |T_i|$ for every $i \in [m]$, and a weight function $w : U \to \mathbb{Z}_+$.

Task: Find a subset $S \subseteq U$ with minimum total weight $\sum_{j \in S} w(j)$ such that $|S \cap T_i| \geq \lambda(i)$ for every $i \in [m]$.

In order to design an FPT algorithm parameterized by vc for (the unweighted) TSS, Banerjee et al. reduced TSS to (the unweighted) MULTI-HITTING SET [1]. One can see that the reduction is also applicable to their weighted variants. Thus, we obtain the following lemma.

Lemma 2. *There is an algorithm that, given an instance (G, thr, w) of* WTSS, *runs in $O((2\text{vc} + 1)^{\text{vc}} \cdot n^c)$ time, where G has a vertex cover C of size* vc *and c (≥ 1) is a constant, and outputs a collection of instances $\mathcal{I} = \{I^h = (V \setminus C, \mathcal{T}^h, \lambda^h, w^h) \mid h \in [s]\}$ of* WEIGHTED MULTI-HITTING SET *such that the following conditions hold:*

(i) $s \leq (2\text{vc} + 1)^{\text{vc}}$;
(ii) $|\mathcal{T}^h| = \text{vc}$ for each $h \in [s]$; and
(iii) an instance (G, thr, w) of WTSS *has a solution with total weight W if and only if there exists an integer $h \in [s]$ such that an instance I^h of* WEIGHTED MULTI-HITTING SET *has a solution with total weight W.*

We design an FPT algorithm parameterized by vc for WEIGHTED MULTI-HITTING SET, using Integer Linear Programming (ILP). First, define $\boldsymbol{w} \in \mathbb{Z}_+^n$ as a vector with $w_i = w(i)$ for each $i \in [n]$. Let $\boldsymbol{x} \in \mathbb{Z}^n$ be a vector of variables with $x_j \in \{0, 1\}$, where $x_j = 1$ means that $j \in [n]$ is included in a solution S, otherwise, not included. We denote by $\boldsymbol{b} \in \mathbb{Z}^m$ the vector with $b_i = \lambda(i)$. Moreover, for each pair of $i \in [m]$ and $j \in [n]$, set $a_{ij} = 1$ if $j \in T_i$, otherwise, $a_{ij} = 0$. Then the constraint $|S \cap T_i| \geq \lambda(i)$ for each $i \in [n]$ can be expressed as $\sum_{j \in [n]} a_{ij} x_j \geq b_i$. Denote $\boldsymbol{A} = (a_{ij}) \in \mathbb{Z}^{m \times n}$. By introducing slack variables $\boldsymbol{s} \in \mathbb{Z}^m$, WEIGHTED MULTI-HITTING SET is expressed using ILP as follows:

$$\max \ -\boldsymbol{w}^T \boldsymbol{x}$$
$$\text{subject to } \boldsymbol{A}\boldsymbol{x} - \boldsymbol{s} = \boldsymbol{b}$$
$$\boldsymbol{0} \leq \boldsymbol{x} \leq \boldsymbol{1} \tag{1}$$
$$\boldsymbol{0} \leq \boldsymbol{s} \leq n\boldsymbol{1}$$
$$\boldsymbol{x} \in \mathbb{Z}^n \text{ and } \boldsymbol{s} \in \mathbb{Z}^m,$$

where $\mathbf{1}$ denotes the vector of appropriate length whose entries are 1.

Eisenbrand and Weismantel showed the following theorem.

Theorem 9 ([12], **Theorem 4.1**). *An ILP of the form*

$$\max\{\boldsymbol{w}^T\boldsymbol{x} : \boldsymbol{A}\boldsymbol{x} = \boldsymbol{b}, 0 \le \boldsymbol{x} \le \boldsymbol{u}, \boldsymbol{x} \in \mathbb{Z}^n\}$$

can be solved in time $n \cdot O(m)^{(m+1)^2} \cdot O(\Delta)^{m(m+1)} \cdot \log^2(m\Delta)$, *where* $\boldsymbol{A} \in \mathbb{Z}^{m \times n}$, $\boldsymbol{b} \in \mathbb{Z}^m$, $\boldsymbol{w}, \boldsymbol{u} \in \mathbb{Z}_+^n$, *and* $\Delta \in \mathbb{Z}$ *such that* $|a_{ij}| \le \Delta$ *for every pair of* $i \in [m]$ *and* $j \in [n]$.

For the ILP in Eq. (1), combined with the condition (ii) of Lemma 2, we have $m = \mathsf{vc}$ and $\Delta = 1$. Thus, WEIGHTED MULTI-HITTING SET expressed in Eq. (1) can be solved in time $n \cdot O(\mathsf{vc})^{(\mathsf{vc}+1)^2} \cdot \log^2(\mathsf{vc})$.

By the condition (i) of Lemma 2, we need to solve $s \le (2\mathsf{vc} + 1)^{\mathsf{vc}}$ instances of WEIGHTED MULTI-HITTING SET. By the condition (iii) of Lemma 2, we conclude that the running time of our algorithm is bounded by $2^{O(\mathsf{vc}^2 \log \mathsf{vc})} n^c$, which completes the proof of Theorem 8.

6   Conclusion and Future Work

In this paper, we studied the parameterized complexity of TSS and WTSS. We showed three intractable results and four tractable cases.

The following are some future directions: (i) Can TSS be solved in polynomial time when a trivially perfect graph is given? (ii) Can TSS and WTSS be solved in polynomial time for threshold graphs and chain graphs? (iii) Is there an XP algorithm parameterized by twin cover number tc? As indicated in Figs. 1 and 2, settling these questions would give interesting contrasts.

Acknowledgements. We thank anonymous referees for their valuable comments and suggestions which greatly helped to improve the presentation of this paper.

References

1. Banerjee, S., Mathew, R., Panolan, F.: Target set selection parameterized by vertex cover and more. Theor. Comput. Syst. **66**(5), 996–1018 (2022). https://doi.org/10.1007/s00224-022-10100-0
2. Bazgan, C., Chopin, M., Nichterlein, A., Sikora, F.: Parameterized approximability of maximizing the spread of influence in networks. J. Discret. Algorithms **27**, 54–65 (2014). https://doi.org/10.1016/j.jda.2014.05.001
3. Ben-Zwi, O., Hermelin, D., Lokshtanov, D., Newman, I.: Treewidth governs the complexity of target set selection. Discret. Optim. **8**(1), 87–96 (2011). https://doi.org/10.1016/j.disopt.2010.09.007
4. Brandstädt, A., Le, V.B., Spinrad, J.P.: Graph Classes: A Survey. Society for Industrial and Applied Mathematics, Philadelphia (1999). https://doi.org/10.1137/1.9780898719796

5. Chen, J., Kanj, I.A., Xia, G.: Improved upper bounds for vertex cover. Theor. Comput. Sci. **411**(40), 3736–3756 (2010). https://doi.org/10.1016/j.tcs.2010.06.026

6. Chen, N.: On the approximability of influence in social networks. SIAM J. Discret. Math. **23**(3), 1400–1415 (2009). https://doi.org/10.1137/08073617X

7. Chiang, C., Huang, L., Li, B., Wu, J., Yeh, H.: Some results on the target set selection problem. J. Comb. Optim. **25**(4), 702–715 (2013). https://doi.org/10.1007/S10878-012-9518-3

8. Chopin, M., Nichterlein, A., Niedermeier, R., Weller, M.: Constant thresholds can make target set selection tractable. In: Even, G., Rawitz, D. (eds.) MedAlg 2012. LNCS, vol. 7659, pp. 120–133. Springer, Heidelberg (2012). https://doi.org/10.1007/978-3-642-34862-4_9

9. Cicalese, F., Cordasco, G., Gargano, L., Milanič, M., Vaccaro, U.: Latency-bounded target set selection in social networks. Theor. Comput. Sci. **535**, 1–15 (2014). https://doi.org/10.1016/j.tcs.2014.02.027

10. Cygan, M., et al.: Parameterized Algorithms, 1st edn. Springer, Cham (2015). https://doi.org/10.1007/978-3-319-21275-3

11. Dvořák, P., Knop, D., Toufar, T.: Target set selection in dense graph classes. SIAM J. Discret. Math. **36**(1), 536–572 (2022). https://doi.org/10.1137/20M1337624

12. Eisenbrand, F., Weismantel, R.: Proximity results and faster algorithms for integer programming using the Steinitz lemma. ACM Trans. Algorithms **16**(1) (2019). https://doi.org/10.1145/3340322

13. Hartmann, T.A.: Target set selection parameterized by clique-width and maximum threshold. In: Tjoa, A.M., Bellatreche, L., Biffl, S., van Leeuwen, J., Wiedermann, J. (eds.) SOFSEM 2018. LNCS, vol. 10706, pp. 137–149. Springer, Cham (2018). https://doi.org/10.1007/978-3-319-73117-9_10

14. Lampis, M.: Structural Graph Parameters, Fine-Grained Complexity, and Approximation. Habilitation à diriger des recherches, Université Paris Dauphine (2022). https://hal.science/tel-03848575

15. McConnell, R.M., Spinrad, J.P.: Modular decomposition and transitive orientation. Discret. Math. **201**(1), 189–241 (1999). https://doi.org/10.1016/S0012-365X(98)00319-7

16. Nichterlein, A., Niedermeier, R., Uhlmann, J., Weller, M.: On tractable cases of target set selection. Soc. Netw. Anal. Min. **3**(2), 233–256 (2013). https://doi.org/10.1007/S13278-012-0067-7

17. Raghavan, S., Zhang, R.: Weighted target set selection on trees and cycles. Networks **77**(4), 587–609 (2021). https://doi.org/10.1002/NET.21972

Mechanism Design for Building Optimal Bridges Between Regions

Zining Qin[2], Hau Chan[3], Chenhao Wang[1,2(✉)], and Ying Zhang[4]

[1] Beijing Normal University, Zhuhai 519087, China
chenhwang@bnu.edu.cn
[2] BNU-HKBU United International College, Zhuhai 519087, China
[3] University of Nebraska-Lincoln, Lincoln, NE 68508, USA
[4] Beijing Electronic Science and Technology Institute, Beijing 100070, China

Abstract. We study the bridge-building problem from the mechanism design perspective. In this problem, a social planner is tasked with building a bridge to connect two regions separated by an obstacle (e.g., a river or valley). Each agent in a region has a private location and is interested in traveling to a facility (e.g., a transportation hub) in the other region. The cost of an agent with respect to a bridge is the distance from their location to the facility of interest via the bridge. Our goal is to design strategy-proof mechanisms that elicit truthful locations from the agents and approximately optimize an objective by determining a location for building a bridge. We consider the social cost and maximum cost objectives, which are the total cost and maximum cost of agents, respectively. For the social cost objective, we characterize an optimal solution and show that it is strategy-proof. For the maximum cost objective, any optimal solution is no longer strategy-proof. We present deterministic $\frac{5}{3}$-approximation and randomized $\frac{3}{2}$-approximation strategy-proof mechanisms. We complement the results by providing tight lower bounds.

1 Introduction

In many urban planning infrastructure projects, a social planner is often tasked with building a bridge to connect two different regions that are in between an obstacle (e.g., a body of water/river, a road, or a valley) [11]. Not surprisingly, building a bridge can lead to many societal benefits. For example, building a bridge over a river can help agents in a region to cross the river to reach the other region more directly. Building a bridge between two mountains can shorten the travel distances of agents compared to using spiral roads on either mountain. Building a viaduct can help to carry a road or railway and reduce the travel distances of agents between different regions. As a result, a bridge would enable agents, at their starting locations, to travel from one region to a point of interest (e.g., an access point, a transit station, or a region center) in another region more efficiently and safely without going through the obstacle directly.

In Fig. 1, we provide an example in which agents are in two regions (modeled simply as a line) separated by a river, which divides the whole region into A

X. Chen and B. Li (Eds.): TAMC 2024, LNCS 14637, pp. 332–343, 2024.
https://doi.org/10.1007/978-981-97-2340-9_28

and B. The agents from Region A need to be connected to a fixed access point in Region B, and agents in Region B need to access the fixed access point in Region A. The access points (referred to as facilities) can be viewed as a hub, such as a center of the community or transit station (determined by the social planner), which the agents can use to reach other places in different regions. The agents' starting locations are alongside the regions (denoted as dots), and the access points are denoted in squares. The social planner would like to improve the distances of the agents from one region to the access point of another region (e.g., from Region A to the access point in Region B) by building a bridge (denoted as a green line in Fig. 1).

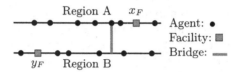

Fig. 1. An example for bridging two regions. The agents are represented as dots. The two access points (facilities) are at x_F and y_F (represented as squares). (Color figure online)

Existing studies have examined building optimal bridges between two different regions as an optimization problem, aiming to minimize the maximum distance between any two points from the two regions (see, e.g., [2,9,10,16,17]). These existing studies have designed polynomial algorithms for building (approximately) optimal bridges between different types of convex polygons (see related work for more details).

Unfortunately, there are two main assumptions that make existing optimization literature not ideal for capturing real-world situations. First, existing literature assumes that all of the points in the regions are the starting locations of some agents. However, in many real-world situations, agents' starting positions consist only of a subset of locations in the regions. Second, existing literature assumes that each agent is required to connect to or access other agents' starting locations in another region. However, agents may not necessarily be concerned about other agents' starting locations and would only be interested in connecting or accessing a given access point (e.g., a shopping mall, region center, or station) Therefore, our goal is to build optimal bridges to account for agents' starting locations and the access points in regions.

1.1 Our Contribution

In this paper, we initiate the mechanism design study of building (approximately) optimal bridges between two regions to connect agents (at their starting locations in their regions) to the corresponding facilities in other regions.

In such a setting, the agent's starting locations in the regions might not be publicly known or visible to the social planner. Therefore, our main goal is to design strategy-proof mechanisms that elicit agents' starting locations truthfully and output bridges that approximately minimize objectives based on the agents' starting locations in their regions to the facilities in other regions.

More specifically, we focus on a basic setting where the two regions are represented by two separate parallel real lines $\mathbb{R}_1$ and $\mathbb{R}_2$ divided by an obstacle (see Fig. 1). Agents' starting locations are points in the real lines (i.e., $x_i \in \mathbb{R}_1$ or $y_j \in \mathbb{R}_2$ for any agent i in $\mathbb{R}_1$ and any agent j in $\mathbb{R}_2$), and the facilities correspond to fixed points (i.e., $x_F \in \mathbb{R}_1$ and $y_F \in \mathbb{R}_2$) in the regions.

We aim to build a bridge that connects the two real lines $\mathbb{R}_1$ and $\mathbb{R}_2$, where one endpoint is in $\mathbb{R}_1$ and the other is in $\mathbb{R}_2$. The bridge must be perpendicular to both $\mathbb{R}_1$ and $\mathbb{R}_2$, and we use a single point s to denote the bridge location. Given a bridge location s, the cost of an agent at $x_i \in \mathbb{R}_1$ is the distance from their starting location to the endpoint of the bridge in $\mathbb{R}_1$, plus the distance from the facility location to the endpoint of the bridge in $\mathbb{R}_2$, i.e., $|x_i - s| + |y_F - s|$.[3] The cost of an agent at $y_j \in \mathbb{R}_2$ is defined similarly. We study two objectives that aim to minimize the social cost and the maximum cost.

For the social cost objective, we characterize an optimal solution and show that an optimal solution is strategy-proof. For the maximum cost objective, we characterize an optimal solution and show that any optimal solution is not strategy-proof.

We provide a deterministic strategy-proof mechanism that has an approximation ratio of $\frac{5}{3}$. We complement this result by providing a tight matching lower bound. We also design a randomized strategy-proof mechanism that has an approximation ratio of $\frac{3}{2}$. We also provide a tight matching lower bound for any randomized strategy-proof mechanisms.

We note that our setting can be reduced to a special setting of Fukui et al. [8], who studied a variant of facility location problems. Fukui et al. [8] proposed a group strategy-proof mechanism that minimizes the social cost, but they did not consider the maximum cost objective. Our characterization of an optimal solution provides a more succinct mechanism result.

1.2   Related Work

The optimal bridge-building problem has been widely studied in the literature [1,2,9,10,16,17]. Cai et al. [2] introduced the problem of adding a line segment to connect two disjoint convex polygonal regions in a plane, such that the length of the longest path from a point in one polygon, passing through the bridge, to a point in another region is minimized. They proposed an $O(n^2 \log n)$-time algorithm, where n is the maximum number of extreme points of the polygons. Later, Bhattacharya and Benkoczi [1] proposed a linear-time algorithm that improves the $O(n^2 \log n)$-time algorithm in [2]. Tan [16] independently presented an alternate linear-time algorithm for the above setting and further generalized it to an $O(n^2)$-time algorithm for bridging two convex polyhedra in space. Kim and Shin

---

[3] We assume that the bridge has zero cost for the agent using the bridge because it is a constant term that each agent would incur. A positive bridge cost can only help improve our approximation ratios.

[10] provided algorithms to find an optimal bridge between two convex polygons, two simple non-convex polygons, and one convex and one simple non-convex polygons in $O(n)$, $O(n^2)$, and $O(n \log n)$, respectively. Later, Tan [17] provided an $O(n \log^3 n)$-time algorithm for the settings of two simple non-convex polygons. The most related setting is by Kim et al. [9], who proposed a linear-time algorithm to compute an optimal bridge between two parallel lines separated by an obstacle to minimize the length of the longest path connecting two points on the lines. On a related note, there are works that focus on minimizing the diameter or average shortest distances between pairs of nodes of a network using new edges (see, e.g., [5,14]). However, all of the works mentioned focusing on *all* points in the regions. Our work focuses on a subset of points, which are the agents' starting points, and aims to bridge agents to their corresponding facilities in other regions from the mechanism design perspective.

Our work is within the paradigm of approximate mechanism design without money. The paradigm of approximate mechanism design without money is initialized by Procaccia and Tennenholtz [15], who used facility location problems as case studies. This paradigm investigates strategy-proof mechanisms through the lens of the approximation ratio. In a typical setting of facility location problems, the agents report their private locations on the real line to a mechanism. The mechanism determines the locations for building facilities, where the cost of agents is the distance to the facilities. Following this work, variations of facility location problems have been introduced (see, e.g., [6,7,12,13]). We refer readers to a survey on models and results for mechanism design for facility location [3]. The most relevant mechanism design work to ours is by Fukui et al. [8], which considers a more general setting called *pit-stop facility game*, where all agents are in a real line, and each agent i reports an interval $[x_i, y_i]$. A mechanism determines a point $s \in \mathbb{R}$. The cost of agent i is $|s - x_i| + |s - y_i|$. It is easy to see that our setting is the special case when the agents in $\mathbb{R}_1$ has 0 (resp. the agents in $\mathbb{R}_2$ has 1) as an endpoint of their intervals. Fukui et al. [8] proposed a deterministic group strategy-proof mechanism that minimizes the social cost (called *lowest balanced mechanism*), but they did not consider the maximum cost objective. Another most relevant work to ours is the work of [4] in which they considered modifying the structure of regions by adding a shuttle or road to improve the distances of the agents to a prelocated facility in a real line. However, they do not consider two regions separated by an obstacle.

2   Model

There are two parallel real lines, denoted by $\mathbb{R}_1$ and $\mathbb{R}_2$. Assume that $\mathbb{R}_1$ is above $\mathbb{R}_2$, where $\mathbb{R}_1$ and $\mathbb{R}_2$ are regions A and B, respectively, as shown in Fig. 1. A set of $M = \{1, \ldots, m\}$ agents is located in $\mathbb{R}_1$, and a set of $N = \{1, \ldots, n\}$ agents is located in $\mathbb{R}_2$. Each agent $i \in M$ has a location $x_i \in \mathbb{R}_1$, and each agent $j \in N$ has a location $y_j \in \mathbb{R}_2$. Let $\mathbf{x} = (x_1, \ldots, x_m) \in \mathbb{R}_1^m$ be the location profile of the agents in M, and $\mathbf{y} = (y_1, \ldots, y_n) \in \mathbb{R}_2^n$ be the location profile of the agents in N. There are two facilities F_1 and F_2 located at $\mathbb{R}_1$ and $\mathbb{R}_2$, respectively. Both facilities have fixed locations. Let $x_F \in \mathbb{R}_1$ be the location of facility F_1, and $y_F \in \mathbb{R}_2$ be the location of facility F_2.

The agents in M want to access facility F_2, and the agents in N want to access facility F_1. To this end, we aim to build a *bridge* to connect the two real lines $\mathbb{R}_1$ and $\mathbb{R}_2$, where one endpoint of this bridge is in $\mathbb{R}_1$ and the other endpoint is in $\mathbb{R}_2$. The bridge is perpendicular to both $\mathbb{R}_1$ and $\mathbb{R}_2$. Hence, we can use a single point s to denote the location of a bridge, where the two endpoints are $s \in \mathbb{R}_1$ and $s \in \mathbb{R}_2$, respectively. Given a bridge s, the cost of each agent $i \in M$ is the length of the shortest path from this agent to facility F_2,

$$c_1(x_i, s) := |x_i - s| + |s - y_F|.$$

Similarly, the cost of each agent $j \in N$ is defined as

$$c_2(y_j, s) := |y_j - s| + |s - x_F|.$$

A (deterministic) mechanism $f : \mathbb{R}_1^m \times \mathbb{R}_2^n \to \mathbb{R}$ is a function that maps the location profiles $\mathbf{x}, \mathbf{y}$ of agents into a real value as the bridge location. A randomized mechanism is a function f from $R_1^m \times R_2^n$ to probability distributions over $\mathbb{R}$. If $f(\mathbf{x}, \mathbf{y}) = P$ is a probability distribution, the cost of agent $i \in M$ is defined as the expectation $c_1(x_i, P) = \mathbb{E}_{s \sim P}[c_1(x_i, s)]$, and the cost of $j \in N$ is defined similarly.

A mechanism is *strategy-proof* if no agent can decrease their cost by misreporting their location, where we restrict the misreporting on their own side, that is, an agent in $\mathbb{R}_1$ cannot report a location in $\mathbb{R}_2$, and vice versa.

Definition 1. *A mechanism f is* strategy-proof *if for any location profiles $\mathbf{x}, \mathbf{y}$, it satisfies two conditions: (1) $c_1(x_i, f(\mathbf{x}, \mathbf{y})) \leq c_1(x_i, f(x_i', \mathbf{x}_{-i}, \mathbf{y}))$ for any agent $i \in M$ and $x_i' \in \mathbb{R}_1$, where $\mathbf{x}_{-i}$ is the location profile of the agents in $M \setminus \{i\}$, and (2) $c_2(y_j, f(\mathbf{x}, \mathbf{y})) \leq c_2(y_j, f(\mathbf{x}, y_j', \mathbf{y}_{-j}))$ for any agent $j \in N$ and $y_j' \in \mathbb{R}_2$, where $\mathbf{y}_{-j}$ is the location profile of the agents in $N \setminus \{j\}$.*

We study two objective functions, minimizing the social cost, and minimizing the maximum cost. Given profiles $\mathbf{x}, \mathbf{y}$ and bridge location s, the *social cost* is the total cost of all agents $SC(s, \mathbf{x}, \mathbf{y}) = \sum_{i \in M} c_1(x_i, s) + \sum_{j \in N} c_2(y_j, s)$, and the *maximum cost* is the maximum value among the cost of all agents $MC(s, \mathbf{x}, \mathbf{y}) = \max\{\max_{i \in M} c_1(x_i, s), \max_{j \in N} c_2(y_j, s)\}$. We say that a mechanism f has an approximation ratio α or is α-approximation for the objective $S \in \{SC, MC\}$, if for any instance $(\mathbf{x}, \mathbf{y}, x_F, y_F)$, we have $\frac{S(f(\mathbf{x}, \mathbf{y}), \mathbf{x}, \mathbf{y})}{\min_{s \in \mathbb{R}} S(s, \mathbf{x}, \mathbf{y})} \leq \alpha$.

Within the agenda of approximate mechanism design, the goal is to design strategy-proof mechanisms with good approximation ratios. Note that when $x_F = y_F$, a trivial mechanism that always returns x_F as bridge location is clearly strategy-proof and optimal for both objectives (as all agents attain their best possible cost). Hence, we focus on the situation $x_F \neq y_F$, and assume that $x_F = 1$, $y_F = 0$ throughout the remainder of this paper. This assumption is without loss of generality because we can scale the locations in real lines.

As a preliminary result, the following lemma says that an optimal bridge location is between the two facilities (public locations 0 and 1).

Lemma 1. *For both social cost and maximum cost objectives, there exists an optimal solution $s^* \in [0, 1]$.*

Proof. Let s be an optimal solution. If $s \in [0, 1]$, the lemma is proved. Assume without loss of generality that $s < 0$. Compared with the solution $s^* = 0$, we show that the cost of every agent under s^* is no more than the cost under s. For every agent $i \in M$, we have

$$c_1(x_i, s) = |x_i - s| + |s - 0| \geq |x_i| = c_1(x_i, 0).$$

For every agent $j \in N$, we have

$$c_2(y_j, s) = |y_j - s| + (1 - s) = |y_j - s| + |s| + 1 \geq |y_j| + 1 = c_2(y_j, 0).$$

Therefore, $s^* = 0$ is also an optimal solution, establishing the proof. □

3 Social Cost

In this section, we study the objective of minimizing the social cost $SC(s, \mathbf{x}, \mathbf{y}) = \sum_{i \in M} c_1(x_i, s) + \sum_{j \in N} c_2(y_j, s)$. We first present an intuitive algorithm for computing an optimal solution and then show that it is strategy-proof.

Given location profiles $\mathbf{x}, \mathbf{y}$, for any $s \in [0, 1]$, define $M_l(s) = \{i \in M | x_i \leq s\}$ to be the set of agents in M whose locations are on the left of s, and $M_r(s) = \{i \in M | x_i > s\}$ to be the set of agents in M on the right of s. Similarly, define $N_l(s) = \{j \in N | y_j \leq s\}$ and $N_r(s) = \{j \in N | y_j > s\}$. The algorithm moves the bridge location s from 0 to 1 continuously until it cannot decrease the social cost or it reaches 1. During the process of moving s to $s + \epsilon$ for any $\epsilon > 0$, if there is no agent located at interval $(s, s + \epsilon)$, then the cost of any agents in $M_r(s) \cup N_l(s)$ does not change, the cost of the agents in $M_l(s)$ increases by 2ϵ, and the cost of agents in $N_r(s)$ decreases by 2ϵ. The algorithm stops when the number of agents in $M_l(s)$ is no less than that in $N_r(s)$. Formally, the algorithm returns a bridge location

$$b(\mathbf{x}, \mathbf{y}) := \begin{cases} 1, & \text{if } |M_l(1)| < |N_r(1)| \\ \min\{s \in [0, 1] \mid |M_l(s)| \geq |N_r(s)|\}, & \text{otherwise.} \end{cases}$$

Theorem 1. *For the social cost, bridge location $b(\mathbf{x}, \mathbf{y})$ is optimal. The mechanism that returns $b(\mathbf{x}, \mathbf{y})$ is strategy-proof.*

Proof. By Lemma 1, there is an optimal solution $s^* \in [0, 1]$. We prove the optimality of $b(\mathbf{x}, \mathbf{y})$ by discussing two cases $s^* < b$ and $s^* > b$ (we write $b(\mathbf{x}, \mathbf{y})$ as b when no confusion arises). If $s^* < b$, by the definition of b, it must be $|M_l(s^*)| < |N_r(s^*)|$. When moving the bridge location from s^* to b, the cost of each agent in $M_r(b) \cup N_l(s^*)$ does not change, the cost of each agent in $M_l(b)$ increases, and the cost of each agent in $N_r(s^*)$ decreases. Let $\epsilon > 0$ be a sufficiently small value so that $s < b - \epsilon$ and no agent lies in interval $(b - \epsilon, b)$. Then, in particular, the cost of each agent in $M_l(b - \epsilon)$ increases by at most

$2(b - s^*)$, and the cost of each agent in $N_r(b - \epsilon) \subseteq N_r(s^*)$ decreases by exactly $2(b - s^*)$. Hence,

$$SC(b, \mathbf{x}, \mathbf{y}) \leq SC(s^*, \mathbf{x}, \mathbf{y}) - 2(b - s^*)|N_r(b - \epsilon)| + 2(b - s^*)|M_l(b - \epsilon)|$$
$$< SC(s^*, \mathbf{x}, \mathbf{y}),$$

where the last inequality comes from the fact that $|M_l(b - \epsilon)| < |N_r(b - \epsilon)|$ by the definition of b.

If $s^* > b$, it must be $|M_l(s^*)| \geq |N_r(s^*)|$. When moving the bridge location from b to s^*, the cost of each agent in $M_r(s^*) \cup N_l(b)$ does not change, the cost of each agent in $M_l(s^*)$ is non-decreasing, and the cost of each agent in $N_r(b)$ decreases. In particular, the cost of each agent in $M_l(b) \subseteq M_l(s^*)$ increases by exactly $2(s^* - b)$, and the cost of each agent in $N_r(b)$ decreases by at most $2(s^* - b)$. Hence, we have

$$SC(s^*, \mathbf{x}, \mathbf{y}) \geq SC(b, \mathbf{x}, \mathbf{y}) - 2(s^* - b)|N_r(b)| + 2(s^* - b)|M_l(b)|$$
$$\geq SC(b, \mathbf{x}, \mathbf{y}),$$

where the last inequality comes from the fact that $|M_l(b)| \geq |N_r(b)|$ by the definition of b. Therefore, b must be an optimal solution for the social cost.

Next, we prove the strategy-proofness. Note that the agents in $M_r(b) \cup N_l(b)$ have no incentive to lie because they attain the best possible cost. The agent $i \in M_l(b)$ with $x_i = b$ also has the best possible cost and, thus, will not misreport. For each agent $i \in M_l(b)$ with $x_i < b$, the only way to change the solution is to misreport a location $x_i' > b$, which leads to a solution $b(x_i', \mathbf{x}_{-i}, \mathbf{y}) \geq b$; however, this can only increase the cost of agent i. For each agent $j \in N_r(b)$, the only way to change the solution is to misreport a location $y_j' < b$, which leads to a solution $b(\mathbf{x}, y_j', \mathbf{y}_{-j}) \leq b$ and cannot bring any benefit. Therefore, no agent has an incentive to misreport. □

4   Maximum Cost

In this section, we consider the objective of minimizing the maximum cost. Before presenting the mechanisms, we first characterize the optimal maximum cost. Given location profiles $\mathbf{x}, \mathbf{y}$, define $x_l = \min_{i \in M} x_i$, $x_r = \max_{i \in M} x_i$, $y_l = \min_{j \in N} y_j$, and $y_r = \max_{j \in N} y_j$. The maximum cost in any solution must be attained by one of the four extreme agents. Intuitively, a good bridge location should balance the cost of these extreme agents. Recall from Lemma 1 that there is an optimal solution in $[0, 1]$. Indeed, the bridge location in $[0, 1]$ that minimizes the difference between the cost of x_l and y_r is optimal, that is,

$$\arg \min_{w \in [0,1]} |c_1(x_l, w) - c_2(y_r, w)| = |\,||x_l - w| - |y_r - w| - 1 + 2w|.$$

Proposition 1. *For the maximum cost, $s^* \in \arg \min_{w \in [0,1]} |c_1(x_l, w) - c_2(y_r, w)|$ is an optimal solution.*

Proof. Let $s \in [0,1]$ be an optimal solution. If $c_1(x_l, s) = c_2(y_r, s)$, the proof is done. Assume without loss of generality that $c_1(x_l, s) > c_2(y_r, s)$. Note that for any solution $s' > s$, the difference $|c_1(x_l, s') - c_2(y_r, s')|$ is no less than $c_1(x_l, s) - c_2(y_r, s)$. If the optimal solution s does not minimize the difference between the cost of x_l and y_r, then there is a solution s' with $0 \le s' < s$ that minimizes the difference. Indeed, compared with s, the cost of the agent at x_l under s' is non-increasing, i.e., $c_1(x_l, s') \le c_1(x_l, s)$, and the cost of the agent at y_r is non-decreasing, i.e., $c_2(y_r, s') \le c_2(y_r, s)$. It follows that the difference between these two agents is non-increasing. Further, while the cost of the agent at x_r is non-increasing, the cost of the agent at y_l may be increasing. If it increases, then it must be $0 \le y_l \le y_r$, and y_l cannot be responsible for the maximum cost. Hence, the maximum cost is non-increasing, and the solution s' is optimal.   □

Define a rounding function $r : \mathbb{R} \to [0,1]$ that maps any real number $x \in \mathbb{R}$ to the nearest point in $[0,1]$ from it, i.e.,

$$r(x) = \arg \min_{w \in [0,1]} |x - w|.$$

For example, $r(1.5) = 1, r(-2) = 0$ and $r(0.3) = 0.3$. We call $r(x_l)$ and $r(y_r)$ the *rounding extremes*, and as a corollary of Proposition 1, there exists an optimal solution between them.

Corollary 1. *For the maximum cost, there exists an optimal solution that lies in $[r(x_l), r(y_r)]$ or $[r(y_r), r(x_l)]$.*

Proof. If an optimal solution $s^* \in [0,1]$ is larger than both $r(x_l)$ and $r(y_r)$, then we can move it to the location $\max\{r(x_l), r(y_r)\}$, in which no agent would increase their cost. Thus, $\max\{r(x_l), r(y_r)\}$ is also optimal. Similarly, if s^* is smaller than $r(x_l)$ and $r(y_r)$, the location $\min\{r(x_l), r(y_r)\}$ is optimal.   □

In contrast to the social cost objective that admits an optimal strategy-proof mechanism, the mechanism that returns an optimal solution for maximum cost is not strategy-proof. Consider the instance with $\mathbf{x} = (0)$ and $\mathbf{y} = (1)$. The unique optimal solution is $\frac{1}{2}$, where both agents have a cost equal to 1. Now, suppose that the agent in N misreports the location as $y_1' = 3$. Then the mechanism takes $\mathbf{x} = (0)$ and $\mathbf{y}' = (3)$ as input, and outputs the unique optimal solution 1. After misreporting, this agent with true location $y_1 = 1$ decreases the cost to 0.

4.1 Deterministic Strategy-Proof Mechanisms

In this subsection, we provide upper and lower bounds on the approximation ratio of deterministic strategy-proof mechanisms for the maximum cost objective. Intuitively, based on Proposition 1, a good solution should balance the cost of the two extremes x_l and y_r. However, to guarantee the strategy-proofness, a mechanism cannot always achieve such a balance perfectly. Instead, we consider specific locations between $r(x_l)$ and $r(y_r)$ by Corollary 1. The following deterministic mechanism returns the bridge location $\frac{1}{2}$ if it lies between the two rounding extremes $r(x_l)$ and $r(y_r)$ and returns the rounding extreme that is closer to $\frac{1}{2}$ otherwise.

Mechanism 1. *Given location profiles* $\mathbf{x}, \mathbf{y}$, *if* $r(x_l) \leq \frac{1}{2} \leq r(y_r)$ *or* $r(y_r) \leq \frac{1}{2} \leq r(x_l)$, *then return* $\frac{1}{2}$. *Otherwise, return* $\arg\min_{s \in \{r(x_l), r(y_r)\}} |s - \frac{1}{2}|$.

Theorem 2. *Mechanism 1 is strategy-proof and* $\frac{5}{3}$-*approximation for the maximum cost.*

Proof. We first prove the strategy-proofness. If $r(x_l) \leq \frac{1}{2} \leq r(y_r)$ and the mechanism returns $\frac{1}{2}$, the misreports from the agents in M can only lead to a situation where both rounding extremes are on the right of $\frac{1}{2}$, and the outcome solution is larger than $\frac{1}{2}$. Thus, the agents in M cannot decrease their cost. Similarly, the misreports from the agents in N can only lead to an outcome smaller than $\frac{1}{2}$, and these agents cannot decrease their cost. If $r(x_l) \geq \frac{1}{2} \geq r(y_r)$, all agents have their best possible cost under the solution $\frac{1}{2}$, and thus they have no incentive to misreport. When both rounding extremes $r(x_l)$ and $r(y_r)$ are larger than $\frac{1}{2}$, if $r(x_l) \geq r(y_r) \geq \frac{1}{2}$, all agents have their best possible cost. If $r(y_r) \geq r(x_l) \geq \frac{1}{2}$, at the solution $r(x_l)$, all agents in M and those agents in N located on the left of $r(x_l)$ already achieve their best possible cost, and only the agents in N located on the right of $r(x_l)$ have the potential incentive to misreport. However, their misreporting cannot benefit them. Finally, the symmetric argument/analysis works for the case when both $r(x_l)$ and $r(y_r)$ are smaller than $\frac{1}{2}$.

Next, we prove the approximation ratio. When $r(y_r) \leq \frac{1}{2} \leq r(x_l)$, it must be that $y_r \leq \frac{1}{2} \leq x_l$, and the mechanism outputs $\frac{1}{2}$. In this case, all agents achieve their best possible cost, and thus the solution $\frac{1}{2}$ is optimal.

When $r(x_l) \leq \frac{1}{2} \leq r(y_r)$, it must be that $x_l \leq \frac{1}{2} \leq y_r$, and the mechanism outputs $\frac{1}{2}$. If the maximum cost $MC(\frac{1}{2}, \mathbf{x}, \mathbf{y})$ is attained by the agents at x_r or y_l, they already achieve the best possible cost, and an optimal solution should have a maximum cost equal to $MC(\frac{1}{2}, \mathbf{x}, \mathbf{y})$. Hence, we only need to focus on the case when the maximum cost is attained by the agents at x_l or y_r, which is

$$MC(\frac{1}{2}, \mathbf{x}, \mathbf{y}) = \max\{c_1(x_l, \frac{1}{2}), c_2(y_r, \frac{1}{2})\} = \max\{\frac{1}{2} - x_l + \frac{1}{2}, y_r - \frac{1}{2} + (1 - \frac{1}{2})\}$$
$$= \max\{1 - x_l, y_r\}.$$

Since an optimal solution s^* lies in interval $[r(x_l), r(y_r)]$ by Corollary 1, the optimal maximum cost is

$$MC(s^*, \mathbf{x}, \mathbf{y}) \geq \max\{c_1(x_l, s^*), c_2(y_r, s^*), \frac{c_1(x_l, s^*) + c_2(y_r, s^*)}{2}\}$$
$$= \max\{c_1(x_l, s^*), c_2(y_r, s^*), \frac{(2s^* - x_l) + (y_r - s^* + 1 - s^*)}{2}\}$$
$$\geq \max\{-x_l, y_r - 1, \frac{y_r + 1 - x_l}{2}\}.$$

Therefore, the approximation ratio is

$$\frac{MC(\frac{1}{2}, \mathbf{x}, \mathbf{y})}{MC(s^*, \mathbf{x}, \mathbf{y})} \leq \frac{\max\{1 - x_l, y_r\}}{\max\{-x_l, y_r - 1, \frac{y_r + 1 - x_l}{2}\}}.$$

Taking into account the constraint that $x_l \leq \frac{1}{2} \leq y_r$, it is easy to verify that this ratio is no more than $\frac{5}{3}$, as desired. It equals $\frac{5}{3}$ when $(x_l, y_r) = (-\frac{3}{2}, \frac{1}{2})$ or $(x_l, y_r) = (\frac{1}{2}, \frac{5}{2})$.

When $r(x_l), r(y_r) \leq \frac{1}{2}$ or $r(x_l), r(y_r) \geq \frac{1}{2}$, by symmetry we can only consider the case when $r(x_l), r(y_r) \leq \frac{1}{2}$. If $r(x_l) \geq r(y_r)$, then the mechanism will output $r(x_l)$, in which all agents achieve their best possible cost. If $r(x_l) \leq r(y_r)$, the mechanism returns $r(y_r)$. All agents in N, and the agents in M who are located on the right of $r(y_r)$, already achieve their best possible cost. Hence, we only need to consider the case when the induced maximum cost of $r(y_r)$ is attained by x_l, that is, $MC(r(y_r), \mathbf{x}, \mathbf{y}) = c_1(x_l, y_r) = 2y_r - x_l$. The optimal solution s^* has a maximum cost

$$MC(s^*, \mathbf{x}, \mathbf{y}) \geq \max\{c_1(x_l, s^*), \frac{c_1(x_l, s^*) + c_2(y_r, s^*)}{2}\}$$

$$= \max\{c_1(x_l, s^*), \frac{|s^* - x_l| + |s^*| + |y_r - s^*| + |1 - s^*|}{2}\}$$

$$\geq \max\{|x_l|, \frac{y_r - x_l + 1}{2}\}.$$

Therefore, the approximation ratio is

$$\frac{MC(\frac{1}{2}, \mathbf{x}, \mathbf{y})}{MC(s^*, \mathbf{x}, \mathbf{y})} \leq \frac{2y_r - x_l}{\max\{|x_l|, \frac{y_r - x_l + 1}{2}\}}.$$

Taking into account the constraint that $x_l \leq y_r \leq \frac{1}{2}$, it is easy to verify that this ratio is no more than $\frac{5}{3}$. It equals $\frac{5}{3}$ when $(x_l, y_r) = (-\frac{3}{2}, \frac{1}{2})$. □

The following theorem provides a matching lower bound $\frac{5}{3}$ for deterministic mechanisms, indicating that Mechanism 1 has tight approximation ratio.

Theorem 3. *For the maximum cost objective, no deterministic strategy-proof mechanism has an approximation ratio less than $\frac{5}{3}$.*

4.2 Randomized Strategy-Proof Mechanisms

While the bounds in Sect. 4.1 for deterministic mechanisms are tight, in this section, we consider randomized mechanisms and also derive tight bounds. Inspired by Corollary 1 that an optimal solution lies between $r(x_l)$ and $r(y_r)$, the following randomized mechanism returns a random point between them. Precisely, it outputs $r(x_l)$, $r(y_r)$, and the point that is between them and is closest to $\frac{1}{2}$, with specified probabilities.

Mechanism 2. *Given location profiles* $\mathbf{x}, \mathbf{y}$, *with probability* $\frac{1}{4}$, *return* $r(x_l)$; *with probability* $\frac{1}{4}$, *return* $r(y_r)$; *with probability* $\frac{1}{2}$, *return*

$$mid(x_l, y_r) = \begin{cases} \frac{1}{2}, & if\, x_l \leq \frac{1}{2} \leq y_r\, or\, x_l \geq \frac{1}{2} \geq y_r \\ \arg\min_{s \in \{r(x_l), r(y_r)\}} |s - \frac{1}{2}|, & otherwise \end{cases}$$

The point $mid(x_l, y_r)$ coincides with $r(x_l)$ or $r(y_r)$ when both x_l and y_r are larger than $\frac{1}{2}$ or both are smaller than $\frac{1}{2}$. For example, when $x_l = 1.2$ and $y_r = 0.6$, the mechanism returns $r(x_l) = 1$ with probability $\frac{1}{4}$, and returns $r(y_r) = mid(x_l, y_r) = 0.6$ with probability $\frac{1}{4} + \frac{1}{2} = \frac{3}{4}$.

Lemma 2. *Mechanism 2 is strategy-proof.*

Proof. We show that any agent i in M cannot decrease the cost by misreporting. The analysis for the agents in N is the same. Clearly, agent i cannot change y_r. When the location of agent i is $x_i > x_l$, the only way to change the solution is to misreport a location $x_i' < x_l$. Then the realizations of the random point returned by the mechanism become $r(x_i')$, $r(y_r)$ and $mid(x_i', y_r)$. Since $x_i > x_l$ and $0 \leq r(x_i') \leq r(x_l)$, we have $c_1(x_i, r(x_i')) = c_1(x_i, r(x_l)) = x_i$. On the other hand, if $x_i \geq mid(x_l, y_r)$, then we have $cost(x_i, mid(x_l, y_r)) = cost(x_i, mid(x_i', y_r)) = x_i$. If $x_i < mid(x_l, y_r)$, then it must be $x_l < x_i < mid(x_l, y_r) \leq r(y_r)$, and the misreport of agent i would not change $mid(x_l, y_r)$, that is, $mid(x_i', y_r) = mid(x_l, y_r) = \min\{\frac{1}{2}, r(y_r)\}$. Hence, agent i cannot benefit from misreporting.

When agent i's location is $x_i = x_l$, if agent i misreports a location $x_i' < x_l$, the analysis is the same as above. If agent i misreports $x_i' > x_l$, as both functions $r(\cdot)$ and $mid(\cdot, \cdot)$ are non-decreasing with x_l, it cannot decrease the cost. $\square$

We remark that the above proof for the strategy-proofness does not rely on the probabilities of the mechanism. Therefore, any constant probabilities assigned to the candidate locations $r(x_l), r(y_r), mid(x_l, y_r)$ can induce a strategy-proof mechanism. Nevertheless, we show that the probabilities specified in Mechanism 2 lead to the best possible approximation ratio.

Theorem 4. *Mechanism 2 is a randomized strategy-proof mechanism and is $\frac{3}{2}$-approximation for the maximum cost.*

Theorem 5. *For the maximum cost objective, no randomized strategy-proof mechanism has an approximation ratio less than $\frac{3}{2}$.*

5   Conclusion

We studied a novel mechanism design setting for building a bridge to connect two regions separated by an obstacle under the social and maximum cost objectives. For both objectives, we characterized their optimal solutions. While any optimal solution for the social cost objective is strategy-proof, it is not strategy-proof for the maximum cost objective. Therefore, for the maximum cost objective, we provided a deterministic $\frac{5}{3}$-approximation mechanism and a randomized $\frac{3}{2}$-approximation mechanism. Furthermore, we derived tight lower bounds, showing that no strategy-proof mechanisms can have better approximation ratios.

For the future directions, we note that our model is just a starting point of the mechanism design for bridge-building problems. While we model the two regions as two real lines, the regions could be a network, Euclidean plane, or other metric spaces. It is interesting to study how the existing and new methods can be applied to these regions. More generally, we can also further consider building multiple non-perpendicular bridges among multiple regions.

References

1. Bhattacharya, B., Benkoczi, R.: On computing the optimal bridge between two convex polygons. Inf. Process. Lett. **79**(5), 215–221 (2001)
2. Cai, L., Xu, Y., Zhu, B.: Computing the optimal bridge between two convex polygons. Inf. Process. Lett. **69**(3), 127–130 (1999)
3. Chan, H., Filos-Ratsikas, A., Li, B., Li, M., Wang, C.: Mechanism design for facility location problems: a survey. In: Proceedings of the 30th International Joint Conference on Artificial Intelligence (IJCAI), pp. 4356–4365 (2021)
4. Chan, H., Wang, C.: Mechanism design for improving accessibility to public facilities. In: Proceedings of the 22nd International Conference on Autonomous Agents and Multiagent Systems (AAMAS), pp. 2116–2124 (2023)
5. Demaine, E.D., Zadimoghaddam, M.: Minimizing the diameter of a network using shortcut edges. In: Kaplan, H. (ed.) SWAT 2010. LNCS, vol. 6139, pp. 420–431. Springer, Heidelberg (2010). https://doi.org/10.1007/978-3-642-13731-0_39
6. Dokow, E., Feldman, M., Meir, R., Nehama, I.: Mechanism design on discrete lines and cycles. In: Proceedings of the 13th ACM Conference on Electronic Commerce (EC), pp. 423–440 (2012)
7. Filos-Ratsikas, A., Voudouris, A.A.: Approximate mechanism design for distributed facility location. In: Caragiannis, I., Hansen, K.A. (eds.) SAGT 2021. LNCS, vol. 12885, pp. 49–63. Springer, Cham (2021). https://doi.org/10.1007/978-3-030-85947-3_4
8. Fukui, Y., Shurbevski, A., Nagamochi, H.: Group strategy-proof mechanisms for shuttle facility games. J. Inf. Process. **28**, 976–986 (2020)
9. Kim, B.J., Shin, C.S., Chwa, K.Y.: Linear algorithms for computing a variant segment center. In: Proceedings of Korean Information Science Society Conference (KISS), vol. A, pp. 708–710 (1998). (in Korean)
10. Kim, S.K., Shin, C.S.: Computing the optimal bridge between two polygons. Theory Comput. Syst. **34**(4), 337–352 (2001)
11. McCartney, G., Whyte, B., Livingston, M., Crawford, F.: Building a bridge, transport infrastructure and population characteristics: explaining active travel into glasgow. Transp. Policy **21**, 119–125 (2012)
12. Mei, L., Li, M., Ye, D., Zhang, G.: Facility location games with distinct desires. Discret. Appl. Math. **264**, 148–160 (2019)
13. Meir, R.: Strategyproof facility location for three agents on a circle. In: Fotakis, D., Markakis, E. (eds.) SAGT 2019. LNCS, vol. 11801, pp. 18–33. Springer, Cham (2019). https://doi.org/10.1007/978-3-030-30473-7_2
14. Meyerson, A., Tagiku, B.: Minimizing average shortest path distances via shortcut edge addition. In: Proceedings of the 12th International Workshop and 13th International Workshop on Approximation, Randomization, and Combinatorial Optimization. Algorithms and Techniques (APPROX/RANDOM), vol. 5687, pp. 272–285 (2009)
15. Procaccia, A.D., Tennenholtz, M.: Approximate mechanism design without money. ACM Trans. Econ. Comput. (TEAC) **1**(4), 1–26 (2013)
16. Tan, X.: On optimal bridges between two convex regions. Inf. Process. Lett. **76**(4–6), 163–168 (2000)
17. Tan, X.: Finding an optimal bridge between two polygons. Int. J. Comput. Geom. Appl. **12**(03), 249–261 (2002)

Joint Bidding in Ad Auctions

Yuchao Ma[1], Weian Li[2], Wanzhi Zhang[3], Yahui Lei[3], Zhicheng Zhang[1],
Qi Qi[1(✉)], Qiang Liu[3], and Xingxing Wang[3]

[1] Gaoling School of Artificial Intelligence, Renmin University of China,
Beijing, China
{yc.ma,mzhangzhicheng,qi.qi}@ruc.edu.cn
[2] Center on Frontiers of Computing Studies, Peking University, Beijing, China
weian_li@pku.edu.cn
[3] Meituan.Inc., Beijing, China
{zhangwanzhi,leiyahui,liuqiang43,wangxingxing04}@meituan.com

Abstract. In traditional advertising auctions, commodity suppliers as
advertisers compete for adverting positions to display commodities. As
e-commerce platforms become more prevalent, offline retailers are also
opening online virtual shops, and retailers are starting to charge a fee
for extra exposure of their shops. This has led to situations where a
single commodity may be sponsored by both the retailer and the supplier, offering opportunities for more profit. In order to explore this novel
advertising pattern, we propose a new model called the joint advertising
system (JAS), where retailers and suppliers jointly bid for advertising
positions. In the context of this realistic scenario, conventional mechanisms such as GFP, GSP and Myerson auction cannot be applied directly.
Besides, the VCG mechanism results in negative revenue in JAS. To solve
this issue, we modify the payment rule of VCG to create a revised VCG
mechanism that guarantees incentive compatibility, individual rationality and weak budget-balance. Additionally, we leverage the structure of
the affine maximizer auction (AMA) and technique of automated mechanism design to train joint AMA. Finally, we conduct several experiments
to demonstrate the performance of the joint AMA. It turns out that our
mechanism maintains good economic properties and outperforms other
mechanisms in various settings.

Keywords: Joint Auction System · Mechanism Design · VCG · Affine
Maximizer Auctions

1 Introduction

Over the past few decades, online advertising has become the main source of
revenue for internet platforms like Google and Amazon. According to the latest
report from the IAB, online advertising revenue surged to an impressive $209.7

Y. Ma and W. Li—These authors contribute equally to this work.

billion in 2022[1]. A prominent method to allocate advertisement slots is through a sponsored search auction (SSA) where advertisers submit their bids to the platform. The platform then decides which advertisements should be displayed and how much they should be charged based on certain auction mechanisms.

Traditionally, the commodity suppliers have been the focus group of SSA as they hope to get ad positions and improve the exposure of their commodities by charging extra fees to platforms (see Fig. 1(a)). However, with the rapid growth of the internet and online two-sided e-commerce platforms such as Uber Eats and Instacart, more and more traditional offline retailers are opening online stores on these platforms. For example, Walmart has established an online shop on Instacart to cater to nearby residents. As a result of this trend, retailers are also willing to pay to enhance their store ranks and be displayed in more appealing positions, attracting more customers. This leads to a new pattern of online advertising where one sponsored ad item is contributed by a bundle of two sides: the retailer (emerging advertiser) and the supplier (traditional advertiser). We call this pattern the *joint advertising system* (JAS) (see Fig. 1(b)), which breaks the simple competitive relationship among bidders in the supplier-side and introduces the concept of *joint bidding*, satisfying the demands of both retailers and suppliers.

In contrast to traditional online advertising, JAS holds the potential to generate more revenue since it charges payments from both retailers and suppliers. However, the pattern of joint bidding also makes auction design more complex. Mechanisms commonly used in traditional settings may not apply seamlessly to this setting. For instance, even though we can use the allocation rule to rank items in a generalized second-price auction, it is difficult to determine the separate payment of the retailer and supplier for an item, which makes equilibrium analysis more complicated. Furthermore, the Myerson auction may suit the traditional pattern of SSA from a theoretical perspective, but it turns out inadequate within the joint bidding environment of JAS. In JAS, a retailer or supplier may appear in multiple slots with different bundles, which means that the bidding prices of bundles are dependent, violating the assumption of the Myerson auction. Therefore, it is important to investigate how to design suitable mechanisms to match the JAS, how to ensure desirable properties under joint bidding, such as incentive compatibility (IC), individual rationality (IR), and weak budget balance (WBB), and how to evaluate the proposed mechanisms.

1.1   Our Contributions

This paper aims to tackle the aforementioned challenges by proposing a new model for JAS. The model consists of two distinct groups of bidders: retailers and suppliers with a given partnership. In departure from the traditional SSA, in JAS, each ad item is jointly contributed by a bundle consisting of one retailer

---

[1] Internet Advertising Revenue Report: Full Year 2022. https://www.iab.com/insights/internet-advertising-revenue-report-full-year-2022/.

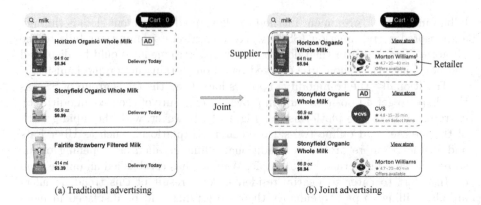

Fig. 1. Examples of traditional advertising and joint advertising.

and one supplier. The platform then determines the allocation of these bundles and the corresponding payments made by each retailer and supplier.

Our model is designed based on an in-depth understanding of JAS. Not only does it contribute to the enrichment of mechanism design theory, but it also offers solutions for more complex industrial e-commerce advertising environments.

Drawing on the structure of JAS, we seek to design mechanisms that satisfy IC, IR and WBB, from both theoretical and experimental perspectives. Initially, we show how the classic VCG mechanism can be adapted for JAS. However, we highlight that this mechanism may yield negative revenues in certain instances. To address this issue, we introduce an innovative mechanism called the revised VCG mechanism (RVCG), which modifies the VCG payment rule while meeting the IC, IR, and WBB criteria.

To further enhance revenue, we focus on a class of auctions referred to as affine maximizer auctions (AMA). We generalize AMA to the joint bidding setting (called joint AMA (JAMA)) and employ the technique of automated mechanism design to train the parameters of the joint AMA and improve revenue.

Specifically, to overcome the poor scalability of the classic AMA mechanism, we preset the size of allocation menu and train different parameters within the range. In addition, it is necessary to consider feasibility constraints in joint advertising environment. Therefore, we further restrict the search space of parameters through normalization and adjusting the structure of neural network. In return, we are capable to enhance scalability of our model.

1.2 Related Work

With the prevalence of Internet, the internet ad auction has been on influential direction in mechanism design and auction theory. Traditional ad auctions can be traced back to [1,25]. The generalized second price (GSP) auction [9] and its variants [13,19,24] have been widely applied in ad systems. In recent years, the rapid development of machine learning and deep learning theory has led to the diversification of ad auction. [27] and [18] propose different methods to optimize different metrics within the GSP framework. [15] come up with the neural multi-slot auction to maximize the revenue of platform with the consideration of externality. Moreover, [14] employ matching-based allocation algorithm to implement learning-based ad auction structure that optimizes profit while ensuring economic constraints in the presence of externalities.

Joint ad auction has been studied before as [3–5,12] mainly focus on the scenario that suppliers could coordinate with retailers aiming to share advertising costs. Their problems are to choose the optimal retailer subset to work with. However, in our JAS, instead of selecting the best retailer subset for suppliers to optimize the profit, our model is based on the premise that the partnership is known in advance and fixed. Thus, our goal is to find the appropriate mechanism to optimize overall platform revenue in JAS within certain economic constraints.

Our work is partly based on the affine maximizer auction [16,17,22], a weighted version of VCG mechanism. VCG mechanism [6,10,26] can maximize the social welfare, while AMA maximizes a weighted and boosted version of social welfare. Adjusting the parameters can lead to a higher revenue than VCG while keeping IC and IR. There are a few of other works that discuss different variants of parameterized generalization of VCG mechanism, such as virtual valuations combinatorial auctions [16,17,22], λ-auctions [11], bundling boosted auction [2] and mixed-bundling auctions with reserve prices [23]. [7] parameterize the allocation menus together with bidder weights and utilize neural network to generate boost variable to pursue the optimal revenue. Based on this, [8] build a transformer-based permutation-equivariant architecture with stronger scalability and better revenue performance. Different from these works, we consider the allocation menus and boost variable in the joint advertising system with more feasibility restrictions.

2 Model and Preliminaries

In this section, we formally propose our model, called *joint advertising system* (JAS). In a JAS, there are n retailers and m suppliers. Denote $R = \{r_1, \ldots, r_n\}$ and $S = \{s_1, \ldots, s_m\}$ by the set of retailers and suppliers, respectively. The partnership between retailers and suppliers can be given by a bipartite graph $G = (V, E)$, where vertex set $V = R \cup S$ contains all retailers and suppliers and edge set E represents the collaborative relation between retailers and suppliers. Specifically, an edge $e = (r_i, s_j) \in E$ means that supplier j provides commodity for retailer i, or retailer i sells the commodity of supplier j. Let $D(r_i) = \{s_j\}_{(r_i, s_j) \in E}$ be the set of suppliers in cooperation with retailer s_i and $C(s_j) = \{r_i\}_{(r_i, s_j) \in E}$ be the set of supplier s_j's cooperative retailers.

In a pageview, suppose that there are K slots provided for displaying advertisements, indexed by $k \in \{1, 2, \ldots, K\}$. Assume that the click-through-rate (CTR) of k-th slot is λ_k and the slot with less index has a larger CTR, i.e., $\lambda_1 \geq \lambda_2 \geq \ldots, \geq \lambda_K \geq 0$.

Each retailer r_i (or supplier s_j) has a private value v_{r_i} (or v_{s_j}) for each click. More specifically, v_{r_i} (or v_{s_j}) represent how much retailer r_i (or supplier s_j)

wants to charge for an extra click. Assume that v_{r_i} (or v_{s_j}) is independently and identically drawn from a publicly known distribution $F(v)$ whose respective p.d.f is $f(v)$. Let $\mathbf{v}^R = (v_{r_1}, \ldots, v_{r_n})$ and $\mathbf{v}^S = (v_{s_1}, \ldots, v_{s_m})$ be the value profiles of retailers and suppliers, respectively. Use $\mathbf{v} = (\mathbf{v}^R, \mathbf{v}^S)$ and $\mathbf{v}_{-r_i}$ (or $\mathbf{v}_{-s_j}$) to represent the value profile of all advertisers and the value profile of all advertisers except retailer i (or supplier j). Similarly, for the bidding profiles, we use the similar notations, e.g., $\mathbf{b}^R$, $\mathbf{b}^S$, $\mathbf{b}$, $\mathbf{b}_{-r_i}$ and $\mathbf{b}_{-s_j}$. In our model, one advertisement item is sponsored by two sides: one retailer and one supplier. Therefore, we define the joint bidding of such a bundle as $b_{ij} = b_{r_i} + b_{s_j}$ for all $e = (r_i, s_j) \in E$. Sometimes we also use "bundle" to represent one ad item.

Having these notations, we can define the auction mechanisms in the JAS. A mechanism $\mathcal{M} = (\mathbf{a}(\mathbf{b}), \mathbf{p}(\mathbf{b}))$ consists two parts: allocation rule and payment rule, where $\mathbf{a}(\mathbf{b}) = (a_i(\mathbf{b}))_{i \in R \cup S}$ and $\mathbf{p}(\mathbf{b}) = (p_i(\mathbf{b}))_{i \in R \cup S}$. Herein, for any advertiser i, $a_i(\mathbf{b}) = \sum_{k=1}^{K} a_{ik}(\mathbf{b})\lambda_k$ represent the total CTR, where $a_{ik}(\mathbf{b})$ is an indicator function to decide whether the advertisement item allocated to slot k contains advertiser i.

Since one slot should be allocated to one bundle and one bundle should be assigned to at most one slot, $a_{ik}(\mathbf{b})$ should satisfy the following feasibility constraints, denoted by $\mathcal{A}$:

$$\sum_{k=1}^{K}[a_{ik}(\mathbf{b}) \cdot a_{jk}(\mathbf{b})] \leq 1, \qquad \forall e = (r_i, s_j) \in E,$$

$$\sum_{i \in R} a_{ik}(\mathbf{b}) + \sum_{j \in S} a_{jk}(\mathbf{b}) = 2, \qquad \forall k \in K,$$

$$\sum_{i \in R} a_{ik}(\mathbf{b}) \leq 1, \quad \sum_{j \in S} a_{jk}(\mathbf{b}) \leq 1, \qquad \forall k \in K,$$

$$a_{ik}(b) \in \{0, 1\}, \qquad \forall i \in R \cup S, k \in K.$$

Given the allocation rule and the payment rule, the utility of any retailer or supplier i can be expressed as

$$u_i(\mathbf{b}) = v_i \cdot a_i(\mathbf{b}) - p_i(\mathbf{b}).$$

The revenue of auction mechanism is the total payment of advertisers, i.e.,

$$\text{Rev}(\mathbf{b}) = \sum_{i \in R} p_i(\mathbf{b}) + \sum_{j \in S} p_j(\mathbf{b}).$$

The social welfare of auction mechanism is defined as the total value of advertisers, that is,

$$\text{SW}(\mathbf{b}) = \sum_{i \in R \cup S} v_i \cdot a_i(\mathbf{b}).$$

In this paper, we mainly focus on the feasible mechanisms with some desirable properties, like *incentive compatibility*, *individual rationality* and *weakly budget-balance*, which are defined formally as follows.

Definition 1 (Incentive Compatibility). *In a JAS, a mechanism is incentive compatible, if for any advertiser i, any bid b'_i and any bidding profile $\mathbf{b}_{-i}$, it holds that*

$$u_i(v_i, \mathbf{b}_{-i}) \geq u_i(b'_i, \mathbf{b}_{-i}).$$

Definition 2 (Individual Rationality). *In a JAS, a mechanism is individual rationality, if for any advertiser i and any bidding profile $\mathbf{b}_{-i}$, it holds that*

$$u_i(v_i, \mathbf{b}_{-i}) \geq 0.$$

Definition 3 (Weakly Budget-Balance). *In a JAS, a mechanism is weakly budget-balanced, if the revenue is always non-negative, that is,*

$$Rev(\mathbf{b}) = \sum_{i \in R} p_i(\mathbf{b}) + \sum_{j \in S} p_j(\mathbf{b}) \geq 0.$$

Namely, IC and IR guarantee that all advertisers will truthfully bid and gain a non-negative utility. WBB ensures that the revenue is always non-negative. In the rest of this paper, we mainly focus on designing the feasible mechanism with IC, IR and WBB.

3   Classic Auction Mechanisms

In this section, we first mainly focus on the classic VCG mechanism and show that it works in JAS, but does not guarantee WBB. Then, we modify the payment rule of VCG mechanism to satisfy WBB and call it revised VCG.

3.1   Revised VCG Mechanism

In a JAS, since one slot is occupied by a bundle of one retailer r_i and one supplier s_j who has cooperative relation with r_i, the allocation rule concentrates more on the bids of feasible bundles. First, we introduce the allocation rule (Algorithm 1) which can achieve the optimal social welfare when all advertisers bid truthfully.

Lemma 1. *In the joint advertising system, the allocation output by Algorithm 1 can achieve the optimal social welfare when all advertisers bid truthfully.*

Now, we elaborate the VCG mechanism in a JAS. In fact, the allocation rule of the VCG mechanism follows the steps of Algorithm 1. Intuitively, the payment rule of the VCG mechanism charges the externality of each advertiser i, i.e., the difference on all other advertisers' social welfare with and without advertiser i. Denote $SW^*(\mathbf{b})$ by the optimal social welfare under bid profile $\mathbf{b}$. The payment of VCG mechanism for advertiser i can be defined as

$$p_i(\mathbf{b}) = SW^*(\mathbf{b}_{-i}) - [SW^*(\mathbf{b}) - a_i(\mathbf{b}) \cdot b_i].$$

It is not difficult to find that VCG mechanism is IR and IC, since the payment of any advertiser is unrelated to her own bid. However, based on the mode of

---

Algorithm 1. Optimal SW* Allocation Rule

Input: Edge set: E, bid profile: $\mathbf{b}^R$, $\mathbf{b}^S$, CTRs of slots: $\{\lambda_1, \ldots, \lambda_K\}$.
Output: $\mathbf{a}(\mathbf{b}) = (\mathbf{a}^R(\mathbf{b}), \mathbf{a}^S(\mathbf{b}))$.

1: **for** $e_t = (r_i, s_j) \in E$ **do**
2:     $b_{e_t} \leftarrow b_{r_i} + b_{s_j}$
3: **end for**
4: Sort and re-index bundle bids $\{b_{e_t}\}_{e_t \in E}$ in a non-increasing order.
5: **for** $k \leftarrow 1$ to K **do**
6:     Allocate k-th slot to e_k which consists of retailer r_i and supplier s_j, $a_{ik}(\mathbf{b}) \leftarrow 1$
       and $a_{jk}(\mathbf{b}) \leftarrow 1$.
7: **end for**
8: $\mathbf{a}^R(\mathbf{b}) \leftarrow \{a_{r_i} = \sum_{k=1}^K a_{ik}(\mathbf{b})\lambda_k\}$
9: $\mathbf{a}^S(\mathbf{b}) \leftarrow \{a_{s_j} = \sum_{k=1}^K a_{jk}(\mathbf{b})\lambda_k\}$

---

joint bidding, when an bidder does not join the auction, her partners will lose a chance to form a bundle and compete for a slot, which may bring a loss on the optimal social welfare. In the following example, we show that when one bidder quits the auction, the optimal total value of other advertisers decreases. It means that the payment of one advertiser may be negative and violates WBB.

Example 1. Assume that there are two retailers $\{r_1, r_2\}$ and two suppliers $\{s_1, s_2\}$ with only two cooperative relationships $\{(r_1, s_1), (r_2, s_2)\}$. Let the values of retailers and suppliers be $(v_{r_1}, v_{r_2}) = (3, 2)$ and $(v_{s_1}, v_{s_2}) = (5, 1)$, respectively. There is only one slot with $\lambda_1 = 1$.

In this setting, VCG mechanism will allocate the slot to the bundle (r_1, s_1) with a social welfare 8. The payment of r_1 is $p_{r_1} = \text{SW}^*_{-r_1} - (\text{SW}^* - \lambda_1 v_{r_1}) = 1 \times (2+1) - (8 - 1 \times 3) = -2$. Similarly, the payment of s_1 is $p_{s_1} = \text{SW}^*_{-s_1} - (\text{SW}^* - \lambda_1 v_{s_1}) = 1 \times (2+1) - (8 - 1 \times 5) = 0$. It is easy to check that $p_{s_2} = p_{r_2} = 0$. Therefore, the total revenue is -2, which violates the constraint of WBB.

Theorem 1. *In the joint advertising system, VCG mechanism is IC, IR and can achieve the optimal social welfare, but is not weakly budget-balanced.*

The main reason leading to a negative revenue of Example 1 is that without retailer r_1, supplier s_1 with a good value loses the chance of competition, which causes a loss on the optimal social welfare. Intuitively, if we can keep the right of supplier s_1 to contribute to the social welfare, the payment of retailer r_1 will not be negative and consequently, WBB will be guaranteed.

Following the above idea, we modify the payment rule of VCG mechanism: we assume that the pivotal advertiser bids 0 and the cooperative relationships still remain rather than delete her directly, when she does not join the auction. After this modification, the payment rule can be written as

$$p_i(\mathbf{b}) = \text{SW}^*(0, \mathbf{b}_{-i}) - [\text{SW}^*(\mathbf{b}) - a_i(\mathbf{b}) \cdot b_i].$$

We define the VCG mechanism with the above payment rule as the revised VCG and prove that revised VCG can guarantee the WBB.

Theorem 2. *In the joint advertising system, revised VCG mechanism is IC, IR, WBB and can achieve the optimal social welfare.*

4   Joint Affine Maximizer Auction

In this section, we first introduce the affine maximizer auction (AMA), the generalized version of VCG mechanism. Then we extend the traditional AMA according to revised VCG in Sect. 3 in the bidding environment of JAS, which is called joint AMA (JAMA) and outline the process of JAMA.

4.1   Overview of Affine Maximizer Auction

As discussed above, the revised VCG mechanism can make sure that the revenue of platform is non-negative. However, the platforms still hope to further enhance the revenue when the desirable properties like IC and IR can be maintained. Therefore, we consider a broader type of mechanisms called *affine maximizer auction* [21], which can be viewed as a generalization of VCG mechanisms.

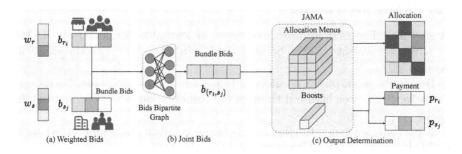

Fig. 2. A schematic view of JAMA. We take bids of retailers and suppliers and their weights as input. Then we calculate the joint bids according to the partnership between retailers and suppliers. Through the training of allocation menus and corresponding boosts, we get the optimal parameter set that optimizes the platform's revenue. Then we output the allocation result and the payment of each retailer and supplier.

Different from VCG mechanism, AMA has two groups of parameters: weights and boosts, where each bidder i has a positive weight w_i and any allocation scheme $\mathbf{a}$ is associated with a boost $u(\mathbf{a})$. The main idea of AMA is to allocate items to maximize the affine social welfare, which is the summation of the total weighted value of bidders and the boost of allocation, and the payment of one bidder is defined as the total decrease in all other bidders' affine social welfare. Intuitively, AMA can increase revenue, since it sacrifices the optimality of social welfare and use a weighted social welfare to measure the bidder's externality. Formally, given weights and boosts, weighted social welfare can be defined as

$$\text{WSW}(\mathbf{b}) = \sum_{i \in R \cup S} w_i v_i \cdot a_i(\mathbf{b}) + u(\mathbf{a}(\mathbf{b})). \tag{1}$$

The allocation of AMA is $\mathbf{a}^*(\mathbf{b}) = \arg\max_{\mathbf{a(b)}} \mathrm{WSW}(\mathbf{b})$ and the payment of any advertiser is

$$p_i(\mathbf{b}) = \left[\mathrm{WSW}^*(\mathbf{b}_{-i}) - (\mathrm{WSW}^*(\mathbf{b}) - w_i b_i \cdot a_i(\mathbf{b}))\right]/w_i. \qquad (2)$$

It is well-known that AMA is IC and IR when given the weights and boosts. Note that VCG mechanism is a special case of AMA with unit weight and zero boost. Therefore, if we can assign the appropriate values to the parameters of AMA, the revenue of AMA will be higher than the revenue of VCG. However, the literature [16] proves that finding the optimal parameters such that AMA achieves the maximum revenue is NP-hard.

Recently, for the multi-item setting, [7] provide a novel approach of automated mechanism design to train the architecture of AMA and illustrate the superiority of their trained mechanism in revenue. Motivated by this approach, in the rest of this section, we aim to extend AMA to our joint advertising system and come up with the corresponding training method. We call our architecture of AMA as joint AMA. Figure 2 illustrates the basic structure of JAMA.

4.2   JAMA Architecture

As shown in Fig. 2, there are three types of learnable parameters in JAMA: the advertiser weights w_i, the boost variable $\mu(\mathbf{a}^*)$ together with the allocation menus $\mathbf{a}^*$.

During the input phase, retailers and suppliers submit their bids. Then we get each advertiser's weighted bid through the combination of her bid and bidding weight, respectively.

After that, we generate joint bids between retailers and suppliers according to bids bipartite graph which represents the cooperative relationship.

Allocation Menus. In order to improve JAMA's revenue performance as much as possible, we initialize multiple allocation menus. Moreover, we normalize the allocation menus to meet the feasible constraints defined in Sect. 2.

Boosts. To get the boost variables, we utilize a multi-layer perceptron (MLP), which is a fully-connected neural network.

After we get the output JAMA parameters w_i, $\mu(\mathbf{a}^*)$ and $\mathbf{a}^*$, we can compute the allocation result according to Eq. (1) and the payment result according to Eq. (2) defined in Sect. 4.

Optimization and Training. Since JAMA is perfectly IC and IR, our goal is to maximize the total revenue of platform. Specifically, we set the loss function as the negative total revenue, i.e., $-[\sum_{i \in R} p_i(\mathbf{b}) + \sum_{j \in S} p_j(\mathbf{b})]$. Then, we learn these parameters together via the method of gradient descent.

For the initialization on the allocation probability matrix of bundle, as the methods in [7], we exploit two methods: the first one is to keep all feasible allocations initially. In cases where the number of advertisers is limited, or when the partnership structure is relatively uncomplicated, this approach can easily obtain the optimal result. The other one is to randomly initialize a large set of allocations to find the best allocation which brings in the most revenue, though few of those allocations are actually used. Besides, in our experiment, the winning allocation matrix is always deterministic, which is in keeping with the experiment setting.

Different from normal auctions, we need to consider the feasible constraints (defined in Sect. 2) in the joint advertising, while allocating ad slots, that is, a bundle cannot appear in more than one slot and one single slot cannot be double-allocated. Therefore, we consider to keep the allocation matrix doubly stochastic while training. We follow the approach used in [20] to normalize the matrix row- and column-wise through softplus operation and take the minimum of the result.

Besides, during training process, we follow the approach used in [7] and utilize softmax function as differentiable surrogate for the max and argmax operations. At test time, we use the regular max operator.

5   Conclusion

In this paper, we introduce a novel online advertising system, joint advertising system, where retailer and supplier can form a bundle and jointly bid for an ad slot. From the perspective of mechanism design, we extend the traditional VCG to JAS, while it may causes a negative revenue. Thus, we propose the revised VCG which keeps IC, IR and weakly budget-balanced. Moreover, to further enhance revenue, we train a AMA-based mechanism, called joint AMA, by using the technique of automated mechanism design.

Acknowledgments. This work is supported by the Fundamental Research Funds for the Central Universities, and the Research Funds of Renmin University of China (22XNKJ07) and Meituan.

References

1. Aggarwal, G., Goel, A., Motwani, R.: Truthful auctions for pricing search keywords. In: Proceedings of the 7th ACM Conference on Electronic Commerce, pp. 1–7 (2006)
2. Balcan, M.F., Prasad, S., Sandholm, T.: Learning within an instance for designing high-revenue combinatorial auctions. In: IJCAI Annual Conference (2021)
3. Bergen, M., John, G.: Understanding cooperative advertising participation rates in conventional channels. J. Mark. Res. **34**(3), 357–369 (1997)
4. Berger, P.D.: Vertical cooperative advertising ventures. J. Mark. Res. **9**(3), 309–312 (1972)
5. Cao, X., Ke, T.T.: Cooperative search advertising. Mark. Sci. **38**(1), 44–67 (2019)
6. Clarke, E.H.: Multipart pricing of public goods. Public Choice 17–33 (1971)
7. Curry, M., Sandholm, T., Dickerson, J.: Differentiable economics for randomized affine maximizer auctions. arXiv preprint arXiv:2202.02872 (2022)

8. Duan, Z., Sun, H., Chen, Y., Deng, X.: A scalable neural network for DSIC affine maximizer auction design. arXiv preprint arXiv:2305.12162 (2023)
9. Edelman, B., Ostrovsky, M., Schwarz, M.: Internet advertising and the generalized second-price auction: selling billions of dollars worth of keywords. Am. Econ. Rev. **97**(1), 242–259 (2007)
10. Groves, T.: Incentives in teams. Econometrica: J. Econometric Soc. 617–631 (1973)
11. Jehiel, P., Meyer-Ter-Vehn, M., Moldovanu, B.: Mixed bundling auctions. J. Econ. Theory **134**(1), 494–512 (2007)
12. Kim, S.Y., Staelin, R.: Manufacturer allowances and retailer pass-through rates in a competitive environment. Mark. Sci. **18**(1), 59–76 (1999)
13. Lahaie, S., Pennock, D.M.: Revenue analysis of a family of ranking rules for keyword auctions. In: Proceedings of the 8th ACM Conference on Electronic Commerce, pp. 50–56 (2007)
14. Li, N., et al.: Learning-based ad auction design with externalities: the framework and a matching-based approach. In: Proceedings of the 29th ACM SIGKDD Conference on Knowledge Discovery and Data Mining, pp. 1291–1302 (2023)
15. Liao, G., et al.: NMA: neural multi-slot auctions with externalities for online advertising. arXiv preprint arXiv:2205.10018 (2022)
16. Likhodedov, A., Sandholm, T.: Methods for boosting revenue in combinatorial auctions. In: AAAI, pp. 232–237 (2004)
17. Likhodedov, A., Sandholm, T., et al.: Approximating revenue-maximizing combinatorial auctions. In: AAAI, vol. 5, pp. 267–274 (2005)
18. Liu, X., et al.: Neural auction: end-to-end learning of auction mechanisms for e-commerce advertising. In: Proceedings of the 27th ACM SIGKDD Conference on Knowledge Discovery & Data Mining, pp. 3354–3364 (2021)
19. Ostrovsky, M., Schwarz, M.: Reserve prices in internet advertising auctions: a field experiment. In: Proceedings of the 12th ACM Conference on Electronic Commerce, pp. 59–60 (2011)
20. Ravindranath, S.S., Feng, Z., Li, S., Ma, J., Kominers, S.D., Parkes, D.C.: Deep learning for two-sided matching. arXiv preprint arXiv:2107.03427 (2021)
21. Roberts, K.: The characterization of implementable choice rules. Aggregation Revelation Preferences **12**(2), 321–348 (1979)
22. Sandholm, T., Likhodedov, A.: Automated design of revenue-maximizing combinatorial auctions. Oper. Res. **63**(5), 1000–1025 (2015)
23. Tang, P., Sandholm, T.: Mixed-bundling auctions with reserve prices. In: AAMAS, pp. 729–736 (2012)
24. Thompson, D.R., Leyton-Brown, K.: Revenue optimization in the generalized second-price auction. In: Proceedings of the Fourteenth ACM Conference on Electronic Commerce, pp. 837–852 (2013)
25. Varian, H.R.: Position auctions. Int. J. Ind. Organ. **25**(6), 1163–1178 (2007)
26. Vickrey, W.: Counterspeculation, auctions, and competitive sealed tenders. J. Financ. **16**(1), 8–37 (1961)
27. Zhang, Z., et al.: Optimizing multiple performance metrics with deep GSP auctions for e-commerce advertising. In: Proceedings of the 14th ACM International Conference on Web Search and Data Mining, pp. 993–1001 (2021)

Lower Bounds for the Sum of Small-Size Algebraic Branching Programs

C. S. Bhargav$^{(\boxtimes)}$ [iD], Prateek Dwivedi[iD], and Nitin Saxena[iD]

Indian Institute of Technology, Kanpur, Kanpur, India
{bhargav,pdwivedi,nitin}@cse.iitk.ac.in

Abstract. We observe that proving strong enough lower bounds for the sum of set-multilinear Algebraic Branching Programs (smABPs) in the *low-degree* regime implies Valiant's conjecture (i.e. it implies general ABP lower bounds). Using this connection, we obtain lower bounds for the sum of small-sized general ABPs. In particular, we show that the sum of poly(n) ABPs, each of size (:= number of vertices) $(nd)^{o(1)}$, cannot compute the family of Iterated Matrix Multiplication polynomials $\text{IMM}_{n,d}$ for any arbitrary function $d = d(n)$.

We also give a dual version of our result for the sum of *low-variate* ROABPs (read-once oblivious ABPs) and read-k oblivious ABPs. Both smABP and ROABP are very well-studied 'simple' models; our work puts them at the forefront of understanding Valiant's conjecture.

Keywords: Algebraic Circuits · Algebraic Branching Programs · Polynomials · Lower Bounds

1 Introduction

In a pioneering work, Leslie Valiant proposed [38] an *algebraic* framework to study efficient ways of computing multivariate polynomials. The computational model was that of *algebraic circuits* – layered directed acyclic graphs with vertices in intermediate layers alternately labeled by addition ($+$) or multiplication ($\times$), and leaves at the bottom layer labeled with variables $x_1, \ldots, x_n$ or constants of the underlying field $\mathbb{F}$. The circuit inductively computes a multivariate polynomial $f \in \mathbb{F}[x_1, \ldots, x_n]$. Each vertex (gate) performs its corresponding operation ($+$ or $\times$) on the inputs it receives until finally, a designated output vertex computes the polynomial. A measure of efficiency is the *size* of the circuit, that is, the number of vertices in the graph. The *depth* of the circuit is the length of the longest path from the input leaves to the output vertex and measures the amount of *parallelism* in the circuit. For a general survey of algebraic complexity, see [7, 24, 35].

Valiant hypothesized that there are *explicit* polynomials that do not have small algebraic circuits computing them, which we now call the VP $\neq$ VNP hypothesis. As algebraic circuits are *non-uniform* models of computation, computing a polynomial more precisely refers to computing a *family* $\{f_n\}_{n \geq 0}$ of

© The Author(s), under exclusive license to Springer Nature Singapore Pte Ltd. 2024
X. Chen and B. Li (Eds.): TAMC 2024, LNCS 14637, pp. 355–366, 2024.
https://doi.org/10.1007/978-981-97-2340-9_30

polynomials, one for each n. The class VP consists of families of polynomials whose degree and circuit size are both polynomially bounded in the number of variables n (denoted poly(n) from now on). On the other hand, if a polynomial has degree poly(n) and the coefficient of any given monomial can be computed in #P/poly, then the polynomial is in VNP[1]. It is not difficult to see that VP $\subseteq$ VNP.

Much like Cook's original P vs. NP hypothesis in the boolean world, very little is known in general about Valiant's hypothesis. A result of Strassen [36] and Baur-Strassen [5] gives a lower bound of $\Omega(n \log n)$ against general circuits. A slightly better lower bound of $\Omega(n^2)$ is known if the directed acyclic graph underlying the circuit is a *tree* – also known as an *Algebraic Formula*. All polynomials that have formulas of size poly(n) form the class VF. We refer the interested reader to the excellent book of Bürgisser [6] for more details on Valiant's hypothesis and connections to the Boolean world.

Intermediate in power, and in between circuits and formulas lie Algebraic Branching Programs (ABPs). An ABP is a layered directed acyclic graph with edges labeled by *affine linear forms*. There is a *source* vertex (s) of in-degree 0 in the first layer and a *sink* vertex (t) of out-degree 0 in the last layer, and edges connect vertices in adjacent layers. The maximum number of vertices in any layer is the *width* of the ABP and the number of layers is its *length*. Each path from s to t computes a polynomial that is the product of the edge labels along the path. The polynomial computed by the ABP is the sum of the polynomials computed by all the $s \rightsquigarrow t$ paths.

An ABP of length ℓ with n_i vertices in the i-th layer can be written as a product of $\ell-1$ matrices $\prod_{i=1}^{\ell-1} M_i$ in a natural way: the matrix M_i is of dimension $n_i \times n_{i+1}$ and contains the edge labels between layers i and $i+1$ as entries. The size of the ABP is the total number of vertices in the graph (or equivalently, the sum of the number of rows of the matrices in matrix representation). Similar to circuits and formulas, the class of polynomials that have ABPs of size poly(n) is denoted VBP.

It is known that VF $\subseteq$ VBP $\subseteq$ VP, and conjectured that all the inclusions are strict. Valiant's hypothesis is considered more generally as the problem of separating any of the classes VF, VBP or VP from VNP. Unfortunately (although probably not surprisingly), general lower bounds in any of these models is hard to come by. In a recent work, Chatterjee, Kumar, She and Volk [8] proved a lower bound of $\Omega(n^2)$ for ABPs. Evidently, the state of affairs is quite similar to that of circuits. In fact, the polynomial $\sum_{i=1}^{n} x_i^n$ used in the lower bound is the same one that Baur and Strassen [5] used for their circuit lower bound.

In this work, we will mainly be interested in *set-multilinear* polynomials, of which the Iterated Matrix Multiplication polynomial is an excellent example. The polynomial $\text{IMM}_{n,d}$ is defined on $N = dn^2$ variables. The variable set X is partitioned into d sets $(X_1, \ldots, X_d)$ of n^2 variables each (viewed as $n \times n$

---

[1] This is simply a sufficient condition for a polynomial to be in VNP, but is enough for our purpose. A precise definition can be found in [35, Definition 1.3].

matrices). The polynomial is defined as the $(1,1)$-th entry of the matrix product $X_1 \cdot X_2 \cdots X_d$:

$$\mathrm{IMM}_{n,d} = \left(\begin{bmatrix} x_{1,1} & \cdots & x_{1,n} \\ \vdots & \ddots & \vdots \\ x_{1,n^2-n+1} & \cdots & x_{1,n^2} \end{bmatrix} \cdots \cdots \begin{bmatrix} x_{d,1} & \cdots & x_{d,n} \\ \vdots & \ddots & \vdots \\ x_{d,n^2-n+1} & \cdots & x_{d,n^2} \end{bmatrix} \right)_{(1,1)} .$$

As all monomials are of the same degree d, the polynomial is *homogeneous*. It is also *multilinear* since every variable has individual degree at most 1. Additionally, every monomial has exactly one variable from each of the d sets of the partition. Thus it is *set-multilinear*. Henceforth, by a set-multilinear polynomial $P_{n,d}$ over the variable set $X = X_1 \sqcup \ldots \sqcup X_d$ (with $|X_i| \leq n$ for all $i \in [d]$), we mean a homogeneous multilinear polynomial with the following property: every monomial m (seen as a set) in $P_{n,d}$ satisfies $|m \cap X_i| = 1$ for all $i \in [d]$.

1.1 Our Results

Our first result is a lower bound against the sum of general *small-size* algebraic branching programs.

Theorem 1 ($\sum$ ABP lower bound). *Let $d < n^{o(1)}$. The polynomial $\mathrm{IMM}_{n,d}$ cannot be computed by the sum of $\mathrm{poly}(n,d)$ ABPs, each of size $(nd)^{o(1)}$.*

Note that the polynomial $\mathrm{IMM}_{n,d}$ has an ABP of size $O(nd)$. The above theorem shows that this is almost optimal: we cannot reduce the size significantly, even by using a sum of polynomially many ABPs.

Remark 1. When $d > n^{o(1)}$, ABPs of size $(nd)^{o(1)}$ cannot produce monomials of degree d. Hence, the theorem statement is obtained trivially (in general, a lower bound of d is trivial for ABPs). But when $d < n^{o(1)}$, the model is quite powerful. In fact, for $d < n^{o(1)}$, the power sum polynomial $\sum_{i=1}^n x_i^d$, that was used in previous ABP lower bounds, can be computed efficiently using a sum of n ABPs, each of size $(nd)^{o(1)}$.

A lower bound of n is not trivial for ABPs (unlike circuits and formulas). Moreover, each edge label can be a general affine linear form, allowing a single path to generate exponentially many monomials. Notwithstanding that, ABPs of size $(nd)^{o(1)}$ are still an incomplete model of computation. Nevertheless, the sum of such ABPs *is* a complete model – every polynomial of degree less than $n^{o(1)}$ can be written as a (exponential) sum of width-1 ABPs (monomials).

The lower bound of Theorem 1 also holds if we replace IMM with an appropriate polynomial from the family of Nisan-Wigderson design-based polynomials.

Our next result is a reformulation of Valiant's conjecture in terms of a different model: the sum of *set-multilinear* ABPs (smABPs) on the set of variables $X = X_1 \sqcup \ldots \sqcup X_d$. An smABP in the *natural order* is a $(d+1)$ layered ABP with edges between layers i and $i+1$ labeled by *linear forms* in X_i. The most natural ABP for the polynomial $\text{IMM}_{n,d}$ is also set-multilinear: each layer (other than the first and the last) has n nodes and the edge connecting the p-th node in layer i to the q-th node in layer $i+1$ is labeled by $x_{i,pq}$.

More generally, for a permutation $\pi \in S_d$ of the variable sets, we say that an smABP is in the order π if the edges between i-th and $(i+1)$-th layer are labeled by *linear forms* in $X_{\pi(i)}$.[2]

We denote by $\sum$ smABP the sum of set-multilinear ABPs, each in a possibly different order. The *width* of a $\sum$ smABP is the sum of the widths of the constituent smABPs.

We show that in the *low-degree* regime, superpolynomial lower bounds against $\sum$ smABP imply superpolynomial ABP lower bounds.

Theorem 2 (Hardness bootstrapping).   *Let n, d be integers such that $d = O(\log n / \log \log n)$. Let $P_{n,d}$ be a set-multilinear polynomial in* VNP *of degree d. If $P_{n,d}$ cannot be computed by a $\sum$ smABP of width* poly(n), *then* VBP $\neq$ VNP.

The above theorem shows that the sum of set-multilinear ABPs, which looks quite restrictive, is surprisingly powerful. This is a recurring theme in algebraic complexity. Interestingly, analogous reductions to the set-multilinear case were known for formulas [29, Theorem 3.1] and circuits [26, Lemma 2.11]. A series of works [2,17,19,37,39] on reducing the *depth* of algebraic circuits culminated in the rather surprising fact that good enough lower bounds for depth-3 circuits imply general circuit lower bounds. The above theorem is in a similar vein. The model of $\sum$ smABP is particularly appealing to study since smABPs are one of the most well-understood objects in algebraic complexity.

Recently, [22] proved near-optimal lower bounds against set-multilinear formulas for a polynomial in VBP. Surprisingly, if the polynomial were computable by an smABP, we would obtain general formula lower bounds. This further illustrates the need to study smABPs.

Non-commuting Matrices Make it Powerful

Note that if the matrices in the smABP were commutative, we can treat $\sum$ smABP as a *single* smABP, against which we know how to prove lower bounds (see Sect. 1.2). So in order to lift the lower bound to VNP, it is essential that we understand the sum of smABPs with non-commuting matrices (see Sect. 1.3 for a detailed discussion).

Arbitrarily Low Degree Suffices

The low-degree regime has recently gained a lot of attention. In a breakthrough work, Limaye Srinivasan and Tavenas [23] showed how to prove superpolynomial

---

[2] This definition differs slightly from that of Forbes [13] as it does not allow *affine* linear forms as edge labels. We use this definition as the ABPs we encounter are of this more restricted form and proving lower bounds for them is sufficient.

lower bounds for constant-depth set-multilinear formulas when the degree is small (set-multilinear lower bounds against arbitrary depth were known before [26,28,31], but degenerated to trivial bounds when the degree was small). They were able to then escalate the low-degree, set-multilinear lower bounds to *general* constant-depth circuit lower bounds. The theorem above shows that the low-degree regime can be helpful in proving lower bounds for ABPs as well.

A Spectrum of Hardness Escalation

We also give a smooth generalization of Theorem 2 using more general versions of both set-multilinear polynomials and smABPs. The variable set is partitioned as before: $X = X_1 \sqcup \ldots \sqcup X_d$ with $|X_i| \le n$ for all i.

A polynomial g is called *set-multi-k-ic* with respect to X if every monomial of g has exactly k variables (with multiplicity) from each of the d sets. That is, for a monomial m (seen as a multiset) in the support of g, $|m \cap X_i| = k$. When $k = 1$, the polynomial g is set-multilinear.

We call an ABP of length kd a *set-multi-k-ic* ABP (denoted $\mathrm{sm}(k)$ABP) if every layer has edges labeled by linear forms from exactly one of the sets X_i, and there are exactly k layers corresponding to each X_i. As a special case, an $\mathrm{sm}(1)$ABP is just a set-multilinear ABP as defined before.

Theorem 3 (Hardness bootstrapping spectrum). *Let* n, d, k *be integers such that* $\min(d^{kd}, (kd)^d) = \mathrm{poly}(n)$, *and let* $P_{n,d,k}$ *be a set-multi-k-ic polynomial in* VNP *of degree kd. If* $P_{n,d,k}$ *cannot be computed by a* $\sum \mathrm{sm}(k)$ABP *of width* $\mathrm{poly}(n)$, *then* VBP $\ne$ VNP.

Remark 2. We note that Theorem 2 is an immediate consequence of Theorem 3 when $k = 1$. An added advantage of this generalization is the flexibility with the degree of the hard polynomial. For example, if $k = d = O(\log n / \log \log n)$, the degree of the polynomial we are allowed is $O(\log^2 n / (\log \log n)^2)$. In contrast, Theorem 2 could only work when the degree is $O(\log n / \log \log n)$.

The *set-multi-k-ic* ABP is inspired from the well-studied *multi-k-ic* depth-restricted circuits and formulas, initiated by Kayal and Saha [18]. We encourage readers to refer [32, Chapter 14] and references therein for a comprehensive discussion.

1.2   The Sum of ROABPs Perspective: The Arbitrarily Low Variate Case

One can also view Theorem 2 through the lens of another well-studied model in the literature, first defined by Forbes and Shpilka [12]. An algebraic branching program over the variables $(x_1, \ldots, x_n)$ is said to be *oblivious* if, for every layer, all the edge labels are univariate polynomials in a single variable. It is further called a *read-once* oblivious ABP (or a ROABP) if every variable appears in at most one layer.

A ROABP in the *natural order* is $n+1$ layered ABP where the edges between layers i and $i+1$ are labeled by univariate polynomials in x_i of degree d. If, instead, the labels were univariate polynomials in $x_{\pi(i)}$ for some permutation $\pi \in S_d$ of the variables, then we say that the ROABP is in the order π.

The computation that a ROABP (or equivalently, an smABP) performs is essentially non-commutative since the variables along a path get multiplied in the same order π as that of the ROABP (smABP). Nisan [25] introduced the powerful technique of using spaces of partial derivatives to study lower bound questions in non-commutative models. This technique can be used to calculate the exact width of the ROABP computing a polynomial.

Following our definition for smABPs, we denote by $\sum$RO the sum of ROABPs, each possibly in a different order. The width of a $\sum$RO is the sum of the widths of the constituent ROABPs. A version of Theorem 2 can also be stated for this model. In contrast to the case of smABPs, we will be interested in the dual *low-variate* regime.

Corollary 1 (Low variate $\sum$RO). *Let n, d be integers such that $n = O(\log d / \log \log d)$. Let $f \in$ VNP be a polynomial on n variables of individual degree d. If f cannot be computed by a $\sum$RO of width* poly(d)*, then* VBP $\neq$ VNP.

The low-variate regime has also recently been shown to be extremely important. The Polynomial Identity Testing (PIT) problem asks to efficiently test whether a polynomial (given as an algebraic circuit, for example) is identically *zero*. In the black-box setting, we are only allowed to evaluate the polynomial (circuit) at various points. Hence, PIT algorithms are equivalent to the construction of *hitting sets* – a collection of points that witness the (non)zeroness of the polynomial computed by the circuit (see [33,34] for a survey of PIT and techniques used).

Recently, several surprising results [1,16,21] essentially conclude that hitting sets for circuits computing extremely low-variate polynomials can be "bootstrapped" to obtain hitting sets for general circuits. See the survey of Kumar and Saptharishi [20] for an exposition of the ideas involved.

We now state a corollary of Theorem 3 analogous to Corollary 1. An *oblivious* ABP is said to be *read-k* if each variable x_i appears in at most k layers. We denote the sum of *read-k* oblivious ABPs as $\sum$R(k)O. Once again, the width of a $\sum$R(k)O is the sum of the widths of the constituent branching programs.

Corollary 2. *Let n, d, k be integers such that $\min(n^{kn}, (kn)^n) =$ poly(d). Let $f \in$ VNP be a polynomial on n variables of individual degree d. If f cannot be computed by a $\sum$R(k)O of width* poly(d)*, then* VBP $\neq$ VNP.

1.3 Proof Techniques and Previous Work

Simulating ABPs Using Sum of smABPs. Unlike the boolean world, *both* the degree d of the polynomial, and the number of variables n are important parameters in algebraic complexity. Often times, it is reasonable and useful to impose restrictions on one of them. Even in the definitions VP and VNP, we require that the degree d be restricted by a polynomial in n (see [15] for more discussion on the motivation behind this choice). Further restrictions on the degree help in proving better structural results which would otherwise be prohibitively costly to perform.

In order to prove Theorem 2, we perform a sequence of structural transformations to the algebraic branching program to obtain a $\sum$ smABP. We first *homogenize* the ABP (Lemma 2), i.e., we alter the ABP so that every vertex in the ABP computes a homogeneous polynomial. In addition, we will ensure that the ABP has d layers and all the edge labels are *linear forms*. The homogenization of ABPs to this form was folklore. Subsequently, we set-multilinearize the branching program (Lemma 1). This step is only efficient in the low-degree regime since what we obtain is a sum of $d^{O(d)}$ set-multilinear ABPs.

With the reduction in place, superpolynomial lower bounds for $\sum$ smABP imply the same for ABPs, albeit in the low-degree regime. The proof of Theorem 3 is similar.

Lower Bounds for the Sum of ABPs. Our proof of the $\sum$ ABP lower bound (Theorem 1) uses the implicit reduction of Theorem 2 to $\sum$ smABP. Using Nisan's characterization [25] mentioned before, we can prove exponential lower bounds against single smABPs (ROABPs), but the characterization does not extend to their sums. There has been progress in handling the sums in recent years, which we now briefly describe.

Arvind and Raja [4] proved a superpolynomial lower bound for the Permanent polynomial against the sum of sub-linear many ROABPs (the bound is exponential if the number of ROABPs is bounded by a constant). Ramya and Rao [27] showed that a sum of sub-exponential size ROABPs computing the multilinear polynomial defined by Raz and Yehudayoff [30] needs exponentially many summands. Ghosal and Rao [14] showed an exponential lower bound for the sum of ROABPs computing the multilinear polynomial defined by Dvir, Malod, Perifel and Yehudayoff [11], provided each of the constituent ABPs is polynomial in size.

Unfortunately, these results do not imply general ABP lower bounds using our hardness escalation theorems, as they only work in regimes where the degree and number of variables are comparable. Viewed differently, they cannot handle a sum of $d!$ smABPs (or $n!$ ROABPs) which is necessary to prove lower bounds in our low-degree (low-variate) regime. In a very recent work Chatterjee, Kush, Saraf and Shpilka [9] improve the bounds in the above works and also prove superpolynomial lower bounds against the sum of smABPs when the degree is $d = \omega(\log n)$. Improving this to work for $d = O(\log n / \log \log n)$ would have dramatic consequences.

Fewer results are known about *read-k* oblivious ABPs. They were studied in [3] as a natural generalization of ROABPs and a lower bound of $\exp(n/k^{O(k)})$ for a single *read-k* oblivious ABP was shown. It remains open to improve this result to prove non-trivial lower bounds when k is large, as well as to prove lower bounds for sums of *read-k* oblivious ABPs. When k is small, the results of Ramya and Rao [27] extend to the sum of multilinear k-pass ABPs, a restriction of *read-k* oblivious ABPs in which the variables are read k times in sequence, each time in a possibly different order.

We demonstrate a way to handle our low-degree regime in certain cases. To prove lower bounds for the sum of smABPs, we use the *partial derivative method*, introduced in the highly influential work of Nisan and Wigderson [26]. We show that the partial derivative measure $\mu(\cdot)$ is large for our hard polynomial but small for the model. In fact, a majority of the lower bounds in algebraic complexity (including the results described above) use modifications and extensions of this measure. For a comprehensive survey of lower bounds and the use of partial derivative measure in algebraic complexity, see [10, 32].

We work with the polynomial $\text{IMM}_{n,d}$, which gives us more flexibility in independently choosing n and d. Unfortunately, this choice creates a two-fold problem. The fundamental one is that $\text{IMM}_{n,d}$ has a small smABP, as we saw before. So we can never prove a superpolynomial lower bound for even a single $\text{poly}(n,d)$ sized smABP (let alone their sum). One might try to avoid this by choosing a different hard polynomial that gives similar flexibility, perhaps something from the family of Nisan-Wigderson design-based polynomials. But in fact, the complexity measure μ is also *maximal* for $\text{IMM}_{n,d}$. Hence, the usual partial derivative method cannot be used to prove lower bounds against any model that efficiently computes $\text{IMM}_{n,d}$. Be that as it may, it might still be possible to use the same technique to prove lower bounds for restrictions of the model. We are able to do this when the smABPs are sub-polynomial in size. It also enables us to handle extremely large sums of smABPs (including those that occur from considering sums of multiple ABPs).

This approach works in the low-degree regime, since our reductions are efficient if the degree is very small. To handle higher degrees, we note that $\text{IMM}_{n,d'}$ with d' small can be obtained as a set-multilinear restriction of $\text{IMM}_{n,d}$. Therefore, our lower bounds translate to higher degrees to finally give superpolynomial lower bounds against sums of small-sized general ABPs.

2   Hardness Bootstrapping Spectrum

We begin by showing that in the low-degree regime, a small sized ABP can be simulated by a $\sum$ smABP of small width. This is very much in the spirit of the set-multilinearization result of Limaye, Srinivasan and Tavenas ([23], Proposition 9) for small-depth circuits. Due to space constraints, we omit detailed proofs which can be found in the full version.[3]

Lemma 1 (ABP set-multilinearization). *Let $P_{n,d}$ be a polynomial of degree d, set-multilinear with respect to the partition $X = X_1 \sqcup \ldots \sqcup X_d$ where $|X_i| \leq n$ for all $i \in [d]$. If $P_{n,d}$ can be computed by an ABP of size s, then there is a $\sum$ smABP of width $d^{O(d)}s$ computing the same polynomial.*

---

[3] Full version - https://www.cse.iitk.ac.in/users/nitin/papers/sumRO.pdf.

We immediately obtain Theorem 2 as an easy consequence. We omit the proof. In order to prove Lemma 1, we first homogenize the ABP (similar to the approach of Raz [29] and LST [23]). Any vertex v in an ABP can be thought of as computing a polynomial corresponding to the 'sub-ABP' between the source s and the vertex v. An ABP is homogenous if the polynomial computed at every vertex is homogenous.

Lemma 2 (ABP homogenization). *Let $f(x_1, \ldots, x_n)$ be a degree d polynomial. Suppose that f can be computed by an* ABP *of size s. Then there is a homogeneous ABP of width s and length d that can compute the same polynomial. Furthermore, all the edge labels are* linear forms.

The above lemma is "folklore" with the proof idea already present in [25]. As our central argument, we show that this homogeneous ABP can be *efficiently* set-multilinearized.

Proposition 1. *Consider a set-multilinear polynomial $P_{n,d}$ over the variable set $X = X_1 \sqcup \ldots \sqcup X_d$ (with $|X_i| \leq n$ for all $i \in [d]$) computed by a homogeneous* ABP *of width w and length d. Then, there is a $\sum$ smABP of width d!w computing $P_{n,d}$.*

With this transformation in hand, we can complete the reduction and obtain Lemma 1. The proof of Theorem 3 follows the template of Theorem 2. We begin with ABP homogenization, followed by a structural transformation to the sum of *set-multi-k-ic* ABP. The superpolynomial lower bound assumption on $\sum$ sm(k)ABP gives the desired separation result. The following lemma is analogous to Lemma 1.

Lemma 3 (ABP to $\sum$ sm(k)ABP). *Let P be a set-multi-k-ic polynomial with respect to the partition $X = X_1 \sqcup \ldots \sqcup X_d$ where $|X_i| \leq n$ for all $i \in [d]$. If P can be computed by an* ABP *of size s, then there is a $\sum$ sm(k)ABP of width $s \cdot \binom{d+kd}{d}$ computing the same polynomial.*

It is straightforward to prove Theorem 3 using the above lemma. The proof is similar to Theorem 2 and we omit it.

3 Lower Bound for the Sum of ABPs

We are now ready to show that in the low degree regime, the Iterated Matrix Multiplication polynomial $\text{IMM}_{n,d}$ cannot be computed even by a polynomially large sum of ABPs, provided that each of the ABPs is small in size. We begin by stating a lower bound for $\sum$ smABP in the *low-degree* regime. Note that in this regime, IMM has an smABP of width $O(nd)$. The lemma shows that even using the sum of multiple smABPs cannot help in reducing the width.

Lemma 4. *Any $\sum$ smABP computing the polynomial $\text{IMM}_{n,d}$ with $d = O(\log n / \log \log n)$, must have width at least $n^{\Omega(1)}$.*

Suppose we had to prove the lower bound of Theorem 1 for a single ABP computing IMM. We could then use Lemma 4 above in conjunction with Lemma 1 to conclude the result. But when we are dealing with a sum of ABPs, we need to be more careful in how we set-multilinearize since the ABPs no longer need to compute set-multilinear or even homogenous polynomials.

4   Discussion and Open Problems

In order to separate VBP from VNP, we need to prove super-polynomial lower bounds against $\sum$ smABP for a polynomial in VNP that we expect to be hard. As noted above, the IMM polynomial is in VBP (in fact, it is a canonical way to define the class VBP) and cannot be used for such a separation. Our Theorem 1 also holds for a polynomial from the Nisan-Wigderson family of design-based polynomials that is in VNP and is a better candidate.

A first step toward proving ABP lower bounds would be to prove any non-trivial lower bounds against the sum of smABPs in the low degree regime, i.e. prove some lower bound for the sum of $d!$ smABPs. Another interesting direction is to show a reduction from ABPs to the sum of fewer than $d!$ smABPs, with a possibly super polynomial blow up in the smABP size. This would still lead to ABP lower bounds if we can prove strongly exponential lower bounds against the sum of (fewer) smABPs. This question remains open as well.

References

1. Agrawal, M., Ghosh, S., Saxena, N.: Bootstrapping variables in algebraic circuits. Proc. Natl. Acad. Sci. U.S.A. **116**(17), 8107–8118 (2019). https://doi.org/10.1073/pnas.1901272116
2. Agrawal, M., Vinay, V.: Arithmetic circuits: a chasm at depth four. In: 2008 49th Annual IEEE Symposium on Foundations of Computer Science, pp. 67–75 (2008). https://doi.org/10.1109/FOCS.2008.32
3. Anderson, M., Forbes, M.A., Saptharishi, R., Shpilka, A., Volk, B.L.: Identity testing and lower bounds for read-k oblivious algebraic branching programs. ACM Trans. Comput. Theory **10**(1), 3:1–3:30 (2018). https://doi.org/10.1145/3170709
4. Arvind, V., Raja, S.: Some lower bound results for set-multilinear arithmetic computations. Chic. J. Theor. Comput. Sci. pp. Art. 6, 26 (2016). https://doi.org/10.4086/cjtcs.2016.006
5. Baur, W., Strassen, V.: The complexity of partial derivatives. Theor. Comput. Sci. **22**(3), 317–330 (1983). https://doi.org/10.1016/0304-3975(83)90110-X
6. Bürgisser, P.: Completeness and Reduction in Algebraic Complexity Theory, Algorithms and Computation in Mathematics, vol. 7. Springer, Berlin (2000). https://doi.org/10.1007/978-3-662-04179-6
7. Bürgisser, P., Clausen, M., Shokrollahi, M.A.: Algebraic complexity theory, Grundlehren der mathematischen Wissenschaften [Fundamental Principles of Mathematical Sciences], vol. 315. Springer, Berlin (1997). https://doi.org/10.1007/978-3-662-03338-8, with the collaboration of Thomas Lickteig

8. Chatterjee, P., Kumar, M., She, A., Volk, B.L.: Quadratic lower bounds for algebraic branching programs and formulas. Comput. Complexity **31**(2), Paper No. 8, 54 (2022). https://doi.org/10.1007/s00037-022-00223-8

9. Chatterjee, P., Kush, D., Saraf, S., Shpilka, A.: Lower bounds for set-multilinear branching programs (2024). https://arxiv.org/abs/2312.15874, preprint

10. Chen, X., Kayal, N., Wigderson, A.: Partial derivatives in arithmetic complexity and beyond. Found. Trends Theor. Comput. Sci. **6**(1-2), 1–138 (2010). https://doi.org/10.1561/0400000043

11. Dvir, Z., Malod, G., Perifel, S., Yehudayoff, A.: Separating multilinear branching programs and formulas. In: Proceedings of the 2012 ACM Symposium on Theory of Computing, STOC 2012, pp. 615–623. ACM, New York (2012). https://doi.org/10.1145/2213977.2214034

12. Forbes, M.A., Shpilka, A.: Quasipolynomial-time identity testing of non-commutative and read-once oblivious algebraic branching programs. In: 2013 IEEE 54th Annual Symposium on Foundations of Computer Science, FOCS 2013, Los Alamitos, CA, pp. 243–252. IEEE Computer Society (2013). https://doi.org/10.1109/FOCS.2013.34

13. Forbes, M.A.: Polynomial Identity Testing of Read-Once Oblivious Algebraic Branching Programs. ProQuest LLC, Ann Arbor, MI, thesis (Ph.D.)–Massachusetts Institute of Technology (2014)

14. Ghosal, P., Rao, B.V.R.: Limitations of sums of bounded read formulas and ABPs. In: Santhanam, R., Musatov, D. (eds.) CSR 2021. LNCS, vol. 12730, pp. 147–169. Springer, Cham (2021). https://doi.org/10.1007/978-3-030-79416-3_9

15. Grochow, J.: Degree restriction for polynomials in VP. Theoretical Computer Science Stack Exchange (2013). https://cstheory.stackexchange.com/q/19268

16. Guo, Z., Kumar, M., Saptharishi, R., Solomon, N.: Derandomization from algebraic hardness. SIAM J. Comput. **51**(2), 315–335 (2022). https://doi.org/10.1137/20M1347395

17. Gupta, A., Kamath, P., Kayal, N., Saptharishi, R.: Arithmetic circuits: a chasm at depth 3. SIAM J. Comput. **45**(3), 1064–1079 (2016). https://doi.org/10.1137/140957123

18. Kayal, N., Saha, C.: Multi-k-ic depth three circuit lower bound. Theory Comput. Syst. **61**(4), 1237–1251 (2017). https://doi.org/10.1007/S00224-016-9742-9

19. Koiran, P.: Arithmetic circuits: the chasm at depth four gets wider. Theor. Comput. Sci. **448**, 56–65 (2012). https://doi.org/10.1016/j.tcs.2012.03.041

20. Kumar, M., Saptharishi, R.: Hardness-randomness tradeoffs for algebraic computation. Bull. Eur. Assoc. Theor. Comput. Sci. EATCS **3**(129), 56–87 (2019). http://bulletin.eatcs.org/index.php/beatcs/article/view/591/599

21. Kumar, M., Saptharishi, R., Tengse, A.: Near-optimal bootstrapping of hitting sets for algebraic circuits. In: Proceedings of the Thirtieth Annual ACM-SIAM Symposium on Discrete Algorithms, Philadelphia, PA, pp. 639–646. SIAM (2019). https://doi.org/10.1137/1.9781611975482.40

22. Kush, D., Saraf, S.: Near-optimal set-multilinear formula lower bounds. In: 38th Computational Complexity Conference, LIPIcs. Leibniz Int. Proc. Inform., vol. 264, pp. Art. No. 15, 33. Schloss Dagstuhl. Leibniz-Zent. Inform., Wadern (2023). https://doi.org/10.4230/lipics.ccc.2023.15

23. Limaye, N., Srinivasan, S., Tavenas, S.: Superpolynomial lower bounds against low-depth algebraic circuits. In: 2021 IEEE 62nd Annual Symposium on Foundations of Computer Science, FOCS 2021, Los Alamitos, CA, pp. 804–814. IEEE Computer Society (2021). https://doi.org/10.1109/FOCS52979.2021.00083

24. Mahajan, M.: Algebraic complexity classes. In: Perspectives in Computational Complexity, Progr. Comput. Sci. Appl. Logic, vol. 26, pp. 51–75. Birkhäuser/Springer, Cham (2014). https://arxiv.org/abs/1307.3863

25. Nisan, N.: Lower bounds for non-commutative computation. In: Proceedings of the Twenty-Third Annual ACM Symposium on Theory of Computing, STOC 1991, pp. 410–418. Association for Computing Machinery, New York (1991). https://doi.org/10.1145/103418.103462

26. Nisan, N., Wigderson, A.: Lower bounds on arithmetic circuits via partial derivatives. Comput. Complex. **6**(3), 217–234 (1996). https://doi.org/10.1007/BF01294256

27. Ramya, C., Raghavendra Rao, B.V.: Lower bounds for special cases of syntactic multilinear ABPs. Theor. Comput. Sci. **809**, 1–20 (2020). https://doi.org/10.1016/j.tcs.2019.10.047

28. Raz, R.: Multi-linear formulas for permanent and determinant are of super-polynomial size. J. ACM **56**(2), Art. 8, 17 (2009). https://doi.org/10.1145/1502793.1502797

29. Raz, R.: Tensor-rank and lower bounds for arithmetic formulas. J. ACM **60**(6), Art. 40, 15 (2013). https://doi.org/10.1145/2535928

30. Raz, R., Yehudayoff, A.: Balancing syntactically multilinear arithmetic circuits. Comput. Complexity **17**(4), 515–535 (2008). https://doi.org/10.1007/s00037-008-0254-0

31. Raz, R., Yehudayoff, A.: Lower bounds and separations for constant depth multilinear circuits. Comput. Complexity **18**(2), 171–207 (2009). https://doi.org/10.1007/s00037-009-0270-8

32. Saptharishi, R.: A survey of lower bounds in arithmetic circuit complexity. Github Survey (2021). https://github.com/dasarpmar/lowerbounds-survey

33. Saxena, N.: Progress on polynomial identity testing. Bull. Eur. Assoc. Theor. Comput. Sci. EATCS (99), 49–79 (2009). https://www.cse.iitk.ac.in/users/nitin/papers/pit-survey09.pdf

34. Saxena, N.: Progress on polynomial identity testing-II. In: Perspectives in Computational Complexity, Progr. Comput. Sci. Appl. Logic, vol. 26, pp. 131–146. Birkhäuser/Springer, Cham (2014). https://arxiv.org/abs/1401.0976

35. Shpilka, A., Yehudayoff, A.: Arithmetic circuits: a survey of recent results and open questions. Found. Trends Theor. Comput. Sci. **5**(3–4), 207–388 (2009). https://doi.org/10.1561/0400000039

36. Strassen, V.: Die Berechnungskomplexität von elementarsymmetrischen Funktionen und von Interpolationskoeffizienten. Numer. Math. **20**, 238–251 (1972/73). https://doi.org/10.1007/BF01436566

37. Tavenas, S.: Improved bounds for reduction to depth 4 and depth 3. Inform. Comput. **240**, 2–11 (2015). https://doi.org/10.1016/j.ic.2014.09.004

38. Valiant, L.G.: Completeness classes in algebra. In: Conference Record of the Eleventh Annual ACM Symposium on Theory of Computing (Atlanta, Ga., 1979), pp. pp 249–261. ACM, New York (1979). https://doi.org/10.1145/800135.804419

39. Valiant, L.G., Skyum, S., Berkowitz, S., Rackoff, C.: Fast parallel computation of polynomials using few processors. SIAM J. Comput. **12**(4), 641–644 (1983). https://doi.org/10.1137/0212043

Author Index

Printed in the United States
by Baker & Taylor Publisher Services